A Dictionary of
Physics

FIFTH EDITION

OXFORD
UNIVERSITY PRESS

Great Clarendon Street, Oxford OX2 6DP

Oxford University Press is a department of the University of Oxford.
It furthers the University's objective of excellence in research, scholarship,
and education by publishing worldwide in

Oxford New York
Auckland Cape Town Dar es Salaam Hong Kong Karachi Kuala Lumpur
Madrid Melbourne Mexico City Nairobi New Delhi Shanghai Taipei Toronto

With offices in
Argentina Austria Brazil Chile Czech Republic France Greece
Guatemala Hungary Italy Japan South Korea Poland Portugal
Singapore Switzerland Thailand Turkey Ukraine Vietnam

Oxford is a registered trade mark of Oxford University Press
in the UK and in certain other countries

Published in the United States
by Oxford University Press Inc., New York

First edition (published under the title) *Concise Dictionary of Physics* 1985
Second edition 1990
Third edition 1996
Fourth edition 2000 retitled *Dictionary of Physics*
Reissued with new covers 2003
Fifth edition 2005

British Library Cataloguing in Publication Data
Data available

Library of Congress Cataloging in Publication Data
Data available

Typeset by Market House Books Ltd
Printed in Great Britain by Clays Ltd, St Ives plc

ISBN 0-19-280628-9   978-0-19-280628-4

1

Preface

This dictionary was originally derived from the *Concise Science Dictionary*, first published by Oxford University Press in 1984 (fifth edition, retitled *A Dictionary of Science*, 2005). It consisted of all the entries relating to physics in the *Concise Science Dictionary*, together with those entries relating to astronomy that are required for an understanding of astrophysics and many entries that relate to physical chemistry. It also included a selection of the words used in mathematics that are relevant to physics, as well as the key words in metal science, computing, and electronics. Subsequent editions have been expanded by the addition of many more entries, including short biographies of important physical scientists; several chronologies tracing the history of some of the key areas in physics; and a number of special one- or two-page feature articles on important topics. For this fifth edition the text has been revised, many entries have been expanded, and over 200 new entries have been added covering all branches of the subject. The more chemical aspects of physical chemistry and the chemistry itself will be found in *A Dictionary of Chemistry*, and biological aspects of biophysics are more fully covered in *A Dictionary of Biology*, which are companion volumes to this dictionary.

An asterisk placed before a word used in an entry indicates that this word can be looked up in the dictionary and will provide further explanation or clarification. However, not every word that appears in the dictionary has an asterisk placed before it. Some entries simply refer the reader to another entry, indicating either that they are synonyms or abbreviations or that they are most conveniently explained in one of the dictionary's longer articles or features. Synonyms and abbreviations are usually placed within brackets immediately after the headword. Terms that are explained within an entry are highlighted by being printed in boldface type.

SI units are used throughout this book and its companion volumes.

JD

2005

Credits

Editor
John Daintith BSc, PhD

Advisers
B. S. Beckett BSc, BPhil, MA(Ed)
R. A. Hands BSc
Michael Lewis MA

Contributors
John Clark BSc
H. M. Clarke MA, MSc
Derek Cooper PhD, FRIC
John Cullerne DPhil

D. E. Edwards BSc, MSc
Richard Rennie BSc, MSc, PhD
David Eric Ward BSc, MSc, PhD

Contents

ab- A prefix attached to the name of a practical electrical unit to provide a name for a unit in the electromagnetic system of units (*see* ELECTROMAGNETIC UNITS), e.g. abampere, abcoulomb, abvolt. The prefix is an abbreviation of the word 'absolute' as this system is also known as the **absolute system**. *Compare* STAT-. In modern practice both absolute and electrostatic units have been replaced by *SI units.

Abelian group *See* GROUP THEORY.

aberration **1.** (in optics) A defect in the image formed by a lens or curved mirror. In **chromatic aberration** the image formed by a lens (but not a mirror) has coloured fringes as a result of the different extent to which light of different colours is refracted by glass. It is corrected by using an *achromatic lens. In **spherical aberration**, the rays from the object come to a focus in slightly different positions as a result of the curvature of the lens or mirror. For a mirror receiving light strictly parallel with its axis, this can be corrected by using a parabolic surface rather than a spherical surface. Spherical aberration in lenses is minimized by making both surfaces contribute equally to the ray deviations, and can (though with reduced image brightness) be reduced by

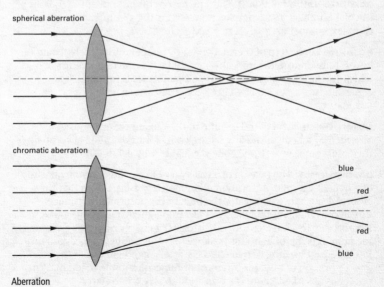

Aberration

the use of diaphragms to let light pass only through the centre part of the lens. *See also* ASTIGMATISM; COMA. **2.** (in astronomy) The apparent displacement in the position of a star as a result of the earth's motion round the sun. Light appears to come from a point that is slightly displaced in the direction of the earth's motion. The angular displacement $\alpha = v/c$, where v is the earth's orbital velocity and c is the speed of light.

abscissa *See* CARTESIAN COORDINATES.

absolute 1. Not dependent on or relative to anything else, e.g. *absolute zero. **2.** Denoting a temperature measured on an **absolute scale**, a scale of temperature based on absolute zero. The usual absolute scale now is that of thermodynamic *temperature; its unit, the kelvin, was formerly called the degree absolute (°A) and is the same size as the degree Celsius. In British engineering practice an absolute scale with Fahrenheit-size degrees has been used: this is the Rankine scale.

absolute expansivity *See* EXPANSIVITY.

absolute humidity *See* HUMIDITY.

absolute permittivity *See* PERMITTIVITY.

absolute pitch (perfect pitch) The ability of a person to identify and reproduce a note without reference to a tuned musical instrument.

absolute temperature *See* ABSOLUTE; TEMPERATURE.

absolute value (modulus) The square root of the sum of the squares of the real numbers in a *complex number, i.e. the absolute value of the complex number $z = x + iy$ is $|z| = \sqrt{(x^2 + y^2)}$.

absolute zero Zero of thermodynamic *temperature (0 kelvin) and the lowest temperature theoretically attainable. It is the temperature at which the kinetic energy of atoms and molecules is minimal. It is equivalent to −273.15°C or −459.67°F. *See also* ZERO-POINT ENERGY; CRYOGENICS.

absorptance Symbol α. The ratio of the radiant or luminous flux absorbed by a body to the flux falling on it. Formerly called **absorptivity**, the absorptance of a *black body is equal to 1 by definition.

absorption 1. The take-up of a gas by a solid or liquid, or the take-up of a liquid by a solid. Absorption differs from *adsorption in that the absorbed substance permeates the bulk of the absorbing substance. **2.** The conversion of the energy of electromagnetic radiation, sound, streams of particles, etc., into other forms of energy on passing through a medium. A beam of light, for instance, passing through a medium, may lose intensity because of two effects: scattering of light out of the beam, and absorption of photons by atoms or molecules in the medium. When a photon is absorbed, there is a transition to an excited state.

absorption coefficient 1. (in physics) *See* LAMBERT'S LAWS. **2.** (in chemistry) The volume of a given gas, measured at standard temperature and pressure, that will dissolve in unit volume of a given liquid.

absorption spectrum *See* SPECTRUM.

absorptivity *See* ABSORPTANCE.

abundance 1. The ratio of the total mass of a specified element in the earth's crust to the total mass of the earth's crust, often expressed as a percentage. For example, the abundance of aluminium in the earth's crust is about 8%. **2.** The ratio of the number of atoms of a particular isotope of an element to the total number of atoms of all the isotopes present, often expressed as a percentage. For example, the abundance of uranium–235 in natural uranium is 0.71%. This is the **natural abundance**, i.e. the abundance as found in nature before any enrichment has taken place.

a.c. *See* ALTERNATING CURRENT.

acceleration Symbol a. The rate of increase of speed or velocity. It is measured in m s^{-2}. For a body moving linearly with constant acceleration a from a speed u to a speed v,

$$a = (v - u)/t = (v^2 - u^2)/2s$$

where t is the time taken and s the distance covered.

If the acceleration is not constant it is given by $dv/dt = d^2s/dt^2$. If the motion is not linear the vector character of displacement, velocity, and acceleration must be considered. *See also* ROTATIONAL MOTION.

acceleration of free fall Symbol g. The acceleration experienced by any massive object falling freely in the earth's gravitational field. Experimentally this is almost constant for all positions near the earth's surface, independent of the nature of the falling body (provided air resistance is eliminated). This is taken to indicate the strict proportionality of *weight (the force causing the acceleration) and inertial *mass, on the basis of *Newton's second law of motion. There is some variation of g with latitude, because of the earth's rotation and because the earth is not completely spherical. The standard value is taken as 9.806 65 m s^{-2}. The acceleration of free fall is also called the **acceleration due to gravity**.

accelerator An apparatus for increasing the kinetic energies of charged particles, used for research in nuclear and particle physics. *See* CYCLOTRON; LINEAR ACCELERATOR; SYNCHROCYCLOTRON; SYNCHROTRON.

acceptor A substance that is added as an impurity to a *semiconductor because of its ability to accept electrons from the valence bands, causing p-type conduction by the mobile positive holes left.

acceptor levels Energy levels of an acceptor atom in a *semiconductor, such as aluminium, in silicon. These energy levels are very near the top of the valence band, and therefore cause *p*-type conduction. *See also* ENERGY BANDS.

accommodation The process by which the focal length of the *lens of the eye is changed so that clear images of objects at a range of distances are displayed on the retina. In man and some other mammals accommodation is achieved by reflex adjustments in the shape of the lens brought about by relaxation and contraction of muscles within the ciliary body.

accretion disc A disc-shaped rotating mass formed by gravitational attraction. *See* BLACK HOLE; NEUTRON STAR; WHITE DWARF.

accumulator (secondary cell; storage battery) A type of *voltaic cell or battery that can be recharged by passing a current through it from an external d.c. supply. The charging current, which is passed in the opposite direction to that in which the cell supplies current, reverses the chemical reactions in the cell. The common types are the *lead–acid accumulator and the *nickel–iron accumulator and nickel–cadmium accumulator. *See also* SODIUM–SULPHUR CELL.

achromatic lens A lens that corrects for chromatic *aberration by using a combination of two lenses, made of different kinds of glass, such that their *dispersions neutralize each other although their *refractions do not. The aberration can be reduced further by using an **apochromatic lens**, which consists of three or more different kinds of glass.

acoustics **1.** The study of sound and sound waves. **2.** The characteristics of a building, especially an auditorium, with regard to its ability to enable speech and music to be heard clearly within it. For this purpose there should be no obtrusive echoes or resonances and the reverberation time should be near the optimum for the hall. Echoes are reduced by avoiding sweeping curved surfaces that could focus the sound and by breaking up large plane surfaces or covering them with sound-absorbing materials. Resonance is avoided by avoiding simple ratios for the main dimensions of the room, so that no one wavelength of sound is a factor of more than one of them. If the reverberation time is too long, speech will sound indistinct and music will be badly articulated, with one note persisting during the next. However, if it is too short, music sounds dead. Reverberation time is long in a bare room with hard walls, and can be deliberately reduced by carpets, soft furnishings, and sound-absorbent ('acoustic') felt. Reverberation times tend to be reduced by the presence of an audience and this must be taken into account in the design of the building.

acoustoelectronic devices (electroacoustic devices) Devices in which electronic signals are converted into acoustic waves. Acoustoelectronic devices are used in constructing *delay lines and also in converting digital data from computers for transmission by telephone lines.

actinic radiation Electromagnetic radiation that is capable of initiating a chemical reaction. The term is used especially of ultraviolet radiation and also to denote radiation that will affect a photographic emulsion.

actinium series *See* RADIOACTIVE SERIES.

actinoid contraction A smooth decrease in atomic or ionic radius with increasing proton number found in the actinoids.

actinometer Any of various instruments for measuring the intensity of electromagnetic radiation. Modern actinometers use the *photoelectric effect; earlier instruments depended either on the fluorescence produced by the radiation on a screen or on the amount of chemical change induced in some suitable substance.

action at a distance A direct and instantaneous interaction between bodies that are not in physical contact with each other. This type of interaction is not consistent with the special theory of *relativity, which states that nothing (including interactions) can travel through space faster than the *speed of light in a vacuum. For this reason it is more logical to describe interactions between bodies by *field theories or by the exchange of virtual particles (*see* VIRTUAL STATE) rather than theories based on action at a distance.

action potential The change in electrical potential that occurs across a cell membrane during the passage of a nerve impulse. As an impulse travels in a wavelike manner along the axon of a nerve, it causes a localized and transient switch in electric potential across the cell membrane from –60 mV (millivolts; the resting potential) to +45 mV. The change in electric potential is caused by an influx of sodium ions. Nervous stimulation of a muscle fibre has a similar effect.

action spectrum A graphical plot of the efficiency of electromagnetic radiation in producing a photochemical reaction against the wavelength of the radiation used. For example, the action spectrum for photosynthesis using light shows a peak in the region 670–700 nm. This corresponds to a maximum absorption in the absorption spectrum of chlorophylls in this region.

activation analysis An analytical technique that can be used to detect elements when present in a sample. In **neutron activation analysis** the sample is exposed to a flux of thermal neutrons in a nuclear reactor. Some of these neutrons are captured by nuclides in the sample to form nuclides of the same atomic number but a higher mass number. These newly formed nuclides emit gamma radiation, which can be used to identify the element present by means of a gamma-ray spectrometer. Activation analysis has also been employed using high-energy charged particles, such as protons or alpha particles.

active device 1. An electronic component, such as a transistor, that is

capable of amplification. **2.** An artificial *satellite that receives
information and retransmits it after amplification. **3.** A radar device that
emits microwave radiation and provides information about a distant body
by receiving a reflection of this radiation. *Compare* PASSIVE DEVICE.

activity 1. Symbol a. A thermodynamic function used in place of
concentration in equilibrium constants for reactions involving nonideal
gases and solutions. For example, in a reaction

$$A \rightleftharpoons B + C$$

the true equilibrium constant is given by

$$K = a_B a_C / a_A$$

where a_A, a_B, and a_C are the activities of the components, which function
as concentrations (or pressures) corrected for nonideal behaviour. **Activity
coefficients** (symbol γ) are defined for gases by $\gamma = a/p$ (where p is pressure)
and for solutions by $\gamma = aX$ (where X is the mole fraction). Thus, the
equilibrium constant of a gas reaction has the form

$$K_p = \gamma_B p_B \gamma_C p_C / \gamma_A p_A$$

The equilibrium constant of a reaction in solution is

$$K_c = \gamma_B X_B \gamma_C X_C / \gamma_A X_A$$

The activity coefficients thus act as correction factors for the pressures or
concentrations. *See* FUGACITY. **2.** Symbol A. The number of atoms of a
radioactive substance that disintegrate per unit time. The **specific activity**
(a) is the activity per unit mass of a pure radioisotope. *See* RADIATION
UNITS.

additive process *See* COLOUR.

adiabatic approximation An approximation used in *quantum
mechanics when the time dependence of parameters, such as the internuclear distance between atoms in a molecule is slowly varying. This
approximation means that the solution of the *Schrödinger equation at
one time goes continuously over to the solution at a later time. This
approximation was formulated by Max Born and the Soviet physicist
Vladimir Alexandrovich Fock (1898–1974) in 1928. The *Born–
Oppenheimer approximation is an example of the adiabatic
approximation.

adiabatic demagnetization A technique for cooling a paramagnetic
salt, such as potassium chrome alum, to a temperature near *absolute
zero. The salt is placed between the poles of an electromagnet and the
heat produced during magnetization is removed by liquid helium. The
salt is then isolated thermally from the surroundings and the field is
switched off; the salt is demagnetized adiabatically and its temperature
falls. This is because the demagnetized state, being less ordered, involves
more energy than the magnetized state. The extra energy can come only
from the internal, or thermal, energy of the substance. It is possible to
obtain temperatures of about 0.005 K in this way.

adiabatic process Any process that occurs without heat entering or leaving a system. In general, an adiabatic change involves a fall or rise in temperature of the system. For example, if a gas expands under adiabatic conditions, its temperature falls (work is done against the retreating walls of the container). The **adiabatic equation** describes the relationship between the pressure (p) of an ideal gas and its volume (V), i.e. $pV^\gamma = K$, where γ is the ratio of the principal specific *heat capacities of the gas and K is a constant.

admittance Symbol Y. The reciprocal of *impedance. It is measured in siemens.

ADSL (asymmetric digital subscriber line) A mechanism by which *broadband communication via the Internet can be made available using pre-existing telephone lines, while allowing simultaneous use of the line for normal telephone calls. Data communication by ADSL is asymmetric in that upstream (transmitting) communication is slower than downstream (receiving) communication, typically half as fast. ADSL coexists with standard telephone operation on the same line by the use of band separation filters at each telephone socket.

adsorbate A substance that is adsorbed on a surface.

adsorption The formation of a layer of gas, liquid, or solid on the surface of a solid or, less frequently, of a liquid. There are two types depending on the nature of the forces involved. In **chemisorption** a single layer of molecules, atoms, or ions is attached to the adsorbent surface by chemical bonds. In **physisorption** adsorbed molecules are held by the weaker *van der Waals' forces. Adsorption is an important feature of surface reactions, such as corrosion, and heterogeneous catalysis. The property is also utilized in adsorption chromatography.

advanced gas-cooled reactor (AGR) *See* NUCLEAR REACTOR.

aerial (antenna) The part of a radio or television system from which radio waves are transmitted into the atmosphere or space (**transmitting aerial**) or by which they are received (**receiving aerial**). A **directional** or **directive aerial** is one in which energy is transmitted or received more effectively from some directions than others, whereas an **omnidirectional aerial** transmits and receives equally well in all directions.

aerodynamics The study of the motion of gases (particularly air) and the motion of solid bodies in air. Aerodynamics is particularly concerned with the motion and stability of aircraft. Another application of aerodynamics is to the flight of birds and insects. The branch of aerodynamics concerned with the flow of gases through compressors, ducts, fans, orifices, etc., is called **internal aerodynamics**.

   Aerodynamic drag is the force that opposes the motion of a body moving relative to a gas and is a function of the density of the gas, the square of the relative velocity, the surface area of the body, and a quantity

a

called the **drag coefficient**, which is a function of the *Reynolds number. **Aerodynamic lift** is an upward force experienced by a body moving through a gas and is a function of the same variables as aerodynamic drag.

aerogenerator *See* WIND POWER.

aeronautics The branch of *aerodynamics concerned with the design, construction, and operation of aircraft and rockets.

aerosol A colloidal dispersion of a solid or liquid in a gas. The commonly used aerosol sprays contain an inert propellant liquefied under pressure. Halogenated alkanes containing chlorine and fluorine (chlorofluorocarbons, or CFCs) have been used in aerosol cans. This use has been criticized on the grounds that these compounds persist in the atmosphere and lead to depletion of the ozone layer.

aerospace The earth's atmosphere and the space beyond it.

AFM *See* ATOMIC FORCE MICROSCOPE.

after-heat Heat produced by a nuclear reactor after it has been shut down. The after-heat is generated by radioactive substances formed in the fuel elements.

age of the earth The time since the earth emerged as a planet of the sun, estimated by *dating techniques to be about 4.6×10^9 years. The oldest known rocks on earth are estimated by their *radioactive age to be about 3.5×10^9 years old. The earth is older than this because of the long time it took to cool. An estimate for the cooling time is included in the estimate for the age of the earth.

age of the universe A time determined by the reciprocal of the value of the *Hubble constant to be about 13.7 billion years. The calculation of the Hubble constant, and hence the age of the universe, depends on which theory of *cosmology is used. Usually, the age of the universe is calculated by assuming that the *expansion of the universe can be described by the *big-bang theory.

AGR Advanced gas-cooled reactor. *See* NUCLEAR REACTOR.

AI *See* ARTIFICIAL INTELLIGENCE.

air *See* EARTH'S ATMOSPHERE.

albedo 1. The ratio of the radiant flux reflected by a surface to that falling on it. **2.** The probability that a neutron entering a body of material will be reflected back through the same surface as it entered.

algebraic sum The total of a set of quantities paying due regard to sign, e.g. the algebraic sum of 3 and –4 is –1.

algorithm A method of solving a problem, involving a finite series of steps. In computing practice the algorithm denotes the expression on paper of the proposed computing process (often by means of a flowchart) prior to the preparation of the program. If no algorithm is possible a *heuristic solution has to be sought.

allowed bands *See* ENERGY BANDS.

allowed transitions *See* SELECTION RULES.

alloy A material consisting of two or more metals (e.g. brass is an alloy of copper and zinc) or a metal and a nonmetal (e.g. steel is an alloy of iron and carbon, sometimes with other metals included). Alloys may be compounds, *solid solutions, or mixtures of the components.

alloy steels *See* STEEL.

Alnico A tradename for a series of alloys, containing iron, aluminium, nickel, cobalt, and copper, used to make permanent magnets.

alpha particle A helium–4 nucleus emitted by a larger nucleus during the course of the type of radioactive decay known as **alpha decay**. As a helium–4 nucleus consists of two protons and two neutrons bound together as a stable entity the loss of an alpha particle involves a decrease in *nucleon number of 4 and decrease of 2 in the *atomic number, e.g. the decay of a uranium–238 nucleus into a thorium–234 nucleus. A stream of alpha particles is known as an **alpha-ray** or **alpha-radiation**.

alternating current (a.c.) An electric current that reverses its direction with a constant *frequency (f). If a graph of the current against time has the form of a *sine wave, the current is said to be **sinusoidal**. Alternating current, unlike direct current, is therefore continuously varying and its magnitude is either given as its peak value (I_0) or its *root-mean-square value ($I_0/\sqrt{2}$ for a sinusoidal current). This r.m.s. value is more useful as it is comparable to a d.c. value in being a measure of the ability of the current to transmit power. The instantaneous value of a sinusoidal current (I) is given by $I = I_0\sin2\pi ft$.

If a direct current is supplied to a circuit the only opposition it encounters is the circuit's *resistance. However, an alternating current is opposed not only by the resistance of the circuit but also by its *reactance. This reactance is caused by *capacitance and *inductance in the circuit. In a circuit consisting of a resistance (R), an inductance (L), and a capacitance (C) all in series, the reactance (X) is equal to $(2\pi fL) - (1/2\pi fC)$. The total opposition to the current, called the *impedance (Z), is then equal to the ratio of the r.m.s. applied p.d. to the r.m.s. current and is given by $\sqrt{(R^2 + X^2)}$.

alternator An *alternating-current generator consisting of a coil or coils that rotate in the magnetic field produced by one or more permanent magnets or electromagnets. The electromagnets are supplied by an independent direct-current source. The frequency of the alternating

a

current produced depends on the speed at which the coil rotates and the number of pairs of magnetic poles. In the large alternators of power stations the electromagnets rotate inside fixed coils; many bicycle dynamos are alternators with rotating permanent magnets inside fixed coils.

altimeter A device used to measure height above sea level. It usually consists of an aneroid *barometer measuring atmospheric pressure. Aircraft are fitted with altimeters, which are set to the atmospheric pressure at a convenient level, usually sea level, before take off. The height of the aircraft can then be read off the instrument as the aircraft climbs and the pressure falls.

ALU (arithmetic/logic unit) The part of the central processor of a *computer in which simple arithmetic and logical operations are performed electronically. For example, the ALU can add, subtract, multiply, compare two numbers, or negate a number.

Alvarez, Luis Walter (1911–88) US physicist most of whose working life was spent at the University of California, Berkeley. After working on radar and the atomic bomb during World War II, he concentrated on particle physics. In 1959 he built the first large *bubble chamber and developed the technique for using it to study charged particles, for which he was awarded the 1968 Nobel Prize for physics. He later became interested in the extinction of the dinosaurs.

AM (amplitude modulation) *See* MODULATION.

amalgam An alloy of mercury with one or more other metals. Most metals form amalgams (iron and platinum are exceptions), which may be liquid or solid. Some contain definite intermetallic compounds, such as $NaHg_2$.

americium Symbol Am. A radioactive metallic transuranic element belonging to the actinoids; a.n. 95; mass number of most stable isotope 243 (half-life 7.95×10^3 years); r.d. 13.67 (20°C); m.p. 994 ± 4°C; b.p. 2607°C. Ten isotopes are known. The element was discovered by G. T. Seaborg and associates in 1945, who obtained it by bombarding uranium–238 with alpha particles.

ammeter An instrument that measures electric current. The main types are the **moving-coil** ammeter, the **moving-iron** ammeter, and the **thermoammeter**. The moving-coil instrument is a moving-coil *galvanometer fitted with a *shunt to reduce its sensitivity. It can only be used for d.c., but can be adapted for a.c. by using a *rectifier. In moving-iron instruments, a piece of soft iron moves in the magnetic field created when the current to be measured flows through a fixed coil. They can be used with a.c. or d.c. but are less accurate (though more robust) than the moving-coil instruments. In thermoammeters, which can also be used with a.c. or d.c., the current is passed through a resistor, which heats up as the current passes. This is in contact with a thermocouple, which is connected to a galvanometer. This indirect system is mainly used for measuring high frequency a.c. In the **hot-wire** instrument the wire is clamped at its ends and its elongation as it is heated causes a pointer to move over a scale.

ammonia clock A form of atomic clock in which the frequency of a quartz oscillator is controlled by the vibrations of excited ammonia molecules (*see* EXCITATION). The ammonia molecule (NH_3) consists of a pyramid with a nitrogen atom at the apex and one hydrogen atom at each corner of the triangular base. When the molecule is excited, once every 20.9 microseconds the nitrogen atom passes through the base and forms a pyramid the other side: 20.9 microseconds later it returns to its original position. This vibration back and forth has a frequency of 23 870 hertz and ammonia gas will only absorb excitation energy at exactly this frequency. By using a *crystal oscillator to feed energy to the gas and a suitable feedback mechanism, the oscillator can be locked to exactly this frequency.

amorphous Describing a solid that is not crystalline; i.e. one that has no long-range order in its lattice. Many powders that are described as 'amorphous' in fact are composed of microscopic crystals, as can be demonstrated by X-ray diffraction. Glasses are examples of true amorphous solids.

amount of substance Symbol n. A measure of the number of entities present in a substance. The specified entity may be an atom, molecule, ion, electron, photon, etc., or any specified group of such entities. The amount of substance of an element, for example, is proportional to the number of atoms present. For all entities, the constant of proportionality is the inverse *Avogadro constant. The SI unit of amount of substance is the *mole.

ampere Symbol A. The SI unit of electric current. The constant current that, maintained in two straight parallel infinite conductors of negligible cross section placed one metre apart in a vacuum, would produce a force between the conductors of 2×10^{-7} N m^{-1}. This definition replaced the earlier international ampere defined as the current required to deposit 0.001 118 00 gram of silver from a solution of silver nitrate in one second. The unit is named after A. M. Ampère.

Ampère, André Marie (1775–1836) French physicist who from 1809 taught at the Ecole Polytechnique in Paris. He is best known for putting electromagnetism (which he called 'electrodynamics') on a mathematical basis. In 1825 he formulated *Ampère's law. The *ampere is named after him.

ampere-hour A practical unit of electric charge equal to the charge flowing in one hour through a conductor passing one ampere. It is equal to 3600 coulombs.

Ampère's law A law of the form

$$dB = (\mu_o \, I \sin \theta \, dl)/4\pi r^2,$$

where dB is the infinitesimal element of the magnitude of the magnetic flux density (*see* MAGNETIC FIELD) at a distance r at a point P from the element length dl of a conductor, μ_o is the magnetic permeability of free space, I is the current flowing through the conductor, and θ is the angle between the direction of the current and the line joining the element of the conductor and P. It is also called the **Ampère–Laplace law** after the French mathematician Pierre-Simon de Laplace (1749–1827). There are several mathematically equivalent statements of this law.

Ampère's rule A rule that relates the direction of the electric current passing through a conductor and the magnetic field associated with it. The rule states that if the electric current is moving away from an observer, the direction of the lines of force of the magnetic field surrounding the conductor is clockwise and that if the electric current is moving towards an observer, the direction of the lines of force is counter-clockwise. An equivalent statement to Ampère's rule is known as the **corkscrew rule**. A corkscrew, or screwdriver, is said to be right-handed if turning the corkscrew in a clockwise direction drives the screw into the object (such as the cork of a bottle). The corkscrew rule states that a right-handed corkscrew is analogous to an electric current and its magnetic field with the direction of the screw being analogous to electric current; the direction in which the corkscrew is being turned is analogous to the direction of lines of force of the field.

ampere-turn The SI unit of *magnetomotive force equal to the magnetomotive force produced when a current of one ampere flows through one turn of a magnetizing coil.

amplifier A device that increases the strength of an electrical signal by drawing energy from a separate source to that of the signal. The original device used in electronic amplifiers was the *triode valve, in which the cathode–anode current is varied in accordance with the low-voltage signal applied to the valve's control grid. In the more recent *transistor, the emitter–collector current is controlled in much the same way by the signal applied to the transistor's base region. In the most modern devices the complete amplifier circuit is manufactured as a single *integrated circuit. The ratio of the output amplitude (of p.d. or current) of an amplifier (or stage of an amplifier) to the corresponding input amplitude is called the **gain** of the amplifier.

amplitude *See* WAVE.

amplitude modulation (AM) *See* MODULATION; RADIO.

a.m.u. *See* ATOMIC MASS UNIT.

analyser A device, used in the *polarization of light, that is placed in the eyepiece of a *polarimeter to observe plane-polarized light. The analyser, which may be a *Nicol prism or *Polaroid, can be oriented in different directions to investigate in which plane an incoming wave is polarized or if the light is plane polarized. If there is one direction from which light does not emerge from the analyser when it is rotated, the incoming wave is plane polarized. If the analyser is horizontal when extinction of light takes place, the polarization of light must have been in the vertical plane. The intensity of a beam of light transmitted through an analyser is proportional to $\cos^2\theta$, where θ is the angle between the plane of polarization and the plane of the analyser. Extinction is said to be produced by 'crossing' the *polarizer and analyser.

analytical geometry (coordinate geometry) A form of geometry in which points are located in a two-dimensional, three-dimensional, or higher dimensional space by means of a system of coordinates. Curves are represented by an equation for a set of such points. The geometry of figures can thus be analysed by algebraic methods. *See* CARTESIAN COORDINATES; POLAR COORDINATES.

anastigmatic lens 1. An objective lens for an optical instrument in which all *aberrations, including *astigmatism, are reduced greatly. **2.** A spectacle lens designed to correct astigmatism. It has different radii of curvature in the vertical and horizontal planes.

anchor ring *See* TORUS.

AND circuit *See* LOGIC CIRCUITS.

Anderson, Carl David (1905–91) US physicist who became a professor at the California Institute of Technology, where he worked mainly in particle physics. In 1932 he discovered the *positron in *cosmic radiation, and four years later was awarded the Nobel Prize. In 1937 he discovered the mu-meson (muon).

anechoic Having a low degree of reverberation with little or no reflection of sound. An **anechoic chamber** is one designed for experiments in acoustics. The walls are covered with small pyramids to avoid the formation of stationary waves between facing surfaces and the whole of the interior surface is covered with an absorbent material to avoid reflections.

anemometer An instrument for measuring the speed of the wind or any other flowing fluid. The simple **vane anemometer** consists of a number of cups or blades attached to a central spindle so that the air, or other fluid, causes the spindle to rotate. The instrument is calibrated to give a wind speed directly from a dial. The instrument can be mounted to rotate about a vertical axis and in this form it also gives an indication of the direction of the wind. A **hot-wire anemometer** consists of an electrically heated wire that is cooled by the flow of fluid passing round it. The faster the flow the lower the temperature of the wire and the lower its resistance. Thus the rate of flow can be calculated by measuring the resistance of the wire.

aneroid barometer *See* BAROMETER.

angle modulation *See* MODULATION.

angle of incidence **1.** The angle between a ray falling on a surface and the perpendicular (normal) to the surface at the point at which the ray strikes the surface. **2.** The angle between a wavefront and a surface that it strikes.

angle of reflection **1.** The angle between a ray leaving a reflecting surface and the perpendicular (normal) to the surface at the point at which the ray leaves the surface. **2.** The angle between a wavefront and a surface that it leaves.

angle of refraction **1.** The angle between a ray that is refracted at a surface between two different media and the perpendicular (normal) to the surface at the point of refraction. **2.** The angle between a wavefront and a surface at which it has been refracted.

angstrom Symbol Å. A unit of length equal to 10^{-10} metre. It was formerly used to measure wavelengths and intermolecular distances but has now been replaced by the nanometre. 1 Å = 0.1 nanometre. The unit is named after A. J. Ångström.

Ångström, Anders Jonas (1814–74) Swedish astronomer and physicist who became professor of physics at the University of Uppsala from 1858 until his death. He worked mainly with emission *spectra, demonstrating the presence of hydrogen in the sun. He also worked out the wavelengths of *Fraunhofer lines. Spectral wavelengths were formerly expressed in *angstroms, but the nanometre is now more usual.

angular displacement, velocity, and acceleration *See* ROTATIONAL MOTION.

angular frequency (pulsatance) A quantity proportional to the *frequency of a periodic phenomenon but having the dimensions of angular velocity. The angular frequency in radians per second = frequency in hertz × 2π radians per cycle.

angular magnification (magnifying power) *See* MAGNIFICATION.

angular momentum Symbol L. The product of the angular velocity of a body and its *moment of inertia about the axis of rotation, i.e. $L = I\omega$.

anharmonic oscillator An oscillating system (in either *classical physics or *quantum mechanics) that is not oscillating in *simple harmonic motion. In general, the problem of an anharmonic oscillator is not exactly soluble, although many systems approximate to harmonic oscillators and for such systems the **anharmonicity** (the deviation of the system from being a *harmonic oscillator) can be calculated using *perturbation theory. If the anharmonicity is large, other approximate or numerical techniques have to be used to solve the problem.

anion A negatively charged *ion, i.e. an ion that is attracted to the *anode in *electrolysis. *Compare* CATION.

anisotropic Denoting a medium in which certain physical properties are different in different directions. Wood, for instance, is an anisotropic material: its strength along the grain differs from that perpendicular to the grain. Single crystals that are not cubic are anisotropic with respect to some physical properties, such as the transmission of electromagnetic radiation. *Compare* ISOTROPIC.

annealing A form of heat treatment applied to a metal to soften it, relieve internal stresses and instabilities, and make it easier to work or machine. It consists of heating the metal to a specified temperature for a specified time, both of which depend on the metal involved, and then allowing it to cool slowly. It is applied to both ferrous and nonferrous metals and a similar process can be applied to other materials, such as glass.

annihilation The destruction of a particle and its *antiparticle as a result of a collision between them. The **annihilation radiation** produced is carried away by *photons or *mesons. For example, in a collision between an electron and a positron the energy produced is carried away by two photons, each having an energy of 0.511 MeV, which is equivalent to the rest-mass energies of the annihilated particles plus their kinetic energies. When nucleons annihilate each other the energy is carried away by mesons.

annulus The plane figure formed between two concentric circles of different radii, R and r. Its area is $\pi(R^2 - r^2)$.

anode A positive electrode. In *electrolysis anions are attracted to the anode. In an electronic vacuum tube it attracts electrons from the *cathode and it is therefore from the anode that electrons flow out of the device. In these instances the anode is made positive by external means; however in a *voltaic cell the anode is the electrode that spontaneously becomes positive and therefore attracts electrons to it from the external circuit.

anomaly 1. An angle used to fix the position of a body, such as a planet, in an elliptical orbit. The **true anomaly** of a planet is the angle between the *perihelion, the sun, and the planet in the direction of the planet's motion. The **mean anomaly** is the angle between the perihelion, the sun, and an imaginary planet having the same period as the real planet, but assumed to be moving at constant speed. **2.** A situation in which a classical theory has a symmetry but the corresponding quantum theory does not. There are several types of quantum anomaly. Examples of the consequences of anomalies are that the number of types of *lepton must equal the number of types of *quark and that only certain groups are viable for *superstring theory.

ANSI American National Standards Institute: a US body that accredits organizations to write industrial standards – publicly available definitions, requirements, criteria, etc. – following the rules established by ANSI.

antenna *See* AERIAL.

anthropic principle The principle that the observable universe has to be as it is, rather than any other way, otherwise we would not be able to observe it. There are many versions of the anthropic principle. The **weak anthropic principle** is specifically concerned with the conditions necessary for conscious life on earth and asserts that numerical relations found for fundamental constants, such as the *gravitational constant, have to hold at the present epoch because at any other epoch there would be no intelligent lifeform to measure the constants. The **strong anthropic principle** is concerned with all possible universes and whether intelligent life could exist in any other universe, including the possibility of different fundamental constants and laws of physics. The anthropic principle is viewed with considerable scepticism by many physicists.

antiatom An atom in which all the particles of an ordinary atom are replaced by their *antiparticles, i.e. electrons by positrons, protons by antiprotons, and neutrons by antineutrons. An antiatom cannot co-exist with an ordinary atom since the atom and the antiatom would annihilate each other with the production of energy in the form of high-energy *photons. Antiatoms have been created artificially in laboratories.

antiferromagnetism *See* MAGNETISM.

antilogarithm *See* LOGARITHM.

antimatter *See* ANTIPARTICLE.

antinode *See* STATIONARY WAVE.

antiparallel spins Neighbouring spinning electrons in which the *spins, and hence the magnetic moments, of the electrons are aligned in the opposite direction. The interaction between the magnetic moments of electrons in atoms is dominated by *exchange forces. Under some circumstances the exchange interactions between magnetic moments favour *parallel spins, while under other conditions they favour antiparallel spins. The case of antiferromagnetism (*see* MAGNETISM) is an example of a system with antiparallel spins.

antiparallel vectors Vectors directed along the same line but in opposite directions.

antiparticle A subatomic particle that has the same mass as another particle and equal but opposite values of some other property or properties. For example, the antiparticle of the electron is the positron, which has a positive charge equal in magnitude to the electron's negative charge. The antiproton has a negative charge equal to the proton's positive charge. The neutron and the antineutron have *magnetic moments opposite in sign relative to their *spins. The existence of antiparticles is predicted by relativistic *quantum mechanics. When a particle and its corresponding antiparticle collide *annihilation takes place. **Antimatter** consists of matter made up of antiparticles. For example, antihydrogen consists of an antiproton with an orbiting positron. Antihydrogen has been artificially created in the laboratory. The spectrum of antihydrogen should be identical to that of hydrogen. It appears that the universe consists overwhelmingly of (normal) matter, and explanations of the absence of large amounts of antimatter have been incorporated into cosmological models that involve the use of *grand unified theories of elementary particles.

anyon *See* QUANTUM STATISTICS.

aperture The effective diameter of a lens or mirror. The ratio of the effective diameter to the focal length is called the **relative aperture**, which is commonly known as the aperture, especially in photographic usage. The reciprocal of the relative aperture is called the **focal ratio**. The numerical value of the focal ratio is known as the **f-number** of a lens. For example, a camera lens with a 40 mm focal length and a 10 mm aperture has a relative aperture of 0.25 and a focal ratio of 4. Its f-number would be f/4, often written f4.

The light-gathering power of a telescope depends on the area of the lens, i.e. it is related to the square of the aperture. However, the larger the relative aperture the greater the *aberrations. In microscopy large-aperture objectives (corrected for aberrations) are preferred, since they reduce the blurring caused by *diffraction of light waves.

aperture synthesis *See* RADIO TELESCOPE.

aphelion The point in the orbit of a planet, comet, or artificial satellite in solar orbit at which it is farthest from the sun. The earth is at aphelion on about July 3. *Compare* PERIHELION.

aplanatic lens A lens that reduces both spherical *aberration and *coma.

apochromatic lens *See* ACHROMATIC LENS.

apocynthion The point in the orbit around the moon of a satellite launched from the earth that is furthest from the moon. For a satellite launched from the moon the equivalent point is the **apolune**. *Compare* PERICYNTHION.

apogee The point in the orbit of the moon, or an artificial earth satellite, at which it is furthest from the earth. At apogee the moon is 406 700 km from the earth, some 42 000 km further away than at **perigee**, the nearest point to the earth.

apolune *See* APOCYNTHION.

apparent expansivity *See* EXPANSIVITY.

Appleton layer *See* EARTH'S ATMOSPHERE.

applications software Computer programs, or collections of programs, designed to meet the needs of the users of computer systems by directly contributing to the performance of specific roles. Examples include a word-processing or spreadsheet program, or a company's payroll package. In contrast, **systems software**, such as an operating system, are the group of programs required for effective use of a computer system.

approximation technique A method used to solve a problem in mathematics, or its physical applications, that does not give an exact solution but that enables an approximate solution to be found.

apsides The two points in an astronomical orbit that lie closest to (**periapsis**) and farthest from (**apoapsis**) the centre of gravitational attraction. The **line of apsides** is the straight line that joins the two apsides. If the orbit is elliptical the line of apsides is the major axis of the ellipse.

aqueous Describing a solution in water.

Arago's spot *See* POISSON'S SPOT.

arccos, arcsin, arctan *See* INVERSE FUNCTIONS.

Archimedes of Syracuse (287–212 BC) Greek mathematician who spent most of his life at his birthplace working on levers and other aspects of mechanics. In hydrostatics he devised a pump (the **Archimedian screw**) and formulated *Archimedes' principle. His method of successive approximations allowed him to determine the value of π to a good approximation. He was killed by a soldier in the Roman siege of Syracuse.

Archimedes' principle The weight of the liquid displaced by a floating body is equal to the weight of the body. The principle was not in fact stated by Archimedes, though it has some connection with his discoveries. The principle is often stated in the form: when a body is (partially or totally) immersed in a fluid, the upthrust on the body is equal to the weight of fluid displaced.

arc lamp *See* ELECTRIC LIGHTING.

arcosh, arsinh, artanh *See* INVERSE FUNCTIONS.

Argand diagram *See* COMPLEX NUMBER.

argument 1. A sequence of logical propositions based on a set of premises and leading to a conclusion. **2.** *See* COMPLEX NUMBER.

arithmetic average (arithmetic mean) *See* AVERAGE.

arithmetic/logic unit *See* ALU.

arithmetic series (arithmetic progression) A series or progression of numbers in which there is a common difference between terms, e.g. 3, 9, 15, 21,… is an arithmetic series with a common difference of 6. The general formula for the nth term is

$$[a + (n - 1)d]$$

and the sum of n terms is

$$n[2a + (n - 1)d]/2.$$

Compare GEOMETRIC SERIES.

armature Any moving part in an electrical machine in which a voltage is induced by a magnetic field, especially the rotating coils in an electric motor or generator and the ferromagnetic bar attracted by an electromagnet in a *relay.

artificial intelligence (AI) A field of computing concerned with the production of programs that perform tasks requiring intelligence when done by people. These tasks include playing games, such as chess or draughts, forming plans, understanding speech and natural languages, interpreting images, reasoning, and learning.

ASCII (pronounced 'asky') American standard code for information interchange: a standard scheme for encoding the letters A–Z, a–z, digits 0–9, punctuation marks, and other special and control characters in binary form. Originally developed in the US, it is widely used in many computers and for interchanging information between computers. Characters are encoded as strings of seven *bits, providing 2^7 or 128 different bit patterns. International 8-bit codes that are extensions of ASCII have been published by the International Standards Organization; these allow the accented Roman letters used in European languages, as well as Cyrillic, Arabic, Greek, and Hebrew characters, to be encoded.

Aspect experiment An experiment conducted by the French physicist Alain Aspect (1947–   ) and his colleagues in the early 1980s to test Bell's inequality (*see* BELL'S THEOREM). The experiment involves producing pairs of photons from a source of excited calcium ions. The photons have different wavelengths and filters are used to ensure that the photons in a pair travel to different detectors in different directions. The photons are circularly polarized and the net angular momentum of the pair is zero. Two polarizing filters are used, each placed at an angle in the path of the

a

photons. These filters either reflect or transmit photons of different linear polarization to one of four detectors (two for each polarizing filter). Coincidence measurements are made using these detectors and the experiment is organized so that the measurements apply only to photons separated to a point at which they cannot communicate by sending a signal at the speed of light. The results are generally believed to show that there are no local *hidden variables in quantum mechanics.

associative law The mathematical law stating that the value of an expression is independent of the grouping of the numbers, symbols, or terms in the expression. The **associative law for addition** states that numbers may be added in any order, e.g. $(x + y) + z = (x + z) + y$. The **associative law for multiplication** states that numbers can be multiplied in any order, e.g. $x(yz) = (xy)z$. Subtraction and division are not associative. *Compare* COMMUTATIVE LAW; DISTRIBUTIVE LAW.

astatic galvanometer A sensitive form of moving-magnet *galvanometer in which any effects of the earth's magnetic field are cancelled out. Two small oppositely directed magnets are suspended at the centres of two oppositely wound coils. As its resultant moment on the magnets is zero, the earth's field has no effect and the only restoring torque on the magnets is that provided by the suspending fibre. This makes a sensitive but delicate instrument.

astatine Symbol At. A radioactive halogen element; a.n. 85; r.a.m. 211; m.p. 302°C; b.p. 337°C. It occurs naturally by radioactive decay from uranium and thorium isotopes. Astatine forms at least 20 isotopes, the most stable astatine–210 has a half-life of 8.3 hours. It can also be produced by alpha bombardment of bismuth–200. Astatine is stated to be more metallic than iodine; at least 5 oxidation states are known in aqueous solutions. It will form interhalogen compounds, such as AtI and AtCl. The existence of At_2 has not yet been established. The element was synthesized by nuclear bombardment in 1940 by D. R. Corson, K. R. MacKenzie, and E. Segrè at the University of California.

asteroids (minor planets; planetoids) A number of small bodies that revolve around the sun between the orbits of Mars and Jupiter in a zone between 1.7 and 4.0 astronomical units from the sun (the **asteroid belt**). The size of the bodies varies from the largest, Ceres (with a diameter of 933 km), to objects less than 1 km in diameter. It is estimated that there are about 10 bodies with diameters in excess of 250 km and some 120 bodies with diameters over 130 km.

asthenosphere A layer of the earth's mantle (*see* EARTH) that underlies the lithosphere at a depth of about 70 km. The velocity of *seismic waves is considerably reduced in the asthenosphere and it is thought to be a zone of partial melting. It extends to a depth of about 250 km where rocks are solid.

astigmatism A lens defect in which when rays in one plane are in focus those in another plane are not. In lenses and mirrors it occurs with objects not on the axis and is best controlled by reducing the *aperture to restrict the use of the lens or mirror to its central portion. The eye can also suffer from astigmatism, usually when the cornea is not spherical. It is corrected by using an *anastigmatic lens.

Aston, Francis William (1877–1945) British chemist and physicist, who until 1910 worked at Mason College (later Birmingham University) and then with J. J. *Thomson at Cambridge University. In 1919 Aston designed the mass spectrograph (*see* MASS SPECTRUM), for which he was awarded the Nobel Prize for chemistry in 1922. With it he discovered the *isotopes of neon, and was thus able to explain nonintegral atomic weights.

astrometry The branch of astronomy concerned with the measurement of the positions of the celestial bodies on the *celestial sphere.

astronomical telescope *See* TELESCOPE.

astronomical unit (AU) The mean distance between the sun and the earth. It is equal to 149 597 870 km (499 light seconds).

astronomy The study of the universe beyond the earth's atmosphere. The main branches are *astrometry, *celestial mechanics, and *astrophysics.

astrophysics The study of the physical and chemical processes involving astronomical phenomena. Astrophysics deals with stellar structure and evolution (including the generation and transport of energy within stars), the properties of the interstellar medium and its interactions with stellar systems, and the structure and dynamics of systems of stars (such as *clusters and *galaxies), and of systems of galaxies. *See also* COSMOLOGY.

asymmetric atom *See* OPTICAL ACTIVITY.

asymptote A line that a curve approaches but only touches at infinity.

asymptotic freedom The consequence of certain *gauge theories, particularly *quantum chromodynamics, that the forces between such particles as quarks become weaker at shorter distances (i.e. higher energies) and vanish as the distance between particles tends to zero. Only non-Abelian gauge theories with unbroken gauge symmetries can have asymptotic freedom (*see* GROUP THEORY). In contrast, *quantum electrodynamics implies that the interaction between particles decreases as a result of dielectric screening; asymptotic freedom for quarks implies that antiscreening occurs. Physically, asymptotic freedom postulates that the *vacuum state for gluons is a medium that has colour paramagnetism, i.e. the vacuum antiscreen colour charges.

Asymptotic freedom explains the successes of the *parton model of pointlike objects inside hadrons and enables systematic corrections to the parton model to be calculated using perturbation theory. That the interaction between quarks increases as the distance between them increases has given rise to the hypothesis of *quark confinement. It appears that if a theory requires the presence of Higgs bosons, asymptotic freedom is destroyed. Thus, *electroweak theory does not have asymptotic freedom.

asymptotic series A series formed by the expansion of a function in the form $a_0 + a_1/x + a_2/x^2 + \ldots + a_n/x^n + \ldots$, such that the error resulting from terminating the series at the term a_n/x^n tends to zero more rapidly than $1/x^n$ as x tends to infinity. An asymptotic series expansion is not necessarily a *convergent series. Asymptotic series are used very extensively in the physical sciences.

atmolysis The separation of a mixture of gases by means of their different rates of diffusion. Usually, separation is effected by allowing the gases to diffuse through the walls of a porous partition or membrane.

atmosphere 1. Symbol atm. A unit of pressure equal to 101 325 pascals. This is equal to 760.0 mmHg. The actual *atmospheric pressure fluctuates around this value. The unit is usually used for expressing pressures well in excess of standard atmospheric pressure, e.g. in high-pressure chemical or physical processes. **2.** *See* EARTH'S ATMOSPHERE.

atmospheric pressure The pressure exerted by the weight of the air above it at any point on the earth's surface. At sea level the atmosphere will support a column of mercury about 760 mm high. This decreases with increasing altitude. The standard value for the atmospheric pressure at sea level in SI units is 101 325 pascals.

atom The smallest part of an element that can exist. Atoms consist of a small dense nucleus of protons and neutrons surrounded by moving electrons. The number of electrons equals the number of protons so the overall charge is zero. The electrons may be thought of as moving in circular or elliptical orbits (*see* BOHR THEORY) or, more accurately, in regions of space around the nucleus (*see* ORBITAL).

The **electronic structure** of an atom refers to the way in which the electrons are arranged about the nucleus, and in particular the *energy levels that they occupy. Each electron can be characterized by a set of four quantum numbers, as follows: (1) The **principal quantum number** n gives the main energy level and has values 1, 2, 3, etc. (the higher the number, the further the electron from the nucleus). Traditionally, these levels, or the orbits corresponding to them, are referred to as **shells** and given letters K, L, M, etc. The K-shell is the one nearest the nucleus. (2) The **orbital quantum number** l, which governs the angular momentum of the electron. The possible values of l are $(n-1)$, $(n-2)$, …, 1, 0. Thus, in the first shell ($n = 1$) the electrons can only have angular momentum zero ($l = 0$). In the second shell ($n = 2$), the values of l can be 1 or 0, giving rise to two **subshells** of slightly different energy. In the third shell ($n = 3$) there are three subshells, with $l = 2$, 1, or 0. The subshells are denoted by letters s ($l = 0$), p ($l = 1$), d ($l = 2$), f ($l = 3$). The orbital quantum number is sometimes called the **azimuthal quantum number**. (3) The **magnetic quantum number** m, which governs the energies of electrons in an external magnetic field. This can take values of $+l$, $+(l - 1)$, ..., 1, 0, -1, ..., $-(l - 1)$, $-l$. In an s-subshell (i.e. $l = 0$) the value of $m = 0$. In a p-subshell

---

ATOMIC THEORY

c.430 BC	Greek natural philosopher Empedocles (d. c. 430 BC) proposes that all matter consists of four elements: earth, air, fire, and water.
c.400 BC	Greek natural philosopher Democritus of Abdera (c. 460–370 BC) proposes that all matter consists of atoms.
306 BC	Greek philosopher Epicurus (c. 342–270 BC) champions Democritus' atomic theory.
1649	French philosopher Pierre Gassendi (1592–1655) proposes an atomic theory (having read Epicurus).
1803	John Dalton proposes Dalton's atomic theory.
1897	J. J. Thomson discovers the electron.
1904	J. J. Thomson proposes his 'plum pudding' model of the atom, with electrons embedded in a nucleus of positive charges. Japanese physicist Hantaro Nagaoka (1865–1950) proposes a 'Saturn' model of the atom with a central nucleus having a ring of many electrons.
1911	Ernest Rutherford discovers the atomic nucleus.
1913	Niels Bohr proposes model of the atom with a central nucleus surrounded by orbiting electrons. British physicist Henry Moseley (1887–1915) equates the positive charge on the nucleus with its atomic number. Frederick Soddy discovers isotopes.
1916	German physicist Arnold Sommerfield (1868–1951) modifies Bohr's model of the atom specifying elliptical orbits for the electrons.
1919	Ernest Rutherford discovers the proton.
1920	Ernest Rutherford postulates the existence of the neutron.
1926	Erwin Schrödinger proposes a wave-mechanical model of the atom (with electrons represented as wave trains).
1932	James Chadwick discovers the neutron. Werner Heisenberg proposes a model of the atomic nucleus in which protons and neutrons exchange electrons to achieve stability.
1936	Niels Bohr proposes a 'liquid drop' model of the atomic nucleus.
1948	German-born US physicist Maria Goeppert-Mayer (1906–72) and German physicist Hans Jensen (1907–73) independently propose the 'shell' structure of the nucleus.
1950	Danish physicist Aage Bohr (1922–) and US physicists Benjamin Mottelson (1926–) and Leo Rainwater (1917–86) combine the 'liquid-drop' and 'shell' models of the nucleus into a single theory.

($l = 1$), m can have values +1, 0, and –1; i.e. there are three p-orbitals in the p-subshell, usually designated p_x, p_y, and p_z. Under normal circumstances, these all have the same energy level. (4) The **spin quantum number** m_s, which gives the spin of the individual electrons and can have the values +½ or –½.

According to the *Pauli exclusion principle, no two electrons in the atom can have the same set of quantum numbers. The numbers define the **quantum state** of the electron, and explain how the electronic structures of atoms occur. See Chronology: Atomic Theory.

atomic bomb *See* NUCLEAR WEAPONS.

atomic clock An apparatus for measuring or standardizing time that is based on periodic phenomena within atoms or molecules. *See* AMMONIA CLOCK; CAESIUM CLOCK.

atomic energy *See* NUCLEAR ENERGY.

atomic force microscope (AFM) A type of microscope in which a small probe, consisting of a tiny chip of diamond, is held on a spring-loaded cantilever in contact with the surface of the sample. The probe is moved slowly across the surface and the tracking force between the tip and the surface is monitored. The probe is raised and lowered so as to keep this force constant, and a profile of the surface is produced. Scanning the probe over the sample gives a computer-generated contour map of the surface. The instrument is similar to the *scanning tunnelling microscope, but uses mechanical forces rather than electrical signals. It can resolve individual molecules and, unlike the scanning tunnelling microscope, can be used with nonconducting samples, such as biological specimens.

atomic mass unit (a.m.u.) A unit of mass used to express *relative atomic masses. It is 1/12 of the mass of an atom of the isotope carbon–12 and is equal to $1.660\ 33 \times 10^{-27}$ kg. This unit superseded both the physical and chemical mass units based on oxygen–16 and is sometimes called the **unified mass unit** or the **dalton**.

atomic number (proton number) Symbol Z. The number of protons in the nucleus of an atom. The atomic number is equal to the number of electrons orbiting the nucleus in a neutral atom.

atomic orbital *See* ORBITAL.

atomic pile An early form of *nuclear reactor using graphite as a *moderator.

atomic volume The relative atomic mass of an element divided by its density.

atomic weight *See* RELATIVE ATOMIC MASS.

attenuation 1. A loss of intensity suffered by sound, radiation, etc., as it passes through a medium. It may be caused by absorption or scattering. **2.** The drop in voltage or current experienced by a signal as it passes through a circuit.

atto- Symbol a. A prefix used in the metric system to denote 10^{-18}. For example, 10^{-18} second = 1 attosecond (as).

attractor The set of points in *phase space to which the representative point of a dissipative system (i.e. one with internal friction) tends as the system evolves. The attractor can be: a single point; a closed curve (a **limit cycle**), which describes a system with periodic behaviour; or a *fractal (or **strange attractor**), in which case the system exhibits *chaos.

AU *See* ASTRONOMICAL UNIT.

audibility The limits of audibility of the human ear are between about 20 hertz (a low rumble) and 20 000 hertz (a shrill whistle). With increased age the upper limit falls quite considerably.

audiofrequency A frequency that is audible to the human ear. *See* AUDIBILITY.

audiometer An instrument that generates a sound of known frequency and intensity in order to measure an individual's hearing ability.

Auger effect The ejection of an electron from an atom without the emission of an X- or gamma-ray photon, as a result of the de-excitation of an excited electron within the atom. This type of transition occurs in the X-ray region of the emission spectrum. The kinetic energy of the ejected electron, called an **Auger electron**, is equal to the energy of the corresponding X-ray photon minus the binding energy of the Auger electron. The effect was discovered by Pierre Auger (1899–1994) in 1925.

aurora The luminous phenomena seen in the night sky in high latitudes, occurring most frequently near the earth's geomagnetic poles. The displays of aurora appear as coloured arcs, rays, bands, streamers, and curtains, usually green or red. The aurora is caused by the interaction of the atoms (mainly atomic oxygen) and molecules in the upper atmosphere (above about 100 km) with charged particles streaming from the sun, attracted to the auroral regions by the earth's magnetic field. The aurora is known as the **aurora borealis** (or northern lights) in the northern hemisphere and as the **aurora australis** (or southern lights) in the southern hemisphere.

austenite *See* STEEL.

autoclave A strong steel vessel used for carrying out chemical reactions, sterilizations, etc., at high temperature and pressure.

autoradiography An experimental technique in which a radioactive specimen is placed in contact with (or close to) a photographic plate, so as to produce a record of the distribution of radioactivity in the specimen. The film is darkened by the ionizing radiation from radioactive parts of the sample. Autoradiography has a number of applications, particularly in the study of living tissues and cells.

autotransformer A *transformer that has a single winding, which is tapped at different points to provide both the primary and secondary circuits.

avalanche A shower of ionized particles created by a single *ionization as a result of secondary ionizations caused by the original electron and ion being accelerated in an electric field. Each ionization leads to the formation of more electrons and ions, which themselves cause further ionizations. Such avalanches occur in a *Geiger counter.

average (mean) **1.** The **arithmetic average** of a set of n numbers is the sum of the numbers divided by n. **2.** The **geometric average** of a set of n numbers is the nth root of their product. *See also* ROOT-MEAN-SQUARE VALUE.

avionics The study and development of electronic circuits and devices as used in either aeronautics or astronautics. Avionics has become of great importance since the advent of the space age.

Avogadro, Amedeo (1776–1856) Italian chemist and physicist. In 1811 he published his hypothesis (*see* AVOGADRO'S LAW), which provided a method of calculating molecular weights from vapour densities. The importance of the work remained unrecognized, however, until championed by Stanislao Cannizzaro (1826–1910) in 1860.

Avogadro constant Symbol N_A or L. The number of atoms or molecules in one *mole of substance. It has the value $6.022\ 1367(36) \times 10^{23}$. Formerly it was called **Avogadro's number**.

Avogadro's law Equal volumes of all gases contain equal numbers of molecules at the same pressure and temperature. The law, often called **Avogadro's hypothesis**, is true only for ideal gases. It was first proposed in 1811 by Amadeo Avogadro.

axial vector (pseudo-vector) A *vector that does not reverse its sign when the coordinate system is changed to a new system by a reflection in the origin (i.e. $x'_i = -x_i$). An example of an axial vector is the *vector product of two *polar vectors, such as $L = r \times p$, where L is the *angular momentum of a particle, r is its position vector, and p is its momentum vector. *Compare* PSEUDO-SCALAR.

axion A hypothetical elementary particle postulated to explain why there is no observed CP violation (*see* CP INVARIANCE) in the strong interaction (*see* FUNDAMENTAL INTERACTIONS). Axions have not been detected experimentally, although it has been possible to put limits on

a

their mass and other properties from the effects that they would have on some astrophysical phenomena (e.g. the cooling of stars). It has also been suggested that they may account for some or all of the missing matter in the universe (*see* MISSING MASS).

axis 1. One of a set of reference lines used to locate points on a graph or in a coordinate system. *See* CARTESIAN COORDINATES; POLAR COORDINATES. **2.** A line about which a figure, curve, or body is symmetrical (**axis of symmetry**) or about which it rotates (**axis of rotation**).

azeotrope (azeotropic mixture; constant-boiling mixture) A mixture of two liquids that boils at constant composition; i.e. the composition of the vapour is the same as that of the liquid. Azeotropes occur because of deviations in Raoult's law leading to a maximum or minimum in the *boiling-point–composition diagram. When the mixture is boiled, the vapour initially has a higher proportion of one component than is present in the liquid, so the proportion of this in the liquid falls with time. Eventually, the maximum and minimum point is reached, at which the two liquids distil together without change in composition. The composition of an azeotrope depends on the pressure.

azimuth *See* POLAR COORDINATES.

azimuthal quantum number *See* ATOM.

Babbit metal Any of a group of related alloys used for making bearings. They consist of tin containing antimony (about 10%) and copper (1–2%), and often lead. The original alloy was invented in 1839 by the US inventor Isaac Babbit (1799–1862).

Babo's law The vapour pressure of a liquid is decreased when a solute is added, the amount of the decrease being proportional to the amount of solute dissolved. The law was discovered in 1847 by the German chemist Lambert Babo (1818–99). *See also* RAOULT'S LAW.

back e.m.f. An electromotive force that opposes the main current flow in a circuit. For example, when the coils of the armature in an electric motor are rotated a back e.m.f is generated in these coils by their interaction with the field magnet (*see* INDUCTANCE). Also, in an electric cell, *polarization causes a back e.m.f. to be set up, in this case by chemical means.

background radiation Low intensity *ionizing radiation present on the surface of the earth and in the atmosphere as a result of *cosmic radiation and the presence of radioisotopes in the earth's rocks, soil, and atmosphere. The radioisotopes are either natural or the result of nuclear fallout or waste gas from power stations. Background counts must be taken into account when measuring the radiation produced by a specified source. *See also* MICROWAVE BACKGROUND RADIATION.

backing store Supplementary computer memory, usually in the form of magnetic disks, in which data and programs are held permanently for reference; small sections of this information can then be copied into the main memory (*RAM) of a computer when required for processing. Backing store is less costly and can hold more information than semiconductor RAM, but the speed of access of information in RAM is considerably faster. Use of a hierarchy of different memory devices, including backing store and RAM, greatly improves performance, efficiency, and economy in a computer.

backup A resource that can be used as a substitute in the event of, say, a fault in a component or system or loss of data from a computer file. A backup file is a copy of a file taken in case the original is destroyed or unintentionally altered and the data lost.

balance An accurate weighing device. The simple **beam balance** consists of two pans suspended from a centrally pivoted beam. Known masses are placed on one pan and the substance or body to be weighed is placed in the other. When the beam is exactly horizontal the two masses are equal. An accurate laboratory balance weighs to the nearest hundredth of a milligram. Specially designed balances can be accurate to a millionth of a milligram. More modern **substitution balances** use the substitution principle. In this calibrated weights are removed from the single lever arm to bring the single pan suspended from it into equilibrium with a fixed counter weight. The substitution balance is more accurate than the two-pan device and enables weighing to be carried out more rapidly. In automatic electronic balances, mass is determined not by mechanical deflection but by electronically controlled compensation of an electric force. A scanner monitors the displacement of the pan support generating a current proportional to the displacement. This current flows through a coil forcing the pan support to return to its original position by means of a magnetic force. The signal generated enables the mass to be read from a digital display. The mass of the empty container can be stored in the balance's computer memory and automatically deducted from the mass of the container plus its contents. *See also* SPRING BALANCE.

ballistic galvanometer A moving-coil *galvanometer designed for measuring charge by detecting a surge of current. It has a heavy coil with minimal damping. When a surge of current is passed through the coil, the initial maximum deflection (the 'throw') is proportional to the total charge that has passed.

ballistic pendulum A device used to measure the velocity of a projectile, such as a bullet. A large mass of relatively soft material is suspended from a horizontal bar and the angle through which this mass is displaced when it is struck by the projectile in flight enables the momentum and hence the velocity of the projectile to be calculated by successive application of the laws of conservation of linear momentum and of energy.

ballistics The study of the flight of projectiles, especially those that have a parabolic flight path from one point on the earth's surface to another.

ball lightning A white or reddish luminous sphere, about 50 cm in diameter, that sometimes appears at ground level in a thunderstorm. Ball lightning is slow moving and usually disappears without detonation. *See also* LIGHTNING.

Balmer series *See* HYDROGEN SPECTRUM.

band spectrum *See* SPECTRUM.

band theory The theory that, in crystals, electrons fall into allowed *energy bands, between which lie forbidden bands. Although *free-electron theory can explain the electrical properties of metals, to understand fully the nature of electrical conduction, free-electron theory must be modified to include the effect of the crystal lattice in which the electrons move. Band theory modifies the free-electron treatment by

b

including a regular periodic potential resulting from the positive ions in the lattice. Although the presence of the lattice alone does not give rise to electron scattering, except under special conditions, the periodic potential generated by the lattice does change the distribution of electron states given by the simple free-electron model.

In principle the original particle-in-a-box problem is modified by the addition of an extra potential term $V(x, y, x)$, where $V(x, y, x)$ has the same periodicity of the lattice it represents. The *Schrödinger equation (*see also* QUANTUM MECHANICS) that must be satisfied by these electron matter waves, ψ, is then:

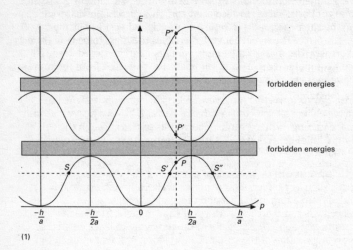

(1)

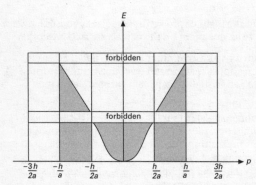

(2)

Band theory

$$- (\hbar/2m)^2(\partial^2\psi/\partial x^2 + \partial^2\psi/\partial y^2 + \partial^2\psi/\partial z^2) + V(x,y,z)\psi = E\psi$$

where E is the energy of the state associated with ψ. This equation is often called the **Bloch equation**; it is difficult to solve even using approximate methods.

As with the free-electron model, the Bloch equation leads to a relationship between the momenta that characterize an electron state, and the energy of that state. However, this relationship is no longer the simple parabolic function:

$$E = p^2/2m,$$

where p is the electron momentum that characterizes the state, and m is the mass of the electron. Instead the function is multivalued and there are certain forbidden bands of energy, i.e. no permitted states exist for them. The relationship between E and p for a one-dimensional crystal is shown in diagram (1).

The main feature is that all the curves repeat themselves over an interval in p of h/a, where h is the Planck constant and a is called the **lattice constant**. The electron states on the lower curve indicated by the points S, S', and S'', are identical states except for their momentum values. The upper curves show that for each momentum value there are several permitted energies; one within each *energy band (e.g. P, P', P''). In order to avoid the confusion that would arise if the complete multivalued set of curves were used, it is conventional to choose certain sections of them in which as p increases so does energy E. The smallest values, starting from $p = 0$, are reserved for the lowest energy band – the lowest curve in diagram (1). The next set from $p = h/2a$ to h/a and $-h/2a$ to $-h/a$ is reserved for the second energy band, (the middle curve in diagram (1), and so on. This zone structure is illustrated in diagram (2) and is called the **extended zone scheme**. It is merely a useful convention; any sets of p values that cover a range of h/a could be used to span all the available energy states.

A comparison of the extended zone scheme and the results of the free-electron model may easily be made by considering the relationship between E and p in the two models. This is shown in diagram (2) with the broken line corresponding to the free-electron parabolic relationship, $E \sim p^2$.

The number of permitted values of p in each of the extended zones is equal to the number of ions in the crystal. As in the free-electron theory, each state corresponding to a value of p can take up two orientations of *spin. So the total number of states in each zone is twice the number of ions in the lattice. This means that if the metal has only one electron per atom contributing to conduction, the electrons in the metal will only occupy all the states in the lower half of the zone – the shaded area in figure (2). No more than one electron can occupy any single state. This restriction on state occupation arises because electrons are **fermions** and therefore obey **Fermi–Dirac statistics**. For such a monovalent metal the energy distribution is virtually the same on the basis of either the free electron or band theory models, because the electrons have not started filling states beyond the first forbidden zone.

When the material has an even number of electrons per atom there would be an occupation of an integral number of zones. If an electric field were to be applied within the material, there would be a shift in the distribution of electron momenta that could only occur if and only if the zones are at close energy proximity. The electrons at the top of the uppermost filled zone can then obtain enough extra energy (e.g. from thermal excitations) to jump over the forbidden energy gap so that they can occupy permitted energy states at the bottom of the next zone. If this energy is very high compared with thermal energies, excitations to the next band cannot occur and there will be no current. The material would be classed as an *insulator.

If the energy gap is not too large, a few electrons can absorb enough thermal energy to cross the forbidden energy gap. At room temperature thermal excitations are sufficiently large to breach this gap and to enable a small current to flow. Because the amount of excitation will increase with temperature, the conductivity will also increase with temperature. This behaviour is typical of a *semiconductor.

Finally, elements with an odd number of electrons per atom are normally electrical conductors. This is because there are always unfilled states available within a given half-filled band. The electron momentum distribution within a band always looks like a small section of the free electron model for that particular extended zone.

bandwidth The frequency range over which a radio signal of specified frequency spreads. For example, in a *modulation system it is the range of frequencies occupied by the modulating signal on either side of the carrier wave. In an amplifier, it is the range of frequencies over which the power amplification falls within a specified fraction of the maximum value. In an aerial it is the range of frequencies that an aerial system can handle without mismatch.

bar A c.g.s. unit of pressure equal to 10^6 dynes per square centimetre or 10^5 pascals (approximately 750 mmHg or 0.987 atmosphere). The **millibar** (100 Pa) is commonly used in meteorology.

Bardeen, John (1908–91) US physicist, who worked at Harvard, Minnesota University, and Bell Telephone Labs before becoming a professor at the University of Illinois in 1951. At Bell, with Walter Brattain (1902–87) and William Shockley (1910–89), he developed the point-contact transistor. The three scientists shared the 1956 Nobel Prize for physics for this work. In 1956, with Leon Cooper (1930–) and John Schrieffer (1931–), he formulated the BCS theory of *superconductivity, for which they shared the 1972 Nobel Prize.

Barkhausen effect The magnetization of a ferromagnetic substance by an increasing magnetic field takes place in discontinuous steps rather than continuously. The effect results from the orientation of magnetic domains (*see* MAGNETISM). It was first observed by H. Barkhausen (1881–1956) in 1919.

Barlow wheel *See* HOMOPOLAR GENERATOR.

barn A unit of area sometimes used to measure *cross sections in nuclear interactions involving incident particles. It is equal to 10^{-28} square metre. The name comes from the phrase 'side of a barn' (something easy to hit).

barograph A meteorological instrument that records on paper variations in atmospheric pressure over a period. It often consists of an aneroid barometer operating a pen that rests lightly on a rotating drum to which the recording paper is attached.

barometer A device for measuring *atmospheric pressure. The **mercury barometer** in its simplest form consists of a glass tube about 80 cm long sealed at one end and filled with mercury. The tube is then inverted and the open end is submerged in a reservoir of mercury; the mercury column is held up by the pressure of the atmosphere acting on the surface of mercury in the reservoir. This type of device was invented by the Italian scientist Evangelista Torricelli (1608–47), who first noticed the variation in height from day to day, and constructed a barometer in 1644.

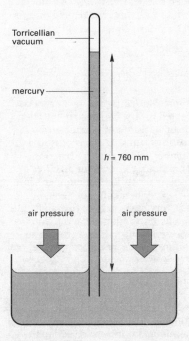

Mercury barometer

In such a device, the force exerted by the atmosphere balanced the weight of the mercury column. If the height of the column is h and the cross-sectional area of the tube is A, then the volume of the mercury in the column is hA and its weight is $hA\rho$, where ρ is the density of mercury. The force is thus $hA\rho g$, where g is the acceleration of free fall and the pressure exerted is this force divided by the area of the tube; i.e. $h\rho g$. Note that the height of the mercury is independent of the diameter of the tube. At standard atmospheric pressure the column is 760 mm high. The pressure is then expressed as 760 mmHg (101 325 pascals).

Mercury barometers of this type, with a reservoir of mercury, are known as **cistern barometers**. A common type is the **Fortin barometer**, in which the mercury is held in a leather bag so that the level in the reservoir can be adjusted. The height is read from a scale along the side of the tube in conjunction with a vernier scale that can be moved up and down. Corrections are made for temperature.

The second main type of barometer is the **aneroid barometer**, in which the cumbersome mercury column is replaced by a metal box with a thin corrugated lid. The air is removed from the box and the lid is supported by a spring. Variations in atmospheric pressure cause the lid to move against the spring. This movement is magnified by a system of delicate levers and made to move a needle around a scale. The aneroid barometer is less accurate than the mercury type but much more robust and convenient, hence its use in *altimeters.

barycentre The *centre of mass of a system.

barye A c.g.s. unit of pressure equal to one dyne per square centimetre (0.1 pascal).

baryon A *hadron with half-integral spin. Nucleons comprise a subclass of baryons. According to currently accepted theory, baryons are made up of three *quarks (**antibaryons** are made up of three antiquarks) held together by gluons (*see* QUANTUM CHROMODYNAMICS). Baryons possess a quantum number, called the **baryon number**, which is +1 for baryons, –1 for antibaryons, 1/3 for quarks, –1/3 for antiquarks, and 0 for all other particles such as electrons, neutrinos, and photons. Baryon number has always appeared to have been conserved experimentally, but *grand unified theories postulate interactions at very high energies that allow it not to be conserved. It is thought that nonconservation of baryon number at the high energies characteristic of the early universe may provide an explanation for the asymmetry between matter and antimatter in the universe. *See* PROTON DECAY.

base 1. (in mathematics) **a.** The number of different symbols in a number system. In the decimal system the base is 10; in *binary notation it is 2. **b.** The number that when raised to a certain power has a *logarithm equal to that power. For example if 10 is raised to the power of 3 it is equal to 1000; 3 is then the (common) logarithm of 1000 to the base 10. In natural or Napierian logarithms the base is e. To change the base from common to natural logarithms the formula used is: $\log_{10}y = \log_{e}y \times \log_{10}e = 0.43429\log_{e}y$. **2.** (in electronics) *See* TRANSISTOR.

base unit A unit that is defined arbitrarily rather than being defined by simple combinations of other units. For example, the ampere is a base unit in the SI system defined in terms of the force produced between two current-carrying conductors, whereas the coulomb is a **derived unit**, defined as the quantity of charge transferred by one ampere in one second.

basic-oxygen process (BOP process) A high-speed method of making high-grade steel. It originated in the **Linz–Donawitz (L–D) process**. Molten pig iron and scrap are charged into a tilting furnace, similar to the Bessemer furnace except that it has no tuyeres. The charge is converted to steel by blowing high-pressure oxygen onto the surface of the metal through a water-cooled lance. The excess heat produced enables up to 30% of scrap to be incorporated into the charge. The process has largely replaced the Bessemer and open-hearth processes.

Basov, Nikolai Gennediyevitch (1922–2001) Russian physicist best known for the development of the *maser, the precursor of the laser. In 1955, while working as a research student with Aleksandr Prokhorov (1916–2000) at the Soviet Academy of Sciences, he devised a microwave amplifier based on ammonia molecules. The two scientists shared the 1964 Nobel Prize with American Charles Townes (1915–), who independently developed a maser.

battery A number of electric cells joined together. The common car battery, or *accumulator, usually consists of six secondary cells connected in series to give a total e.m.f. of 12 volts. A torch battery is usually a dry version of the *Leclanché cell, two of which are often connected in series. Batteries may also have cells connected in parallel, in which case they have the same e.m.f. as a single cell, but their capacity is increased, i.e. they will provide more total charge. The capacity of a battery is usually specified in ampere-hours, the ability to supply 1 A for 1 hr, or the equivalent.

baud A unit for measuring signal speed in a computer or communications system. When the signal is a sequence of *bits, the baud rate is given in bits per second (bps). The unit is named after J. M. E. Baudot (1845–1903).

BBGKY hierarchy A hierarchy of equations for the distribution function in statistical mechanics. This hierarchy leads to kinetic equations that are generalizations of the Boltzmann equation and it has given important insights into the approach to equilibrium in statistical mechanics. The initials BBGKY stand for N. N. Bogoliubov, Max Born, H. S. Green, J. G. Kirkwood, and J. Yvon, who derived this hierarchy of equations in the 1930s and 1940s.

BCS theory *See* SUPERCONDUCTIVITY.

beam A group of rays moving in an organized manner. It may consist of particles (e.g. an electron beam) or of electromagnetic radiation (e.g. a radar beam).

beam balance *See* BALANCE.

beam hole A hole through the shielding of a *nuclear reactor to enable a beam of neutrons or other particles to escape for experimental purposes.

beats A periodic increase and decrease in loudness heard when two notes of slightly different frequency are sounded at the same time. If a note of frequency n is heard at the same time as a note of frequency m, the resulting note will have a frequency of about $(n + m)/2$. However the amplitude of this note will vary from the difference to the sum of the amplitudes of the m and n notes and the frequency (called the **beat frequency**) of this variation will be $(m - n)$. The beating sound produced occurs as the waves successively reinforce and oppose each other as they move in and out of phase. Beating also occurs with radio-frequency waves and is made use of in *heterodyne devices. *See also* INTERFERENCE.

Beckmann thermometer A thermometer for measuring small changes of temperature. It consists of a mercury-in-glass thermometer with a scale covering only 5 or 6°C calibrated in hundredths of a degree. It has two mercury bulbs, the range of temperature to be measured is varied by running mercury from the upper bulb into the larger lower bulb (see

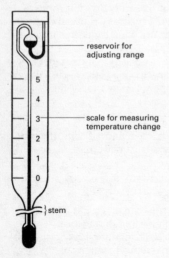

Beckmann thermometer

illustration). It is used particularly for measuring *depression of freezing point or *elevation of boiling point of liquids when solute is added, in order to find relative molecular masses. The instrument was invented by the German chemist E. O. Beckmann (1853–1923).

becquerel Symbol Bq. The SI unit of activity (*see* RADIATION UNITS). The unit is named after the discoverer of radioactivity A. H. Becquerel.

Becquerel, Antoine Henri (1852–1908) French physicist. His early researches were in optics; then, in 1896, he accidentally discovered *radioactivity in fluorescent salts of uranium. Three years later he showed that it consists of charged particles that are deflected by a magnetic field. For this work he was awarded the 1903 Nobel Prize, which he shared with Pierre and Marie *Curie.

bel Ten *decibels.

bell metal A type of *bronze used in casting bells. It consists of 60–85% copper alloyed with tin, often with some zinc and lead included.

Bell's theorem A theorem stating that no local *hidden-variables theory can make predictions in agreement with those of *quantum mechanics. Local hidden-variables theories give rise to a result, called **Bell's inequality**, which is one of many similar results concerning the probabilities of two events both occurring in well-separated parts of a system. The British physicist John S. Bell (1928–90) showed in 1964 that quantum mechanics predicts a violation of the inequalities, which are consequences of local hidden-variables theories. Experiments are in agreement with quantum mechanics rather than local hidden-variables theories by violating Bell's inequality, in accordance with Bell's theorem.

Bénard cell A structure associated with a layer of liquid that is confined by two horizontal parallel plates, in which the lateral dimensions are much larger than the width of the layer. Before heating, the liquid is homogeneous. However, if after heating from below, the temperatures of the plates are T_1 and T_2, at a critical value of the temperature gradient $\Delta T = T_1 - T_2$ the liquid abruptly starts to convect. The liquid spontaneously organizes itself into a set of convection rolls, i.e. the liquid goes round in a series of 'cells', called Bénard cells. The cell was devised by the French scientist Henri Bénard in about 1900. *See also* COMPLEXITY.

bending moment (about any point or section of a horizontal beam under load) The algebraic sum of the moments of all the vertical forces to either side of that point or section. *See* MOMENT OF A FORCE.

Bergius, Friedrich Karl Rudolf (1884–1949) German organic chemist. While working with Fritz Haber in Karlsruhe, he become interested in reactions at high pressures. In 1912 he devised an industrial process for making light hydrocarbons by the high-pressure hydrogenation of coal or heavy oil. The work earned him a share of the 1931 Nobel Prize for chemistry with Carl Bosch (1874–1940). The Bergius process proved important for supplying petrol for the German war effort in World War II.

berkelium Symbol Bk. A radioactive metallic transuranic element belonging to the actinoids; a.n. 97; mass number of the most stable isotope 247 (half-life 1.4×10^3 years); r.d. (calculated) 14. There are eight known isotopes. It was first produced by G. T. Seaborg and associates in 1949 by bombarding americium–241 with alpha particles.

Bernoulli, Daniel (1700–82) Swiss mathematician. In 1724 he published a work on differential equations, which earned him a professorship at St Petersburg. He returned to Basel, Switzerland, in 1733 and began researches on hydrodynamics (*see* BERNOULLI THEOREM), the work for which he is best known. He also initiated the kinetic theory of matter.

Bernoulli theorem At any point in a pipe through which a fluid is flowing the sum of the pressure energy, the kinetic energy, and the potential energy of a given mass of the fluid is constant. This is equivalent to a statement of the law of the conservation of energy. The law was published in 1738 by Daniel Bernoulli.

Berzelius, Jöns Jacob (1779–1848) Swedish chemist. After moving to Stockholm he worked with mining chemists and, with them, discovered several elements, including cerium (1803), selenium (1817), lithium (1818), *thorium (1828), and vanadium (1830). He also worked on atomic weights and electrochemistry and devised the notation for chemical elements.

Bessel function A type of function that occurs as a solution to problems involving waves in systems with cylindrical symmetry. Bessel functions have been extensively studied and tabulated and are used in many branches of mathematical physics. They are named after the German astronomer Friedrich Wilhelm Bessel (1784–1846).

Bessemer process A process for converting *pig iron from a *blast furnace into *steel. The molten pig iron is loaded into a refractory-lined tilting furnace (**Bessemer converter**) at about 1250°C. Air is blown into the furnace from the base and *spiegel is added to introduce the correct amount of carbon. Impurities (especially silicon, phosphorus, and manganese) are removed by the converter lining to form a slag. Finally the furnace is tilted so that the molten steel can be poured off. In the modern VLN (very low nitrogen) version of this process, oxygen and steam are blown into the furnace in place of air to minimize the absorption of nitrogen from the air by the steel. The process is named after the British engineer Sir Henry Bessemer (1813–98), who announced it in 1856. *See also* BASIC-OXYGEN PROCESS.

beta decay A type of weak interaction (*see* FUNDAMENTAL INTERACTIONS) in which an unstable atomic nucleus changes into a nucleus of the same nucleon number (*A*) but different proton number (*Z*). There are three types of beta decay: negative beta decay, positive beta decay, and electron capture.

Negative beta decay:

$$^A_ZX \rightarrow {}^A_{Z+1}Y + {}^0_{-1}e + {}^0_0\bar{\nu}$$

A neutron in the nucleus X has decayed into a proton forming a new nucleus Y with the emission of an electron and antineutrino. This process involves a decrease in mass and is energetically favourable; it can also occur outside the nucleus – free neutrons decay with a mean lifetime of about 15 minutes.

Positive beta decay:

$$^A_ZX \rightarrow {}^A_{Z-1}Y + {}^0_1e + {}^0_0\nu$$

A proton in the nucleus X transforms into a neutron and a new nucleus Y is formed with the emission of an antimatter electron (positron) and neutrino. This process involves an effective increase in mass for the proton and is not energetically favourable. It cannot occur outside the nucleus – free protons do not undergo this kind of interaction. The process is allowed within the environment of the nucleus because when the nucleus as a whole is taken into account the interaction represents an overall decrease in mass.

Electron capture:

$$^A_ZX + {}^0_{-1}e \rightarrow {}^A_{Z-1}Y + {}^0_0\nu$$

A proton in the nucleus X captures an electron from the atomic environment and becomes a neutron, emitting a neutrino in the process. This process also involves an effective increase in mass for the proton and is not energetically favourable; again, it cannot also occur outside the nucleus – free protons do not undergo this kind of interaction. The process is allowed within the environment of the nucleus because, taking into account the whole nucleus, the interaction represents an overall decrease in mass.

beta-iron A nonmagnetic allotrope of iron that exists between 768°C and 900°C.

beta particle An electron or positron emitted during *beta decay. A stream of beta particles is known as **beta radiation**.

betatron A particle *accelerator for producing high-energy electrons (up to 340 MeV) for research purposes, including the production of high-energy X-rays. The electrons are accelerated by electromagnetic induction in a doughnut-shaped (toroidal) ring from which the air has been removed. This type of accelerator was first developed by D. W. Kerst (1911–) in 1939; the largest such machine, at the University of Illinois, was completed in 1950.

Bevatron A colloquial name for the proton *synchrotron at the Berkeley campus of the University of California. It produces energies up to 6 GeV.

biaxial crystal *See* DOUBLE REFRACTION.

biconcave *See* CONCAVE.

bifurcation A phenomenon that can occur in a dynamical system in which some solutions of a certain type for the system, such as laminar flow in a fluid, suddenly change to another type, such as turbulent flow, when one of the parameters that characterizes the system reaches a critical value. Bifurcation is an important phenomenon in *chaos theory.

big-bang theory The cosmological theory that all the matter and energy in the universe originated from a state of enormous density and temperature that exploded at a finite moment in the past. See Feature (pp 42–43).

billion Formerly in the UK, one million million, 10^{12}, but the US meaning of one thousand million, 10^9, has now been adopted worldwide.

bimetallic strip A strip consisting of two metals of different *expansivity riveted or welded together so that the strip bends on heating. If one end is fixed the other end can be made to open and close an electric circuit, as in a *thermostat.

bimorph cell A device consisting of two plates of piezoelectric material, such as Rochelle salt, joined together so that one expands on the application of a potential difference and the other contracts. The cell thus bends as a result of the applied p.d. The opposite effect is also used, in which the mechanical bending of the cell is used to produce a p.d., as in the crystal microphone and some types of record-player pickups.

binary notation A number system using only two different digits, 0 and 1. Instead of units, tens, hundreds, etc., as used in the decimal system, digits in the binary notation represent units, twos, fours, eights, etc. Thus one in decimal notation is represented by 0001, two by 0010, four by 0100, and eight by 1000. Because 0 and 1 can be made to correspond to off and on conditions in an electric circuit, the binary notation is widely used in computers.

binary prefixes A set of prefixes for binary powers designed to be used in data processing and data transmission contexts. They were suggested in 1998 by the International Electrotechnical Commission (IEC) as a way of resolving the ambiguity in use of kilo-, mega-, giga-, etc., in computing. In scientific usage, these prefixes indicate 10^3, 10^6, 10^9, etc. (*see* SI UNITS). In computing, it became common to use the prefix "kilo-" to mean 2^{10}, so one kilobit was 1024 bits (not 1000 bits). This was extended to larger prefixes, so "mega-" in computing is taken to be 2^{20} (1 048 576) rather than 10^6 (1 000 000). However, there is a variation in usage depending on the context. In discussing memory capacities megabyte generally means 2^{20} bytes, but in disk storage (and data transmission) megabyte is often taken to mean 10^6 bytes. (In some contexts, as in the capacity of a floppy disk, it has even been quoted as 1 024 000 bytes, i.e. 1000 times a (binary)

kilobyte.) The IEC attempted to resolve this confusion by introducing
binary prefixes, modelled on the normal decimal prefixes, as follows:

kibi- 2^{10}
mebi- 2^{20}
gebi- 2^{30}
tebi- 2^{40}
pebi- 2^{50}
exbi- 2^{60}

These names are contractions of 'kilobinary', 'megabinary', etc., but are
pronounced so that the second syllable rhymes with 'bee'. Using these
prefixes, one gebibyte would be 1 073 741 824 bytes and one gigabyte
would (unambiguously) be 1 000 000 000 bytes.

binary stars A pair of stars revolving about a common centre of mass.
In a **visual binary** the stars are far enough apart to be seen separately by
an optical telescope. In an **astrometric binary** one component is too faint
to be seen and its presence is inferred from the perturbations in the
motion of the other. In a **spectrosopic binary** the stars cannot usually be
resolved by a telescope, but the motions can be detected by different
Doppler shifts in the spectrum at each side of the binary, according to
whether the components are approaching or receding from the observer.

binding energy The energy equivalent to the *mass defect when
nucleons bind together to form an atomic nucleus. When a nucleus is
formed some energy is released by the nucleons, since they are entering a
more stable lower-energy state. Therefore the energy of a nucleus consists
of the energy equivalent of the mass of its individual nucleons minus the
binding energy. The binding energy per nucleon plotted against the mass
number provides a useful graph showing that up to a mass number of
50–60, the binding energy per nucleon increases rapidly, thereafter it falls
slowly (*see* NUCLEAR STABILITY). Energy is released both by fission of heavy
elements and by fusion of light elements because both processes entail a
rearrangement of nuclei in the lower part of the graph to form nuclei in
the higher part of the graph.

binoculars Any optical instrument designed to serve both the observer's
eyes at once. **Binocular field glasses** consist of two refracting astronomical
*telescopes inside each of which is a pair of prisms to increase the
effective length and produce an upright image. Simpler binoculars, such
as **opera glasses**, consist of two Galilean telescopes which produce upright
images without prisms. Commonly, binoculars are specified by a pair of
numbers, such as 10×50. The first number indicates the angular
*magnification produced. The second is the diameter of the objective lens
in millimetres, and indicates the amount of light gathered by the
instrument. **Binocular microscopes** are used in biology and surgery to
enable the observer to obtain a stereoscopic view of small objects or parts.

binocular vision The ability, found only in animals with forward-facing
eyes, to produce a focused image of the same object simultaneously on

THE BIG-BANG THEORY

Newton's work gave a mathematical basis for the universe on a large scale. However, the data available at the time suggested a static unchanging universe. This could not easily be explained in the context of the law of gravitation, since all bodies in the universe attract all other bodies with the force of gravity. Newton realized that there was only one solution to this problem: in a static universe, matter had to be uniformly spread throughout an infinitely large space. In 1826, Heinrich Olbers published a paper containing what is known as *Olbers' paradox; such a universe would lead to a perpetually bright sky on earth.

Space–time
Cosmologists now believe that Newton's model was based on incorrect assumptions about the structure of space, time, and matter. Einstein in his general theory of *relativity (1915) proposed that the universe exists in four-dimensional space–time. This space–time is curved by the presence of matter, and the matter moves by following the resulting curves.

The expanding universe
The discovery by Hubble in 1929 that the universe is expanding provided a starting point for the ideas on which our present understanding of the universe is based. Hubble made his discovery by analysing the *spectra of light from distant galaxies and noting a persistent *red shift, which he explained in terms of the *Doppler effect; an increase in observed wavelengths of light occurs because the light source is receding from the observer. The larger the speed, of recession, the larger the red shifts. Hubble discovered a pattern in his data: the further away the galaxy, the greater the speed of recession. Known as *Hubble's law, this provided the evidence that the universe is expanding and a resolution to Olbers' paradox. If the galaxies and the earth are moving apart, the radiation falling on the earth from the galaxies is reduced. The further galaxies are away from the earth, the smaller their contribution to the radiation falling on the earth.

This model might seem to place the earth at the centre of the universe again. However, it is space itself that is expanding and the galaxies are imbedded in this space. The ring (space) in the diagram has dots (the galaxies). The expansion of the ring means that the view from any one dot is that the other dots are receding at a speed proportional to their distance away. No single dot is at the centre of the system but all dots see the same thing.

Age of the universe
Hubble's law may be stated in the form: $H_0 = v/d$, where v is the speed of recession of the galaxy, d is the earth–galaxy distance, and H_0 is called the

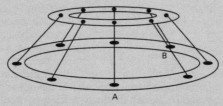

The expansion leads to the recession of B from A along the ring. The speed of recession will be directly proportional to the distance of B from A along the ring.

Hubble constant. Assuming that the galaxies have always been moving apart, the age of the universe (T) can be estimated, i.e. $T = 1/H_0$. On this basis the age of the universe would be between 15–18 billion years.

b

The origin of the universe
This view of the origin of the universe is called the **big-bang theory**. The theory suggests that the universe originated as a minute but very hot body and that the temperature has been falling as the expansion has continued. In 1945 George Gamow predicted that there should be a *microwave background corresponding to a black-body temperature a few degrees above absolute zero. This microwave background was discovered 20 years later. The big-bang theory also explains the amount of helium in the universe.

In 1992, the COBE satellite discovered that there were very small variations in the microwave background. This discovery helped to explain why the universe formed into galaxies and stars. The non-uniformities that began the nucleation of galactic matter in the early universe now appear as the small variations in the microwave background.

Fundamental forces
It is thought that the four *fundamental interactions in the universe are all manifestations of the same force. This force existed when the big bang, occurred at a temperature above 10^{15}K. As the universe cooled the forces separated as the original symmetries were broken. Gravity was the first to separate, followed by the strong nuclear force, and the weak and electromagnetic forces (see table).

Time from big bang	Temperature (K)	State of the universe/forces
0 second	infinite	The universe is infinitesimally small and infinitely dense (i.e. a mathematical singularity).
10^{-12} second	10^{15}	Weak and electromagnetic forces begin to separate.
10^{-6} second	10^{14}	Quarks and leptons begin to form.
10^{-3} second	10^{12}	Quarks form the hadrons; quark confinement begins.
10^{2} second	10^{7}	Helium nuclei formed by fusion.
10^{5} years	10^{4}	Atomic era; atoms form as protons combine with electrons.
10^{6} years	10^{3}	Matter undergoes gravitational collapse.
$.5–1.8 \times 10^{10}$ years	2.7	Present day: cosmic background corresponds to about 2.7K.

The future
Research into the future of the universe is clearly speculative. Whether the universe will continue to expand indefinitely depends on its mean density. Below a critical level (the critical density), gravitational attraction will not be enough to stop the expansion. However, if the mean density is above the critical density the universe is **bound** and an eventual contraction will occur resulting in a **big crunch**. This may precede another big bang initiating the whole cycle again.

the retinas of both eyes. This permits three-dimensional vision and contributes to distance judgment.

binomial theorem (binomial expansion) A rule for the expansion of a binomial expression (expression consisting of the sum of two variables raised to a given power). The general binomial expression $(x + y)^n$ expands to:

$$x^n + nx^{n-1}y + [n(n-1)/2!]x^{n-2}y^2 + \ldots y^n.$$

bioenergetics The study of the flow and the transformations of energy that occur in living organisms. Typically, the amount of energy that an organism takes in (from food or sunlight) is measured and divided into the amount used for growth of new tissues; that lost through death, wastes, and (in plants) transpiration; and that lost to the environment as heat (through respiration).

bioluminescence The emission of light without heat (*see* LUMINESCENCE) by living organisms. The phenomenon occurs in glow-worms and fireflies, bacteria and fungi, and in many deep-sea fish (among others); in animals it may serve as a means of protection (e.g. by disguising the shape of a fish) or species recognition or it may provide mating signals. The light is produced during the oxidation of a compound called **luciferin** (the composition of which varies according to the species), the reaction being catalysed by an enzyme, **luciferase**. Bioluminescence may be continuous (e.g. in bacteria) or intermittent (e.g. in fireflies).

biomechanics The application of the principles of *mechanics to living systems, particularly those living systems that have coordinated movements. Biomechanics also deals with the properties of biological materials, such as blood and bone. For example, biomechanics would be used to analyse the stresses on bones in animals, both when the animals are static and when they are moving. Other types of problems in biomechanics include the *fluid mechanics associated with swimming in fish and the *aerodynamics of birds flying. It is sometimes difficult to perform realistic calculations in biomechanics because of complexity in the shape of animals or the large number of *degrees of freedom that need to be considered (for example, the large number of muscles involved in the movement of a human leg).

biophysics *See* PHYSICS.

biotechnology The development of techniques for the application of biological processes to the production of materials of use in medicine and industry. For example, the production of antibiotics, cheese, and wine rely on the activity of various fungi and bacteria. Genetic engineering can modify bacterial cells to synthesize completely new substances, e.g. hormones, vaccines, monoclonal antibodies, etc.

Biot–Savart law A result in electromagnetism that states that the magnetic flux density B of a magnetic field produced by a current I in a long straight conductor is directly proportional to *I* and inversely proportional to the perpendicular distance *r* from the conductor. This law was discovered empirically in 1820 by the French physicists Jean Baptiste Biot (1774–1862) and Felix Savart (1791–1841). It can be derived from *Ampère's law.

biprism A glass prism with an obtuse angle that functions as two acute-angle prisms placed base-to-base. A double image of a single object is thus formed; the device was used by Fresnel to produce two coherent beams for interference experiments.

birefringence *See* DOUBLE REFRACTION.

bistable circuit *See* FLIP-FLOP.

bit (binary digit) Either of the digits 0 or 1 as used in the *binary notation. Bits are therefore the basic unit of information in a computer system.

Bitnet A computer network originally linking IBM mainframe systems located in North America and with backing from IBM. The network has been substantially extended to other parts of the world, usually on a region-by-region basis, and has been implemented on other computer systems. Complete messages of any length are transmitted from one computer system to the next, until the destination is reached.

Bitter pattern A microscopic pattern that forms on the surface of a ferromagnetic material that has been coated with a colloidal suspension of small iron particles. The patterns outline the boundaries of the magnetic domains (*see* MAGNETISM). They were first observed by F. Bitter in 1931.

Black, Joseph (1728–99) British chemist and physician, born in France. He studied at Glasgow and Edinburgh, where his thesis (1754) contained the first accurate description of the chemistry of carbon dioxide. In 1757 he discovered latent heat, and was the first to distinguish between heat and temperature.

black body A hypothetical body that absorbs all the radiation falling on it. It thus has an *absorptance and an *emissivity of 1. While a true black body is an imaginary concept, a small hole in the wall of an enclosure at uniform temperature is the nearest approach that can be made to it in practice.

  Black-body radiation is the electromagnetic radiation emitted by a black body. It extends over the whole range of wavelengths and the distribution of energy over this range has a characteristic form with a maximum at a certain wavelength. The position of the maximum depends on temperature, moving to shorter wavelengths with increasing temperature. *See* STEFAN'S LAW; WIEN'S DISPLACEMENT LAW.

black hole An object in space that has collapsed under its own gravitational forces to such an extent that its *escape velocity is equal to the speed of light. Black holes are believed to be formed in the gravitational collapse of very large stars at the ends of their lives (*see* DEATH OF A STAR; STELLAR EVOLUTION; SUPERNOVA). If the mass of an evolved stellar core is greater than the *Chandrasekhar limit for neutron stars then neutron degeneracy pressure is unable to prevent contraction until the gravitational field is sufficiently strong to prevent the escape of electromagnetic radiation. The boundary of the black hole, known as the **event horizon**, is the surface in space at which the gravitational field reaches this critical value. Events occurring within this horizon (i.e. in the interior of the black hole) cannot be observed from outside.

The theoretical study of black holes involves the use of general *relativity. It has been shown that a black hole can be characterized uniquely by just three properties: its mass, angular momentum, and electrical charge (this is known as the **no-hair theorem**). Mathematical expressions have been derived for describing black holes; these are the **Schwarzschild solution** (uncharged nonrotating hole), the **Reissner–Nordstrøm solution** (charged nonrotating hole), the **Kerr solution** (uncharged rotating hole), and the **Kerr–Newman solution** (charged rotating hole).

The ultimate fate of matter inside the black hole's event horizon is as yet unknown. General relativity predicts that at the centre of the hole there is a **singularity**, a point at which the density becomes infinite and the presently understood laws of physics break down. It is possible that a successful quantum theory of gravity could resolve this problem. However, since any singularity is hidden within the event horizon, it cannot influence the outside universe, so the normal laws of physics, including general relativity, can be used to describe processes outside the black hole.

Observational evidence of objects thought to be black holes comes from their effect on surrounding matter. Thus, if a black hole is part of a binary system with another star it will attract and capture matter from this star. The material leaving the star first forms a rotating **accretion disc** around the black hole, in which the matter becomes compressed and heated to such an extent that it emits X-rays. In the constellation Cygnus there is an X-ray source, Cygnus X-1, which consists of a supergiant star revolving around a small invisible companion with a mass of about ten times that of the sun, and therefore well above the Chandrasekhar limit. The companion is thought to be a black hole. Black holes have also been postulated as the power sources of *quasars and as possible generators of *gravitational waves. It appears that there are very large black holes at the centre of all galaxies.

It has been suggested that the formation of a black hole is responsible for gamma-ray bursts. Theoreticians have also postulated the existence of 'mini' black holes (with masses of about 10^{12} kilogram and radii about 10^{-15} metre). Such entities might have been formed shortly after the big bang when the universe was created. Quantum-mechanical effects are important for mini black holes, which emit Hawking radiation (*see* HAWKING PROCESS). *See also* SCHWARZSCHILD RADIUS.

Blandford–Znajek process An astrophysical process in which an external magnetic field is able to 'tap' the rotational energy of a rotating black hole, thereby making the black hole a powerful source of energy. There is some evidence that the process occurs around certain types of black hole. It was proposed by Roger Blandford and Roman Znajek in 1977. *See also* PENROSE PROCESS.

blast furnace A furnace for smelting iron ores, such as haematite (Fe_2O_3) or magnetite (Fe_3O_4), to make *pig iron. The furnace is a tall refractory-lined cylindrical structure that is charged at the top with the dressed ore, coke, and a flux, usually limestone. The conversion of the iron oxides to metallic iron is a reduction process in which carbon monoxide and hydrogen are the reducing agents. The overall reaction can be summarized thus:

$$Fe_3O_4 + 2CO + 2H_2 \rightarrow 3Fe + 2CO_2 + 2H_2O$$

The CO is obtained within the furnace by blasting the coke with hot air from a ring of tuyeres about two-thirds of the way down the furnace. The reaction producing the CO is:

$$2C + O_2 \rightarrow 2CO$$

In most blast furnaces hydrocarbons (oil, gas, tar, etc.) are added to the blast to provide a source of hydrogen. In the modern **direct-reduction process** the CO and H_2 may be produced separately so that the reduction process can proceed at a lower temperature. The pig iron produced by a blast furnace contains about 4% carbon and further refining is usually required to produce steel or cast iron.

Bloch's theorem A theorem relating to the *quantum mechanics of crystals stating that the wave function ψ for an electron in a periodic potential has the form $\psi(r) = \exp(ik \cdot r)U(r)$, where k is the *wave vector, r is a position vector, and $U(r)$ is a periodic function that satisfies $U(r + R) = U(r)$, for all vectors R of the Bravais lattice of the crystal. Bloch's theorem is interpreted to mean that the wave function for an electron in a periodic potential is a plane wave modulated by a periodic function. This explains why a free-electron model has some success in describing the properties, such as electrical and thermal *conductivity, of certain metals and why the free-electron model is inadequate to give a quantitative description of the properties of most metals. Bloch's theorem was formulated by the Swiss-born US physicist Felix Bloch (1905–83) in 1928. *See also* ENERGY BAND.

blooming The process of depositing a transparent film of a substance, such as magnesium fluoride, on a lens to reduce (or eliminate) the reflection of light at the surface. The film is about one quarter of a wavelength thick and has a lower *refractive index than the lens. The anti-reflection effect is achieved by destructive interference.

B-meson Symbol B^0. A meson that consists of a down quark and an anti-bottom quark. It is electrically neutral, has spin zero, and a mass of 5.279

GeV. The antiparticle consists of a bottom quark and an anti-down quark. It is hoped that study of the decays of B-mesons will shed light on the problem of CP violation (*see* CP INVARIANCE).

body-centred cubic (b.c.c.) *See* CUBIC CRYSTAL.

Bohr, Niels Henrik David (1885–1962) Danish physicist. In 1913 he published his explanation of how atoms, with electrons orbiting a central nucleus, achieve stability by assuming that their angular momentum is quantized. Movement of electrons from one orbit to another is accompanied by the absorption or emission of energy in the form of light, thus accounting for the series of lines in the emission *spectrum of hydrogen. For this work Bohr was awarded the 1922 Nobel Prize. He also explained the periodic table in terms of shells of electrons and developed the liquid-drop model of the nucleus. *See* BOHR THEORY.

bohrium Symbol Bh. A radioactive *transactinide element; a.n. 107. It was first made in 1981 by Peter Armbruster and a team in Darmstadt, Germany, by bombarding bismuth-209 nuclei with chromium-54 nuclei. Only a few atoms of bohrium have ever been detected.

Bohr theory The theory published in 1913 by Niels Bohr to explain the line spectrum of hydrogen. He assumed that a single electron of mass m travelled in a circular orbit of radius r, at a velocity v, around a positively charged nucleus. The *angular momentum of the electron would then be mvr. Bohr proposed that electrons could only occupy orbits in which this angular momentum had certain fixed values, $h/2\pi, 2h/2\pi, 3h/2\pi, \ldots nh/2\pi$, where h is the Planck constant. This means that the angular momentum is quantized, i.e. can only have certain values, each of which is a multiple of n. Each permitted value of n is associated with an orbit of different radius and Bohr assumed that when the atom emitted or absorbed radiation of frequency v, the electron jumped from one orbit to another; the energy emitted or absorbed by each jump is equal to hv. This theory gave good results in predicting the lines observed in the spectrum of the hydrogen atom and simple ions such as He^+, Li^{2+}, etc. The idea of quantized values of angular momentum was later explained by the wave nature of the electron. Each orbit has to have a whole number of wavelengths around it; i.e. $n\lambda = 2\pi r$, where λ is the wavelength and n a whole number. The wavelength of a particle is given by h/mv, so $nh/mv = 2\pi r$, which leads to $mvr = nh/2\pi$. Modern atomic theory does not allow subatomic particles to be treated in the same way as large objects, and Bohr's reasoning is somewhat discredited. However, the idea of quantized angular momentum has been retained.

boiling point (b.p.) The temperature at which the saturated vapour pressure of a liquid equals the external atmospheric pressure. As a consequence, bubbles form in the liquid and the temperature remains constant until all the liquid has evaporated. As the boiling point of a liquid depends on the external atmospheric pressure, boiling points are usually quoted for standard atmospheric pressure (760 mmHg = 101 325 Pa).

boiling-point–composition diagram A graph showing how the boiling point and vapour composition of a mixture of two liquids depends on the composition of the mixture. The abscissa shows the range of compositions from 100% A at one end to 100% B at the other. The diagram has two curves: the lower one gives the boiling points (at a fixed pressure) for the different compositions. The upper one is plotted by taking the composition of vapour at each temperature on the boiling-point curve. The two curves would coincide for an ideal mixture, but generally they are different because of deviations from Raoult's law. In some cases, they may show a maximum or minimum and coincide at some intermediate composition, explaining the formation of *azeotropes.

boiling-water reactor (BWR) *See* NUCLEAR REACTOR.

bolometer A sensitive instrument used to measure radiant heat. The original form consists of two elements, each comprising blackened platinum strips (about 10^{-3} mm thick) arranged in series on an insulated frame to form a zigzag. The two elements are connected into the adjacent arms of a *Wheatstone bridge; one element is exposed to the radiation, the other is shielded from it. The change in the resistance of the exposed element, as detected by the bridge galvanometer, enables the heat reaching it to be calculated.

Modern semiconductor bolometers are now common, in which the platinum is replaced by a strip of semiconductor: this has a much greater (though usually negative) *temperature coefficient of resistance, and makes the system more sensitive.

Boltzmann, Ludwig Eduard (1844–1906) Austrian physicist. He held professorships in Graz, Vienna, Munich, and Leipzig, where he worked on the kinetic theory of gases (*see* MAXWELL–BOLTZMANN DISTRIBUTION) and on thermodynamics (*see* BOLTZMANN EQUATION). He suffered from depression and committed suicide.

Boltzmann constant Symbol k or k_B. The ratio of the universal gas constant (R) to the Avogadro constant (N_A). It may be thought of therefore as the gas constant per molecule:

$$k = R/N_A = 1.380\ 658(12) \times 10^{-23}\ \text{J K}^{-1}$$

It is named after Ludwig Boltzmann.

Boltzmann equation An equation used in the study of a collection of particles in *nonequilibrium statistical mechanics, particularly their transport properties. The Boltzmann equation describes a quantity called the **distribution function**, f, which gives a mathematical description of the state and how it is changing. The distribution function depends on a position vector $\mathbf{r}$, a velocity vector $\mathbf{v}$, and the time t; it thus provides a statistical statement about the positions and velocities of the particles at any time. In the case of one species of particle being present, Boltzmann's equation can be written:

$$\partial f/\partial t + \boldsymbol{a}.(\partial f/\partial \boldsymbol{v}) + \boldsymbol{v}.(\partial f/\partial \boldsymbol{r}) = (\partial f/\partial t)_{\text{coll}},$$

where $\boldsymbol{a}$ is the acceleration of bodies between collisions and $(\partial f/\partial t)_{\text{coll}}$ is the rate of change of $f(\boldsymbol{r},\boldsymbol{v},t)$ due to collisions. The Boltzmann equation can be used to calculate *transport coefficients, such as *conductivity. The Boltzmann equation was proposed by Ludwig Boltzmann in 1872. It is the first equation in a hierarchy of equations in nonequilibrium statistical mechanics.

bomb calorimeter An apparatus used for measuring heats of combustion (e.g. calorific values of fuels and foods). It consists of a strong container in which the sample is sealed with excess oxygen and ignited electrically. The heat of combustion at constant volume can be calculated from the resulting rise in temperature.

Bondi, Hermann *See* HOYLE, SIR FRED.

Boolean algebra A form of symbolic logic, devised by George Boole (1815–64) in the middle of the 19th century, which provides a mathematical procedure for manipulating logical relationships in symbolic form. For example in Boolean algebra $a + b$ means a or b, while ab means a and b. It makes use of *set theory and is extensively used by the designers of computers to enable the bits 0 and 1, as used in the binary notation, to relate to the logical functions the computer needs in carrying out its calculations.

Born, Max (1882–1970) German-born British physicist who was awarded the 1954 Nobel Prize for physics (with W. Bothe, 1891–1957) for his work on the statistical interpretation of quantum mechanics. With *Heisenberg he also developed matrix mechanics. He also made major contributions to the theory of lattice dynamics.

Born–Haber cycle A cycle of reactions used for calculating the lattice energies of ionic crystalline solids. For a compound MX, the lattice energy is the enthalpy of the reaction

$$M^+(g) + X^-(g) \rightarrow M^+X^-(s) \; \Delta H_{\text{L}}$$

The standard enthalpy of formation of the ionic solid is the enthalpy of the reaction

$$M(s) + \tfrac{1}{2}X_2(g) \rightarrow M^+X^-(s) \; \Delta H_{\text{f}}$$

The cycle involves equating this enthalpy (which can be measured) to the sum of the enthalpies of a number of steps proceeding from the elements to the ionic solid. The steps are:

(1) Atomization of the metal:

$$M(s) \rightarrow M(g) \; \Delta H_1$$

(2) Atomization of the nonmetal:

$$\tfrac{1}{2}X_2(g) \rightarrow X(g) \; \Delta H_2$$

(3) Ionization of the metal:

   $M(g) \rightarrow M^+(g) + e\ \Delta H_3$

This is obtained from the ionization potential.
(4) Ionization of the nonmetal:

   $X(g) + e \rightarrow X^-(g)\ \Delta H_4$

This is the electron affinity.
(5) Formation of the ionic solids:

   $M^+(g) + X^-(g) \rightarrow M^+X^-(s)\ \Delta H_L$

Equating the enthalpies gives:

   $\Delta H_f = \Delta H_1 + \Delta H_2 + \Delta H_3 + \Delta H_4 + \Delta H_L$

from which ΔH_L can be found. It is named after Max Born and the German chemist Fritz Haber (1868–1934).

Born–Oppenheimer approximation An *adiabatic approximation used in molecular and solid-state physics in which the motion of atomic nuclei is taken to be so much slower than the motion of electrons that, when calculating the motions of electrons, the nuclei can be taken to be in fixed positions. This approximation was justified using *perturbation theory by Max *Born and the US physicist Julius Robert Oppenheimer (1904–67) in 1927.

boron counter A *counter tube containing a **boron chamber**, used for counting slow neutrons. The boron chamber is lined with boron or a boron compound or is filled with the gas boron trifluoride (BF_3). As natural boron contains about 18% of the isotope boron–10, and as this isotope absorbs neutrons with the emission of an alpha particle, the chamber can be coupled with a scaler to count the alpha particles emitted when neutrons enter the chamber.

Bose–Einstein condensation A phenomenon occurring in a macroscopic system consisting of a large number of *bosons at a sufficiently low temperature, in which a significant fraction of the particles occupy a single quantum state of lowest energy (the ground state). Bose–Einstein condensation can only take place for bosons whose total number is conserved in collisions. Because of the Pauli exclusion principle, it is impossible for two or more fermions to occupy the same quantum state, and so there is no analogous condensation phenomenon for such particles. Bose–Einstein condensation is of fundamental importance in explaining the phenomenon of *superfluidity. At very low temperatures (around 2×10^{-7} K) a Bose–Einstein condensate can form, in which several thousand atoms become a single entity (a **superatom**). This effect has been observed with atoms of rubidium, lithium, and other atomic systems at extremely low temperatures. The effect is named after the Indian physicist Satyendra Nath Bose (1894–1974) and Albert *Einstein.

Bose–Einstein statistics *See* QUANTUM STATISTICS.

boson An *elementary particle (or bound state of an elementary particle, e.g. an atomic nucleus or an atom) with integral spin; i.e. a particle that conforms to Bose–Einstein statistics (*see* QUANTUM STATISTICS), from which it derives its name. *Compare* FERMION.

bosonization The procedure used in *quantum field theory and the many-body problem in *quantum mechanics in which theories involving fermions have the fermions replaced by an effective field theory with bosons. In one-space-dimensional systems the transformation from fermion fields to boson fields is exact. For higher-dimensional systems, bosonization is a procedure that, in general, can only be carried out approximately; it is, for example, only valid as a low-energy approximation. The derivation of an effective field theory for mesons, starting from *quantum chromodynamics, is an example of an approximate bosonization applicable to low energies. The transformation to the description of an electron gas in terms of *plasmon variables is another example of approximate bosonization.

bottom quark *See* ELEMENTARY PARTICLES.

boundary layer The thin layer of fluid formed around a solid body or surface relative to which the fluid is flowing. Adhesion between the molecules of the fluid and that of the body or surface causes the molecules of the fluid closest to the solid to be stationary relative to it. The transfer of heat or mass between a solid and a fluid flowing over it is largely controlled by the nature of the boundary layer.

bound state A system in which two (or more) parts are bound together in such a way that energy is required to split them. An example of a bound state is a molecule formed from two (or more) atoms.

Bourdon gauge A pressure gauge consisting essentially of a C-shaped or spiral tube with an oval cross section. One end of the tube is connected to the fluid whose pressure is to be measured and the other end is sealed. As the pressure inside the tube is increased, the oval tube tends to become circular and this causes the tube to straighten. The movement of the end of the tube is transferred by a simple mechanism to a needle moving round a dial or to a digital display. With suitable design, Bourdon gauges can be used for high-pressure measurement and also for low pressures. It was invented by Eugène Bourdon (1804–88).

Boyle, Robert (1627–91) English chemist and physicist, born in Ireland. After moving to Oxford in 1654 he worked on gases, using an air pump made by Robert *Hooke. With it he proved that sound does not travel in a vacuum. In 1662 he discovered *Boyle's law. In chemistry he worked on flame tests and acid-base indicators.

Boyle's law The volume (V) of a given mass of gas at a constant temperature is inversely proportional to its pressure (p), i.e. pV = constant. This is true only for an *ideal gas. This law was discovered in 1662 by

Robert Boyle. On the continent of Europe it is known as **Mariotte's law** after E. Mariotte (1620–84), who discovered it independently in 1676. *See also* GAS LAWS.

Brackett series *See* HYDROGEN SPECTRUM.

Bragg, Sir William Henry (1862–1942) British physicist, who with his son **Sir (William) Lawrence Bragg** (1890–1971), was awarded the 1915 Nobel Prize for physics for their pioneering work on *X-ray crystallography. He also constructed an X-ray spectrometer for measuring the wavelengths of X-rays. In the 1920s, while director of the Royal Institution in London, he initiated X-ray diffraction studies of organic molecules.

Bragg's law When a beam of X-rays (wavelength λ) strikes a crystal surface in which the layers of atoms or ions are separated by a distance d, the maximum intensity of the reflected ray occurs when $\sin\theta = n\lambda/2d$, where θ (known as the **Bragg angle**) is the complement of the angle of incidence and n is an integer. The law enables the structure of many crystals to be determined. It was discovered in 1912 by Sir Lawrence Bragg.

brass A group of alloys consisting of copper and zinc. A typical yellow brass might contain about 67% copper and 33% zinc.

Brattain, Walter *See* BARDEEN, JOHN.

Braun, Karl Ferdinand (1850–1918) German physicist, who became professor of physics at Strasbourg in 1895. In the early 1900s he used crystals as diodes (later employed in crystal-set radios) and developed the *cathode-ray tube for use as an oscilloscope. He also worked on radio and in 1909 shared the Nobel Prize for physics with *Marconi.

breakdown The sudden passage of a current through an insulator. The voltage at which this occurs is the **breakdown voltage**.

breaking stress *See* ELASTICITY.

breeder reactor *See* NUCLEAR REACTOR.

Bremsstrahlung (German: braking radiation) The X-rays emitted when a charged particle, especially a fast electron, is rapidly slowed down, as when it passes through the electric field around an atomic nucleus. The X-rays cover a whole continuous range of wavelengths down to a minimum value, which depends on the energy of the incident particles. Bremsstrahlung are produced by a metal target when it is bombarded by electrons.

Brewster's law The extent of the polarization of light reflected from a transparent surface is a maximum when the reflected ray is at right angles to the refracted ray. The angle of incidence (and reflection) at which this maximum polarization occurs is called the **Brewster angle** or **polarizing angle**. For this angle i_B, the condition is that $\tan i_B = n$, where n is the refractive index of the transparent medium. The law was discovered in 1811 by the British physicist David Brewster (1781–1868).

bridge rectifier *See* RECTIFIER.

Brinell hardness A scale for measuring the hardness of metals introduced around 1900 by the Swedish metallurgist J. A. Brinell (1849–1925). A small chromium-steel ball is pressed into the surface of the metal by a load of known weight. The ratio of the mass of the load in kilograms to the area of the depression formed in square millimetres is the **Brinell number**.

Britannia metal A silvery alloy consisting of 80–90% tin, 5–15% antimony, and sometimes small percentages of copper, lead, and zinc. It is used in bearings and some domestic articles.

British thermal unit (Btu) The Imperial unit of heat, being originally the heat required to raise the temperature of 1 lb of water by 1°F. 1 Btu is now defined as 1055.06 joules.

broadband Communication by a system that supports a wide range of frequencies, so that identical messages can be carried simultaneously. *See also* ADSL.

broken symmetry A situation in which the lowest-energy state of a many-body system or *vacuum state of a relativistic *quantum field theory has a lower symmetry than the equations defining the system. Examples in solid-state physics include ferromagnetism, antiferromagnetism, and superconductivity. In particle physics, the Weinberg–Salam model (*see* ELECTROWEAK THEORY) is an important example of a relativistic quantum field theory with broken symmetry.

A result associated with broken symmetry is **Goldstone's theorem**. This states that a relativistic quantum field theory having continuous symmetry that is broken must include the existence of massless particles called **Goldstone bosons**. In many-body theory Goldstone bosons are *collective excitations. An exception to Goldstone's theorem is provided in the case of broken *gauge theories, such as the Weinberg–Salam model, in which the Goldstone bosons become massive bosons known as *Higgs bosons. In many-body theory, long-range forces provide the analogous exception to Goldstone's theorem, with the Higgs bosons being excitations with a nonzero gap.

bronze Any of a group of alloys of copper and tin, sometimes with lead and zinc present. The amount of tin varies from 1% to 30%. The alloy is hard and easily cast and extensively used in bearings, valves, and other machine parts. Various improved bronzes are produced by adding other elements; for instance, **phosphor bronzes** contain up to 1% phosphorus. In addition certain alloys of copper and metals other than tin are called bronzes – **aluminium bronze** is a mixture of copper and aluminium. Other special bronzes include *bell metal and *gun metal.

brown dwarf An astronomical object with a mass intermediate between the mass of a planet and that of a small star. The mass of a brown dwarf is large enough to generate energy by gravitational pressure, but not large enough to sustain nuclear fusion. The energy is radiated as electromagnetic radiation. Brown dwarfs are faint objects, which are expected to shine for about 100 million years before cooling. Their masses lie between a few times the mass of Jupiter and 80 times the mass of Jupiter. It has been suggested that brown dwarfs may contribute to the *missing mass of the universe.

Brownian movement The continuous random movement of microscopic solid particles (of about 1 micrometre in diameter) when suspended in a fluid medium. First observed by the botanist Robert Brown (1773–1858) in 1827 when studying pollen particles, it was originally thought to be the manifestation of some vital force. It was later recognized to be a consequence of bombardment of the particles by the continually moving molecules of the liquid. The smaller the particles the more extensive is the motion. The effect is also visible in particles of smoke suspended in a still gas.

brush An electrical contact to a moving commutator on a motor or generator. It is made of a specially prepared form of carbon and is kept in contact with the moving part by means of a spring.

brush discharge A luminous discharge from a conductor that takes the form of luminous branching threads that penetrate into the surrounding gas. It is a form of *corona and it occurs when the electric field near the surface of the conductor exceeds a certain value but is not sufficiently high for a spark to appear.

bubble chamber A device for detecting ionizing radiation. It consists of a chamber containing a liquid, often hydrogen, kept at slightly above its boiling point at a preliminary pressure that is high enough to prevent boiling. Immediately before the passage of the ionizing particles the pressure is reduced, and the particles then act as centres for the formation of bubbles, which can be photographed to obtain a record of the particles' tracks. The device was invented in 1952 by the US physicist Donald Arthur *Glaser. *Compare* CLOUD CHAMBER.

buckminsterfullerene A form of carbon composed of clusters of 60 carbon atoms bonded together in a polyhedral structure composed of pentagons and hexagons. Originally it was identified in 1985 in products obtained by firing a high-power laser at a graphite target. It can be made by an electric arc struck between graphite electrodes in an inert atmosphere. The molecule, C_{60}, was named after the US architect Richard Buckminster Fuller (1895–1983) because of the resemblance of the structure to the geodesic dome, which Fuller invented. The molecules are informally called **buckyballs**; more formally, the substance itself is also called **fullerene**. The substance is a yellow crystalline solid (**fullerite**), soluble in benzene.

Various fullerene derivatives are known in which organic groups are attached to carbon atoms on the sphere. In addition, it is possible to produce novel enclosure compounds by trapping metal ions within the C_{60} cage. Some of these have semiconducting properties. The electric-arc method of producing C_{60} also leads to a smaller number of fullerenes such as C_{70}, which have less symmetrical molecular structures. It is also possible to produce forms of carbon in which the atoms are linked in a cylindrical, rather than spherical, framework with a diameter of a few nanometres. They are known as **buckytubes** (or **nanotubes**).

buckyball *See* BUCKMINSTERFULLERENE.

buckytube *See* BUCKMINSTERFULLERENE.

bulk modulus *See* ELASTIC MODULUS.

bumping Violent boiling of a liquid caused by superheating so that bubbles form at a pressure above atmospheric pressure. It can be prevented by putting pieces of porous pot in the liquid to enable bubbles of vapour to form at the normal boiling point.

Bunsen, Robert Wilhelm (1811–99) German chemist, who held professorships at Kassel, Marburg, and Heidelberg. His early researches on arsenic-containing compounds cost him an eye in an explosion. He then turned to gas analysis and spectroscopy, enabling him and *Kirchhoff to discover the elements caesium (1860) and rubidium (1861). He also popularized the use of the *Bunsen burner and developed the *Bunsen cell.

Bunsen burner A laboratory gas burner having a vertical metal tube into which the gas is led, with a hole in the side of the base of the tube to admit air. The amount of air can be regulated by a sleeve on the tube. When no air is admitted the flame is luminous and smoky. With air, it has a faintly visible hot outer part (the oxidizing part) and an inner blue cone where combustion is incomplete (the cooler reducing part of the flame). The device is named after Robert Bunsen, who used a similar device (without a regulating sleeve) in 1855.

Bunsen cell A *primary cell developed by Robert Bunsen consisting of a zinc cathode immersed in dilute sulphuric acid and a carbon anode immersed in concentrated nitric acid. The electrolytes are separated by a porous pot. The cell gives an e.m.f. of about 1.9 volts.

buoyancy The upward thrust on a body immersed in a fluid. This force is equal to the weight of the fluid displaced (*see* ARCHIMEDES' PRINCIPLE).

Burgers vector A vector that gives a measure of the size and direction of a dislocation in a crystal. It is defined in terms of a circuit inside a crystal. In the case of a perfect crystal the circuit closes. If there is a dislocation inside the circuit it fails to close, with the Burgers vector being given by the measure of the failure to close. It was introduced by the Dutch physicist Johannes Martinus Burgers (1895–1981) in 1939.

bus (short for **busbars**) A set of conducting paths – wires or optical fibres – connecting several components of a computer system and allowing the components to send signals to each other. The components take it in turns to transmit.

busbar 1. An electrical conductor held at a constant voltage, used to carry high currents between different circuits in a system. **2.** *See* BUS.

butterfly effect *See* CHAOS.

BWR (boiling-water reactor) *See* NUCLEAR REACTOR.

bypass capacitor A capacitor providing a low *impedance path in an electronic circuit, over a predetermined frequency range.

byte A subdivision of a *word in a computer, it usually consists of eight *bits. A kilobyte is 1024 bytes (not 1000 bytes). A gigabyte is $2^{30} = 1\ 073\ 741\ 824$ bytes.

cadmium cell *See* WESTON CELL.

caesium clock An *atomic clock that depends on the energy difference between two states of the caesium–133 nucleus when it is in a magnetic field. In one type, atoms of caesium–133 are irradiated with *radio-frequency radiation, whose frequency is chosen to correspond to the energy difference between the two states. Some caesium nuclei absorb this radiation and are excited to the higher state. These atoms are deflected by a further magnetic field, which causes them to hit a detector. A signal from this detector is fed back to the radio-frequency oscillator to prevent it drifting from the resonant frequency of 9 192 631 770 hertz. In this way the device is locked to this frequency with an accuracy better than 1 part in 10^{13}. The caesium clock is used in the *SI unit definition of the second.

Cahn–Hilliard model A model describing critical dynamics in which the dynamics is dissipative and the *order parameter is conserved. The Cahn–Hilliard model was introduced by J. W. Cahn and J. E. Hilliard in 1958. *See also* GLAUBER MODEL.

calculus A series of mathematical techniques developed independently by Isaac *Newton and Gottfried Leibniz (1646–1716). **Differential calculus** treats a continuously varying quantity as if it consisted of an infinitely large number of infinitely small changes. For example, the velocity v of a body at a particular instant can be regarded as the infinitesimal distance, written ds, that it travels in the vanishingly small time interval, dt; the instantaneous velocity v is then ds/dt, which is called the **derivative** of s with respect to t. If s is a known function of t, v at any instant can be calculated by the process of *differentiation. The differential calculus is a powerful technique for solving many problems concerned with rate processes, maxima and minima, and similar problems.

   Integral calculus is the opposite technique. For example, if the velocity of a body is a known function of time, the infinitesimal distance ds travelled in the brief instant dt is given by $ds = vdt$. The measurable distance s travelled between two instants t_1 and t_2 can then be found by a process of summation, called *integration, i.e.

$$s = \int_{t_2}^{t_1} vdt$$

The technique is used for finding areas under curves and volumes and other problems involving the summation of infinitesimals.

californium Symbol Cf. A radioactive metallic transuranic element belonging to the actinoids; a.n. 98; mass number of the most stable isotope 251 (half-life about 700 years). Nine isotopes are known; californium–252 is an intense neutron source, which makes it useful in neutron *activation analysis and potentially useful as a radiation source in medicine. The element was first produced by G. T. Seaborg (1912–99) and associates in 1950.

caloric theory A former theory concerning the nature of heat, which was regarded as a weightless fluid (called **caloric**). It was unable to account for the fact that friction could produce an unlimited quantity of heat and it was abandoned when Joule showed that heat is a form of energy.

calorie The quantity of heat required to raise the temperature of 1 gram of water by 1°C (1 K). The calorie, a c.g.s. unit, is now largely replaced by the *joule, an *SI unit. 1 calorie = 4.186 8 joules.

Calorie (kilogram calorie; kilocalorie) 1000 calories. This unit is still in limited use in estimating the energy value of foods, but is obsolescent.

calorific value The heat per unit mass produced by complete combustion of a given substance. Calorific values are used to express the energy values of fuels; usually these are expressed in megajoules per kilogram (MJ kg^{-1}). They are also used to measure the energy content of foodstuffs; i.e. the energy produced when the food is oxidized in the body. The units used in this context are kilojoules per gram (kJ g^{-1}), although Calories (kilocalories) are often still used in nontechnical contexts. Calorific values are measured using a *bomb calorimeter.

calorimeter Any of various devices used to measure thermal properties such as *calorific value, *heat capacity, *latent heat, etc. *See* BOMB CALORIMETER.

calx A metal oxide formed by heating an ore in air.

camera **1.** An optical device for obtaining still photographs or for exposing cinematic film. It consists of a light-proof box with a lens at one end and a plate or film at the other. To make an exposure the shutter is opened and an image of the object to be photographed is formed on the light-sensitive film. The length of the exposure is determined by the intensity of light available, the film speed, and the *aperture of the lens. In the simpler cameras the shutter speed and aperture are controlled manually, but in automatic cameras the iris over the lens or the shutter is adjusted on the basis of information provided by a built-in *exposure meter. In ciné cameras the shutter automatically opens as the film comes to rest behind the lens for each frame; the film passes through the camera so that a set number (commonly 16, 18, or 24) of frames are exposed every second. **2.** The part of a television system that converts optical images into electronic signals. It consists of a lens system, which focuses the image to be televised on the photosensitive mosaic of the camera tube, causing localized discharge of those of its elements that are illuminated. This mosaic is scanned from behind by an electron beam so that the beam current is varied as it passes over areas of light and shade. The signal so picked up by the scanning beam is preamplified in the camera and passed to the transmitter with sound and synchronization signals. In *colour television three separate camera tubes are used, one for each *primary colour.

Canada balsam A yellow-tinted resin used for mounting specimens in optical microscopy. It has similar optical properties to glass.

canal rays Streams of positive ions produced in a *discharge tube by boring holes (canals) in the cathode. The positive ions attracted to the cathode pass through the holes and emerge on the other side as positive rays.

candela Symbol Cd. The *SI unit of luminous intensity equal to the luminous intensity in a given direction of a source that emits monochromatic radiation of frequency 540×10^{12} Hz and has a radiant intensity in that direction of 1/683 watt per steradian.

candle power Luminous intensity as formerly expressed in terms of the international candle but now expressed in candela.

capacitance The property of a conductor or system of conductors that describes its ability to store electric charge. The capacitance (C) is given by Q/V, where Q is stored charge on one conductor and V the potential difference between the two conductors (or between a single conductor and earth); it is measured in farads, although microfarads are frequently a more convenient unit for practical purposes.

An isolated sphere has a capacitance of $4\pi\varepsilon r$, where r is the radius and ε the *permittivity of the medium surrounding it. Capacitance is more commonly applied to systems of conductors (or semiconductors) separated by insulators (*see* CAPACITOR).

capacitor An arrangement of conductors separated by an insulator (dielectric) used to store charge or introduce *reactance into an alternating-current circuit. The earliest form was the *Leyden jar. Capacitors used as circuit elements have two conducting plates separated by the dielectric. The dielectric may be air, paper impregnated with oil or wax, plastic film, or ceramic. The simplest form has two parallel rectangular conducting plates (area A) separated by a dielectric (thickness d, permittivity ε). The capacitance of such a capacitor is $A\varepsilon/d$. **Electrolytic capacitors** are devices in which a thin layer of an oxide is deposited on one of the electrodes to function as the dielectric.

capacitor microphone A microphone consisting of a *capacitor with a steady voltage applied across its parallel plates. One plate is fixed, the other is a thin diaphragm that is moved by the pressure of the sound waves. The movements of the diaphragm cause a variation in the spacing and therefore in the *capacitance of the device. This variation in capacitance is, in turn, reflected in a similar variation in the charge carried by each plate. The consequent current to and from one plate is carried by a resistor, the varying potential difference across which constitutes the device's output signal. It was formerly known as a **condenser microphone**.

capillarity *See* SURFACE TENSION.

capillary A tube of small diameter, such as the narrowest type of blood vessel in the vertebrate circulatory system.

capture Any of various processes in which a system of particles absorbs an extra particle. There are several examples in atomic and nuclear physics. For instance, a positive ion may capture an electron to give a neutral atom or molecule. Similarly, a neutral atom or molecule capturing an electron becomes a negative ion. An atomic nucleus may capture a neutron to produce a different (often unstable) nucleus. Another type of nuclear capture is the process in which the nucleus of an atom absorbs an electron from the innermost orbit (the K shell) to transform into a different nucleus. In this process (called **K capture**) the atom is left in an excited state and generally decays by emission of an X-ray photon.

  Radiative capture is any such process in which the capture results in an excited state that decays by emission of photons. A common example is neutron capture to yield an excited nucleus, which decays by emission of a gamma ray.

carat 1. A measure of fineness (purity) of gold. Pure gold is described as 24-carat gold. 14-carat gold contains 14 parts in 24 of gold, the remainder usually being copper. **2.** A unit of mass equal to 0.200 gram, used to measure the masses of diamonds and other gemstones.

carbon cycle A series of nuclear reactions in which four hydrogen nuclei combine to form a helium nucleus with the liberation of energy, two positrons, and two neutrinos. The process is believed to be the source of energy in many stars and to take place in six stages. In this series carbon–12 acts as if it were a catalyst, being reformed at the end of the series:

$$^{12}_{6}\text{C} + ^{1}_{1}\text{H} \rightarrow ^{13}_{7}\text{N} + \gamma$$

$$^{13}_{7}\text{N} \rightarrow ^{13}_{6}\text{C} + e^{+} + \nu_{e}$$

$$^{13}_{6}\text{C} + ^{1}_{1}\text{H} \rightarrow ^{14}_{7}\text{N} + \gamma$$

$$^{14}_{7}\text{N} + ^{1}_{1}\text{H} \rightarrow ^{15}_{8}\text{O} + \gamma$$

$$^{15}_{8}\text{O} \rightarrow ^{15}_{7}\text{N} + e^{+} + \nu_{e}$$

$$^{15}_{7}\text{N} + ^{1}_{1}\text{H} \rightarrow ^{12}_{6}\text{C} + ^{4}_{2}\text{He}$$

See STELLAR EVOLUTION.

carbon dating (**radiocarbon dating**) A method of estimating the ages of archaeological specimens of biological origin. As a result of *cosmic radiation a small number of atmospheric nitrogen nuclei are continuously being transformed by neutron bombardment into radioactive nuclei of carbon–14:

$$^{14}_{7}N + n \rightarrow {}^{14}_{6}C + p$$

Some of these radiocarbon atoms find their way into living trees and other plants in the form of carbon dioxide, as a result of photosynthesis. When the tree is cut down photosynthesis stops and the ratio of radiocarbon atoms to stable carbon atoms begins to fall as the radiocarbon decays. The ratio $^{14}C/^{12}C$ in the specimen can be measured and enables the time that has elapsed since the tree was cut down to be calculated. The method has been shown to give consistent results for specimens up to some 40 000 years old, though its accuracy depends upon assumptions concerning the past intensity of the cosmic radiation. The technique was developed by Willard F. Libby (1908–80) and his coworkers in 1946–47.

carbon fibres Fibres of carbon in which the carbon has an oriented crystal structure. Carbon fibres are made by heating textile fibres and are used in strong composite materials for use at high temperatures.

Carnot, Nicolas Léonard Sadi (1796–1832) French physicist, who first worked as a military engineer. He then turned to scientific research and in 1824 published his analysis of the efficiency of heat engines. The key to this analysis is the thermodynamic *Carnot cycle. He died at an early age of cholera.

Carnot cycle The most efficient cycle of operations for a reversible *heat engine. Published in 1824 by N. L. S. Carnot, it consists of four operations on the working substance in the engine (see illustration): (1) Isothermal expansion at thermodynamic temperature T_1 with heat q_1 taken in. (2) Adiabatic expansion with a fall of temperature to T_2. (3) Isothermal compression at temperature T_2 with heat q_2 given out. (4) Adiabatic compression with a rise of temperature back to T_1. According to the **Carnot principle**, the efficiency of any reversible heat engine depends only on the temperature range through which it works, rather than the properties of the working substances. In any reversible engine,

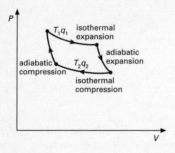

Carnot cycle

the efficiency (η) is the ratio of the work done (W) to the heat input (q_1), i.e. $\eta = W/q_1$. As, according to the first law of *thermodynamics, $W = q_1 - q_2$, it follows that $\eta = (q_1 - q_2)/q_1$. For the Kelvin temperature scale, $q_1/q_2 = T_1/T_2$ and $\eta = (T_1 - T_2)/T_1$. For maximum efficiency T_1 should be as high as possible and T_2 as low as possible.

carrier **1.** *See* CARRIER WAVE. **2.** *See* CHARGE CARRIER. **3.** *See* CARRIER GAS.

carrier gas The gas that carries the sample in *gas chromatography.

carrier wave An electromagnetic wave of specified frequency and amplitude that is emitted by a radio transmitter in order to carry information. The information is superimposed onto the carrier by means of *modulation.

Cartesian coordinates A system used in analytical geometry to locate a point P, with reference to two or three **axes** (see graphs). In a two-dimensional system the vertical axis is the y-axis and the horizontal axis is the x-axis. The point at which they intersect is called the **origin**, O. Values of $y < 0$ fall on the y-axis below the origin, values of $x < 0$ fall on the x-axis to the left of the origin. Any point P is located by its perpendicular distances from the two axes. The distance from the x-axis is called the **ordinate**; the distance from the y-axis is the **abscissa**. The position is indicated numerically by enclosing the values of the abscissa and the ordinate in parentheses and separating the two by means of a comma, e.g. (x,y). In three dimensions the system can be used to locate a point with reference to a third, z-axis. It is named after René Descartes (1596–1650).

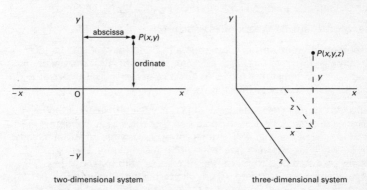

two-dimensional system   three-dimensional system

Cartesian coordinates

cascade liquefier An apparatus for liquefying a gas of low *critical temperature. Another gas, already below its critical temperature, is liquified and evaporated at a reduced pressure in order to cool the first gas to below its critical temperature. In practice a series of steps is often used, each step enabling the critical temperature of the next gas to be reached.

cascade process Any process that takes place in a number of steps, usually because the single step is too inefficient to produce the desired result. For example, in various uranium-enrichment processes the separation of the desired isotope is only poorly achieved in a single stage; to achieve better separation the process has to be repeated a number of times, in a series, with the enriched fraction of one stage being fed to the succeeding stage for further enrichment. Another example of cascade process is that operating in a *cascade liquefier.

case hardening The hardening of the surface layer of steel, used for tools and certain mechanical components. The commonest method is to carburize the surface layer by heating the metal in a hydrocarbon or by dipping the red hot metal into molten sodium cyanide. Diffusion of nitrogen into the surface layer to form nitrides is also used.

Casimir effect The force between two *macroscopic conducting surfaces in a volume that only contains an electromagnetic field. The *zero-point energy of the electromagnetic field depends on the *mode frequencies of the field, which in turn depend on the boundary conditions on the field. Thus, changing the positions of the surfaces changes the mode frequencies and zero-point energy of the field. This energy change can be expressed as a potential energy per unit area of the plates (as a function of distance between the plates), which leads to a force between the plates. The existence of these forces was predicted by the Dutch physicist Hendrik B. G. Casimir (1909–) in the 1940s and detected experimentally by M. J. Sparnaay in 1958. Both the sign and magnitude of the Casimir effect depend critically on the geometry of the surface.

Cassegrainian telescope *See* TELESCOPE.

cast iron A group of iron alloys containing 1.8 to 4.5% of carbon. It is usually cast into specific shapes ready for machining, heat treatment, or assembly. It is sometimes produced direct from the *blast furnace or it may be made from remelted *pig iron.

catalytic converter A device used in the exhaust systems of motor vehicles to reduce atmospheric pollution. The three main pollutants produced by petrol engines are: unburnt hydrocarbons, carbon monoxide produced by incomplete combustion of hydrocarbons, and nitrogen oxides produced by nitrogen in the air reacting with oxygen at high engine temperatures. Hydrocarbons and carbon monoxide can be controlled by a higher combustion temperature and a weaker mixture. However, the higher temperature and greater availability of oxygen arising from these measures encourage formation of nitrogen oxides. The use of three-way catalytic converters solves this problem by using platinum and palladium catalysts to oxidize the hydrocarbons and the CO and rhodium catalysts to reduce the nitrogen oxides back to nitrogen. These three-way catalysts require that the air–fuel ratio is strictly stochiometric. Some catalytic converters promote oxidation reactions only, leaving the nitrogen oxides unchanged. Three-way converters can reduce hydrocarbons and CO emissions by some 85%, at the same time reducing nitrogen oxides by 62%.

catastrophe theory A branch of mathematics dealing with the sudden emergence of discontinuities, in contrast to *calculus, which is concerned with continuous quantities. Catastrophe theory originated in *topology in work by the French mathematician René Thom (1923–2002) and was developed by Thom and the Russian mathematician Vladimir Igorevich Arnold (1937–). There are physical applications of catastrophe theory in *optics and in systems involving *complexity, including biological systems.

catenary A curve formed when a chain or rope of uniform density hangs from two fixed points. If the lowest point on the curve passes through the origin, the equation is $y = c(\cosh x/c)$, where c is the distance between the x-axis and the directrix.

cathetometer A telescope or microscope fitted with crosswires in the eyepiece and mounted so that it can slide along a graduated scale. Cathetometers are used for accurate measurement of lengths without mechanical contact. The microscope type is often called a **travelling microscope**.

cathode A negative electrode. In *electrolysis cations are attracted to the cathode. In vacuum electronic devices electrons are emitted by the cathode and flow to the *anode. It is therefore from the cathode that electrons flow into these devices. However, in a primary or secondary cell the cathode is the electrode that spontaneously becomes negative during discharge, and from which therefore electrons emerge.

cathode-ray oscilloscope (CRO) An instrument based on the *cathode-ray tube that provides a visual image of electrical signals. The horizontal deflection is usually provided by an internal *timebase, which causes the beam to sweep across the screen at a specified rate. The signal to be investigated is fed to the vertical deflection plates after amplification. Thus the beam traces a graph of the signal amplitude against time.

cathode rays Streams of electrons emitted at the cathode in an evacuated tube containing a cathode and an anode. They were first observed in gas *discharge tubes operated at low pressure. Under suitable conditions electrons produced by secondary emission at the cathode are accelerated down the tube to the anode. In such devices as the *cathode-ray tube the electrons are produced by *thermionic emission from a hot cathode in a vacuum.

cathode-ray tube (CRT) The device that provides the viewing screen in the television tube, the radar viewer, and the *cathode-ray oscilloscope. The cathode-ray tube consists of an evacuated tube containing a heated cathode and two or more ring-shaped anodes through which the cathode rays can pass so that they strike the enlarged end of the tube (see illustration). This end of the tube is coated with fluorescent material so that it provides a screen. Any point on the screen that is struck by the cathode ray becomes luminous. A *control grid between the cathode and the anode enables the intensity of the beam to be varied, thus controlling the brightness of the illumination on the screen. The assembly of cathode, control grid, and anode is called the *electron gun. The beam emerging from the electron gun is focused and deflected by means of plates providing an electric field or coils providing a magnetic field. This enables the beam to be focused to a small point of light and deflected to produce the illusion of an illuminated line as this point sweeps across the tube.

The television tube is a form of cathode-ray tube in which the beam is made to scan the screen 625 times to form a frame, with 25 new frames being produced every second. (These are the figures for standard television tubes in the UK). Each frame creates a picture by variations in the intensity of the beam as it forms each line.

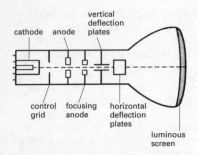

Cathode-ray tube

cathodic protection *See* SACRIFICIAL PROTECTION.

cation A positively charged ion, i.e. an ion that is attracted to the cathode in *electrolysis. *Compare* ANION.

causality The principle that effect cannot precede cause. The principle is particularly useful when combined with the principle that the highest attainable speed in the universe is the *speed of light in a vacuum. Causality is used to analyse the results of scattering experiments and in optics.

caustic (in optics) The curve or surface formed by the reflection of parallel rays of light in a large-aperture concave mirror. The apex of the caustic lies at the principal focus of the mirror. Such a curve can sometimes be seen on the surface of the liquid in a cup as a result of reflection by the curved walls of the cup. A similar curve is formed by a convex lens with spherical surfaces refracting parallel rays of light.

Cavendish, Henry (1731–1810) British chemist and physicist, born in France. Although untrained, his inheritance from his grandfather, the Duke of Devonshire, enabled him to live as a recluse and study science. In his experiments with gases (1766), he correctly distinguished between hydrogen and carbon dioxide, and in 1781 synthesized water by exploding hydrogen in oxygen. He also constructed a torsion balance in 1798, with which he measured the mean density (and hence mass) of the earth.

cavitation The formation of gas- or vapour-filled cavities in liquids in motion when the pressure is reduced to a critical value while the ambient temperature remains constant. If the velocity of the flowing liquid exceeds a certain value, the pressure can be reduced to such an extent that *Bernoulli's theorem breaks down. It is at this point that cavitation occurs, causing a restriction on the speed at which hydraulic machinery can be run without noise, vibration, erosion of metal parts, or loss of efficiency.

cavity resonator *See* RESONANT CAVITY.

CD *See* COMPACT DISK.

CD-I CD interactive: a variant of *CD-ROM in which data, sound, and images can be interleaved on the same disk, i.e. it is a *multimedia disk. It was designed as a 'buy and play' system for the home.

CD-ROM CD read-only memory: a device that is based on the audio *compact disk and provides read-only access to a large amount of data (up to 640 megabytes) for use on computer systems. The term also refers to the medium in general. A **CD-ROM drive** must be used with the computer system to read the data from disk; the data cannot normally be rewritten. Most drives can also play CD audio disks, but audio disk players cannot handle CD-ROMs. The data may be in any form – text, sound, images, or binary data, or a mixture – and various CD-ROM format standards exist to handle these. CD-ROM is widely used for the distribution of data, images, and software and for archiving data.

CD-RW CD-rewritable: a CD format launched around 1997 that enabled recording and re-use of CDs. CD-RW uses a phase change to record data. The recording layer is a special alloy (typically silver/indium/antimony/tellurium). The laser in the CD drive has three power levels. The highest level melts small regions of the recording layer and these cool quickly to an amorphous form, thereby creating small pits in the recording surface. This level is used for writing data to the disk. The intermediate power level heats the surface to a temperature below the melting point, but high enough to cause recrystallization of the amorphous pits. This is used for erasing data. The lowest power level is used for reading data from the disk in the same way that data is read from a *CD-ROM.

celestial equator *See* EQUATOR.

celestial mechanics The study of the motions of and forces between the celestial bodies. It is based on *Newton's laws of motion and *Newton's law of gravitation. Refinements based on the general theory of *relativity are also included, although the differences between the two theories are only important in a few cases.

celestial sphere The imaginary sphere of infinite radius within which celestial bodies appear to lie. The earth, and the observer, are visualized as being at the centre of the sphere and the sphere as rotating once every sidereal *day (see illustration). The sphere is used to describe the position of celestial bodies with respect to the earth.

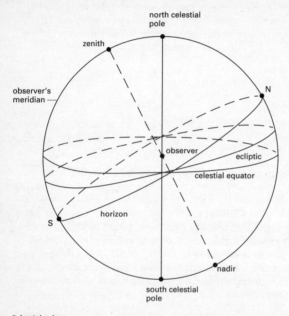

Celestial sphere

cell (in physical chemistry) **1.** A system in which two electrodes are in contact with an electrolyte. The electrodes are metal or carbon plates or rods or, in some cases, liquid metals (e.g. mercury). In an *electrolytic cell a current from an outside source is passed through the electrolyte to produce chemical change (*see* ELECTROLYSIS). In a *voltaic cell, spontaneous reactions between the electrodes and electrolyte(s) produce a potential difference between the two electrodes.

Voltaic cells can be regarded as made up of two *half cells, each composed of an electrode in contact with an electrolyte. For instance, a zinc rod dipped in zinc sulphate solution is a $Zn|Zn^{2+}$ half cell. In such a system zinc atoms dissolve as zinc ions, leaving a negative charge on the electrode

$$Zn(s) \rightarrow Zn^{2+}(aq) + 2e$$

The solution of zinc continues until the charge build-up is sufficient to prevent further ionization. There is then a potential difference between the zinc rod and its solution. This cannot be measured directly, since measurement would involve making contact with the electrolyte, thereby introducing another half cell (*see* ELECTRODE POTENTIAL). A rod of copper in copper sulphate solution comprises another half cell. In this case the spontaneous reaction is one in which copper ions in solution take electrons from the electrode and are deposited on the electrode as copper atoms. In this case, the copper acquires a positive charge.

The two half cells can be connected by using a porous pot for the liquid junction (as in the Daniell cell) or by using a salt bridge. The resulting cell can then supply current if the electrodes are connected through an external circuit. The cell is written

$$Zn(s)|Zn^{2+}(aq)|Cu^{2+}(aq)|Cu$$

$$E = 1.10 \, V$$

Here, E is the e.m.f. of the cell equal to the potential of the right-hand electrode minus that of the left-hand electrode for zero current. Note that 'right' and 'left' refer to the cell as written. Thus, the cell could be written

$$Cu(s)|Cu^{2+}(aq)|Zn^{2+}(aq)|Zn(s)$$

$$E = -1.10 \, V$$

The overall reaction for the cell is

$$Zn(s) + Cu^{2+}(aq) \rightarrow Cu(s) + Zn^{2+}(aq)$$

This is the direction in which the cell reaction occurs for a positive e.m.f.

The cell above is a simple example of a **chemical cell**; i.e. one in which the e.m.f. is produced by a chemical difference. **Concentration cells** are cells in which the e.m.f. is caused by a difference of concentration. This may be a difference in concentration of the electrolyte in the two half cells. Alternatively, it may be an electrode concentration difference (e.g. different concentrations of metal in an amalgam, or different pressures of gas in two gas electrodes). Cells are also classified into cells **without transport** (having a single electrolyte) and **with transport** (having a liquid junction across which ions are transferred). Various types of voltaic cell exist, used as sources of current, standards of potential, and experimental set-ups for studying electrochemical reactions. *See also* DRY CELL; PRIMARY CELL; SECONDARY CELL. **2.** *See* PHOTOELECTRIC CELL. **3.** *See* SOLAR CELL. **4.** *See* KERR EFFECT (for Kerr cell).

Celsius scale A *temperature scale in which the fixed points are the temperatures at standard pressure of ice in equilibrium with water (0°C) and water in equilibrium with steam (100°C). The scale, between these two temperatures, is divided in 100 degrees. The degree Celsius (°C) is equal in magnitude to the *kelvin. This scale was formerly known as the centigrade scale; the name was officially changed in 1948 to avoid confusion with a hundredth part of a grade. It is named after the Swedish astronomer Anders Celsius (1701–44), who devised the inverted form of this scale (ice point 100°, steam point 0°) in 1742.

cementation Any metallurgical process in which the surface of a metal is impregnated by some other substance, especially an obsolete process for making steel by heating bars of wrought iron to red heat for several days in a bed of charcoal. *See also* CASE HARDENING.

cementite *See* STEEL.

centi- Symbol c. A prefix used in the metric system to denote one hundredth. For example, 0.01 metre = 1 centimetre (cm).

centigrade scale *See* CELSIUS SCALE.

central processing unit *See* COMPUTER; CPU.

centre of curvature The centre of the sphere of which a *lens surface or curved *mirror forms a part. The **radius of curvature** is the radius of this sphere.

centre of gravity *See* CENTRE OF MASS.

centre of mass The point at which the whole mass of a body may be considered to be concentrated. This is the same as the **centre of gravity**, the point at which the whole weight of a body may be considered to act, if the body is situated in a uniform gravitational field.

centrifugal force *See* CENTRIPETAL FORCE.

centrifugal pump *See* PUMP.

centrifuge A device in which solid or liquid particles of different densities are separated by rotating them in a tube in a horizontal circle. The denser particles tend to move along the length of the tube to a greater radius of rotation, displacing the lighter particles to the other end.

centripetal force A force acting on a body causing it to move in a circular path. If the mass of the body is m, its constant speed v, and the radius of the circle r, the magnitude of the force is mv^2/r and it is directed towards the centre of the circle. Even though the body is moving with a constant speed v, its velocity is changing, because its direction is constantly changing. There is therefore an acceleration v^2/r towards the centre of the circle. For example, when an object is tied to a string and swung in a horizontal circle there is a tension in the string equal to mv^2/r. If the string breaks, this restraining force disappears and the object will move off in a straight line along the tangent to the circle in which it was previously moving.

In the case of a satellite (mass m) orbiting the earth (mass M), the centripetal force holding the satellite in orbit is the gravitational force,

GmM/d^2, where G is the gravitational constant and d is the height of the satellite above the centre of the earth. Therefore $GmM/d^2 = mv^2/d$. This equation enables the height of the orbit to be calculated for a given orbital velocity.

Another way of looking at this situation, which was once popular, is to assume that the centripetal force is balanced by an equal and opposite force, acting away from the centre of the circle, called the **centrifugal force**. One could then say that the satellite stays in orbit when the centrifugal force balances the gravitational force. This is, however, a confusing and misleading argument because the centrifugal force is fictitious – it does not exist. The gravitational force is not balanced by the centrifugal force: it *is* the centripetal force.

Another example is that of a car rounding a bend. To an observer in the car, a tennis ball lying on the back shelf will roll across the shelf as if it was acted on by an outward centrifugal force. However, to an observer outside the car it can be seen that the ball, because of its almost frictionless contact with the car, is continuing in its straight line motion, uninfluenced by the centripetal force. Occasionally the concept of a centrifugal force can be useful, as long as it is recognized as a fictitious force. A true centrifugal force is exerted, as a *reaction, by the rotating object on whatever is providing its centripetal force.

centroid The point within an area or volume at which the centre of mass would be if the surface or body had a uniform density. For a symmetrical area or volume it coincides with the centre of mass. For a nonsymmetrical area or volume it has to be found by integration.

Cerenkov, Pavel Alekseyevich (1904–90) Soviet physicist, who became a professor at the Lebedev Institute of Physics in Moscow. In 1934, while observing radioactive radiation underwater, he discovered *Cerenkov radiation. The explanation of the phenomenon was provided by Igor Tamm (1895–1971) and Ilya Frank (1908–90), and in 1958 the three scientists shared the Nobel Prize for physics.

Cerenkov counter (Cerenkov detector) A type of *counter for detecting and counting high-energy charged particles. The particles pass through a liquid and the light emitted as *Cerenkov radiation is registered by a *photomultiplier tube.

Cerenkov radiation Electromagnetic radiation, usually bluish light, emitted by a beam of high-energy charged particles passing through a transparent medium at a speed greater than the speed of light in that medium. It was discovered in 1934 by Pavel Cerenkov. The effect is similar to that of a *sonic boom when an object moves faster than the speed of sound; in this case the radiation is a shock wave set up in the electromagnetic field. Cerenkov radiation is used in the *Cerenkov counter.

cermet A composite material consisting of a ceramic in combination with a sintered metal, used when a high resistance to temperature, corrosion, and abrasion is needed.

CERN (Conseil Européen pour la Recherche Nucléaire) The European Organization for Nuclear Research, which is situated close to Geneva in Switzerland and is supported by a number of European nations. It runs the **Super Proton Synchrotron** (SPS), which has a 7-kilometre underground tunnel enabling protons to be accelerated to 400 GeV, and the **Large Electron-Positron Collider** (LEP), in which 50 GeV electron and positron beams are collided.

cetane number A number that provides a measure of the ignition characteristics of a Diesel fuel when it is burnt in a standard Diesel engine. It is the percentage of cetane (hexadecane) in a mixture of cetane and 1-methylnaphthalene that has the same ignition characteristics as the fuel being tested. *Compare* OCTANE NUMBER.

c.g.s. units A system of *units based on the centimetre, gram, and second. Derived from the metric system, it was not well suited for use with thermal quantities (based on the inconsistently defined *calorie) and with electrical quantities (in which two systems, based respectively on unit permittivity and unit permeability of free space, were used). For many scientific purposes c.g.s. units have now been replaced by *SI units.

Chadwick, Sir James (1891–1974) British physicist. After working at Manchester University under *Rutherford, he went to work with Hans *Geiger in Leipzig in 1913. Interned for the duration of World War I, he joined Rutherford in Cambridge after the war. In 1932 he discovered the *neutron, as predicted by Rutherford. In 1935 he was awarded the Nobel Prize, the same year in which he built Britain's first *cyclotron at Liverpool University.

chain reaction A reaction that is self-sustaining as a result of the products of one step initiating a subsequent step.

In nuclear chain reactions the succession depends on production and capture of neutrons. Thus, one nucleus of the isotope uranium–235 can disintegrate with the production of two or three neutrons, which cause similar fission of adjacent nuclei. These in turn produce more neutrons. If the total amount of material exceeds a *critical mass, the chain reaction may cause an explosion.

Chemical chain reactions usually involve free radicals as intermediates. An example is the reaction of chlorine with hydrogen initiated by ultraviolet radiation. A chlorine molecule is first split into atoms:

$$Cl_2 \rightarrow Cl\bullet + Cl\bullet$$

These react with hydrogen as follows

$$Cl\bullet + H_2 \rightarrow HCl + H\bullet$$

$$H\bullet + Cl_2 \rightarrow HCl + Cl\bullet \text{ etc.}$$

Combustion and explosion reactions involve similar free-radical chain reactions.

Chandrasekhar limit The maximum possible mass of a star that is prevented from collapsing under its own gravity by the *degeneracy pressure of electrons (a *white dwarf). For white dwarfs the **Chandrasekhar mass** is about 1.4 times the mass of the sun. There is an analogue of the Chandrasekhar limit for neutron stars. For neutron stars its value is less precisely known because of uncertainties regarding the equation of state of neutron matter, but it is generally taken to be in the range of 1.5 to 3 (and almost certainly no more than 5) times the mass of the sun. It is named after the Indian-born US astrophysicist Subrahmanyan Chandrasekhar (1910–95).

change of phase (change of state) A change of matter in one physical *phase (solid, liquid, or gas) into another. The change is invariably accompanied by the evolution or absorption of energy, even if it takes place at constant temperature (*see* LATENT HEAT).

channel 1. The region between the source and the drain in a field-effect *transistor. The conductivity of the channel is controlled by the voltage applied to the gate. **2.** A path, or a specified frequency band, along which signals, information, or data flow.

chaos Unpredictable and seemingly random behaviour occurring in a system that should be governed by deterministic laws. In such systems, the equations that describe the way the system changes with time are nonlinear and involve several variables. Consequently, they are very sensitive to the initial conditions, and a very small initial difference may make an enormous change to the future state of the system. Originally, the theory was introduced to describe unpredictability in meteorology, as exemplified by the **butterfly effect**. It has been suggested that the dynamical equations governing the weather are so sensitive to the initial data that whether or not a butterfly flaps its wings in one part of the world may make the difference between a tornado occurring or not occurring in some other part of the world. Chaos theory has subsequently been extended to other branches of science; for example to turbulent flow, planetary dynamics, and electrical oscillations in physics, and to combustion processes and oscillating reactions in chemistry. *See also* ATTRACTOR; FRACTAL.

characteristic *See* LOGARITHM.

charge A property of some *elementary particles that gives rise to an interaction between them and consequently to the host of material phenomena described as electrical. Charge occurs in nature in two forms, conventionally described as **positive** and **negative** in order to distinguish between the two kinds of interaction between particles. Two particles that have similar charges (both negative or both positive) interact by repelling each other; two particles that have dissimilar charges (one positive, one negative) interact by attracting each other. The size of the interaction is determined by *Coulomb's law.

The natural unit of negative charge is the charge on an *electron, which is equal but opposite in effect to the positive charge on the proton. Large-scale matter that consists of equal numbers of electrons and protons is electrically neutral. If there is an excess of electrons the body is negatively charged; an excess of protons results in a positive charge. A flow of charged particles, especially a flow of electrons, constitutes an electric current. Charge is measured in coulombs, the charge on an electron being 1.602×10^{-19} coulombs.

charge carrier The entity that transports electric charge in an electric current. The nature of the carrier depends on the type of conductor: in metals, the charge carriers are electrons; in *semiconductors the carriers are electrons (n-type) or positive *holes (p-type); in gases the carriers are positive ions and electrons; in electrolytes they are positive and negative ions.

charge conjugation Symbol C. A property of elementary particles that determines the difference between a particle and its *antiparticle. The property is not restricted to electrically charged particles (i.e. it applies to neutral particles such as the neutron). *See* CP INVARIANCE.

charge density **1.** The electric charge per unit volume of a medium or body (**volume charge density**). **2.** The electric charge per unit surface area of a body (**surface charge density**).

Charles, Jacques Alexandre César (1746–1823) French chemist and physicist, who became professor of physics at the Paris Conservatoire des Arts et Métiers. He is best remembered for discovering *Charles' law (1787), relating to the volume and temperature of a gas. In 1783 he became the first person to make an ascent in a hydrogen balloon.

Charles' law The volume of a fixed mass of gas at constant pressure expands by a constant fraction of its volume at 0°C for each Celsius degree or kelvin its temperature is raised. For any *ideal gas the fraction is approximately 1/273. This can be expressed by the equation $V = V_0(1 + t/273)$, where V_0 is the volume at 0°C and V is its volume at t°C. This is equivalent to the statement that the volume of a fixed mass of gas at constant pressure is proportional to its thermodynamic temperature, $V = kT$, where k is a constant. The law resulted from experiments begun around 1787 by J. A. C. Charles but was properly established only by the more accurate results published in 1802 by Joseph Gay-Lussac. Thus the law is also known as **Gay-Lussac's law**. An equation similar to that given above applies to pressures for ideal gases: $p = p_0(1 + t/273)$, a relationship known as **Charles' law of pressures**. *See also* GAS LAWS.

charm A property of certain *elementary particles that is expressed as a

quantum number (*see* ATOM) and is used in the quark model. It was
suggested to account for the unusually long lifetime of the *psi particle.
In this theory the three original quark–antiquark pairs were
supplemented by a fourth pair – the charmed quark and its antiquark.
The psi particle itself is a meson having zero charm as it consists of the
charmed pair. However, charmed *hadrons do exist; they are said to
possess **naked charm**. Charm is thought to be conserved in strong and
electromagnetic interactions.

chemical cell *See* CELL.

chemical dating An absolute *dating technique that depends on
measuring the chemical composition of a specimen. Chemical dating can
be used when the specimen is known to undergo slow chemical change at
a known rate. For instance, phosphate in buried bones is slowly replaced
by fluoride ions from the ground water. Measurement of the proportion of
fluorine present gives a rough estimate of the time that the bones have
been in the ground. Another, more accurate, method depends on the fact
that amino acids in living organisms are L-optical isomers. After death,
these racemize and the age of bones can be estimated by measuring the
relative amounts of D- and L-amino acids present.

chemical potential Symbol: μ. For a given component in a mixture,
the coefficient $\partial G/\partial n$, where G is the Gibbs free energy and n the amount
of substance of the component. The chemical potential is the change in
Gibbs free energy with respect to change in amount of the component,
with pressure, temperature, and amounts of other components being
constant. Components are in equilibrium if their chemical potentials are
equal.

chemiluminescence *See* LUMINESCENCE.

chemisorption *See* ADSORPTION.

chip *See* SILICON CHIP.

chirality The property of existing in left- and right-handed structural
forms. *See* OPTICAL ACTIVITY.

Chiron A minor planet discovered in 1977. It has an orbit of 50.68 years
that, unlike other known minor planets, lies almost entirely outside that
of Saturn. Its diameter is uncertain, but seems to be of the order of 300 km.

Chladni figures *Stationary wave patterns produced in metal or glass
plates. Vibrations of plates can be set up by stroking the edge with a
violin bow. It was shown in 1787, in experiments by the German lawyer
and physicist Ernst F. F. Chladni (1756–1827), that the vibrations of plates
could be made visible by sprinkling sand on the plate before applying the
bow to it. Salt can also be used to illustrate vibrations in plates with the
particles accumulating in drifts at *nodes and moving away from
antinodes. The nodes thus appear as white lines. The vibrations associated

with these patterns are also associated with sounds given off by these plates (an example of this being a drumhead).

choke A coil of wire with high inductance and low resistance. It is used in radio circuits to impede the passage of audio-frequency or radio-frequency currents or to smooth the output of a rectifying circuit.

cholesteric crystal *See* LIQUID CRYSTAL.

chromatic aberration *See* ABERRATION.

chromaticity An objective description of the colour quality of a visual stimulus that does not depend on its luminance but which, together with its luminance, completely specifies the colour. The colour quality is defined in terms of **chromaticity coordinates**, x,y, and z, where

$x = X/(X + Y + Z)$

$y = Y/(X + Y + Z)$

and $z = Z/(X + Y + Z)$

X, Y, and Z are the **tristimulus values** of a light, i.e. they are the amounts of three reference stimuli needed to match exactly the light under consideration in a trichromatic system.

chromium steel Any of a group of *stainless steels containing 8–25% of chromium. A typical chromium steel might contain 18% of chromium, 8% of nickel, and 0.15% of carbon. Chromium steels are highly resistant to corrosion and are used for cutlery, chemical plant, ball bearings, etc.

chromosphere The layer of the *sun's atmosphere immediately above the *photosphere. The chromosphere is normally only visible when the photosphere is totally eclipsed by the moon. The chromosphere is about 10 000 kilometres thick and the temperature in it rises from 4000 K, where it merges with the photosphere, to about 50 000 K, where it reaches the transition region below the *corona.

chronology protection conjecture A conjecture put forward by Stephen *Hawking in the early 1990s asserting that the fundamental laws of physics should forbid *time travel. There is some theoretical evidence in favour of this idea.

circle A closed curve every point on which is a fixed distance (the **radius**) from a point (the **centre**) within the curve (see illustration). The **diameter** is a line that joins two points on the **circumference** and passes through the centre: the diameter is twice the radius (r). The circumference of a circle is equal to $2\pi r$; the area is πr^2, where π is a constant with the value 3.141 592. In analytical geometry the equation of a circle, centred at the origin, is $x^2 + y^2 = r^2$.

circular measure A method of measuring angles by treating them as the angle formed by a sector of a circle at the circle's centre. The unit of

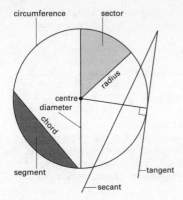

A circle

measure is the **radian**, the angle subtended at the centre of a circle by an arc of equal length to the radius. Since an arc of length r subtends an angle of 1 radian, the whole circumference, length $2\pi r$, will subtend an angle of $2\pi r/r = 2\pi$ radians. Thus, $360° = 2\pi$ radians; 1 radian = 57.296°.

circular polarization　*See* POLARIZATION OF LIGHT.

cladding 1. A thin coating of an expensive metal rolled on to a cheaper one. **2.** A thin covering of a metal around a fuel element in a nuclear reactor to prevent corrosion of the fuel elements by the coolant.

Clapeyron–Clausius equation　A differential equation that describes the relationship between variables when there is a change in the state of a system. In a system that has two *phases of the same substance, for example solid and liquid, heat is added or taken away very slowly so that one phase can change reversibly into the other while the system remains at equilibrium. If the two phases are denoted A and B, the Clapeyron–Clausius equation is:

$$dp/dT = L/T(V_B - V_A),$$

where p is the pressure, T is the *thermodynamic temperature, L is the heat absorbed per mole in the change from A to B, and V_B and V_A are the volumes of B and A respectively. In the case of a transition from liquid to vapour, the volume of the liquid can be ignored. Taking the vapour to be an *ideal gas, the Clapeyron–Clausius equation can be written:

$$d\log_e p \,/\, dT = L/RT^2.$$

The Clapeyron–Clausius equation is named after the French engineer Benoit-Pierre-Émile Clapeyron (1799–1864) and the German physicist Rudolf Julius Emmanuel Clausius (1822–88).

Clark cell　A type of *voltaic cell consisting of an anode made of zinc amalgam and a cathode of mercury both immersed in a saturated solution of zinc sulphate. The Clark cell was formerly used as a standard of e.m.f.; the e.m.f. at 15°C is 1.4345 volts. It is named after the British scientist Hosiah Clark (d. 1898).

classical field theory A theory that describes a *field in terms of *classical physics rather than *quantum mechanics. Examples of classical field theories include classical *electrodynamics, described by *Maxwell's equations, and the general theory of *relativity, describing classical gravitation. A classical field theory emerges as a limit of the corresponding *quantum field theory. In order for a classical field theory to apply on a macroscopic scale it is necessary for the interactions to be long range, as they are in electrodynamics and gravitation, rather than short range, as in nuclear forces. Classical field theory is also used for mathematical convenience to describe the physics of continuous media, such as fluids.

classical physics Theoretical physics up to approximately the end of the 19th century, before the concepts of *quantum theory (1900) and special *relativity (1905). Classical physics relied largely on *Newtonian mechanics and James Clerk *Maxwell's theory of electromagnetism. It may still be applied with high precision to large-scale phenomena.

Claude process A process for liquefying air on a commercial basis. Air under pressure is used as the working substance in a piston engine, where it does external work and cools adiabatically. This cool air is fed to a counter-current *heat exchanger, where it reduces the temperature of the next intake of high-pressure air. The same air is re-compressed and used again, and after several cycles eventually liquefies. The process was perfected in 1902 by the French scientist Georges Claude (1870–1960).

Clausius, Rudolf Julius Emmanuel (1822–88) German physicist, who held teaching posts in Berlin and Zurich, before going to Würzburg in 1869. He is best known for formulating the second law of *thermodynamics in 1850, independently of William Thomson (Lord *Kelvin). In 1865 he introduced the concept of *entropy, and later contributed to electrochemistry and electrodynamics (see CLAUSIUS–MOSSOTTI EQUATION).

Clausius–Mossotti equation A relation between the *polarizability α of a molecule and the dielectric constant (see PERMITTIVITY), ε of a dielectric substance made up of molecules with this polarizability. The Clausius–Mossotti equation is usually written in the form

$\alpha = (3/4\pi N)/[(\varepsilon - 1)/(\varepsilon - 2)]$,

where N is the number of molecules per unit volume. The equation provides a link between a microscopic quantity (the polarizability) and a macroscopic quantity (the dielectric constant); it was derived using macroscopic electrostatics by the Italian physicist Ottaviano Fabrizio Mossotti (1791–1863) in 1850 and independently by R. Clausius in 1879. It works best for gases and is only approximately true for liquids or solids, particularly if the dielectric constant is large. *Compare* LORENTZ–LORENZ EQUATION.

close packing The packing of spheres so as to occupy the minimum amount of space. In a single plane, each sphere is surrounded by six close neighbours in a hexagonal arrangement. The spheres in the second plane fit into depressions in the first layer, and so on. Each sphere has 12 other touching spheres. There are two types of close packing. In **hexagonal close packing** the spheres in the third layer are directly over those in the first, etc., and the arrangement of planes is ABAB…. In **cubic close packing** the spheres in the third layer occupy a different set of depressions than those in the first. The arrangement is ABCABC…. *See also* CUBIC CRYSTAL.

cloud chamber A device for making visible the paths of particles of *ionizing radiation. The **Wilson (expansion) cloud chamber** consists of a container containing air and ethanol vapour, which is cooled suddenly by adiabatic expansion, causing the vapour to become supersaturated. The excess moisture in the vapour is then deposited in drops on the tracks of ions created by the passage of the ionizing radiation. The resulting row of droplets can be photographed. If the original moving particle was being deflected by electric or magnetic fields, the extent of the deflection provides information on its mass and charge. This device was invented in 1911 by C. T. R. *Wilson.

A simpler version of this apparatus is the **diffusion cloud chamber**, developed by Cowan, Needels, and Nielsen in 1950, in which supersaturation is achieved by placing a row of felt strips soaked in a suitable alcohol at the top of the chamber. The lower part of the chamber is cooled by solid carbon dioxide. The vapour continuously diffuses downwards, and that in the centre (where it becomes supersaturated) is almost continuously sensitive to the presence of ions created by the radiation.

Clusius column A device for separating isotopes by *thermal diffusion. One form consists of a vertical column some 30 metres high with a heated electric wire running along its axis. The lighter isotopes in a gaseous mixture of isotopes diffuse faster than the heavier isotopes. Heated by the axial wire, and assisted by natural convection, the lighter atoms are carried to the top of the column, where a fraction rich in lighter isotopes can be removed for further enrichment.

cluster *See* GALAXY CLUSTER; STAR CLUSTER.

coagulation The process in which colloidal particles come together irreversibly to form larger masses. Coagulation can be brought about by adding ions to change the ionic strength of the solution and thus destabilize the colloid. Ions with a high charge are particularly effective (e.g. alum, containing Al^{3+}, is used in styptics to coagulate blood). Another example of ionic coagulation is in the formation of river deltas, which occurs when colloidal silt particles in rivers are coagulated by ions in sea water. Alum and iron(III) sulphate are also used for coagulation in sewage treatment. Heating is another way of coagulating certain colloids (e.g. boiling an egg coagulates the albumin).

coal A brown or black carbonaceous deposit derived from the accumulation and alteration of ancient vegetation, which originated largely in swamps or other moist environments. As the vegetation decomposed it formed layers of peat, which were subsequently buried (for example, by marine sediments following a rise in sea level or subsidence of the land). Under the increased pressure and resulting higher temperatures the peat was transformed into coal. Two types of coal are recognized: **humic** (or **woody**) **coals**, derived from plant remains; and **sapropelic coals**, which are derived from algae, spores, and finely divided plant material.

As the processes of coalification (i.e. the transformation resulting from the high temperatures and pressures) continue, there is a progressive transformation of the deposit: the proportion of carbon relative to oxygen rises and volatile substances and water are driven out. The various stages in this process are referred to as the **ranks** of the coal. In ascending order, the main ranks of coal are: **lignite** (or **brown coal**), which is soft, brown, and has a high moisture content; **subbituminous coal**, which is used chiefly by generating stations; **bituminous coal**, which is the most abundant rank of coal; **semibituminous coal**; **semianthracite coal**, which has a fixed carbon content of between 86% and 92%; and **anthracite coal**, which is hard and black with a fixed carbon content of between 92% and 98%.

Most deposits of coal were formed during the Carboniferous and Permian periods. More recent periods of coal formation occurred during the early Jurassic and Tertiary periods. Coal deposits occur in all the major continents; the leading producers include the USA, China, Ukraine, Poland, UK, South Africa, India, Australia, and Germany. Coal is used as a fuel and in the chemical industry; by-products include coke and coal tar.

coaxial cable A cable consisting of a central conductor surrounded by an insulator, which is in turn contained in an earthed sheath of another conductor. The central conductor and the outer conductor are coaxial (i.e. have the same axis). They are used to transmit high-frequency signals as they produce no external fields and are not influenced by them.

cobalt steel Any of a group of alloy *steels containing 5–12% of cobalt, 14–20% of tungsten, usually with 4% of chromium and 1–2% of vanadium. They are very hard but somewhat brittle. Their main use is in high-speed tools.

COBE Cosmic Background Explorer; an orbiting satellite launched in November 1989 for cosmological research. In 1992, statistical studies of measurements on the *microwave background radiation indicated the presence of weak temperature fluctuations thought to be imprints of quantum fluctuations in the *early universe. *See also* WMAP.

Cockcroft, Sir John Douglas (1897–1967) British physicist, who joined
*Rutherford at the Cavendish Laboratory in Cambridge, where with
Ernest Walton (1903–95) he built a *linear accelerator. In 1932, using the
apparatus to bombard lithium nuclei with protons, they produced the first
artificial nuclear transformation. For this work they were awarded the
1951 Nobel Prize.

Cockcroft–Walton generator The first proton accelerator; a simple
*linear accelerator producing a potential difference of some 800 kV (d.c.)
from a circuit of rectifiers and capacitors fed by a lower (a.c.) voltage. The
experimenters, Sir John Cockcroft and E. T. S. Walton (1903–95), used this
device in 1932 to achieve the first artificially induced nuclear reaction by
bombarding lithium with protons to produce helium:

$$^1_1H + ^7_3Li = ^4_2He + ^4_2He$$

coefficient **1.** (in mathematics) A number or other known factor by
which a variable quantity is multiplied, e.g. in $ax^2 + bx + c = 0$, a is the
coefficient of x^2 and b is the coefficient of x. **2.** (in physics) A measure of a
specified property of a particular substance under specified conditions, e.g.
the coefficient of *friction of a substance.

coefficient of expansion *See* EXPANSIVITY.

coefficient of friction *See* FRICTION.

coelostat A device that enables light from the same area of the sky to
be continuously reflected into the field of view of an astronomical
telescope or other instrument. It consists of a plane mirror driven by a
clockwork or electrical mechanism so that it rotates from east to west to
compensate for the apparent west-to-east rotation of the *celestial sphere.

coercive force (coercivity) The magnetizing force necessary to reduce
the flux density in a magnetic material to zero. *See* HYSTERESIS.

coherence length A length-scale giving the range within which
electrons in a superconductor can interact to form Cooper pairs. The
concept was put forward by Sir Brian Pippard (1920–   ) in 1953.

coherent radiation Electromagnetic radiation in which two or more
sets of waves have a constant phase relationship, i.e. with peaks and
troughs always similarly spaced.

coherent scattering Scattering for which there is a well-defined
relationship between the phase of the incoming wave and the phase of
the outgoing wave. Scattering for which there is no well-defined such
relationship is called **incoherent scattering**.

coherent units A system of *units of measurement in which derived
units are obtained by multiplying or dividing base units without the use
of numerical factors. *SI units form a coherent system; for example the
unit of force is the newton, which is equal to 1 kilogram metre per second

squared (kg m s^{-2}), the kilogram, metre, and second all being base units of the system.

coinage metals A group of three malleable ductile transition metals forming group 11 (formerly IB) of the periodic table: copper (Cu), silver (Ag), and gold (Au). Their outer electronic configurations have the form $nd^{10}(n+1)s^1$. Although this is similar to that of alkali metals, the coinage metals all have much higher ionization energies and higher (and positive) standard electrode potentials. Thus, they are much more difficult to oxidize and are more resistant to corrosion. In addition, the fact that they have d-electrons makes them show variable valency (Cu^I, Cu^{II}, and Cu^{III}; Ag^I and Ag^{II}; Au^I and Au^{III}) and form a wide range of coordination compounds. They are generally classified with the transition elements.

coincidence circuit An electronic logic device that gives an output only if two input signals are fed to it simultaneously or within a specified time of each other. A **coincidence counter** is an electronic counter incorporating such a device.

cold emission The emission of electrons by a solid without the use of high temperature (thermal emission), either as a result of field emission (*see* FIELD-EMISSION MICROSCOPE) or *secondary emission.

cold fusion *See* NUCLEAR FUSION.

Coleman–Gross theorem The theory that non-Abelian *gauge theories with unbroken gauge theories are the only renormalizable (*see* RENORMALIZATION) *quantum field theories able to have *asymptotic freedom. The Coleman–Gross theorem was stated by the US physicists Sidney Coleman and David James Gross in 1973.

collective excitation A quantized mode in a many-body system, occurring because of cooperative motion of the whole system as a result of interactions between particles. *Plasmons and *phonons in solids are examples of collective excitations. Collective excitations obey Bose–Einstein statistics (*see* QUANTUM STATISTICS).

collective model *See* UNIFIED MODEL.

collective oscillation A *mode of oscillation in a many-body system in which there is a cooperative motion of the system as a result of interactions between the particles in the system. A *plasma oscillation is an example of a collective oscillation. It is possible for collective oscillations to exist both in *classical physics and *quantum mechanics. In the quantum theory of many-body systems, collective oscillations are called *collective excitations.

collector *See* TRANSISTOR.

colliding-beam experiments *See* PARTICLE-BEAM EXPERIMENTS.

colligative properties Properties that depend on the concentration of particles (molecules, ions, etc.) present in a solution, and not on the nature of the particles. Examples of colligative properties are osmotic pressure (*see* OSMOSIS), *lowering of vapour pressure, *depression of freezing point, and *elevation of boiling point.

collimator **1.** Any device for producing a parallel beam of radiation. A common arrangement used for light consists of a convex achromatic lens fitted to one end of a tube with an adjustable slit at the other end, the slit being at the principal focus of the lens. Light rays entering the slit leave the lens as a parallel beam. Collimators for particle beams and other types of electromagnetic radiation utilize a system of slits or apertures. **2.** A small fixed telescope attached to a large astronomical telescope to assist in lining up the large one onto the desired celestial body.

collision density The number of collisions that occur in unit volume in unit time when a given neutron flux passes through matter.

colloids Colloids were originally defined by Thomas Graham in 1861 as substances, such as starch or gelatin, which will not diffuse through a membrane. He distinguished them from **crystalloids** (e.g. inorganic salts), which would pass through membranes. Later it was recognized that colloids were distinguished from true solutions by the presence of particles that were too small to be observed with a normal microscope yet were much larger than normal molecules. Colloids are now regarded as systems in which there are two or more phases, with one (the **dispersed phase**) distributed in the other (the **continuous phase**). Moreover, at least one of the phases has small dimensions (in the range 10^{-9}–10^{-6} m). Colloids are classified in various ways.

    Sols are dispersions of small solid particles in a liquid. The particles may be macromolecules or may be clusters of small molecules. **Lyophobic sols** are those in which there is no affinity between the dispersed phase and the liquid. An example is silver chloride dispersed in water. In such colloids the solid particles have a surface charge, which tends to stop them coming together. Lyophobic sols are inherently unstable and in time the particles aggregate and form a precipitate. **Lyophilic sols**, on the other hand, are more like true solutions in which the solute molecules are large and have an affinity for the solvent. Starch in water is an example of such a system. **Association colloids** are systems in which the dispersed phase consists of clusters of molecules that have lyophobic and lyophilic parts. Soap in water is an association colloid.

    Emulsions are colloidal systems in which the dispersed and continuous phases are both liquids, e.g. oil-in-water or water-in-oil. Such systems require an emulsifying agent to stabilize the dispersed particles.

    Gels are colloids in which both dispersed and continuous phases have a three-dimensional network throughout the material, so that it forms a jelly-like mass. Gelatin is a common example. One component may sometimes be removed (e.g. by heating) to leave a rigid gel (e.g. silica gel).

Other types of colloid include *aerosols (dispersions of liquid or solid particles in a gas, as in a mist or smoke) and foams (dispersions of gases in liquids or solids). Colloids are analysed theoretically in terms of intermolecular forces.

cologarithm The logarithm of the reciprocal of a number.

colorimeter Any instrument for comparing or reproducing colours. Monochromatic colorimeters match a *colour with a mixture of monochromatic and white lights. Trichromatic colorimeters use a mixture of three *primary colours.

colour The sensation produced when light of different wavelengths falls on the human eye. Although the visible spectrum covers a continuously varying range of colours from red to violet it is usually split into seven colours (the **visible spectrum**) with the following approximate wavelength ranges:

red 740–620 nm
orange 620–585 nm
yellow 585–575 nm
green 575–500 nm
blue 500–445 nm
indigo 445–425 nm
violet 425–390 nm

A mixture of all these colours in the proportions encountered in daylight gives white light; other colours are produced by varying the proportions or omitting components.

A coloured light has three attributes: its **hue**, depending on its wavelength; its **saturation**, depending on the degree to which it departs from white light; and its *luminosity. Coloured objects that owe their colour to pigments or dyes absorb some components of white light and reflect the rest. For example, a red book seen in white light absorbs all the components except the red, which it reflects. This is called a **subtractive process** as the final colour is that remaining after absorption of the rest. This is the basis of the process used in *colour photography. Combining coloured lights, on the other hand, is an **additive process** and this is the method used in *colour television. *See also* PRIMARY COLOURS.

colour charge *See* ELEMENTARY PARTICLES.

colour photography Any of various methods of forming coloured images on film or paper by photographic means. One common process is a subtractive reversal system that utilizes a film with three layers of light-sensitive emulsion, one responding to each of the three *primary colours. On development a black image is formed where the scene is blue. The white areas are dyed yellow, the *complementary colour of blue, and the blackened areas are bleached clean. A yellow filter between this emulsion layer and the next keeps blue light from the second emulsion, which is green-sensitive. This is dyed magenta where no green light has fallen. The final emulsion is red-sensitive and is given a cyan (blue-green) image on the negative after dying. When white light shines through the three dye layers the cyan dye subtracts red where it does not occur in the scene, the magenta subtracts green, and the yellow subtracts blue. The light projected by the negative therefore reconstructs the original scene either as a transparency or for use with printing paper.

colour television A television system in which the camera filters the light from the scene into the three *primary-colour components, red, blue, and green, detected by separate camera tubes. The separate information so obtained relating to the colour of the image is combined with the sound and synchronization signals and transmitted using one of three systems, the American, British, or French. At the receiver, the signal is split again into red, blue, and green components, each being fed to a separate *electron gun in the cathode-ray tube of the receiver. By an additive process (see COLOUR) the picture is reconstituted by the beam from each gun activating a set of phosphor dots of that colour on the screen.

colour temperature The temperature of a nonblack body as indicated by the temperature of a black body having approximately the same spectral distribution.

coma 1. A nebulous cloud of gas and dust that surrounds the nucleus of a *comet. **2.** An *aberration of a lens or mirror in which the image of a point lying off the axis has a comet-shaped appearance.

combinations See PERMUTATIONS AND COMBINATIONS.

combined cycle See FLUIDIZATION.

combustion A chemical reaction in which a substance reacts rapidly with oxygen with the production of heat and light. Such reactions are often free-radical chain reactions. See also FLAME.

comet A small body that travels around the sun in an eccentric orbit. **Short-period comets** have orbital periods of less than 150 years. The others have very long periods, some exceeding 100 000 years. Typical comets have three components: the **nucleus** of ice and dust, the *coma of gas and dust, and the **comet tail**, which only appears when the comet is near the sun (it, too, consists of gas and dust). The nuclei of most comets are thought to be 'dirty snowballs' about one kilometre in diameter, although the solar system has a few comets with nuclei exceeding 10 km in diameter. The coma may be 10^4–10^5 km in diameter, and the tail can be 10^7 km in length. See also HALLEY'S COMET.

commensurate lattice A lattice that can be divided into two or more sublattices, with the basis vectors of the lattice being a rational multiple of the basis vectors of the sublattice. The *phase transition between a commensurate lattice and an *incommensurate lattice can be analysed using the Frenkel–Kontorowa model.

common-collector connection A technique used in the operation of some *transistors, in which the *collector is common to both the input and output circuits, the input terminal is the *base, and the output terminal is the *collector.

common logarithm *See* LOGARITHM.

communication satellite An unmanned artificial satellite sent by rocket into a geostationary orbit (*see* SYNCHRONOUS ORBIT) around the earth to enable television broadcasts and telephone communications to be made between points on the earth's surface that could not otherwise communicate by radio owing to the earth's curvature. Modulated *microwaves are transmitted to the satellite, which amplifies them and retransmits them at a different frequency to the receiving station. The satellites are powered by *solar cells. Three or more satellites in equatorial orbits can provide a world-wide communications linkage. The satellites are placed well above the ionosphere (*see* EARTH'S ATMOSPHERE) and therefore the carrier waves used have to be in the microwave region of the spectrum in order to pass through the ionosphere.

commutative law The mathematical law stating that the value of an expression is independent of the order of combination of the numbers, symbols, or terms in the expression. The **commutative law for addition** applies if $x + y = y + x$. The **commutative law of multiplication** applies if $x \times y = y \times x$. Subtraction and division are not commutative. *Compare* ASSOCIATIVE LAW; DISTRIBUTIVE LAW.

commutator The part of the armature of an electrical motor or generator through which connections are made to external circuits. It consists of a cylindrical assembly of insulated copper conductors, each of which is connected to one point in the armature winding. Spring-loaded carbon brushes are positioned around the commutator to carry the current to or from it.

compact disk (CD) A 120 mm metal disk on which there is a *digital recording of audio information, providing high-quality recording and reproduction of music, speech, etc. The recording is protected by a layer of clear plastic. The information is permanently encoded in the form of a spiral track of minute pits impressed on one surface of the disk during manufacture; these impressions correspond to a changing sequence of *bits. The CD is rotated at constant linear velocity (CLV) in a CD player; the rotation rate varies according to the radius of the track accessed. Data are retrieved from the rotating disk by means of a low-power laser focused on the track and modulated by the binary code impressed on the track.

compass A small magnet pivoted at its central point to revolve in a horizontal plane. In the earth's magnetic field the magnet (called the compass needle) aligns itself so that its north-seeking end points to the earth's magnetic north pole. A scale (called a compass card) is placed below the needle for use in navigation. In some navigation compasses the entire card is pivoted, indicating direction by a fixed mark on the casing. Such compasses are often filled with alcohol to provide damping. Magnetic compasses suffer from being affected by magnetic metals in their vicinity and to a large extent they have been replaced by *gyrocompasses. The compass was invented in ancient China.

complementarity The concept that a single model may not be adequate to explain all the observations made of atomic or subatomic systems in different experiments. For example, *electron diffraction is best explained by assuming that the electron is a wave (*see* DE BROGLIE WAVELENGTH), whereas the *photoelectric effect is described by assuming that it is a particle. The idea of two different but complementary concepts to treat quantum phenomena was first put forward by the Danish physicist Niels *Bohr in 1927. *See also* LIGHT.

complementary colours A pair of coloured lights of specific hue (*see* COLOUR) that produce the sensation of white when mixed in appropriate intensities. There is an infinite number of such pairs, an example (with wavelengths) is orange (608 nm) and blue (490 nm).

complex conjugate Symbol Z^*. The quantity given by $Z^* = x - iy$, when the *complex number Z, is given by $Z = x + iy$. The polar form of Z^* is $r\cos\theta - ir\sin\theta$. In the Argand diagram the complex conjugate of a complex number is a reflection of the complex number about the real axis. The sum and product of Z and Z^* satisfy $Z + Z^* = 2x$, $ZZ^* = x^2 + y^2$.

complexity The levels of *self-organization of a system. In physical systems, complexity is associated with *broken symmetry and the ability of a system to have different states between which it can make *phase transitions. It is also associated with having coherence in space over a long range. Examples of complexity include *superconductivity, *superfluidity, *lasers, and ordered phases that arise when a system is driven far from thermal equilibrium (*see* BÉNARD CELL). It is not necessary for a system to have a large number of degrees of freedom in order for complexity to occur. The study of complexity is greatly aided by computers in systems that cannot be described analytically. Complexity is also very important in theoretical biology.

complex number A number that has a real part, x, and an imaginary part, iy, where $i = \sqrt{-1}$ and x and y are real (x can also equal 0). The complex number therefore has the form $x + iy$, which can also be written in the polar form $r\cos\theta + ir\sin\theta$, where r is the **modulus** and θ is the **argument** (or **amplitude**). A complex number can be represented on an **Argand diagram**, devised by J. R. Argand (1768–1822), in which the horizontal axis represents the real part of the number and the vertical axis the imaginary part (see illustration). In the polar form the modulus is the line joining the origin to the point representing the complex number and the argument is the angle between the modulus and the x-axis.

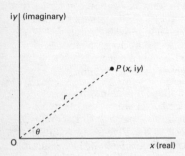

iy (imaginary)

P (x, iy)

r

θ

O

x (real)

Argand diagram

component vectors Two or more vectors that produce the same effect as a given vector; the vectors that combine to produce the effect of a resultant vector. A component vector in a given direction is the projection of the given vector (**V**) along that direction, i.e. $V\cos\theta$, where θ is the angle between the given vector and the direction.

compound microscope *See* MICROSCOPE.

compressibility The reciprocal of bulk modulus (*see* ELASTIC MODULUS). The compressibility (k) is given by $-V^{-1}\mathrm{d}V/\mathrm{d}p$, where $\mathrm{d}V/\mathrm{d}p$ is the rate of change of volume (V) with pressure.

compression ratio The ratio of the total volume enclosed in the cylinder of an *internal-combustion engine at the beginning of the compression stroke to the volume enclosed at the end of the compression stroke. For petrol engines the compression ratio is 8.5–9:1, with a recent tendency to the lower end of the range in order to make use of unleaded petrols. For Diesel engines the compression ratio is in the range 12–25:1.

Compton, Arthur Holly (1892–1962) US physicist, who became professor of physics at the University of Chicago in 1923. He is best known for his discovery (1923) of the *Compton effect, for which he shared the 1927 Nobel Prize with C. T. R. *Wilson. In 1938 he demonstrated that *cosmic radiation consists of charged particles.

Compton effect The reduction in the energy of high-energy (X-ray or gamma-ray) photons when they are scattered by free electrons, which thereby gain energy. The phenomenon, which was first observed in 1923 by the US physicist A. H. Compton, occurs when the photon collides with an electron; some of the photon's energy is transferred to the electron and consequently the photon loses energy $h(v_1 - v_2)$, where h is the *Planck constant and v_1 and v_2 are the frequencies before and after collision. As $v_1 > v_2$, the wavelength of the radiation increases after the collision. This type of inelastic scattering is known as **Compton scattering** and is similar to the *Raman effect. *See also* INVERSE COMPTON EFFECT.

Compton wavelength The length scale below which a particle's quantum-mechanical properties become evident in relativistic *quantum mechanics. For a particle of rest mass m the Compton wavelength is $\hbar/mc$, where $\hbar$ is the rationalized Planck constant and c is the speed of light. The Compton wavelength is so named because of its occurrence in the theory of the *Compton effect, where its value for the electron is 3.8616×10^{-13} m. The Compton wavelength is sometimes defined as h/mc, with h being the Planck constant, in which case the electron value is 2.4263×10^{-12} m.

computer An electronic device that processes information according to a set of instructions, called the **program**. The most versatile type of computer is the **digital computer**, in which the input is in the form of characters, represented within the machine in *binary notation. Central to the operation of a computer is the **central processing unit** (**CPU**), which contains circuits for manipulating the information (*see* LOGIC CIRCUITS). The CPU contains the arithmetic/logic unit (ALU), which performs operations, and a control unit. It is supported by a short-term **memory**, in which data is stored in electronic circuits (*see* RAM). Associated storage usually involves *magnetic disks or *CD-ROM. There are also various peripheral input and output devices, such as a keyboard, visual-display unit (VDU), magnetic tape unit, and *printer. Computers range in size from the **microprocessor** with a few thousand logic elements, to the large **mainframe computer** with millions of logic circuits.

The **analog computer** is used in scientific experiments, industrial control, etc. In this type of device the input and output are continuously varying quantities, such as a voltage, rather than the discrete digits of the more commercially useful digital device. **Hybrid computers** combine the properties of both digital and analog devices. Input is usually in analog form, but processing is carried out digitally in a CPU.

Computer **hardware** consists of the actual electronic or mechanical devices used in the system; the **software** consists of the programs and data. *See also* ROM.

concave Curving inwards. A **concave mirror** is one in which the reflecting surface is formed from the interior surface of a sphere or paraboloid. A **concave lens** has at least one face formed from the interior surface of a sphere. A **biconcave lens** has both faces concave and is therefore thinnest at its centre. The **plano-concave lens** has one plane face and one concave face. The **concavo-convex lens** (also called a **meniscus**) has one concave face and one *convex face. *See* LENS.

concavo-convex *See* CONCAVE.

concentration cell *See* CELL.

conchoidal fracture Fracture of a solid in which the surface of the material is curved and marked by concentric rings. It occurs particularly in amorphous materials.

condensation The change of a vapour or gas into a liquid. The change of phase is accompanied by the evolution of heat (*see* LATENT HEAT).

condensation pump *See* DIFFUSION PUMP.

condensed-matter physics *See* SOLID-STATE PHYSICS.

condenser **1.** A mirror or set of lenses used in optical instruments, such as a microscope or film projector, to concentrate the light diverging from a compact source. A common form consists of two plano-convex lenses with the plane faces pointing outwards. **2.** A device used to cool a vapour to cause it to condense to a liquid. In a steam engine the condenser acts as a reservoir that collects the part of the steam's internal energy that has not been used in doing work on the piston. The cooling water passed through the condenser is warmed and is used as fresh feedwater for the boiler. **3.** *See* CAPACITOR.

condenser microphone *See* CAPACITOR MICROPHONE.

conductance The reciprocal of electrical resistance in a direct-current circuit. The ratio of the resistance to the square of the *impedance in an alternating-current circuit. The SI unit is the siemens, formerly called the mho or reciprocal ohm.

conduction **1. (thermal conduction)** The transmission of heat through a substance from a region of high temperature to a region of lower temperature. In gases and most liquids, the energy is transmitted mainly by collisions between atoms and molecules with those possessing lower kinetic energy. In solid and liquid metals, heat conduction is predominantly by migration of fast-moving electrons, followed by collisions between these electrons and ions. In solid insulators the absence of *free electrons restricts heat transfer to the vibrations of atoms and molecules within crystal lattices. *See* CONDUCTIVITY.
2. (electrical conduction) The passage of electric charge through a substance under the influence of an electric field. *See also* CHARGE CARRIER; ENERGY BANDS.

conduction band *See* ENERGY BANDS.

conductivity **1. (thermal conductivity)** A measure of the ability of a substance to conduct heat. For a block of material of cross section A, the energy transferred per unit time E/t, between faces a distance, l, apart is given by $E/t = \lambda A(T_2 - T_1)/l$, where λ is the conductivity and T_2 and T_1 are the temperatures of the faces. This equation assumes that the opposite faces are parallel and that there is no heat loss through the sides of the block. The SI unit is therefore $J s^{-1} m^{-1} K^{-1}$. **2. (electrical conductivity)** The reciprocal of the *resistivity of a material. It is measured in siemens per metre in SI units. When a fluid is involved the electrolytic conductivity is given by the ratio of the current density to the electric field strength.

conductor **1.** A substance that has a high thermal *conductivity. Metals are good conductors on account of the high concentration of *free electrons they contain. Energy is transmitted through a metal predominantly by means of collisions between electrons and ions. Most nonmetals are poor conductors (good **thermal insulators**) because there are relatively few free electrons. **2.** A substance that has a high electrical conductivity. Again conduction results from the movement of free electrons. *See* ENERGY BANDS.

cone 1. (in optics) A type of light-sensitive receptor cell, found in the *retinas of all diurnal vertebrates. Cones are specialized to transmit information about colour and they function best in bright light. They are not evenly distributed on the retina, being concentrated in the fovea and absent on the margin of the retina. *Compare* ROD. **2.** (in mathematics) A solid figure generated by a line (the **generator**) joining a point on the perimeter of a closed plane curve (the **directrix**) to a point (the **vertex**) outside this plane, as the line moves round the directrix. If the directrix is a circle, the figure is a **circular cone** standing on a circular **base**. If the line joining the vertex to the centre of the base (the **axis**) is perpendicular to the base the figure is a **right circular cone**, which has a volume $\pi r^2 h/3$, where r is the radius of the base and h the height of the vertex above the base. If the axis of the cone is not perpendicular to the base, the figure is an **oblique cone**. In general, the volume of any cone is one third of its base area multiplied by the perpendicular distance of the vertex from the base.

configuration 1. The arrangement of atoms or groups in a molecule. **2.** The arrangement of electrons in atomic *orbitals in an atom.

configuration space The n-dimensional space with coordinates $(q_1, q_2, \ldots, q_n)$ associated with a system that has n *degrees of freedom, where the values q describe the degrees of freedom. For example, in a gas of N atoms each atom has three positional coordinates, so the configuration space is $3N$-dimensional. If the particles also have internal degrees of freedom, such as those caused by vibration and rotation in a molecule, then these must be included in the configuration space, which is consequently of a higher dimension. *See also* PHASE SPACE.

confinement *See* QUANTUM CHROMODYNAMICS; QUARK CONFINEMENT.

conic A figure formed by the intersection of a plane and a *cone. If the intersecting plane is perpendicular to the axis of a right circular cone, the figure formed is a *circle. If the intersecting plane is inclined to the axis at an angle in excess of half the apex angle of the cone it is an *ellipse. If the plane is parallel to the sloping side of the cone, the figure is a *parabola. If the plane cuts both halves of the cone a *hyperbola is formed.

A conic can be defined as a plane curve in which for all points on the curve the ratio of the distance from a fixed point (the **focus**) to the perpendicular distance from a straight line (the **directrix**) is a constant called the **eccentricity** e. For a parabola $e = 1$, for an ellipse $e < 1$, and for a hyperbola $e > 1$.

conjugate points Two points in the vicinity of a *lens or *mirror such that a bright object placed at one will form an image at the other.

conjunction The alignment of two celestial bodies within the solar system so that they have the same longitude as seen from the earth. A planet that orbits between the sun and the earth (Venus and Mercury) is in **superior conjunction** when it is in line with the sun and the earth but on the opposite side of the sun to the earth. It is in **inferior conjunction** when it lies between the earth and the sun. Conjunction may also occur between two planets or a moon and a planet. *Compare* OPPOSITION.

conservation law A law stating that the total magnitude of a certain physical property of a system, such as its mass, energy, or charge, remains unchanged even though there may be exchanges of that property between components of the system. For example, imagine a table with a bottle of salt solution (NaCl), a bottle of silver nitrate solution ($AgNO_3$), and a beaker standing on it. The mass of this table and its contents will not change even when some of the contents of the bottles are poured into the beaker. As a result of the reaction between the chemicals two new substances (silver chloride and sodium nitrate) will appear in the beaker:

$$NaCl + AgNO_3 \rightarrow AgCl + NaNO_3,$$

but the total mass of the table and its contents will not change. This **conservation of mass** is a law of wide and general applicability, which is true for the universe as a whole, provided that the universe can be considered a closed system (nothing escaping from it, nothing being added to it). According to Einstein's mass–energy relationship, every quantity of energy (E) has a mass (m), which is given by E/c^2, where c is the speed of light. Therefore if mass is conserved, the law of **conservation of energy** must be of equally wide application. The laws of **conservation** of **linear momentum** and **angular momentum** also are believed to be universally true.

Because no way is known of either creating or destroying electric charge, the law of **conservation of charge** is also a law of universal application. Other quantities are also conserved in reactions between elementary particles.

conservative field A field of force in which the work done in moving a body from one point to another is independent of the path taken. The force required to move the body between these points in a conservative field is called a **conservative force**.

consistent histories An interpretation of quantum mechanics that makes use of the concept of *decoherence to explain how the classical world emerges from quantum mechanics. The consistent-histories interpretation avoids the problem of observers and has greatly clarified our understanding of the problem of measurement in quantum mechanics.

consolute temperature The temperature at which two partially miscible liquids become fully miscible as the temperature is increased.

constant 1. A component of a relationship between variables that does not change its value, e.g. in $y = ax + b$, b is a constant. **2.** A fixed value that has to be added to an indefinite integral. Known as the **constant of integration**, it depends on the limits between which the integration has been performed. **3.** *See* FUNDAMENTAL CONSTANTS.

constantan An alloy having an electrical resistance that varies only very slightly with temperature (over a limited range around normal room temperatures). It consists of copper (50–60%) and nickel (40–50%) and is used in resistance wire, thermocouples, etc.

constant-boiling mixture *See* AZEOTROPE.

constitutive equations The equations $D = \varepsilon E$ and $B = \mu H$, where D is the electric displacement, ε is the *permittivity of the medium, E is the electric field intensity, B is the magnetic flux density, μ is the *permeability of the medium, and H is the magnetic field strength (*see* MAGNETIC FIELD).

contact potential difference The potential difference that occurs between two electrically connected metals or between the base regions of two semiconductors. If two metals with work functions ϕ_1 and ϕ_2 are brought into contact, their Fermi levels will coincide. If $\phi_1 > \phi_2$ the first metal will acquire a positive surface charge with respect to the other at the area of contact. As a result, a contact potential difference occurs between the two metals or semiconductors.

containment 1. The prevention of the escape of radioactive materials from a *nuclear reactor. **2.** The process of preventing the plasma in a *thermonuclear reactor from touching the walls of the vessel by means of magnetic fields.

continuous function A function $f(x)$ is continuous at $x = a$ if the limit of $f(x)$ as x approaches a is $f(a)$. A function that does not satisfy this condition is said to be a **discontinuous function**.

continuous phase *See* COLLOIDS.

continuous spectrum *See* SPECTRUM.

continuous wave A wave that is transmitted continuously rather than in pulses.

continuum A system of axes that form a *frame of reference. The three dimensions of space and the dimension of time together can be taken to form a four-dimensional continuum; this was suggested by Minkowski in connection with special *relativity.

control grid A wire-mesh electrode placed between the cathode and anode in a *thermionic valve or a *cathode-ray tube to control the flow of electrons from one to the other. A fluctuating potential signal fed to the control grid produces at the anode a current signal with similar but amplified fluctuations. It thus forms the basis of the electronic valve amplifier. In a cathode-ray tube the grid controls the intensity of the electron beam and hence the brightness of the image on the screen.

control rod One of a number of rods of a material, such as boron or cadmium, that absorbs neutrons. Control rods can be moved into or out of the core of a *nuclear reactor to control the rate of the reaction taking place within it.

control unit (CU) The part of the central processor of a *computer that supervises the execution of a computer program.

convection A process by which heat is transferred from one part of a fluid to another by movement of the fluid itself. In **natural convection** the movement occurs as a result of gravity; the hot part of the fluid expands, becomes less dense, and is displaced by the colder denser part of the fluid as this drops below it. This is the process that occurs in most domestic hot-water systems between the boiler and the hot-water cylinder. A natural convection current is set up transferring the hot water from the boiler up to the cylinder (always placed above the boiler) so that the cold water from the cylinder can move down into the boiler to be heated. In some modern systems, where small-bore pipes are used or it is inconvenient to place the cylinder above the boiler, the circulation between boiler and hot-water cylinder relies upon a pump. This is an example of **forced convection**, where hot fluid is transferred from one region to another by a pump or fan.

conventional current A 19th-century convention, still in use, that treats any electrical current as a flow of positive charge from a region of positive potential to one of negative potential. The real motion, however, in the case of electrons flowing through a metal conductor, is in the opposite direction, from negative to positive. In semiconductors *hole conduction is in the direction of the conventional current; electron conduction is in the opposite direction.

convergent series A series $a_1 + a_2 + \ldots + a_i + \ldots$, for which a partial sum $S_n = a_1 + a_2 + \ldots + a_n$ tends to a finite (or zero) limit as n tends to infinity. This limit is the **sum** of the series. For example, the series $1 + 1/2 + 1/4 + 1/8 + \ldots$ (with the general term a_i equal to $(1/2)^{i-1}$) tends to the limit 2. A series that is not convergent is said to be a **divergent series**. In such a series the partial sum tends to plus or minus infinity or may oscillate. For example, the series $1 + 1/2 + 1/3 + 1/4 + \ldots$ (with a_i equal to $1/i$) is divergent. As can be seen from this latter example, a series may be divergent even if the individual terms a_i tend to zero as i tends to infinity.

converging lens or mirror A lens or mirror that can refract or reflect a parallel beam of light so that it converges at a point (the principal focus). Such a mirror is concave; a converging lens is thicker at its centre than at its edges (i.e. it is biconvex, plano-convex, or convexo-concave). *Compare* DIVERGING LENS OR MIRROR.

conversion electron *See* INTERNAL CONVERSION.

converter 1. An electrical machine for converting alternating current into direct current, or less frequently, vice versa. **2.** The reaction vessel in the *Bessemer process or some similar steel-making process. **3.** A computer device for converting information coded in one form into some other form.

converter reactor A *nuclear reactor that converts fertile material (e.g. thorium–232) into *fissile material (e.g. uranium–233). A converter reactor can also be used to produce electrical power.

convex Curving outwards. A **convex mirror** is one in which the reflecting surface is formed from the exterior surface of a sphere or paraboloid. A **convex lens** has at least one face formed from the exterior surface of a sphere. A **biconvex lens** has both faces convex and is therefore thickest at its centre. The **plano-convex lens** has one plane face and one convex face. The **convexo-concave lens** (also called a **meniscus**) has one convex face and one *concave face. *See* LENS.

coolant A fluid used to remove heat from a system by *convection (usually forced), either to control the temperature or to extract energy. In a water-cooled car engine the coolant is water (or water and antifreeze), which is pumped around the engine and cooled in the radiator. In a *nuclear reactor the coolant is used to transfer the heat of the reaction from the core to a heat exchanger or to the steam-raising plant. In gas-cooled reactors the coolant is usually carbon dioxide. Pressurized water or boiling water is used as both coolant and *moderator in several types of reactor. In fast reactors, liquid sodium is used as the coolant.

Cooper, Leon *See* BARDEEN, JOHN; SUPERCONDUCTIVITY.

cooperative phenomenon A phenomenon in which the constituents of a system cannot be regarded as acting independently from each other. Cooperative phenomena result from interactions between the constituents. Phenomena that can be described by the *liquid-drop model of nuclei, such as nuclear fission, are examples of cooperative phenomena because they involve the *nucleus as a whole rather than individual nucleons. Other examples of cooperative phenomena occur when a substance undergoes a *phase transition, as in the phenomena of ferromagnetism (*see* MAGNETISM) or *superconductivity.

Cooper pairs *See* SUPERCONDUCTIVITY.

coordinate *See* CARTESIAN COORDINATES; POLAR COORDINATES.

coordinate geometry *See* ANALYTICAL GEOMETRY.

coordinate system A system that uniquely specifies points in a plane or in three-dimensional space. The simplest coordinate system is the *Cartesian coordinate system. In a plane two coordinates are necessary to specify a point. In three-dimensional space three coordinates are required. Many coordinate systems can be used to specify a point; however, sometimes one particular coordinate system is more convenient than others; indeed, certain problems can be solved in one coordinate system but not in others. For example, the Schrödinger equation for the hydrogen atom can be solved using spherical *polar coordinates but not using Cartesian coordinates.

coordination number The number of groups, molecules, atoms, or ions surrounding a given atom or ion in a complex or crystal. For instance, in a square-planar complex the central ion has a coordination number of four. In a close-packed crystal (*see* CLOSE PACKING) the coordination number is twelve.

Copenhagen interpretation The standard interpretation of *quantum mechanics associated with the ideas of Niels Bohr, who worked at the University of Copenhagen. In this interpretation a system (e.g. a particle) can be described by a *wave function. This is a complex function – i.e. there is no real value, but the physical significance is that the square of the wave function is proportional to the probability of a particular definite state. In the Copenhagen interpretation a particle does not have a definite position or spin, for example, until it is observed – i.e. until a measurement is made. The idea is that the measurement 'collapses the wave function', leading to a definite measurement of the state. However, any prediction of the state of a system can only be probabilistic. The Copenhagen interpretation is the one most generally accepted by physicists but it does imply certain apparent paradoxes. *See* SCHRÖDINGER'S CAT; EPR EXPERIMENT. *See also* MANY-WORLDS INTERPRETATION.

Copernican astronomy The system of astronomy that was proposed by Nicolaus *Copernicus in his book *De revolutionibus orbium coelestium*, which was published in the month of his death and first seen by him on his deathbed. It used some elements of *Ptolemaic astronomy, but rejected the notion, then current, that the earth was a stationary body at the centre of the universe. Instead, Copernicus proposed the apparently unlikely concept that the sun was at the centre of the universe and that the earth was hurtling through space in a circular orbit about it. Galileo's attempts, some 70 years later, to convince the Catholic church that in spite of scriptural authority to the contrary, the Copernican system was correct, resulted in *De revolutionibus* being placed on the Index of forbidden books, where it remained until 1835.

Copernicus, Nicolaus (Mikolaj Kopernik; 1473–1543) Polish astronomer, who studied mathematics and optics. By 1514 he had formulated his proposal that the planets, including the earth, orbit the sun in circular paths, although it was not formally published until the year he died. This refutation of an earth-centred universe raised hostile opposition from the church as well as from other astronomers. *See* COPERNICAN ASTRONOMY.

core **1.** A rod or frame of magnetic material that increases the inductance of a coil through which it passes. Cores are used in transformers, electromagnets, and the rotors and stators of electrical machines. It may consist of laminated metal, ferrite, or compressed ferromagnetic particles in a matrix of an insulating binder (**dust core**). **2.** The inner part of a *nuclear reactor in which the nuclear reaction takes place. **3.** The devices that make up the memory in certain types of computer. **4.** The central region of a star or planet.

Coriolis force A fictitious force sometimes used to simplify calculations involving rotating systems, such as the movement of air, water, and projectiles over the surface of the rotating earth. The concept was first used in 1835 by Gaspard de Coriolis (1792–1843), a French physicist. The daily rotation of the earth means that in 24 hours a point on its equator moves a distance of some 40 000 kilometres, giving it a tangential velocity of about 1670 kilometres per hour. A point at the latitude of, say, Rome, travels a shorter distance in the same time and therefore has a lower tangential velocity – about 1340 km/hr. Air over the equator has the full tangential velocity of 1670 km/hr and as it travels north, say, it will retain this velocity; to an observer outside the earth this would be clear. However, to an observer in Rome it appears to be moving eastwards, because the earth at that point is moving eastwards more slowly than the air. The Coriolis force (which is quite fictitious) is the force that a naive observer thinks is needed to push the air eastwards.

corkscrew rule *See* AMPÈRE'S RULE.

corona **1.** The outer part of the sun's atmosphere. Its two main components are the K-corona (or inner corona), with a temperature of about 2×10^6 K at a height of some 75 000 km, and the F-corona (or outer corona), which is considerably cooler and extends for several million kilometres into space. **2.** A glowing region of the air surrounding a conductor when the potential gradient near it exceeds a critical value. It is caused by ionization of the air and may be accompanied by hissing sounds. **Corona discharge** (or **point discharge**) occurs at sharp points where the surface charge density is high by the attraction, charging, and consequent repulsion of air molecules.

corpuscular theory *See* LIGHT.

corrosion Chemical or electrochemical attack on the surface of a metal. *See also* ELECTROLYTIC CORROSION.

cosine rule In any triangle, with sides of length a, b, and c, $c^2 = a^2 + b^2 - 2ab\cos\theta$, where θ is the angle between sides a and b.

cosmic censorship A hypothesis concerning singularities and *black holes in the general theory of *relativity. It was suggested in 1969 by the British physicist Roger Penrose (1931–). The **cosmic censorship conjecture** asserts that all singularities in general relativity are hidden behind an event horizon (*see* DEATH OF A STAR). The conjecture has never been proved mathematically, although there is some evidence for it in many situations. Even if cosmic censorship is not correct, singularities would not be seen experimentally if the singularities are removed by *quantum gravity. It may be that in classical general relativity the cosmic censorship hypothesis is true for 'reasonable' physical situations but that it is possible to construct counter-examples to it for various special situations.

cosmic radiation High-energy particles that fall on the earth from space. **Primary cosmic rays** consist of nuclei of the most abundant elements, with *protons (hydrogen nuclei) forming by far the highest proportion; electrons, positrons, neutrinos, and gamma-ray photons are also present. The particle energies range from 10^{-11} J to 10 J (10^8 to 10^{20} eV) and as they enter the earth's atmosphere they collide with oxygen and nitrogen nuclei producing **secondary cosmic rays**. The secondary rays consist of elementary particles and gamma-ray photons. A single high-energy primary particle can produce a large **shower** of secondary particles. The sources of the primary radiation are not all known, particularly in the case of very high-energy rays, although the sun is believed to be the principal source of particles with energies up to about 10^{10} eV. It is believed that all particles with energies of less than 10^{18} eV originate within the Galaxy.

cosmic string *See* STRING.

cosmological constant A term that can be added to Einstein's field equation for general *relativity theory. The cosmological constant is independent of space and time. It was put forward by Einstein in 1917 to allow for the possibility of a static universe. Although the discovery of the *expansion of the universe removed the original motivation for the cosmological constant, the discovery that the expansion of the universe is accelerating suggests that the cosmological constant has a non-zero value, albeit smaller by a factor of 10^{120} of what might be expected theoretically. Explaining the small but non-zero value of the cosmological constant is one of the greatest challenges facing theoretical physics at the present time.

cosmological principle The claim that on extremely large scales, i.e. much greater scales than those associated with *large-scale structure, the universe is homogeneous and isotropic. There is some evidence that the cosmological principle is valid, notably from the cosmic microwave background radiation, but it has not been demonstrated conclusively.

cosmology The study of the nature, origin, and evolution of the universe. Various theories concerning the origin and evolution of the universe exist. See Chronology. *See* BIG-BANG THEORY; EARLY UNIVERSE; STEADY-STATE THEORY.

Cottrell precipitator An electrostatic precipitator used to remove dust particles from industrial waste gases, by attracting them to charged grids or wires.

coudé system *See* TELESCOPE.

coulomb Symbol C. The *SI unit of electric charge. It is equal to the charge transferred by a current of one ampere in one second. The unit is named after Charles de Coulomb.

Coulomb, Charles Augustin de (1736–1806) French physicist, who

---

COSMOLOGY

260 BC	Greek astronomer Aristarchus of Samos (c. 320–230 BC) proposes a sun-centred universe.
c.150 AD	Greek-Egyptian astronomer Ptolemy (2nd century AD) proposes an earth-centred universe.
1543	Copernicus publishes his sun-centred theory of the universe (solar system).
1576	English mathematician Thomas Digges (c. 1546–95) proposes that the universe is infinite (because stars are at varying distances).
1584	Italian philosopher Giordano Bruno (1548–1600) states that the universe is infinite.
1633	Galileo champions Copernicus's sun-centred universe, but is forced by the Roman Catholic Inquisition to recant.
1854	Helmholtz predicts the heat death of the universe, based on thermo-dynamics.
1917	Einstein proposes a static universe theory.
1922	Russian astronomer Alexander Friedmann proposes the expanding universe theory.
1927	George Lemaître proposes the big-bang theory of the universe.
1929	Edwin Hubble demonstrates the expansion of the universe.
1948	US physicists George Gamow (1904–68), Ralph Alpher (1921–), and Hans Bethe (1906–2005) develop the big-bang theory, and the α-β-γ theory of the origin of the elements; Alpher also predicts that the big bang would have produced a microwave background. British astronomers Herman Bondi (1919–), Thomas Gold (1920–2004), and Fred Hoyle propose the steady-state theory of the universe.
1965	US astrophysicists Arno Penzias (1933–) and Robert Wilson (1936–) discover the microwave background radiation.
1980	US physicist Allan Guth (1947–) proposes the inflationary theory of the universe.
1992	US COBE astronomical satellite detects ripples in residual cosmic radiation (cited as evidence of the big bang).

served as an army engineer in Martinique before returning to France. He is best known for his 1785 proposal of *inverse-square laws to describe the interaction between electrical charges and between magnets (*see* COULOMB'S LAW), which he proved experimentally using a *torsion balance.

Coulomb field *See* COULOMB'S LAW.

Coulomb force *See* COULOMB'S LAW.

Coulomb's law The force (sometimes called the **Coulomb force**) between two charged particles, regarded as point charges Q_1 and Q_2 a distance d apart, is proportional to the product of the charges and inversely proportional to the square of the distance between them. The law is now usually stated in the form $F = Q_1Q_2/4\pi\varepsilon d^2$, where ε is the absolute *permittivity of the intervening medium. $\varepsilon = \varepsilon_r\varepsilon_0$, where ε_r is the relative permittivity (the dielectric constant) and ε_0 is the electric constant. The electric field surrounding a point charge is called the **Coulomb field** and the scattering of charged particles by the Coulomb field surrounding an atomic nucleus is called **Coulomb scattering**. The law was first published by Charles de Coulomb in 1785. This law was found independently by Henry Cavendish.

counter Any device for detecting and counting objects or events, often incident charged particles or photons. The latter devices usually work by allowing the particle to cause ionization, which creates a current or voltage pulse. The pulses are then counted electronically. *See* CERENKOV COUNTER; CRYSTAL COUNTER; GEIGER COUNTER; PROPORTIONAL COUNTER; SCINTILLATION COUNTER; SPARK CHAMBER. These names are often applied merely to the actual detectors; the ancillary counting mechanism is then called a *scaler.

couple Two equal and opposite parallel forces applied to the same body that do not act in the same line. The forces create a torque, the moment of which is equal to the product of the force and the perpendicular distance between them. *See* MOMENT OF A FORCE.

coupling 1. (in physics) An interaction between two different parts of a system or between two or more systems. Examples of coupling in the *spectra of atoms and nuclei are *Russell–Saunders coupling, *j-j coupling, and *spin–orbit coupling. In the spectra of molecules there are five idealized ways (called the **Hund coupling cases**) in which the different types of angular momentum in a molecule (the electron orbital angular momentum L, the electron spin angular momentum S, and the angular momentum of nuclear rotation N) couple to form a resultant angular momentum J. (In practice, the coupling for many molecules is intermediate between Hund's cases due to interactions, which are ignored in the idealized cases.) In *solid-state physics an example of coupling is electron–phonon coupling, the analysis of which gives the theories of electrical *conductivity and *superconductivity. *See also* COUPLING CONSTANT. **2.** (in chemistry) A type of chemical reaction in which two molecules join together; for example, the formation of an azo compound by coupling of a diazonium ion with a benzene ring.

coupling constant A physical constant that is a measure of the strength of interaction between two parts of a system or two or more systems. In the case of a *field theory, the coupling constant is a measure of the magnitude of the force exerted on a particle by a field. In the case of a *quantum field theory, a coupling constant is not constant but is a function of energy, the dependence on energy being described by the *renormalization group. *See also* COUPLING; FINE STRUCTURE; ASYMPTOTIC FREEDOM.

CP invariance The symmetry generated by the combined operation of changing *charge conjugation (*C*) and *parity (*P*). **CP violation** occurs in weak interactions in kaon decay and also in *B-mesons. *See also* CPT THEOREM; TIME REVERSAL.

CPT theorem The theorem that the combined operation of changing *charge conjugation *C*, *parity *P*, and *time reversal *T*, denoted **CPT**, is a fundamental *symmetry of relativistic *quantum field theory. No violation of the CPT theorem is known experimentally. When *C*, *P*, and *T* (or any two of them) are violated, the principles of relativistic quantum field theory are not affected; however, violation of **CPT invariance** would drastically alter the fundamentals of relativistic quantum field theory. It is not known whether *superstring theory obeys a version of the CPT theorem.

CPU (central processing unit) The main operating part of a *computer; it includes the **control unit** (CU) and the arithmetic/logic unit (*see* ALU). Its function is to fetch instructions from memory, decode them, and execute the program. It also provides timing signals. An *integrated circuit that has a complete CPU on a single silicon chip is called a microprocessor.

crack A flaw in the surface (or just below the surface) of a material. Cracks at the microscopic level can have a large effect on the mechanical properties of a material because they concentrate stress. In **crack growth** there is an increase in energy due to new surfaces being generated but this is more than compensated for by a reduction in energy due to stress relief. This means that if there is sufficient energy stored round the crack it is energetically favourable for the crack to grow, and hence for the material to be ruptured.

creep The continuous deformation of a solid material, usually a metal, under a constant stress that is well below its yield point (*see* ELASTICITY). It usually only occurs at high temperatures and the creep characteristics of any material destined to be used under conditions of high stress at high temperatures must be investigated.

critical angle *See* TOTAL INTERNAL REFLECTION.

critical damping *See* DAMPING.

critical exponents Numbers that quantify how thermodynamic quantities, such as the specific heat capacity, diverge at the critical temperature of a second-order *phase transition. In some model systems, particularly in low dimensions, critical exponents can be calculated exactly. In general, they can be estimated accurately using various approximation techniques, frequently using the *renormalization group.

critical Üeld The critical value of the magnetic flux density of a magnetic field, above which the superconductivity of a superconductor is destroyed. This value depends on both the temperature and the nature of the superconducting substance. It is possible for the critical field to be caused by the superconducting current. This means that at any temperature beneath the transition temperature of the superconductor there must be some critical maximum value for the superconducting current.

critical mass The minimum mass of fissile material that will sustain a nuclear *chain reaction. For example, when a nucleus of uranium–235 disintegrates two or three neutrons are released in the process, each of which is capable of causing another nucleus to disintegrate, so creating a chain reaction. However, in a mass of U–235 less than the critical mass, too many neutrons escape from the surface of the material for the chain reaction to proceed. In the atom bomb, therefore, two or more subcritical masses have to be brought together to make a mass in excess of the critical mass before the bomb will explode.

critical pressure The pressure of a fluid in its *critical state; i.e. when it is at its critical temperature and critical volume.

critical reaction A nuclear *chain reaction in which, on average, one transformation causes exactly one other transformation so that the chain reaction is self-sustaining. If the average number of transformations caused by one transformation falls below one, the reaction is **subcritical** and the chain reaction ceases; if it exceeds one the reaction is **supercritical** and proceeds explosively.

critical state The state of a fluid in which the liquid and gas phases both have the same density. The fluid is then at its *critical temperature, *critical pressure, and *critical volume.

critical temperature 1. The temperature above which a gas cannot be liquefied by an increase of pressure. *See also* CRITICAL STATE. **2.** *See* TRANSITION POINT.

critical volume The volume of a fixed mass of a fluid in its *critical state; i.e. when it is at its critical temperature and critical pressure. The **critical specific volume** is its volume per unit mass in this state: in the past this has often been called the critical volume.

CRO *See* CATHODE-RAY OSCILLOSCOPE.

Crookes, Sir William (1832–1919) British chemist and physicist, who in 1861 used *spectroscopy to discover thallium and in 1875 invented the radiometer. He also developed an improved vacuum tube (**Crookes' tube**) for studying gas discharges.

cross product *See* VECTOR PRODUCT.

cross section 1. A plane surface formed by cutting a solid, especially by cutting at right angles to its longest axis. **2.** The area of such a surface. **3.** A measure of the probability that a collision will occur between a beam of radiation and a particular particle, expressed as the effective area presented by the particle in that particular process. It is measured in square metres or *barns. The **elastic cross section** accounts for all elastic scattering in which the radiation loses no energy to the particle. The **inelastic cross section** accounts for all other collisions. It is further subdivided to account for specific interactions, such as the **absorption cross section, fission cross section, ionization cross section**, etc.

crucible A dish or other vessel in which substances can be heated to a high temperature.

cryogenic pump A *vacuum pump in which pressure is reduced by condensing gases on surfaces maintained at about 20 K by means of liquid hydrogen or at 4 K by means of liquid helium. Pressures down to 10^{-8} mmHg (10^{-6} Pa) can be maintained; if they are used in conjunction with a *diffusion pump, pressures as low as 10^{-15} mmHg (10^{-13} Pa) can be reached.

cryogenics The study of very low temperatures and the techniques for producing them. Objects are most simply cooled by placing them in a bath containing liquefied gas maintained at a constant pressure. In general, a liquefied gas can provide a constant bath temperature from its triple point to its critical temperature and the bath temperature can be varied by changing the pressure above the liquid. The lowest practical temperature for a liquid bath is 0.3 K. Refrigerators (*see* REFRIGERATION) consist essentially of devices operating on a repeated cycle, in which a low-temperature reservoir is a continuously replenished liquid bath. Above 1 K they work by compressing and expanding suitable gases. Below this temperature liquids or solids are used and by *adiabatic demagnetization it is possible to reach 10^{-6} K.

cryohydrate A eutectic mixture of ice and some other substance (e.g. an ionic salt) obtained by freezing a solution.

cryometer A thermometer designed to measure low temperatures. *Thermocouples can be used down to about 1 K and *resistance thermometers can be used at 0.01 K. Below this magnetic thermometers (0.001 K) and nuclear-resonance thermometers (3×10^{-7} K) are required.

cryoscopic constant *See* DEPRESSION OF FREEZING POINT.

cryostat A vessel enabling a sample to be maintained at a very low temperature. The *Dewar flask is generally used for controlling heat leaking in by radiation, conduction, or convection. Cryostats usually consist of two or more Dewar flasks nesting in each other.

cryotron A switch that relies on *superconductivity. It consists of a coil of wire of one superconducting material surrounding a straight wire of another superconducting material; both are immersed in a liquid-helium bath. A current passed through the coil creates a magnetic field, which alters the superconducting properties of the central wire, switching its resistance from zero to a finite value. Cryotron switches can be made very small and take very little current.

crystal A solid with a regular polyhedral shape. All crystals of the same substance grow so that they have the same angles between their faces. However, they may not have the same external appearance because different faces can grow at different rates, depending on the conditions. The external form of the crystal is referred to as the **crystal habit**. The atoms, ions, or molecules forming the crystal have a regular arrangement and this is the **crystal structure**.

crystal counter A type of solid-state *counter in which a potential difference is applied across a crystal; when the crystal is struck by an elementary particle or photon, the electron–ion pairs created cause a transient increase in conductivity. The resulting current pulses are counted electronically.

crystal defect An imperfection in the regular lattice pattern of a crystal. See Feature (pp 106–107).

crystal habit *See* CRYSTAL.

crystal lattice The regular pattern of atoms, ions, or molecules in a crystalline substance. A crystal lattice can be regarded as produced by repeated translations of a **unit cell** of the lattice. *See also* CRYSTAL SYSTEM.

crystallography The study of crystal form and structure. *See also* X-RAY CRYSTALLOGRAPHY.

crystalloids *See* COLLOIDS.

crystal microphone A microphone in which the sound waves fall on a plate of Rochelle salt or similar material with piezoelectric properties, the variation in pressure being converted into a varying electric field by the *piezoelectric effect.

crystal oscillator (piezoelectric oscillator) An oscillator in which a piezoelectric crystal is used to determine the frequency. An alternating electric field applied to two metallic films sputtered onto the parallel faces of a crystal, usually of quartz, causes it to vibrate at its natural frequency; this frequency can be in the kilohertz or megahertz range, depending on how the crystal is cut. The mechanical vibrations in turn create an alternating electric field across the crystal that does not suffer from frequency drift. The device can be used to replace the tuned circuit in an oscillator by providing the resonant frequency or it can be coupled to the oscillator circuit, which is tuned approximately to the crystal frequency. In this type, the crystal prevents frequency drift. The device is widely used in *quartz clocks and watches.

crystal pick-up A pick-up in a record player in which the mechanical vibrations produced by undulations in the record groove are transmitted to a piezoelectric crystal, which produces a varying electric field of the same frequency as the sound. This signal is amplified and fed to loudspeakers in order to recreate the sound.

crystal structure *See* CRYSTAL.

crystal system A method of classifying crystals on the basis of their unit cell. There are seven crystal systems. If the cell is a parallelopiped with sides a, b, and c and if α is the angle between b and c, β the angle between a and c, and γ the angle between a and b, the systems are:
(1) **cubic** $a=b=c$ and $\alpha=\beta=\gamma=90°$
(2) **tetragonal** $a=b\neq c$ and $\alpha=\beta=\gamma=90°$
(3) **rhombic** (or **orthorhombic**) $a\neq b\neq c$ and $\alpha=\beta=\gamma=90°$
(4) **hexagonal** $a=b\neq c$ and $\alpha=\beta=\gamma=90°$
(5) **trigonal** $a=b\neq c$ and $\alpha=\beta=\gamma\neq90°$
(6) **monoclinic** $a\neq b\neq c$ and $\alpha=\gamma=90°\neq\beta$
(7) **triclinic** $a=b=c$ and $\alpha\neq\beta\neq\gamma$

CT scanner (computerized tomography scanner) *See* TOMOGRAPHY.

cubic crystal A crystal in which the unit cell is a cube (*see* CRYSTAL SYSTEM). There are three possible packings for cubic crystals: **simple cubic**, **face-centred cubic**, and **body-centred cubic**. See illustration.

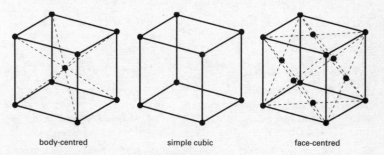

body-centred   simple cubic   face-centred

Cubic crystal structures

CRYSTAL DEFECTS

A crystal *lattice is formed by a repeated arrangement of atoms, ions, or molecules. Within one cubic centimetre of material one can expect to find up to 10^{22} atoms and it is extremely unlikely that all of these will be arranged in perfect order. Some atoms will not be exactly in the right place with the result that the lattice will contain *defects. The presence of defects within the crystal structure has profound consequences for certain bulk properties of the solid, such as the electrical resistance and the mechanical strength.

Point defects
Local crystal defects called **point defects**, appear as either impurity atoms or gaps in the lattice. Impurity atoms can occur in the lattice either at **interstitial sites** (between atoms in a non-lattice site) or at **substitutional sites** (replacing an atom in the host lattice). Lattice gaps are called **vacancies** and arise when an atom is missing from its site in the lattice. Vacancies are sometimes called **Schottky defects**. A vacancy in which the missing atom has moved to an interstitial position is known as a **Frenkel defect**.

Colour centres
In ionic crystals, the ions and vacancies always arrange themselves so that there is no build-up of one type of charge in any small volume of the crystal. If ions or charges are introduced into or removed from the lattice, there will, in general, be an accompanying rearrangement of the ions and their outer valence electrons. This rearrangement is called **charge compensation** and is most dramatically observed in **colour centres**. If certain crystals are irradiated with X-rays, gamma rays, neutrons, or electrons a colour change is observed. For example, diamond may be coloured blue by electron bombardment and quartz may be coloured brown by irradiation with neutrons. The high-energy radiation produces defects in the lattice and, in an attempt to maintain charge neutrality, the crystal undergoes some measure of charge compensation. Just as electrons around an atom have a series of discrete permitted energy levels, so charges residing at point defects exhibit sets of discrete levels, which are separated from one another by energies corresponding to wavelengths in the visible region of the spectrum. Thus light of certain wavelengths can be absorbed at the defect sites, and the material appears to be coloured. Heating the irradiated crystal can, in many cases, repair the irradiation damage and the crystal loses its coloration.

Dislocations
Non-local defects may involve entire planes of atoms. The most important of

Formation of a Schottky defect

Formation of a Frenkel defect

Point defects in a two-dimensional crystal

these is called a **dislocation.** Dislocations are essentially **line-defects**; that is, there is an incomplete plane of atoms in the crystal lattice. In 1934, Taylor, Orowan, and Polanyi independently proposed the concept of the dislocation to account for the mechanical strength of metal crystals. Their microscopic studies revealed that when a metal crystal is plastically deformed, the deformation does not occur by a separation of individual atoms but rather by a slip of one plane of atoms over another plane. Dislocations provide a mechanism for this slipping of planes that does not require the bulk movement of crystal material. The passage of a dislocation in a crystal is similar to the movement of a ruck in a carpet. A relatively large force is required to slide the carpet as a whole. However, moving a ruck over the carpet can inch it forward without needing such large forces. This movement of dislocations is called **plastic flow**.

Strength of materials

In practice most metal samples are **polycrystalline;** that is they consist of many small crystals or grains at different angles to each other. The boundary between two such grains is called a **grain boundary**. The plastic flow of dislocations may be hindered by the presence of grain boundaries, impurity atoms, and other dislocations. Pure metals produced commercially are generally too weak to be of much mechanical use. The weakness of these samples can be attributed to the ease with which the dislocations are able to move within the sample. Slip, and therefore deformation, can then occur under relatively low stresses. Impurity atoms, other dislocations, and grain boundaries can all act as obstructions to the slip of atomic planes. Traditionally, methods of making metals stronger involved introducing defects that provide regions of disorder in the material. For example, in an alloy, such as steel, impurity atoms (e.g. carbon) are introduced into the lattice during the forging process. The perfection of the iron lattice structure is disturbed and the impurities oppose the dislocation motion. This makes for greater strength and stiffness.

The complete elimination of dislocations may seem an obvious way to strengthen materials. However, this has only proved possible for hair-like single crystal specimens called **whiskers.** These whiskers are only a few micrometers thick and are seldom more than a few millimetres long; nevertheless their strength approaches the theoretical value.

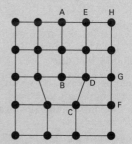 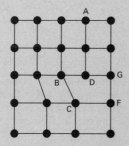

Dislocation in a two-dimensional crystal. The extra plane of atoms AB causes strain at bond CD. On breaking, the bond flips across to form CB. This incremental movement shifts the dislocation across so that the overall effect is to slide the two planes BDG and CF over each other.

cubic equation An equation in which the highest power of the variable is three. It has the general form $ax^3 + bx^2 + cx + d = 0$ and, in general, is satisfied by three values of x.

cubic expansivity *See* EXPANSIVITY.

curie The former unit of *activity (*see* RADIATION UNITS). It is named after Marie Curie.

Curie, Marie (Marya Sklodowska; 1867–1934) Polish-born French chemist, who went to Paris in 1891. She married the physicist Pierre Curie (1859–1906) in 1895 and soon began work on seeking radioactive elements other than uranium in pitchblende (to account for its unexpectedly high radioactivity). By 1898 she had discovered *radium and *polonium, although it took her four years to purify them. In 1903 the Curies shared the Nobel Prize for physics with Henri *Becquerel, who had discovered radioactivity.

Curie point (Curie temperature) The temperature at which a ferromagnetic substance loses its ferromagnetism and becomes only paramagnetic. For iron the Curie point is 760°C and for nickel 356°C. It is named after Pierre Curie.

Curie's law The susceptibility (χ) of a paramagnetic substance is proportional to the thermodynamic temperature (T), i.e. $\chi = C/T$, where C is the Curie constant. A modification of this law, the **Curie–Weiss law**, is more generally applicable. It states that $\chi = C/(T - \theta)$, where θ is the Weiss constant, a characteristic of the material. The law was first proposed by Pierre Curie and modified by another French physicist, Pierre-Ernest Weiss (1865–1940).

curium Symbol Cm. A radioactive metallic transuranic element belonging to the actinoids; a.n. 96; mass number of the most stable isotope 247 (half-life 1.64×10^7 years); r.d. (calculated) 13.51; m.p. 1340±40°C. There are nine known isotopes. The element was first identified by G. T. Seaborg (1912–99) and associates in 1944 and first produced by L. B. Werner and I. Perlman in 1947 by bombarding americium–241 with neutrons. It is named after Marie Curie.

curl (rot) The *vector product of the *gradient operator with a vector. For a vector $\mathbf{u}$ that has components u_1, u_2, and u_3 in the x, y, and z directions (with respective unit vectors $\mathbf{i}$, $\mathbf{j}$, and $\mathbf{k}$), and is a function of x, y, and z, the curl is given by:

$$\text{curl}\,\mathbf{u} = \nabla \times \mathbf{u} = (\partial u_3/\partial y - \partial u_2/\partial z)\mathbf{i} + (\partial u_1/\partial z - \partial u_3/\partial x)\mathbf{j} + (\partial u_2/\partial x - \partial u_1/\partial y)\mathbf{k}.$$

See also DIVERGENCE.

current Symbol I. A flow of electric charge through a conductor. The current at a particular cross section is the rate of flow of charge. The charge may be carried by electrons, ions, or positive holes (*see* CHARGE

CARRIER). The unit of current is the ampere. *See also* CONVENTIONAL CURRENT.

current balance An instrument used to measure a current absolutely, on the basis of the definition of the ampere. An accurate form consists of a beam balance with similar coils attached to the ends of the balance arms. Fixed coils are situated above and below these two coils. The six coils are then connected in series so that a current passing through them creates a torque on the beam, which is restored to the horizontal by means of a rider. From the position and weight of the rider, and the geometry of the system, the current can be calculated.

current density 1. The current flowing through a conductor per unit cross-sectional area, measured in amperes per square metre. **2.** The current flowing through an electrolyte per unit area of electrode.

cusp A point at which two arcs of a curve intersect.

cycle A regularly repeated set of changes to a system that brings back all its parameters to their original values once in every set of changes. The duration of one cycle is called its period and the rate of repetition of cycle, called the *frequency, is measured in *hertz. *See* SIMPLE HARMONIC MOTION.

cycloid The curve traced by a point on the circumference of a circle as it rolls without slipping along a straight line. The length of the arc formed by one revolution of the circle is $8r$, where r is the radius of the circle. The horizontal distance between cusps is $2\pi r$.

cyclotron A cyclic particle *accelerator in which charged particles fed into the centre of the device are accelerated in an outward spiral path inside two hollow D-shaped conductors placed to form a split circle. A magnetic field is applied at right-angles to the plane of the dees and an alternating potential difference is applied between them. The frequency of the alternating p.d. is arranged so that the particles are accelerated each time they reach the evacuated gap between the dees. The magnetic field makes them follow curved paths. After several thousand revolutions inside the dees the particles reach the perimeter of the dees, where a deflecting field directs them onto the target. In this device protons can achieve an energy of 10^{-12} J (10 MeV). The first working cyclotron was produced in 1931 by the US physicist E. O. Lawrence (1901–58). *See also* SYNCHROCYCLOTRON.

cylindrical polar coordinates *See* POLAR COORDINATES.

dalton *See* ATOMIC MASS UNIT.

Dalton, John (1766–1844) British chemist and physicist. In 1801 he formulated his law of partial pressures (*see* DALTON'S LAW), but he is best remembered for *Dalton's atomic theory, which he announced in 1803. Dalton also studied colour blindness (a condition, once called Daltonism, that he shared with his brother).

Dalton's atomic theory A theory of chemical combination, first stated by John Dalton in 1803. It involves the following postulates:
  (1) Elements consist of indivisible small particles (atoms).
  (2) All atoms of the same element are identical; different elements have different types of atom.
  (3) Atoms can neither be created nor destroyed.
  (4) 'Compound elements' (i.e. compounds) are formed when atoms of different elements join in simple ratios to form 'compound atoms' (i.e. molecules).
Dalton also proposed symbols for atoms of different elements (later replaced by the present notation using letters).

Dalton's law The total pressure of a mixture of gases or vapours is equal to the sum of the partial pressures of its components, i.e. the sum of the pressures that each component would exert if it were present alone and occupied the same volume as the mixture of gases. Strictly speaking, the principle is true only for ideal gases. It was discovered by John Dalton.

damping A decrease in the amplitude of an oscillation as a result of energy being drained from the oscillating system to overcome frictional or other resistive forces. For example, a pendulum soon comes to rest unless it is supplied with energy from an outside source; in a pendulum clock, energy is supplied through an *escapement from a wound spring or a falling mass to compensate for the energy lost through friction. Damping is introduced intentionally in measuring instruments of various kinds to overcome the problem of taking a reading from an oscillating needle. A measuring instrument is said to be **critically damped** if the system just fails to oscillate and the system comes to rest in the shortest possible time. If it is **underdamped** it will oscillate repeatedly before coming to rest; if it is **overdamped** it will not oscillate but it will take longer to come to rest than it would if it was critically damped. An instrument, such as a galvanometer, that is critically damped is often called a **deadbeat** instrument.

Daniell cell A type of primary *voltaic cell with a copper positive electrode and a negative electrode of a zinc amalgam. The zinc-amalgam electrode is placed in an electrolyte of dilute sulphuric acid or zinc sulphate solution in a porous pot, which stands in a solution of copper sulphate in which the copper electrode is immersed. While the reaction takes place ions move through the porous pot, but when it is not in use the cell should be dismantled to prevent the diffusion of one electrolyte into the other. The e.m.f. of the cell is 1.08 volts with sulphuric acid and 1.10 volts with zinc sulphate. It was invented in 1836 by the British chemist John Daniell (1790–1845).

dark energy Energy in the universe associated with the fact that the expansion of the universe is accelerating and the *cosmological constant could have a nonzero value. Analysis of data from *WMAP indicates that about 70% of the energy of the universe is in the form of dark energy. The nature of this energy is not known.

dark galaxy A galaxy that is composed largely of dark matter. There is some evidence for the existence of such galaxies. It is thought that they should be very common, particularly since theories of large-scale structure work much better if the existence of plentiful dark galaxies is assumed.

dark matter *See* MISSING MASS.

DAT (digital audio tape) A type of magnetic tape originally designed for audio recording but now adapted for computer storage and backup use. The recording method allows a capacity of about 1 gigabyte.

database A large collection of information that has been coded and stored in a computer in such a way that it can be extracted under a number of different category headings.

dating techniques Methods of estimating the true age of rocks, palaeontological specimens, archaeological sites, etc. **Relative dating techniques** date specimens in relation to one another; for example, stratigraphy is used to establish the succession of fossils. **Absolute** (or **chronometric**) **techniques** give an absolute estimate of the age and fall into two main groups. The first depends on the existence of something that develops at a seasonally varying rate, as in dendrochronology and varve dating. The other uses some measurable change that occurs at a known rate, as in *chemical dating, radioactive (or radiometric) dating (*see* CARBON DATING; FISSION-TRACK DATING; POTASSIUM–ARGON DATING; RUBIDIUM–STRONTIUM DATING; URANIUM–LEAD DATING), and *thermoluminescence.

daughter **1.** A nuclide produced by radioactive *decay of some other nuclide (the **parent**). **2.** An ion or free radical produced by dissociation or reaction of some other (**parent**) ion or radical.

Davisson–Germer experiment *See* ELECTRON DIFFRACTION.

Davy, Sir Humphry (1778–1829) British chemist, who studied gases at the Pneumatic Institute in Bristol, where he discovered the anaesthetic properties of dinitrogen oxide (nitrous oxide). He moved to the Royal Institution, London, in 1801 and five years later isolated potassium and sodium by electrolysis. He also prepared barium, boron, calcium, and strontium as well as proving that chlorine and iodine are elements. In 1816 he invented the Davy lamp.

day The time taken for the earth to complete one revolution on its axis. The **solar day** is the interval between two successive returns of the sun to the *meridian. The **mean solar day** of 24 hours is the average value of the solar day for one year. The **sidereal day** is measured with respect to the fixed stars and is 4.09 minutes shorter than the mean solar day as a result of the imposition of the earth's orbital motion on its rotational motion.

D-brane *See* SUPERSTRING THEORY.

d.c. *See* DIRECT CURRENT.

deadbeat *See* DAMPING.

death of a star The final collapse of a star. The process begins when *red giants with masses less than or equal to about three solar masses do not have internal temperatures that are sufficiently high to ignite any further nuclear fusion reactions after the completion of the helium burning stage. For these relatively small stars, shedding of the outer layers begins and the mass is reduced by some 50%.

The material that was once the outer layers of the star surrounds a small collapsing core. The radiation from this core can be very intense and the resulting glow, caused by the ionization of the surrounding material, is known as a planetary *nebula. The core continues to undergo gravitational collapse until equilibrium is reached with the **Fermi pressure**. This Fermi pressure is generated by the electrons being confined to smaller and smaller spaces as a result of the gravitational collapse. Electrons, which are responsible for most of the volume in ordinary matter, when confined in this way generate a pressure because as *fermions they are subject to the *Pauli exclusion principle and therefore have an appreciable mutual repulsion. The equilibrium between gravitational collapse and the Fermi pressure is reached when the core has a diameter of about 1% of the sun. This implies a very high density of about 1000 kg/cm^3. At this stage the star is classed as a *white dwarf and appears on the bottom-left quarter of the *Hertzsprung–Russell diagram. White dwarfs gradually become dimmer as they cool and eventually reach a temperature equilibrium with the surrounding vacuum.

If at the white dwarf stage a star has a mass exceeding 1.4 solar masses, it is said to have exceeded the *Chandrasekhar limit. For stars of this mass, the electron Fermi pressure is not large enough to counter the continued gravitational contraction. Electrons and protons in the resulting dense matter then combine in an interaction similar to the radioactive transformation known as *electron capture. Neutrons are formed in this electron-proton fusion and further collapse may be halted (if the star's mass is not so large) by a Fermi pressure generated by the tightly packed neutrons (as neutrons are also fermions). The core is then said to consist of **neutronium**, since it is exclusively made of neutrons.

The final collapse to the neutronium stage is very rapid and accompanied by an extreme rise in temperature. This rapid contraction is stopped as suddenly as it began by the neutron Fermi pressure, which generates an intense radiation pressure causing the star to explode. The resulting explosion is called a *supernova. Supernovae can be so intense that over a short period (typically a few days) they have been known to emit as much radiation as an entire galaxy of stable stars. The extreme pressure and temperatures within a supernova explosion are sufficient for thermonuclear fusion of nuclei to form nuclei of elements heavier than iron. The debris from supernova explosions is therefore the source of all the elements with which we are familiar on earth. As a result of density fluctuations in clouds of debris of this type, gravitational collapse can begin once again and form a new generation of stars. At the centre of some supernovae there may be a residual highly dense core. This remnant, composed of neutrons, is called a *neutron star. Theoretically, neutron stars can rotate at a very high rate and in doing so emit very regular radio-frequency pulses. Before the theoretical work that led to the idea of neutron stars, *pulsars – pulsating sources of radio waves – had already been observed. The current explanation of pulsar radiation is that it originates from a rotating neutron star.

The Fermi pressure generated by the tightly packed neutrons within a neutron star is the only influence preventing further gravitational contraction. However, if the mass of the neutron star is above about three solar masses, then the neutron Fermi pressure is not sufficient to prevent a further collapse. As the radius of the star reduces, the gravitational field becomes increasingly strong until eventually the radius of the star falls to lower than a limit, called its **Schwarzschild radius** (τ_s). At this point the equations of general relativity are only valid for an observer outside the Schwarzschild radius. The boundary that delimits this region of validity is called the **event horizon.**

The gravitational fields within the event horizon are so large that not even electromagnetic radiation (including light) can escape. The star is then said to have collapsed to a *black hole. It is thought that the collapse of a large star to a black hole is accompanied by a gamma-ray burst. Black holes are still only theoretical constructs although many observations suggest that they do exist. For example, the intense radiation emitted from *quasars may be produced as a consequence of matter falling into black holes. Other examples that support the existence of black holes include the detection of very massive but invisible partners in binary systems and the rapidly orbiting gases around galactic centres. Both phenomena could be explained by the presence of centrally directed forces resulting from a black hole. Cygnus X-1 is believed to be an example of a binary system in which one member is a *main sequence star and the other a black hole. As charged gases from the main sequence star spiral in towards the black hole, they reach very high speeds and can emit X-rays.

de Broglie, Louis-Victor Pierre Raymond (1892–1987) French physicist, who taught at the Sorbonne in Paris for 34 years. He is best known for his 1923 theory of *wave–particle duality (*see also* DE BROGLIE WAVELENGTH), which reconciled the corpuscular and wave theories of *light and proved important in *quantum theory. For this work he was awarded the 1929 Nobel Prize.

de Broglie wavelength The wavelength of the wave associated with a moving particle. The wavelength (λ) is given by $\lambda = h/mv$, where h is the Planck constant, m is the mass of the particle, and v its velocity. The **de Broglie wave** was first suggested by Louis de Broglie in 1923 on the grounds that, since electromagnetic waves can be treated as particles (*photons), one could therefore expect particles to behave in some circumstances like waves (*see* COMPLEMENTARITY). The subsequent observation of *electron diffraction substantiated this argument and the de Broglie wave became the basis of *wave mechanics.

debye A unit of electric *dipole moment in the electrostatic system, used to express dipole moments of molecules. It is the dipole moment produced by two charges of opposite sign, each of 1 statcoulomb and placed 10^{-18} cm apart, and has the value $3.335\,64 \times 10^{-30}$ coulomb metre. It is named after the Dutch chemist and physicist Peter J. W. Debye (1884–1966).

Debye–Hückel theory A theory to explain the nonideal behaviour of electrolytes, published in 1923 by P. J. W. Debye (1884–1966) and Erich Hückel (1896–1980). It assumes that electrolytes in solution are fully dissociated and that nonideal behaviour arises because of electrostatic interactions between the ions. The theory shows how to calculate the extra free energy per ion resulting from such interactions, and consequently the activity coefficient. It gives a good description of nonideal electrolyte behaviour for very dilute solutions, but cannot be used for more concentrated electrolytes.

Debye–Scherrer method A method of X-ray diffraction in which a beam of X-rays is diffracted by material in the form of powder. Since the powder consists of very small crystals of the material in all possible orientations, the diffraction pattern is a series of concentric circles. This type of pattern allows the unit cell to be found with great precision. This method was first used by Peter Debye and Paul Scherrer in 1916 and independently by Albert Hull in 1917.

Debye temperature Symbol θ. A parameter that appears in the *Debye theory of specific heat. It has the dimensions of temperature and is defined by $\theta = h\nu/k$, where h is the Planck constant, ν is the maximum frequency for the lattice vibrations of the solid, and k is the Boltzmann constant. The Debye temperature is characteristic of a particular material; for example, for sodium θ is 150 K.

Debye theory of speciÜc heat A theory of the specific heat of solids proposed by Peter Debye in 1912. Debye gave an accurate quantitative description of the specific heat capacity by attributing it to the lattice vibrations. He postulated that these vibrations have a range of frequencies up to a maximum frequency. *See also* DEBYE TEMPERATURE; EINSTEIN THEORY OF SPECIFIC HEAT.

Debye–Waller factor A quantity that takes the effects of quantum or thermal fluctuations on the intensities of X-rays diffracted by crystals into account. This factor was calculated by Peter Debye in 1913 and refined by Ivar Waller in 1923.

deca- Symbol da. A prefix used in the metric system to denote ten times. For example, 10 coulombs = 1 decacoulomb (daC).

decalescence *See* RECALESCENCE.

decay The spontaneous transformation of one radioactive nuclide into a daughter nuclide, which may be radioactive or may not, with the emission of one or more particles or photons. The decay of N_0 nuclides to give N nuclides after time t is given by $N = N_0\exp(-\gamma t)$, where γ is called the **decay constant** or the **disintegration constant**. The reciprocal of the decay constant is the **mean life**. The time required for half the original nuclides to decay (i.e. $N = \frac{1}{2}N_0$) is called the **half-life** of the nuclide. The same terms are applied to elementary particles that spontaneously transform into other particles. For example, a free neutron decays into a proton and an electron (*see* BETA DECAY). *See also* ALPHA PARTICLE; Q-VALUE.

decay sequence A plot of neutron number (N) against proton number (Z) for nuclides belonging to a particular *radioactive series. Each point marked on the plot represents a member of the series. The lines joining these points represent the nuclear transformations that have occurred in transmuting the original nucleus to the new one.

Four types of radioactive decay can be conveniently represented as shifts along a decay sequence:

Alpha decay: The emission of an *alpha particle corresponds to the loss of a helium nucleus (two protons and two neutrons) by the parent nucleus. An alpha decay is easily represented as a diagonal arrow made up of two steps down the N axis and two steps towards the left on the Z axis.

Beta (negative) decay: A neutron in the nucleus decays into a proton and an electron; the electron is emitted as a beta particle. This transformation is therefore represented on a decay sequence as the loss of a neutron and the gain of a proton; that is, a diagonal arrow made up of one step down the N axis and one step to the right on the Z axis. *See also* BETA DECAY.

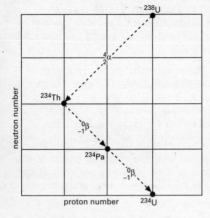

Decay sequence

Beta (positive) decay: In this decay a proton transforms into a neutron with the emission of an antimatter electron (positron). Unlike the beta (negative) decay, which can also occur outside the nucleus, the beta (positive) decay does not occur spontaneously with free protons. This transformation is represented on a decay sequence as a diagonal arrow made up of one step up the N axis and one step to the left on the Z axis.
Electron capture: A nuclear proton captures an electron from the atomic environment. This transformation is represented in the same way as beta (positive) decay. The diagram shows a portion of the decay sequence for the **uranium series**.

deci- Symbol d. A prefix used in the metric system to denote one tenth. For example, 0.1 coulomb = 1 decicoulomb (dC); 0.1 metre = 1 decimetre (dm).

decibel A unit used to compare two power levels, usually applied to sound or electrical signals. Although the decibel is one tenth of a **bel**, it is the decibel, not the bel, that is invariably used. Two power levels P and P_0 differ by n decibels when $n = 10\log_{10}P/P_0$. If P is the level of sound intensity to be measured, P_0 is a reference level, usually the intensity of a note of the same frequency at the threshold of audibility.

The logarithmic scale is convenient as human audibility has a range of 1 (just audible) to 10^{12} (just causing pain) and one decibel, representing an increase of some 26%, is about the smallest change the ear can detect.

decimal system A number system based on the number 10; the number system in common use. All rational numbers can be written as a **finite decimal** (e.g. ¼ = 0.25) or a **repeating decimal** (e.g. 5/27 = 0.185 185 185…). An *irrational number can be written to any number of decimal places, but can never be given exactly (e.g. √3 = 1.732 050 8…).

declination **1.** The angle between the magnetic meridian and the geographic meridian at a point on the surface of the earth. *See* GEOMAGNETISM. **2.** The angular distance of a celestial body north (positive) or south (negative) of the celestial *equator.

decoherence A process in which a quantum mechanical state of a system is altered by the interaction between the system and its environment. The process of decoherence has been detected experimentally. Decoherence was postulated in the 1980s and has been used to clarify discussions of the foundations of quantum mechanics and the problems of measurement.

deconfinement temperature *See* QUARK CONFINEMENT.

deep inelastic scattering *See* INELASTIC COLLISION.

defect **1.** A discontinuity in a crystal lattice. A **point defect** consists either of a missing atom or ion creating a **vacancy** in the lattice (a vacancy is sometimes called a **Schottky defect**) or an extra atom or ion between two normal lattice points creating an **interstitial**. A **Frenkel defect** consists of a vacancy in which the missing atom or ion has moved to an interstitial position. If more than one adjacent point defect occurs in a crystal there may be a slip along a surface causing a **line defect** (or **dislocation**). Defects are caused by strain or, in some cases, by irradiation. All crystalline solids contain an equilibrium number of point defects above absolute zero; this number increases with temperature. The existence of defects in crystals is important in the conducting properties of *semiconductors. **2.** *See* MASS DEFECT.

definite integral *See* INTEGRATION.

degaussing The process of neutralizing the magnetization in an object that has inadvertently become magnetized. For example, ferromagnetic components of TV sets may become magnetized and misdirect the electron beams. A degaussing coil is often provided and fed with a diminishing alternating current each time the set is switched on. Ships can be degaussed by surrounding them with current-carrying cables that set up an equal and opposite field. This prevents the ships from detonating magnetic mines. Degaussing is used to protect scientific and other electronic devices from strong magnetic fields; usually a system of coils is designed to neutralize such fields over the important region or the equipment is surrounded by a shield of suitable alloy (e.g. Mumetal).

degeneracy pressure The pressure in a *degenerate gas of fermions caused by the Pauli exclusion principle and the Heisenberg uncertainty principle. Because of the exclusion principle, fermions at a high density, with small interparticle spacing, must have different momenta; from the uncertainty principle, the momentum difference must be inversely proportional to the spacing. Consequently, in a high-density gas (small spacing) the particles have high relative momenta, which leads to a degeneracy pressure much greater than the thermal pressure. *White dwarfs and *neutron stars are supported against collapse under their own gravitational fields by the degeneracy pressure of electrons and neutrons, respectively.

degenerate gas A gas in which, because of high density, the particle concentration is so high that the *Maxwell–Boltzmann distribution does not apply and the behaviour of the gas is governed by *quantum statistics. Examples of degenerate gases are the conduction electrons in a metal, the electrons in a *white dwarf, and the neutrons in a *neutron star. *See also* DEGENERACY PRESSURE.

degenerate level An *energy level of a quantum-mechanical system that corresponds to more than one *quantum state.

degenerate semiconductor A heavily doped *semiconductor in which the *Fermi level is located in either the valence band or the conduction band (*see* ENERGY LEVELS) causing the material to behave as a metal.

degenerate states *Quantum states of a system that have the same energy.

degree 1. A unit of plane angle equal to 1/360th of a complete revolution. **2.** A division on a *temperature scale. **3.** The power to which a variable is raised. If one expression contains several variables the overall degree of the expression is the sum of the powers. For example, the expression $p^2q^3r^4$ has a degree of 9 overall (it is a second-degree expression in p). The degree of a polynomial is the degree of the variable with the highest power, e.g. $ax^5 + bx^4 + c$ has a degree of 5. **4.** The highest power to which the derivative of the highest order is raised in a *differential equation. For example, $(d^2y/dx^2)^3 + dy/dx = c$ is a differential equation of the third degree (but second order).

degrees of freedom 1. The number of independent parameters required to specify the configuration of a system. This concept is applied in the *kinetic theory to specify the number of independent ways in which an atom or molecule can take up energy. There are however various sets of parameters that may be chosen, and the details of the consequent theory vary with the choice. For example, in a monatomic gas each atom may be allotted three degrees of degrees of freedom, corresponding to the three coordinates in space required to specify its position. The mean energy per atom for each degree of freedom is the same, according to the principle of the *equipartition of energy, and is equal to $kT/2$ for each degree of freedom (where k is the *Boltzmann constant and T is the thermodynamic temperature). Thus for a monatomic gas the total molar energy is $3LkT/2$, where L is the Avogadro constant (the number of atoms per mole). As $k = R/L$, where R is the molar gas constant, the total molar energy is $3RT/2$.

In a diatomic gas the two atoms require six coordinates between them, giving six degrees of freedom. Commonly these are interpreted as six independent ways of storing energy: on this basis the molecule has three degrees of freedom for different directions of translational motion (*see* TRANSLATION), and in addition there are two degrees of freedom for rotation of the molecular axis and one vibrational degree of freedom along the bond between the atoms. The rotational degrees of freedom each contribute $kT/2$, to the total energy; similarly the vibrational degree of freedom has an equal share of kinetic energy and must on average have as much potential energy (*see* SIMPLE HARMONIC MOTION). The total energy per molecule for a diatomic gas is therefore $3kT/2$ (for translational energy of the whole molecule) plus $2kT/2$ (for rotational energy of each atom) plus $2kT/2$ (for vibrational energy), i.e. a total of $7kT/2$. **2.** The least number of independent variables required to define the state of a system in the *phase rule. In this sense a gas has two degrees of freedom (e.g. temperature and pressure).

deionized water Water from which dissolved ionic salts have been removed by *ion-exchange techniques. It is used for many purposes as an alternative to distilled water.

dekatron A neon-filled tube with a central anode surrounded by ten cathodes and associated transfer electrodes. As voltage pulses are received by the tube a glow discharge moves from one set of electrodes to the next, enabling the device to be used as a visual counting tube in the decimal system. The tube can also be used for switching.

del (nabla) Symbol ∇. The differential vector operator $\mathbf{i}\,\partial/\partial x + \mathbf{j}\,\partial/\partial y + \mathbf{k}\,\partial/\partial z$.

delayed neutrons The small proportion of neutrons that are emitted with a measurable time delay in a nuclear fission process. *Compare* PROMPT NEUTRONS.

delay line A component in an electronic circuit that is introduced to provide a specified delay in transmitting the signal. Coaxial cable or inductor-capacitor networks can be used to provide a short delay but for longer delays an **acoustic delay line** is required. In this device the signal is converted by the *piezoelectric effect into an acoustic wave, which is passed through a liquid or solid medium, before reconversion to an electronic signal.

deliquescence The absorption of water from the atmosphere by a hygroscopic solid to such an extent that a concentrated solution of the solid eventually forms.

delta function Symbol $\delta(x)$. A type of function that has the value zero except at the origin where it is infinite, with the integral of this function having the value 1. The delta function is not well defined in terms of simple function theory but it can be justified in a mathematically rigorous way. The function is used in mathematical physics to represent concepts such as point charges and impulsive forces.

demagnetization The removal of the ferromagnetic properties of a body by disordering the domain structure (*see* MAGNETISM). One method of achieving this is to insert the body within a coil through which an alternating current is flowing; as the magnitude of the current is reduced to zero, the domains are left with no predominant direction of magnetization.

demodulation The process of extracting the information from a modulated carrier wave (*see* MODULATION; RADIO). The device used is called a **demodulator** or a **detector**.

denature To add another isotope to a fissile material to make it unsuitable for use in a nuclear weapon.

dendrite A crystal with a branching structure like that of a tree, formed when liquids solidify. The commonest example is the structure of snowflakes. Dendrites are particularly important in determining the properties of metals and alloys. In an alloy, the central part of a dendrite is richer in matter with higher melting points than the matter in the outer regions, which are slow to solidify. The growth of this type of crystal is called **dendritic growth**.

densitometer An instrument used to measure the *photographic density of an image on a film or photographic print. Densitometers work by letting the specimen transmit or reflect a beam of light and monitoring the transmitted or reflected intensity. They originally consisted of visual *photometers but most instruments are now photoelectric. The simplest transmission densitometer consists of a light source, a photosensitive cell, and a microammeter: the density is measured in terms of the meter readings with and without the sample in place. They have a variety of uses, including detecting the sound track on a cinematic film, measuring intensities in spectrographic records, and checking photographic prints.

density **1.** The mass of a substance per unit of volume. In *SI units it is measured in $kg\ m^{-3}$. *See also* RELATIVE DENSITY; VAPOUR DENSITY. **2.** *See* CHARGE DENSITY. **3.** *See* PHOTOGRAPHIC DENSITY.

density functional theory A theory used to describe many-fermion systems in which the energy is a *functional of the density of fermions. Density functional theory has been used extensively in the theory of electrons in atoms, molecules, and solids and in the theory of nucleons in nuclei.

density of states *See* FREE ELECTRON.

depleted Denoting a material that contains less of a particular isotope than it normally contains, especially a residue from a nuclear reactor or isotope-separation plant containing fewer fissile atoms than natural uranium.

depletion layer A region in a *semiconductor that has a lower-than-usual number of mobile charge carriers. A depletion layer forms at the interface between two dissimilar regions of conductivity (e.g. a p–n junction).

depolarization The prevention of *polarization in a *primary cell. For example, maganese(IV) oxide (the **depolarizer**) is placed around the positive electrode of a *Leclanché cell to oxidize the hydrogen released at this electrode.

depression of freezing point The reduction in the freezing point of a pure liquid when another substance is dissolved in it. It is a *colligative property – i.e. the lowering of the freezing point is proportional to the number of dissolved particles (molecules or ions), and does not depend on their nature. It is given by $\Delta t = K_f C_m$, where C_m is the molar concentration of dissolved solute and K_f is a constant (the **cryoscopic constant**) for the solvent used. Measurements of freezing-point depression (using a Beckmann thermometer) can be used for finding relative molecular masses of unknown substances.

depth of field The range of distance in front of and behind an object that is being focused by an optical instrument, such as a microscope or camera, within which other objects will be in focus. The **depth of focus** is the amount by which the distance between the camera and the film can be changed without upsetting the sharpness of the image.

derivative *See* DIFFERENTIATION; CALCULUS.

derived unit *See* BASE UNIT.

desorption The removal of adsorbed atoms, molecules, or ions from a surface.

detailed balance The cancellation of the effect of one process by another process that operates at the same time with the opposite effect. An example of detailed balance is provided by a chemical reaction between two molecular species A and B, which results in the formation of the molecular species C and D. Detailed balance for this chemical reaction occurs if the rate at which the reaction A + B → C + D occurs is equal to the rate at which the reaction C + D → A + B occurs. The equilibrium state in thermodynamics is characterized by detailed balance. When there is detailed balance in a system, the *self-organization far from equilibrium associated with *nonequilibrium statistical mechanics cannot occur.

detector 1. *See* DEMODULATION. **2.** *See* COUNTER.

deuterated compound A compound in which some or all of the hydrogen–1 atoms have been replaced by deuterium atoms.

deuterium (heavy hydrogen) Symbol D. The isotope of hydrogen that has a mass number 2 (r.a.m. 2.0144). Its nucleus contains one proton and one neutron. The abundance of deuterium in natural hydrogen is about 0.015%. It is present in water as the oxide HDO (*see also* HEAVY WATER), from which it is usually obtained by electrolysis or fractional distillation. Its chemical behaviour is almost identical to hydrogen although deuterium compounds tend to react rather more slowly than the corresponding hydrogen compounds. Its physical properties are slightly different from those of hydrogen, e.g. b.p. 23.6 K (hydrogen 20.4 K).

deuterium oxide *See* HEAVY WATER.

deuteron A nucleus of a deuterium atom, consisting of a proton and a neutron bound together.

deviation **1. (angle of deviation)** The angle formed between a ray of light falling on a surface or transparent body and the ray leaving it. **2.** The difference between one of an observed set of values and the true value, usually represented by the mean of all the observed values. The **mean deviation** is the mean of all the individual deviations of the set. *See* STANDARD DEVIATION.

devil's staircase A continuous function that has flat regions. If the flat regions, which by definition have a derivative of zero, are connected by regions with a nonzero derivative, the function is called an **incomplete devil's staircase**. If the derivative of the function is zero almost everywhere, the function is called a **complete devil's staircase** (sometimes called a **singular continuous function**). If there are discontinuous jumps between the flat regions, the function is called a **harmless staircase**. A physical realization of a devil's staircase can occur in the *Frenkel–Kontorowa model of atoms adsorbed on a periodic substrate.

dew A type of *hydrometeor in which water condenses onto grass and other exposed surfaces near the ground when the temperature has fallen below the *dew point. If the dew point is below freezing a **hoarfrost** forms. If dew is formed before the temperature falls below freezing, frozen dew called **white dew**, is formed. Conditions that encourage the formation of dew include: (1) a surface that is insulated from the heat supply for the soil, (2) a clear still atmosphere with a low relative *humidity.

Dewar, Sir James (1842–1923) British chemist and physicist, born in Scotland. In 1875 he became a professor at Cambridge University, while carrying out much of his experimental work at the Royal Institution in London. He began studying gases at low temperatures and in 1872 invented the *Dewar flask. In 1891, together with Frederick Abel (1827–1902), he developed the smokeless propellant explosive cordite, and in 1898 was the first to liquefy hydrogen.

Dewar flask A vessel for storing hot or cold liquids so that they maintain their temperature independently of the surroundings. Heat transfer to the surroundings is reduced to a minimum: the walls of the vessel consist of two thin layers of glass (or, in large vessels, steel) separated by a vacuum to reduce conduction and convection; the inner surface of a glass vessel is silvered to reduce radiation; and the vessel is stoppered to prevent evaporation. It was devised around 1872 by Sir James Dewar and is also known by its first trade name **Thermos flask**. *See also* CRYOSTAT.

dew point The temperature at which the water vapour in the air is saturated. As the temperature falls the dew point is the point at which the vapour begins to condense as droplets of water.

dew-point hygrometer *See* HYGROMETER.

dextrorotatory Denoting a compound that rotates the plane of polarization of plane-polarized light to the right (clockwise as observed by someone facing the oncoming radiation). *See* OPTICAL ACTIVITY.

***d*-form** *See* OPTICAL ACTIVITY.

diamagnetism *See* MAGNETISM.

diamond The hardest known mineral (with a hardness of 10 on Mohs' scale). It is an allotropic form of pure carbon that has crystallized in the cubic system, usually as octahedra or cubes, under great pressure. Diamond crystals may be colourless and transparent or yellow, brown, or black. They are highly prized as gemstones but also have extensive uses in industry, mainly for cutting and grinding tools. Industrial diamonds are increasingly being produced synthetically.

diaphragm An opaque disc with a circular aperture at its centre. Diaphragms of different sizes are used to control the total light flux passing through an optical system or to reduce aberration by restricting the light passing through a system to the central portion. An **iris diaphragm** consists of a number of overlapping crescent-shaped discs arranged so that the central aperture can be continuously varied in diameter.

diastereoisomers Stereoisomers that are not identical but not mirror images. For instance, the *d*-form of tartaric acid and the meso form constitute a pair of diastereoisomers. *See* OPTICAL ACTIVITY.

dichroism The property of some crystals, such as tourmaline, of selectively absorbing light vibrations in one plane while allowing light vibrations at right angles to this plane to pass through. Polaroid is a synthetic dichroic material. *See* POLARIZATION.

dielectric A nonconductor of electric charge in which an applied electric field causes a *displacement of charge but not a flow of charge. Electrons within the atoms of a dielectric are, on average, displaced by an applied field with respect to the nucleus, giving rise to a dipole that has an electric moment in the direction of the field. The resulting stress within the dielectric is known as the **electric polarization** (P) and is defined by $P = D - E\varepsilon_0$, where D is the displacement, E is the electric field strength, and ε_0 is the electric constant.

The **dielectric constant** is now called the relative *permittivity. The **dielectric strength** is the maximum potential gradient that can be applied to a material without causing it to break down. It is usually expressed in volts per millimetre. *See also* CAPACITOR.

dielectric constant *See* PERMITTIVITY.

dielectric heating The heating of a dielectric material, such as a plastic, by applying a radio-frequency electric field to it. The most common method is to treat the material as the dielectric between the plates of a capacitor. The heat produced is proportional to $V^2 f A\phi/t$, where V is the applied potential difference, f its frequency, A is the area of the dielectric, t its thickness, and ϕ is the loss factor of the material (related to its *permittivity).

Diesel engine *See* INTERNAL-COMBUSTION ENGINE.

differential amplifier An *amplifier with two inputs in which the output is a function of the difference between the inputs.

differential calculus *See* CALCULUS.

differential equation An equation in which a derivative of y with respect to x appears as well as the variables x and y. The **order** of a differential equation is the order of its highest derivative. The **degree** of the equation is the highest power present of the highest-order derivative. There are many types of differential equation, each having its own method of solution. The simplest type has separable variables, enabling each side of the equation to be integrated separately.

differential geometry The branch of geometry concerned with the applications of differential calculus to geometry. Differential geometry has important applications in *gauge theories, the general theory of *relativity, and some problems in the theory of condensed matter. In *quantum gravity, radically new physical ideas are needed for which differential geometry may be inadequate for their mathematical formulation.

differentiation The process of finding the **derivative** of a function in differential *calculus. If $y = f(x)$, the derivative of y, written dy/dx or $f'(x)$, is equal to the limit as $\Delta x \to 0$ of $[f(x + \Delta x) - f(x)]/\Delta x$. In general, if $y = x^n$, then $dy/dx = nx^{n-1}$. On a graph of $y = f(x)$, the derivative dy/dx is the gradient of the tangent to the curve at the point x.

diffraction The spreading or bending of waves as they pass through an aperture or round the edge of a barrier. The diffracted waves subsequently interfere with each other (*see* INTERFERENCE) producing regions of reinforcement and weakening. First noticed as occurring with light by

Francesco Grimaldi (1618–63), the phenomenon gave considerable support to the wave theory of light. Diffraction also occurs with streams of particles because of the quantum-mechanical wave nature of such particles. *See also* FRESNEL DIFFRACTION; FRAUNHOFER DIFFRACTION; ELECTRON DIFFRACTION.

diffraction grating A device for producing spectra by diffraction and interference. The usual grating consists of a glass or speculum-metal sheet with a very large number of equidistant parallel lines ruled on it (usually of the order of 1000 per mm). Diffracted light after transmission through the glass or reflection by the speculum produces maxima of illumination (spectral lines) according to the equation $m\lambda = d(\sin i + \sin\theta)$, where d is the distance between grating lines, λ is the wavelength of the light, i is the angle of incidence, θ the direction of the diffracted maximum, and m is the 'order' of the spectral line. Reflection gratings are also used to produce spectra in the ultraviolet region of the electromagnetic spectrum.

diffusion 1. The process by which different substances mix as a result of the random motions of their component atoms, molecules, and ions. In gases, all the components are perfectly miscible with each other and mixing ultimately becomes nearly uniform, though slightly affected by gravity (*see also* GRAHAM'S LAW). The diffusion of a solute through a solvent to produce a solution of uniform concentration is slower, but otherwise very similar to the process of gaseous diffusion. In solids, however, diffusion occurs very slowly at normal temperatures. **2.** The scattering of a beam of light by reflection at a rough surface or by transmission through a translucent (rather than transparent) medium, such as frosted glass. **3.** The passage of elementary particles through matter when there is a high probability of scattering and a low probability of capture.

diffusion cloud chamber *See* CLOUD CHAMBER.

diffusion limited aggregation (DLA) A process of aggregation dominated by particles diffusing and having a nonzero probability of sticking together irreversibly when they touch. The clusters formed by DLA are *fractal in type.

diffusion pump (condensation pump) A *vacuum pump in which oil or mercury vapour is diffused through a jet, which entrains the gas molecules from the container in which the pressure is to be reduced. The diffused vapour and entrained gas molecules are condensed on the cooled walls of the pump. Pressures down to 10^{-7} Pa can be reached by sophisticated forms of the diffusion pump.

digit A symbol used to represent a single number. For example, the number 479 consists of three digits.

digital audio tape *See* DAT.

digital camera A form of camera in which film is replaced by a semiconductor CCD array, which records the picture and stores it within the camera in a (usually) replaceable memory module. Often a viewing screen is built in allowing immediate reviewing of the pictures stored. Pictures can be transferred to a computer for later viewing, editing, and printing.

digital computer *See* COMPUTER.

digital display A method of indicating a reading of a measuring instrument, clock, etc., in which the appropriate numbers are generated on a fixed display unit by the varying parameter being measured rather than fixed numbers on a scale being indicated by a moving pointer or hand. *See* DIGITRON; LIGHT-EMITTING DIODE; LIQUID-CRYSTAL DISPLAY.

digital recording A method of recording or transmitting sound in which the sound itself is not transmitted or recorded. Instead the pressure in the sound wave is sampled at least 30 000 times per second and the successive values represented by numbers, which are then transmitted or recorded. Afterwards they are restored to analogue form in the receiver or player. This method is used for very high fidelity recordings as no distortion or interference occurs during transmission or in the recording process.

digitron An electronic gas-discharge tube that provides a *digital display in calculators, counters, etc. It usually has 10 cold cathodes shaped into the form of the digits 0–9. The cathode selected receives a voltage pulse causing a glow discharge to illuminate the digit. It has now largely been superseded by *light-emitting diodes and *liquid-crystal displays.

dihedral (dihedron) An angle formed by the intersection of two planes (e.g. two faces of a polyhedron). The **dihedral angle** is the angle formed by taking a point on the line of intersection and drawing two lines from this point, one in each plane, perpendicular to the line of intersection.

dilatancy *See* NEWTONIAN FLUID.

dilation (dilatation) 1. An increase in volume. **2.** *See* TIME DILATION.

dilatometer A device for measuring the cubic *expansivities of liquids. It consists of a bulb of known volume joined to a graduated capillary tube, which is closed at the top to prevent evaporation. A known mass of liquid is introduced into the device, which is submerged in a bath maintained at different temperatures t_1 and t_2. The two volumes corresponding to these temperatures, V_1 and V_2, are read off the calibrated stem. The value of the cubic expansivity (γ) is then given by

$$\gamma = (V_2 - V_1)/V_1(t_2 - t_1).$$

dimensional analysis A method of checking an equation or a solution to a problem by analysing the dimensions in which it is expressed. It is also useful for establishing the form, but not the numerical coefficients, of an empirical relationship. If the two sides of an equation do not have the same dimensions, the equation is wrong. If they do have the same dimensions, the equation may still be wrong, but the error is likely to be in the arithmetic rather than the method of solution.

dimensions The product or quotient of the basic physical quantities, raised to the appropriate powers, in a derived physical quantity. The basic physical quantities of a mechanical system are usually taken to be mass (M), length (L), and time (T). Using these dimensions, the derived physical quantity velocity will have the dimensions L/T and acceleration will have the dimensions L/T^2. As force is the product of a mass and an acceleration (*see* NEWTON'S LAW OF MOTION), force has the dimensions MLT^{-2}. In electrical work in *SI units, current, I, can be regarded as dimensionally independent and the dimensions of other electrical units can be found from standard relationships. Charge, for example, is measured as the product of current and time. It therefore has the dimension IT. Potential difference V is related to the current I and the power P by the relationship $P = VI$, where P is power. As power is force × distance ÷ time ($MLT^{-2} \times L \times T^{-1} = ML^2T^3$), voltage V is given by $V = ML^2T^{-3}I^{-1}$.

diode An electronic device with two electrodes. In the obsolescent **thermionic diode** a heated cathode emits electrons, which flow across the intervening vacuum to the anode when a positive potential is applied to it. The device permits flow of current in one direction only as a negative potential applied to the anode repels the electrons. This property of diodes was made use of in the first thermionic radios, in which the diode was used to demodulate the transmitted signal (*see* MODULATION). In the **semiconductor diode**, a p–n junction performs a similar function. The forward current increases with increasing potential difference whereas the reverse current is very small indeed. *See* SEMICONDUCTOR; TRANSISTOR.

dioptre A unit for expressing the power of a lens or mirror equal to the reciprocal of its focal length in metres. Thus a lens with a focal length of 0.5 metre has a power of 1/0.5 = 2 dioptres. The power of a converging lens is usually taken to be positive and that of a diverging lens negative. Because the power of a lens is a measure of its ability to cause a beam to converge, the dioptre is now sometimes called the radian per metre.

dip *See* GEOMAGNETISM.

dipole 1. Two equal and opposite charges that are separated by a distance. The **dipole moment** is the product of either charge and the distance between them. Some molecules behave as dipoles and measurement of the dipole moments can often provide information regarding the configuration of the molecule. **2.** An aerial commonly used for frequencies below 30 megahertz. It consists of a horizontal rod, fed or tapped at its centre. It may be half a wavelength or a full wavelength long.

dipole–dipole interaction The interaction of two systems, such as atoms or molecules, by their *dipole moments. The energy of dipole–dipole interaction depends on the relative orientation and the strength of the dipoles and how far apart they are. A water molecule has a permanent dipole moment, thus causing a dipole–dipole interaction if two water molecules are near each other. Although isolated atoms do not have permanent dipole moments, a dipole moment can be induced by the presence of another atom near it, thus leading to **induced dipole–dipole interactions**. Dipole–dipole interactions are responsible for *van der Waals' forces and *surface tension in liquids.

dipole radiation *See* FORBIDDEN TRANSITIONS.

Dirac, Paul Adrien Maurice (1902–84) British physicist, who shared the 1933 Nobel Prize with Erwin *Schrödinger for developing Schrödinger's non-relativistic wave equations to take account of relativity (*see* DIRAC EQUATION). This modified equation predicted the existence and properties of the *positron. Dirac also invented, independently of Enrico Fermi, the form of *quantum statistics known as Fermi–Dirac statistics.

Dirac constant *See* PLANCK CONSTANT.

Dirac equation An equation derived by P. A. M. Dirac, which can be regarded as an interpretation of the *Schrödinger equation that takes account of relativity. It can be written in the form:

$$i\alpha\nabla\psi + (mc/\hbar)\,\beta\psi = (i/c)\delta\psi/\delta t,$$

where m is the mass of a free particle, c the speed of light, t the time, and $\hbar$ is the Dirac constant. The wave function is ψ and α and β are square matrices; i is $\sqrt{-1}$. The Dirac equation, unlike the *Klein–Gordon equation, can be used with spin-½ particles, such as the electron; it also predicts the existence of *antiparticles.

direct current (d.c.) An electric current in which the net flow of charge is in one direction only. *Compare* ALTERNATING CURRENT.

direct-current motor *See* ELECTRIC MOTOR.

direct motion **1.** The apparent motion of a planet from west to east as seen from the earth against the background of the stars. **2.** The anticlockwise rotation of a planet, as seen from its north pole. *Compare* RETROGRADE MOTION.

directrix **1.** A plane curve defining the base of a *cone. **2.** A straight line from which the distance to any point on a *conic is in a constant ratio to the distance from that point to the focus.

direct transition A transition between two electronic energy levels in a molecule that proceeds directly between the two levels. By contrast, an **indirect transition** between two electronic energy levels in a molecule involves either changes in vibrational energy levels and/or energy level crossing in potential energy curves. Simple electronic spectra of molecules are associated with direct transitions while phenomena such as fluorescence and phosphorescence are associated with indirect transitions.

discharge 1. The conversion of the chemical energy stored in a *secondary cell into electrical energy. **2.** The release of electric charge from a capacitor in an external circuit. **3.** The passage of charge carriers through a gas at low pressure in a **discharge tube**. A potential difference applied between cathode and anode creates an electric field that accelerates any free electrons and ions to their appropriate electrodes. Collisions between electrons and gas molecules create more ions. Collisions also produce excited ions and molecules (*see* EXCITATION), which decay with emission of light in certain parts of the tube.

discontinuous function *See* CONTINUOUS FUNCTION.

disintegration Any process in which an atomic nucleus breaks up spontaneously into two or more fragments in a radioactive decay process or breaks up as a result of a collision with a high-energy particle or nuclear fragment.

disintegration constant *See* DECAY.

diskette *See* FLOPPY DISK.

dislocation *See* DEFECT.

disordered solid A material that neither has the structure of a perfect *crystal lattice nor of a crystal lattice with isolated *defects. In a **random alloy**, one type of disordered solid, the order of the different types of atom occurs at random. Another type of disordered solid is formed by introducing a high concentration of defects, with the defects distributed randomly throughout the solid. In an *amorphous solid, such as glass, there is a random network of atoms with no lattice.

disperse phase *See* COLLOIDS.

dispersion The splitting up of a ray of light of mixed wavelengths by refraction into its components. Dispersion occurs because the *deviation for each wavelength is different on account of the different speeds at which waves of different wavelengths pass through the refracting medium. If a ray of white light strikes one face of a prism and passes out of another face, the white light will be split into its components and the full visible spectrum will be formed. The **dispersive power** of a prism (or other medium) for white light is defined by

$$(n_b - n_r)/(n_y - 1),$$

where n_b, n_r, and n_y are the *refractive indices for blue, red, and yellow light respectively. The term is sometimes applied to the separation of wavelengths produced by a *diffraction grating.

dispersion forces *See* VAN DER WAALS' FORCE.

dispersive power *See* DISPERSION.

displacement **1.** Symbol *s*. A specified distance in a specified direction. It is the vector equivalent of the scalar distance. **2.** *See* ELECTRIC DISPLACEMENT.

distortion The extent to which a system fails to reproduce the characteristics of its input in its output. It is most commonly applied to electronic amplifiers and to optical systems. *See* ABERRATION.

distributive law The mathematical law stating that one operation is independent of being carried out before or after another operation. For example, multiplication is distributive with respect to addition and subtraction, i.e. $x(y + z) = xy + xz$. *Compare* ASSOCIATIVE LAW; COMMUTATIVE LAW.

diurnal Daily; denoting an event that happens once every 24 hours.

divergence (div) The *scalar product of the *gradient operator ∇ with a vector. For a vector u that has components u_1, u_2, and u_3 in the x, y, and z directions, and is a function of x, y, and z, the divergence is given by:

$$\text{div}\,u = \nabla.u = \partial u_1/\partial x + \partial u_2/\partial y + \partial u_3/\partial z.$$

The divergence of a vector at a given point represents the flux of the vector per unit volume in the neighbourhood of that point. *See also* CURL; LAPLACE EQUATION.

divergence theorem A theorem that gives the relation between the total flux of a vector F out of a surface S, which surrounds the volume V, to the vector inside the volume. The divergence theorem states that

$$\int_V \text{div}\,F\,dV = \int_S F \cdot dS.$$

The divergence theorem is also known as **Gauss' theorem** and **Ostrogradsky's theorem** (named after the Russian mathematician Michel Ostrogradsky (1801–61), who stated it in 1831). *Gauss' law for electric fields is a particular case of the divergence theorem.

divergent series *See* CONVERGENT SERIES.

diverging lens or mirror A lens or mirror that can refract or reflect a parallel beam of light into a diverging beam. A diverging lens is predominantly concave; a diverging mirror is convex. *Compare* CONVERGING LENS OR MIRROR.

DLA *See* DIFFUSION LIMITED AGGREGATION.

***dl*-form** *See* OPTICAL ACTIVITY; RACEMIC MIXTURE.

D-lines Two close lines in the yellow region of the visible spectrum of sodium, having wavelengths 589.0 and 589.6 nm. As they are prominent and easily recognized they are used as a standard in spectroscopy.

domain *See* MAGNETISM.

donor *See* SEMICONDUCTOR.

doping *See* SEMICONDUCTOR.

Doppler cooling *See* LASER COOLING.

Doppler effect The apparent change in the observed frequency of a
wave as a result of relative motion between the source and the observer.
For example, the sound made by a low-flying aircraft as it approaches
appears to fall in pitch as it passes and flies away. In fact, the frequency
of the aircraft engine remains constant but as it is approaching more
sound waves per second impinge on the ear and as it recedes fewer sound
waves per second impinge on the ear. The apparent frequency, F, is given
by

$$F = f(c - u_o)/(c - u_s),$$

where f is the true frequency, c is the speed of sound, and u_o and u_s are the
speeds of the observer and the source, respectively.

Although the example of sound is most commonly experienced, the
effect was suggested by Christian Johann Doppler (1803–53), an Austrian
physicist, as an attempt to explain the coloration of stars. In fact the
Doppler effect cannot be observed visually in relation to the stars,
although the effect does occur with electromagnetic radiation and the
*redshift of light from receding stars can be observed spectroscopically.
The Doppler effect is also used in radar to distinguish between stationary
and moving targets and to provide information regarding the speed of
moving targets by measuring the frequency shift between the emitted
and reflected radiation.

For electromagnetic radiation, the speed of light, c, features in the
calculation and as there is no fixed medium to provide a frame of
reference, relativity has to be taken into account, so that

$$F = f\sqrt{[(1 - v/c)/(1 + v/c)]},$$

where v is the speed at which source and observer are moving apart. If
v^2/c^2 is small compared to 1, i.e. if the speed of separation is small
compared to the speed of light, this equation simplifies to

$$F = f(1 - v/c).$$

***d*-orbital** *See* ORBITAL.

dose A measure of the extent to which matter has been exposed to
*ionizing radiation. The **absorbed dose** is the energy per unit mass
absorbed by matter as a result of such exposure. The SI unit is the gray,
although it is often measured in rads (1 rad = 0.01 gray; *see* RADIATION
UNITS). The **maximum permissible dose** is the recommended upper limit
of absorbed dose that a person or organ should receive in a specified
period according to the International Commission on Radiological
Protection. *See also* LINEAR ENERGY TRANSFER.

dosimeter Any device used to measure absorbed *dose of ionizing radiation. Methods used include the *ionization chamber, photographic film, or the rate at which certain chemical reactions occur in the presence of ionizing radiation.

dot product *See* SCALAR PRODUCT.

double refraction The property, possessed by certain crystals (notably calcite), of forming two refracted rays from a single incident ray. The **ordinary ray** obeys the normal laws of refraction. The other refracted ray, called the **extraordinary ray**, follows different laws. The light in the ordinary ray is polarized at right angles to the light in the extraordinary ray. Along an *optic axis the ordinary and extraordinary rays travel with the same speed. Some crystals, such as calcite, quartz, and tourmaline, have only one optic axis; they are **uniaxial crystals**. Others, such as mica and selenite, have two optic axes; they are **biaxial crystals**. The phenomenon is also known as **birefringence** and the double-refracting crystal as a **birefringent crystal**. *See also* POLARIZATION.

doublet 1. A pair of optical lenses of different shapes and made of different materials used together so that the chromatic aberration produced by one is largely cancelled by the reverse aberration of the other. **2.** A pair of associated lines in certain spectra, e.g. the two lines that make up the sodium D line.

drain *See* TRANSISTOR.

drift-tube accelerator *See* LINEAR ACCELERATOR.

dripline One of the limits on a chart in which the number of protons in a nucleus is plotted against the number of neutrons that indicate where nuclei can exist. There are two driplines: the **proton dripline** and the **neutron dripline**. The proton dripline indicates the maximum number of protons a nucleus can have, with the neutron dripline indicating the maximum number of neutrons a nucleus can have. **Dripline nuclei** are nuclei that are close to these limits.

dry cell A primary or secondary cell in which the electrolytes are in the form of a paste. Many torch, radio, and calculator batteries are *Leclanché cells in which the electrolyte is an ammonium chloride paste and the container is the negative zinc electrode (with an outer plastic wrapping).

dry ice Solid carbon dioxide used as a refrigerant. It is convenient because it sublimes at −78°C (195 K) at standard pressure rather than melting.

duality An object or system in which two separate aspects can be isolated. For example, in *quantum mechanics, the particle and wave nature of both light and electrons is a duality (*see* COMPLEMENTARITY). In the theory of *phase transitions, in certain *model systems there is a duality in which quantities in the low-temperature phase are related to quantities in the high-temperature phase to give information about the *transition point. In projective geometry (the type of geometry associated with perspective), there is duality between straight lines and points. Duality is also a key concept in *supersymmetry, particularly in *superstring theory.

dubnium Symbol Db. A radioactive *transactinide element; a.n. 105. It was first reported in 1967 by a group at Dubna near Moscow and was confirmed in 1970 at Dubna and at Berkeley, California. It can be made by bombarding californium–249 nuclei with nitrogen–15 nuclei. Only a few atoms have ever been made.

ductility The ability of certain metals, such as copper, to retain their strength when their shape is changed, especially the ability of such metals to be drawn into a thin wire without cracking or breaking.

Dulong and Petit's law For a solid element the product of the relative atomic mass and the specific heat capacity is a constant equal to about $25 \, J \, mol^{-1} \, K^{-1}$. Formulated in these terms in 1819 by the French scientists Pierre Dulong (1785–1838) and A. T. Petit (1791–1820), the law in modern terms states: the molar heat capacity of a solid element is approximately equal to $3R$, where R is the *gas constant. The law is only approximate but applies with fair accuracy at normal temperatures to elements with a simple crystal structure.

dust core *See* CORE.

DVD Digital versatile disk: a disk format similar to a compact disk (*see* CD-ROM) but containing much more data. It was introduced in 1996. DVD disks are the same 120 mm diameter as CDs with potential capacities of up to 4.7 gigabytes for a single-sided single-layer disk. The technology involved in DVD storage is similar to that in compact disks, but more precise. The extra capacity is achieved in a number of ways. The tracks on a DVD are closer together and the pits are smaller, allowing more pits per unit area. The key to this is the use of a shorter wavelength laser (typically 635 or 650 nm in the red region for DVDs as opposed to 780 nm in the infrared for CDs). Moreover, a DVD can have two layers on the same side of the disk. The top layer is translucent and the bottom layer opaque. Data can be read from either layer by refocusing the laser. In addition DVDs may be double-sided. DVD formats also have a more efficient error-correction system. The potential capacity of a double-sided double-layer DVD is up to 17 gigabytes. DVDs have been increasingly used in computing as a higher-capacity version of compact disks. As with compact disks, there are various types. **DVD-ROM** (DVD read-only memory) is similar to CD-ROM. **DVD-R** (DVD-recordable) is similar to CD-R. There are also different rewritable formats: **DVD-RAM**, **DVD+RW**, and **DVD-RW**.

dwarf star A star, such as the sun, that lies on the main sequence in a *Hertzsprung–Russell diagram. *See also* WHITE DWARF.

dynamical system A system governed by *dynamics (either classical mechanics or quantum mechanics). The evolution of dynamical systems can be very complex, even for systems with only a few *degrees of freedom, sometimes involving such considerations as *ergodicity, which originated in *statistical mechanics. The evolution of dynamical systems can be studied using the *phase space for the system. *Chaos is an example of the complex behaviour that can occur in a dynamical system. The global behaviour of the evolution of dynamical systems in phase space can be analysed using *topology, a notable result of this approach being the *KAM theorem.

dynamic equilibrium *See* EQUILIBRIUM.

dynamics The branch of mechanics concerned with the motion of bodies under the action of forces. Time intervals, distances, and masses are regarded as fundamental and bodies are assumed to possess *inertia. Bodies in motion have an attribute called *momentum (*see* NEWTON'S LAWS OF MOTION), which can only be changed by the application of a force. *Compare* KINETICS; STATICS.

dynamo An electric *generator, especially one designed to provide *direct current. Alternating-current generators can be called dynamos but are more often called alternators.

dynamo action The generation of electrical current and magnetic field by the motion of an electrically conducting fluid. It is generally believed that the magnetic fields of the earth and the sun are produced by dynamo action in the molten iron–nickel core of the earth and in the plasma of the solar interior respectively.

dynamometer 1. An instrument used to measure a force, often a spring balance. **2.** A device used to measure the output power of an engine or motor. **3. (current dynamometer)** A variety of *current balance, for measuring electric current.

dyne The c.g.s unit of force; the force required to give a mass of one gram an acceleration of 1 cm s^{-2}. 1 dyne = 10^{-5} newton.

dystectic mixture A mixture of substances that has a constant maximum melting point.

e The irrational number defined as the limit as n tends to infinity of $(1 + 1/n)^n$. It has the value 2.718 28.... It is used as the base of natural *logarithms and occurs in the *exponential function, e^x.

early universe The study of *cosmology at the time very soon after the *big bang. Theories of the early universe have led to a mutually beneficial interaction between cosmology and the theory of *elementary particles, particularly *grand unified theories.

Because there were very high temperatures in the early universe many of the *broken symmetries in *gauge theories were unbroken symmetries at these temperatures. As the universe cools after the big bang there is thought to be a sequence of transitions to broken symmetry states.

Combining cosmology with grand unified theories helps to explain why the observed universe appears to consist of matter with no antimatter (*see* ANTIPARTICLES). This means that one has a nonzero *baryon number for the universe. This solution relies on the fact that there were nonequilibrium conditions in the early universe due to its rapid expansion after the big bang.

An important idea in the theory of the early universe is that of **inflation** – the idea that the nature of the *vacuum state gave rise, after the big bang, to an exponential expansion of the universe. The hypothesis of the **inflationary universe** solves several long-standing problems in cosmology, such as the flatness and homogeneity of the universe. In particular, it is thought that quantum fluctuations in the early universe were responsible for the emergence of large-scale structure in the universe, such as galaxies.

Earnshaw's theorem A fundamental result concerning bodies that interact according to inverse square laws, such as *Coulomb's law of electrostatics and *Newton's law of gravitation. It states that a system consisting of such bodies cannot be in stable static equilibrium. The Reverend Samuel Earnshaw (1805–1888) proved this result in 1842. Earnshaw's theorem means that neither a planetary system nor an atom can be composed of static bodies.

earth The planet that orbits the sun between the planets Venus and Mars at a mean distance from the sun of 149 600 000 km. It has a mass of about 5.976×10^{24} kg and an equatorial diameter of 12 756 km. The earth consists of three layers: the gaseous atmosphere (*see* EARTH'S ATMOSPHERE), the liquid hydrosphere, and the solid lithosphere. The solid part of the earth also consists of three layers: the **crust** with a mean thickness of about 32 km under the land and 10 km under the seas; the **mantle**, which extends some 2900 km below the crust; and the **core**, part of which is believed to be liquid. The crust has a relative density of about 3 and consists largely of sedimentary rocks overlaying igneous rocks. The composition of the crust is: oxygen 47%, silicon 28%, aluminium 8%, iron 4.5%, calcium 3.5%, sodium and potassium 2.5% each, and magnesium 2.2%. Hydrogen, carbon, phosphorus, and sulphur are all present to an extent of less than 1%. The mantle reaches a relative density of about 5.5 at its maximum depth and is believed to consist mainly of silicate rocks. The core is believed to have a maximum relative density of 13 and a maximum temperature of 6400 K. *See also* GEOMAGNETISM; GEOPHYSICS.

earthquake A sudden movement or fracturing within the earth's lithosphere, causing a series of shocks. This may range from a mild tremor to a large-scale earth movement causing extensive damage over a wide area. The point at which the earthquake originates is known as the **seismic focus**; the point on the earth's surface directly above this is the **epicentre** (or **hypocentre**). *See* SEISMIC WAVES. Earthquakes result from a build-up of stresses within the rocks until they are strained to the point beyond which they will fracture. They occur in narrow continuous belts of activity, which correspond with the junction of lithospheric plates, including the circum-Pacific belt, the Alpine–Himalayan belt, and mid-ocean ridges. The scale of the shock of an earthquake is known as the magnitude; the most commonly used scale for comparing the magnitude of earthquakes is the logarithmic *Richter scale (9.5 is the highest recorded magnitude on the scale).

The mathematical analysis of the geophysics responsible for earthquakes makes it clear that it is very difficult to predict when an earthquake will occur.

earth's atmosphere The gas that surrounds the earth. The composition of dry air at sea level is: nitrogen 78.08%, oxygen 20.95%, argon 0.93%, carbon dioxide 0.03%, neon 0.0018%, helium 0.0005%, krypton 0.0001%, and xenon 0.00001%. In addition to water vapour, air in some localities contains sulphur compounds, hydrogen peroxide, hydrocarbons, and dust particles.

The lowest level of the atmosphere, in which most of the weather occurs, is called the **troposphere**. Its thickness varies from about 7 km at the poles to 28 km at the equator and in this layer temperature falls with increasing height. The next layer is the **stratosphere**, which goes up to about 50 km. Here the temperature remains approximately constant. Above this is the **ionosphere**, which extends to about 1000 km, with the temperature rising and the composition changing substantially. At about 100 km and above most of the oxygen has dissociated into atoms; at above 150 km the percentage of nitrogen has dropped to nil. In the ionosphere the gases are ionized by the absorption of solar radiation. This enables radio transmissions to be made round the curved surface of the earth as the ionized gas acts as a reflector for certain wavelengths. The ionosphere is divided into three layers. The D-layer (50–90 km) contains a low

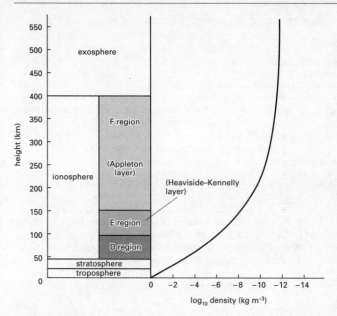

Earth's atmosphere

concentration of free electrons and reflects low-frequency radio waves. The E-layer (90–150 km) is also called the **Heaviside layer** or **Heaviside–Kennelly layer** as its existence was predicted independently by Oliver Heaviside (1850–1925) and Arthur E. Kennelly (1861–1939). This layer reflects medium-frequency waves. The F-layer (150–1000 km) is also called the **Appleton layer** after its discoverer Sir Edward Appleton (1892–1965). It has the highest concentration of free electrons and is the most useful for radio transmission. Wavelengths between 8 mm and 20 m are not reflected by the ionosphere but escape into space. Therefore television transmissions, which utilize this range, require artificial *satellites for reflection (or reception, amplification, and retransmission). From about 400 km, the outermost region of the atmosphere is also called the **exosphere**. See illustration.

earthshine Sunlight reflected from the surface of the earth. An observer in space may see nearby objects dimly illuminated by earthshine, as things on earth may be illuminated by moonlight. Under certain conditions near new moon the dark disc of the moon can be seen faintly illuminated by earthshine – a phenomenon called 'the old moon in the new moon's arms'.

earth's magnetic field *See* GEOMAGNETISM.

ebullioscopic constant *See* ELEVATION OF BOILING POINT.

eccentricity *See* CONIC.

ECG *See* ELECTROCARDIOGRAM.

echelon A form of *interferometer consisting of a stack of glass plates arranged stepwise with a constant offset. It gives a high resolution and is used in spectroscopy to study hyperfine line structure. In the **transmission echelon** the plates are made equal in optical thickness to introduce a constant delay between adjacent parts of the wavefront. The **reflecting echelon** has the exposed steps metallized and acts like an exaggerated *diffraction grating.

echo The reflection of a wave by a surface or object so that a weaker version of it is detected shortly after the original. The delay between the two is an indication of the distance of the reflecting surface. An **echo sounder** is an apparatus for determining the depth of water under a ship. The ship sends out a sound wave and measures the time taken for the echo to return after reflection by the sea bottom. **Sonar** (*so*und *na*vigation *ra*nging) is a technique for locating underwater objects by a similar method. Echoes also occur with radio waves; reflection of waves causes an echo in radio transmission and ghosts in television pictures. *See also* RADAR.

echolocation 1. *See* RADAR; ECHO. **2.** A method used by some animals (such as bats, dolphins, and certain birds) to detect objects in the dark. The animal emits a series of high-pitched sounds that echo back from the object and are detected by the ear or some other sensory receptor. From the direction of the echo and from the time between emission and reception of the sounds the object is located, often very accurately.

ECL *See* EMITTER-COUPLED LOGIC.

eclipse The total (**total eclipse**) or partial (**partial eclipse**) obscuring of

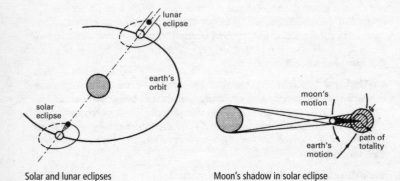

Solar and lunar eclipses   Moon's shadow in solar eclipse

light from a celestial body as it passes behind or through the shadow of another body. A **lunar eclipse** occurs when the sun, earth, and moon are in a straight line and the shadow of the earth falls on the moon. A **solar eclipse** occurs when the shadow of the moon falls on the earth. See illustrations.

ecliptic The *great circle in which the plane of the earth's orbit round the sun intersects the *celestial sphere. It is thus the sun's apparent annual path across the sky.

Eddington limit A limit for the maximum value of the brightness of a star of a given mass. This limit exists because the radiation pressure caused by the nuclear fusion reactions powering the star has to counter, but not exceed, the gravitational force that would cause gravitational collapse of the star. The existence of this limit was first pointed out by the British astrophysicist Sir Arthur Stanley Eddington (1882–1944).

eddy current A current induced in a conductor situated in a changing magnetic field or moving in a fixed one. Any imagined circuit within the conductor will change its magnetic flux linkage, and the consequent induced e.m.f. will drive current around the circuit. In a substantial block of metal the resistance will be small and the current therefore large. Eddy currents occur in the cores of transformers and other electrical machines and represent a loss of useful energy (the **eddy-current loss**). To reduce this loss to a minimum metal cores are made of insulated sheets of metal, the resistance between these laminations reducing the current. In high-frequency circuits *ferrite cores can be used. Eddy currents in a moving conductor interact with the magnetic field producing them to retard the motion of the conductor. This enables some electrical instruments (moving-coil type) to utilize eddy currents to create damping and hence a stable reading in a short time. Eddy currents are also used in *induction heating.

Edison cell *See* NICKEL–IRON ACCUMULATOR.

EEG *See* ELECTROENCEPHALOGRAM.

effective temperature *See* LUMINOSITY.

effective value *See* ROOT-MEAN-SQUARE VALUE.

efficiency A measure of the performance of a machine, engine, etc., being the ratio of the energy or power it delivers to the energy or power fed to it. In general, the efficiency of a machine varies with the conditions under which it operates and there is usually a load at which it operates with the highest efficiency. The **thermal efficiency** of a heat engine is the ratio of the work done by the engine to the heat supplied by the fuel. For a reversible heat engine (*see* REVERSIBLE PROCESS) this efficiency is equal to $(T_1 - T_2)/T_1$, where T_1 is the thermodynamic temperature at which all the heat is taken up and T_2 is the thermodynamic temperature at which it is given out (*see* CARNOT CYCLE). For real engines it is always less than this.

effusion The flow of a gas through a small aperture. The relative rates at which gases effuse, under the same conditions, is approximately inversely proportional to the square roots of their densities.

eigenfunction An allowed *wave function of a system in quantum mechanics. The associated energies are **eigenvalues**.

Einstein, Albert (1879–1955) German-born US physicist, who took Swiss nationality in 1901. A year later he went to work in the Bern patent office. In 1905 he published four enormously influential papers, one on *Brownian movement, one on the *photoelectric effect, one on the special theory of *relativity, and one on energy and inertia (which included the famous result $E = mc^2$). In 1915 he published the general theory of relativity, concerned mainly with gravitation. He made several key contributions to the early development of quantum mechanics, including the quantum theory of radiation and the prediction of *Bose–Einstein condensation. In 1921 he was awarded the Nobel Prize. In 1933, as a Jew, Einstein decided to remain in the USA (where he was lecturing), as Hitler had come to power. For the remainder of his life he sought a unified field theory. In 1939 he informed President Roosevelt that an atom bomb was feasible and that Germany might be able to make one.

Einstein coefficients Coefficients used in the *quantum theory of radiation, related to the probability of a transition occurring between the ground state and an excited state (or vice versa) in the processes of *induced emission and *spontaneous emission. For an atom exposed to *electromagnetic radiation, the rate of absorption R_a is given by $R_a = B\rho$, where ρ is the density of electromagnetic radiation and B is the **Einstein B coefficient** associated with absorption. The rate of induced emission is also given by $B\rho$, with the coefficient B of induced emission being equal to the coefficient of absorption. The rate of spontaneous emission is given by A, where A is the **Einstein A coefficient of spontaneous emission**. The A and B coefficients are related by $A = 8\pi h v^3 B/c^3$, where h is the *Planck constant, v is the frequency of electromagnetic radiation, and c is the speed of light. The coefficients were put forward by Albert Einstein in 1916–17 in his analysis of the quantum theory of radiation.

Einstein equation 1. The mass–energy relationship announced by Einstein in 1905 in the form $E = mc^2$, where E is a quantity of energy, m its mass, and c is the speed of light. It presents the concept that energy possesses mass. *See also* RELATIVITY. **2.** The relationship $E_{max} = hf - W$, where E_{max} is the maximum kinetic energy of electrons emitted in the photoemissive effect, h is the Planck constant, f the frequency of the incident radiation, and W the *work function of the emitter. This is also written $E_{max} = hf - \phi e$, where e is the electronic charge and ϕ a potential difference, also called the work function. (Sometimes W and ϕ are distinguished as **work function energy** and **work function potential**.) The equation can also be applied to photoemission from gases, when it has the form: $E = hf - I$, where I is the ionization potential of the gas.

einsteinium Symbol Es. A radioactive metallic transuranic element belonging to the actinoids; a.n. 99; mass number of the most stable isotope 254 (half-life 270 days). Eleven isotopes are known. The element was first identified by A. Ghiorso and associates in debris from the first hydrogen bomb explosion in 1952. Microgram quantities of the element did not become available until 1961.

Einstein–Podolsky–Rosen experiment *See* EPR EXPERIMENT.

Einstein shift *See* REDSHIFT.

Einstein theory of specific heat A theory of the specific heat of solids proposed by Albert Einstein in 1906. In this theory, Einstein attributed the specific heat of solids to the vibrations of the solid and made the simplifying assumption that all the vibrations have the same frequency. This theory was partially successful since it was able to derive *Dulong and Petit's law at high temperatures and showed that the specific heat capacity goes to zero as the absolute temperature also goes to zero. A better description of the specific heat of solids was given by the more realistic *Debye theory of specific heat.

elastance The reciprocal of *capacitance. It is measured in farad^{-1} (sometimes called a 'daraf').

elastic collision A collision in which the total kinetic energy of the colliding bodies after collision is equal to their total kinetic energy before collision. Elastic collisions occur only if there is no conversion of kinetic energy into other forms, as in the collision of atoms. In the case of macroscopic bodies this will not be the case as some of the energy will become heat. In a collision between polyatomic molecules, some kinetic energy may be converted into vibrational and rotational energy of the molecules, but otherwise molecular collisions appear to be elastic.

elasticity The property of certain materials that enables them to return to their original dimensions after an applied *stress has been removed. In general, if a stress is applied to a wire, the *strain will increase in proportion (see OA on the illustration) until a certain point called the **limit of proportionality** is reached. This is in accordance with *Hooke's law. Thereafter there is at first a slight increase in strain with increased load until a point L is reached. This is the **elastic limit**; up to this point the deformation of the specimen is elastic, i.e. when the stress is removed the specimen returns to its original length. Beyond the point L there is permanent deformation when the stress is removed, i.e. the material has ceased to be **elastic** and has become **plastic**. In the plastic stages individual materials vary somewhat; in general, however, at a point B there is a sudden increase in strain with further increases of stress – this is the **yield**

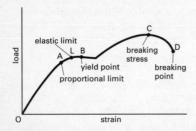

Elasticity

point. Beyond the point C, the **breaking stress**, the wire will snap (which occurs at point D).

elastic modulus The ratio of the *stress applied to a body to the *strain produced. The **Young modulus of elasticity**, named after the British physicist Thomas *Young, refers to longitudinal stress and strain. The **bulk modulus** is the ratio of the pressure on a body to its fractional decrease in volume. The **shear** (or **rigidity**) **modulus** is the tangential force per unit area divided by the angular deformation in radians.

electret A permanently electrified substance or body that has opposite charges at its extremities. They resemble permanent magnets in many ways. An electret can be made by cooling certain waxes in a strong electric field.

electrical conductivity in metals See Feature (pp 144–145).

electrical energy A form of energy related to the position of an electric charge in an electric field. For a body with charge Q and an electric potential V, its electrical energy is QV. If V is a potential difference, the same expression gives the energy transformed when the charge moves through the p.d.

electric arc A luminous discharge between two electrodes. The discharge raises the electrodes to incandescence, the resulting thermal ionization largely providing the carriers to maintain the high current between the electrodes.

electric-arc furnace A furnace used in melting metals to make alloys, especially in steel manufacture, in which the heat source is an electric arc. In the direct-arc furnace, such as the Héroult furnace, an arc is formed between the metal and an electrode. In the indirect-arc furnace, such as the Stassano furnace, the arc is formed between two electrodes and the heat is radiated onto the metal.

electric bell A device in which an electromagnetically operated hammer strikes a bell (see illustration). Pressing the bell-push closes a circuit, causing current to flow from a battery or mains step-down

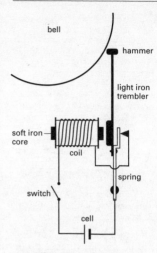

Electric bell

transformer through an electromagnet. The electromagnet attracts a piece of soft iron attached to the hammer, which strikes the bell and at the same time breaks the circuit. The hammer springs back into its original position again, closing the circuit and causing the magnet to attract the soft iron. This process continues until the bell-push is released.

electric charge *See* CHARGE.

electric constant *See* PERMITTIVITY.

electric current *See* CURRENT.

electric displacement (electric flux density) Symbol D. The charge per unit area that would be displaced across a layer of conductor placed across an *electric field. This describes also the charge density on an extended surface that could be causing the field.

electric field A region in which an electric charge experiences a force usually because of a distribution of other charges. The **electric field strength** or **electric intensity** (E) at any point in an electric field is defined as the force per unit charge experienced by a small charge placed at that point. This is equivalent to a potential gradient along the field and is measured in volts per metre. The strength of the field can alternatively be described by its *electric displacement D. The ratio D/E for measurements in a vacuum is the electric constant ε_0. In a substance the observed potential gradient is reduced by electron movement so that D/E appears to increase: the new ratio (ε) is called the *permittivity of the substance. An electric field can be created by an isolated electric charge, in which case the field strength at a distance r from a point charge Q is given by

ELECTRICAL CONDUCTIVITY IN METALS

Elementary conductivity theory is an extension of *free electron theory, in which the electrons in a metal are treated as a gas of negatively charged free particles. In a metallic crystal, the average kinetic energy of electrons is very large compared to the energy associated with thermal vibrations in the lattice. Therefore the thermal vibrations have little or no effect on the *Fermi level, which represents the maximum energy that these electrons can have. The possible energy states that the electrons can occupy are derived from a **particle-in-a-box** analysis of the metal crystal. In this the electron is described as a *matter wave confined within the crystal lattice of the metal, which is identified as the box.

In the absence of an electric field, the electrons are distributed amongst the possible energy states, up to a maximum energy E_{max} (or e_F, the Fermi level). Each of these energy states corresponds to a free electron state of motion at a particular speed and direction. The average velocity of the electrons in the metal will be zero and hence no current will flow.

In an electric field, each electron will experience a force that will accelerate it in the direction of the field. A convenient way of illustrating the electric field is to use **momentum space** diagrams. Each electron state in the particle-in-the-box problem corresponds to a certain momentum possessed by an electron. Imagine a three-dimensional space equipped with a Cartesian coordinate system corresponding to momentum values rather than spatial positions. Each point in this space represents a state of momentum. See diagram (1).

The energy associated with free electrons is restricted to their kinetic energy; therefore the Fermi surface may be represented by the surface of a sphere in the momentum coordinate system, the maximum magnitude of momentum (corresponding to the maximum kinetic energy) being the radius of this sphere.

The momentum space diagram for zero field is therefore a sphere centred at the origin, with a radius P_F, which is the magnitude of the maximum momentum at the Fermi surface. See diagram (2). In an electric field applied in the x-direction, the Fermi surface would be shifted in the direction of the field. However, one might expect that maintaining the field would progressively shift the Fermi surface further along the x-direction, as electrons would be accelerated. This does not happen; after a short time a constant current is established whose value depends on the field. The Fermi surface is shifted on application of the field, but only by an amount proportional to the size of the field. See diagram (3).

This stabilization of the current occurs because electrons in states near the Fermi surface are scattered by two main mechanisms: (1) the thermal vibration of the lattice; and (2) the presence of impurity atoms and other point *defects. The scattered electrons have their momentum changed by the scattering event and take up states on the other side of the Fermi sphere.

A dynamic equilibrium is therefore established between the electrons being accelerated to the right-hand side of the sphere and those scattered back to the left. On average, the electrons travel the **mean free path** (λ) between each scattering event. The velocity v_F at the Fermi surface is very high, ~ 10^6 m/s, compared with any change in velocity produced by the applied field. Therefore, the time between collisions may be considered constant at λ/v_F. This means that the shift in the Fermi surface will be greater for a stronger field; i.e. a greater change in velocity can be achieved for a larger field in the same time λ/v_F. The current will therefore be higher, which is of the basis of *Ohm's law.

An expression for the **resistivity** of the conductor can be obtained from kinetic theory, which neglects scattering by imperfections and lattice vibrations. Such an expression relates the macroscopic to the microscopic and leads to an estimate of the mean free path. For copper at room temperature this expression gives a value for λ ~ 3.0×10^{-8} m. This suggests that electrons in copper can travel up to 100 atomic spacings between collisions. This value of λ is quite surprising as one might expect that a perfect lattice would itself scatter the electrons. However if the lattice is regular, an electron is attracted to the ions as much in one direction as any other. It is only if the lattice is not perfect that the electrons will be scattered.

As lattice vibrations fall with thermodynamic temperature (T), one would expect that their contribution to resistivity (ρ_i) would decrease at low temperatures, eventually becoming zero at 0K. A more detailed calculation shows that at low temperatures $\rho_i \propto T^5$ and changes to a linear dependence on T at higher temperatures. See diagram (4). *See also* BAND THEORY.

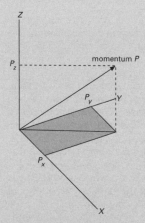

(1) Diagram of momentum space, the point at the end of the arrow corresponds to a state of momentum in an arbitary direction of magnitude P.

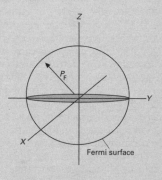

(2) Occupied electron states in momentum space. With no external electric field the sphere is centred at the origin; that is, there is an isotropic distribution of momentum states in all directions.

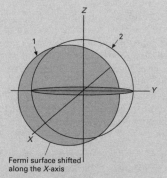

Fermi surface shifted along the X-axis

(3) In an electric field the Fermi surface is shifted. Scattering of electrons occurs at the surface (1) and electrons take up empty states at (2).

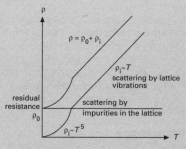

(4)

$E = Q/4\pi r^2 \varepsilon$, where ε is the permittivity of the intervening medium (*see* COULOMB'S LAW). An electric field can also be created by a changing magnetic field.

electric flux Symbol Ψ. In an *electric field, the product of the electric flux density and the relevant area. *See* ELECTRIC DISPLACEMENT.

electric flux density *See* ELECTRIC DISPLACEMENT.

electricity Any effect resulting from the existence of stationary or moving electric charges.

electric lighting Illumination provided by electric currents. The devices used are the **arc lamp**, the **light bulb** (incandescent filament lamp), and the **fluorescent tube**. In the arc lamp, which is no longer used as a general means of illumination, an electric current flows through a gap between two carbon electrodes, between which a high potential difference is maintained. The current is carried by electrons and ions in the vapour produced by the electrodes and a mechanism is required to bring the electrodes closer together as they are vaporized. The device produces a strong white light but has many practical disadvantages. However, arcs enclosed in an inert gas (usually xenon) are increasingly used for such purposes as cinema projectors. The common light bulb is a glass bulb containing a tungsten filament and usually an inert gas. The passage of an electric current through the filament heats it to a white heat. Inert gas is used in the bulb to minimize blackening of the glass by evaporation of tungsten. In the fluorescent tube a glass tube containing mercury vapour (or some other gas) at a low pressure has its inner surface coated with a fluorescent substance. A discharge is created within the tube between two electrodes. Electrons emitted by the cathode collide with gas atoms or molecules and raise them to an excited state (*see* EXCITATION). When they fall back to the *ground state they emit photons of ultraviolet radiation, which is converted to visible light by the coating of phosphor on the inner walls of the tube. In some lamps, such as the *sodium-vapour lamp and *mercury-vapour lamp used in street lighting, no fluorescent substance is used, the light being emitted directly by the excited atoms of sodium or mercury. Vapour lights are more efficient than filament lights as less of the energy is converted into heat. At the present time, a great deal of effort is being devoted to finding materials that are more efficient at producing light than traditional light bulbs, i.e. to create light bulbs in which a higher proportion of electrical energy is converted into light rather than heat.

electric motor A machine for converting electrical energy into mechanical energy. They are quiet, clean, and have a high efficiency (75–95%). They work on the principle that a current passing through a coil within a magnetic field will experience forces that can be used to rotate the coil. In the **induction motor**, alternating current is fed to a stationary coil (the **stator**), which both creates the magnetic field and induces a current in the rotating coil (**rotor**), which it surrounds. The advantage of this kind of motor is that current does not have to be fed through a commutator to a moving part. In the **synchronous motor**, alternating current fed to the stator produces a magnetic field that rotates and locks with the field of the rotor, in this case an independent magnet, causing the rotor to rotate at the same speed as the stator field rotates. The rotor is either a permanent magnet or an electromagnet fed by a direct current through slip rings. In the **universal motor**, current is fed to the stator and, through a commutator, to the rotor. In the series-wound motor the two are in series; in the shunt-wound motor they are in parallel. These motors can be used with either a.c. or d.c. but some small motors use a permanent magnet as the stator and require d.c. for the rotor (via the commutator). *See also* LINEAR MOTOR.

electric polarization *See* DIELECTRIC.

electric potential Symbol V. The energy required to bring unit electric charge from infinity to the point in an electric field at which the potential is being specified. The unit of electric potential is the volt. The **potential difference** (**p.d.**) between two points in an electric field or circuit is the difference in the values of the electric potentials at the two points; in other words, it is the work done in moving unit charge from one point to the other.

electric power The rate of expending energy or doing work in an electrical system. For a direct-current circuit, it is given by the product of the current passing through a system and the potential difference across it. In alternating-current circuits, the power is given by $VI\cos\phi$, where V and I are the RMS values and ϕ is the *phase angle. $\cos\phi$ is called the **power factor** of the circuit.

electric spark The transient passage of an electric current through a gas between two points of high opposite potential, with the emission of light and sound. *Lightning consists of a spark between a cloud and earth or between two oppositely charged parts of the same cloud.

electric susceptibility *See* SUSCEPTIBILITY.

electrocardiogram (**ECG**) A tracing or graph of the electrical activity of the heart. Recordings are made from electrodes fastened over the heart and usually on both arms and a leg. Changes in the normal pattern of an ECG may indicate heart irregularities or disease.

electrochemical cell *See* CELL.

electrochemical equivalent Symbol z. The mass of a given element liberated from a solution of its ions in electrolysis by one coulomb of charge. *See* FARADAY'S LAWS (of electrolysis).

electrochemical series *See* ELECTROMOTIVE SERIES.

electrochemistry The study of chemical properties and reactions involving ions in solution, including electrolysis and electric cells.

electrochromatography *See* ELECTROPHORESIS.

electrode A conductor that emits or collects electrons in a cell, thermionic valve, semiconductor device, etc. The **anode** is the positive electrode and the **cathode** is the negative electrode.

electrodeposition The process of depositing one metal on another by electrolysis, as in *electroforming and *electroplating.

electrode potential The potential difference produced between the electrode and the solution in a *half cell. It is not possible to measure this directly since any measurement involves completing the circuit with the electrolyte, thereby introducing another half cell. **Standard electrode potentials** $E^{\ominus}$ are defined by measuring the potential relative to a standard *hydrogen half cell using 1.0 molar solution at 25°C. The convention is to designate the cell so that the oxidized form is written first. For example,

$$Pt(s)|H_2(g)H^+(aq)|Zn^{2+}(aq)|Zn(s)$$

The e.m.f. of this cell is –0.76 volt (i.e. the zinc electrode is negative). Thus the standard electrode potential of the $Zn^{2+}|Zn$ half cell is –0.76 V. Electrode potentials are also called **reduction potentials**. *See also* ELECTROMOTIVE SERIES.

electrodialysis A method of obtaining pure water from water containing a salt, as in desalination. The water to be purified is fed into a cell containing two electrodes. Between the electrodes is placed an array of semipermeable membranes alternately semipermeable to positive ions and negative ions. The ions tend to segregate between alternate pairs of membranes, leaving pure water in the other gaps between membranes. In this way, the feed water is separated into two streams: one of pure water and the other of more concentrated solution.

electrodynamics The study of electric charges in motion, the forces created by electric and magnetic fields, and the relationship between them. *Compare* ELECTROSTATICS.

electroencephalogram (EEG) A tracing or graph of the electrical activity of the brain. Electrodes taped to the scalp record electrical waves from different parts of the brain. The pattern of an EEG reflects an individual's level of consciousness and can be used to detect such disorders as epilepsy, tumours, or brain damage.

electroforming A method of forming intricate metal articles or parts by *electrodeposition of the metal on a removable conductive mould.

electroluminescence *See* LUMINESCENCE.

electrolysis The production of a chemical reaction by passing an electric current through an electrolyte. In electrolysis, positive ions migrate to the cathode and negative ions to the anode.

electrolyte A liquid that conducts electricity as a result of the presence of positive or negative ions. Electrolytes are molten ionic compounds or solutions containing ions, i.e. solutions of ionic salts or of compounds that ionize in solution. Liquid metals, in which the conduction is by free electrons, are not usually regarded as electrolytes. Solid conductors of ions, as in the sodium–sulphur cell, are also known as electrolytes.

electrolytic capacitor *See* CAPACITOR.

electrolytic cell A cell in which electrolysis occurs; i.e. one in which current is passed through the electrolyte from an external source.

electrolytic corrosion Corrosion that occurs through an electrochemical reaction.

electrolytic gas (detonating gas) The highly explosive gas formed by the electrolysis of water. It consists of two parts hydrogen and one part oxygen by volume.

electrolytic rectifier A *rectifier consisting of two dissimilar electrodes immersed in an electrolyte. By suitable choice of electrodes and electrolyte the cell can be made to pass current easily in one direction but hardly at all in the other. Examples include a lead–aluminium cell with ammonium phosphate(V) electrolyte and a tantalum–lead cell with sulphuric acid as the electrolyte.

electrolytic refining The purification of metals by electrolysis. It is commonly applied to copper. A large piece of impure copper is used as the anode with a thin strip of pure copper as the cathode. Copper(II) sulphate solution is the electrolyte. Copper dissolves at the anode: $Cu \rightarrow Cu^{2+} + 2e$, and is deposited at the cathode. The net result is transfer of pure copper from anode to cathode. Gold and silver in the impure copper form a so-called **anode sludge** at the bottom of the cell, which is recovered.

electrolytic separation A method of separating isotopes by exploiting the different rates at which they are released in electrolysis. It was formerly used for separating deuterium and hydrogen. On electrolysis of water, hydrogen is formed at the cathode more readily than deuterium, thus the water becomes enriched with deuterium oxide.

electromagnet A magnet consisting of a soft ferromagnetic core with a coil of insulated wire wound round it. When a current flows through the wire the core becomes magnetized; when the current ceases to flow the core loses its magnetization. Electromagnets are used in switches, solenoids, electric bells, metal-lifting cranes, and many other applications.

electromagnetic induction The production of an electromotive force in a conductor when there is a change of magnetic flux linkage with the conductor or when there is relative motion of the conductor across a magnetic field. The magnitude of the e.m.f. is proportional (and in modern systems of units equal) to the rate of change of the flux linkage or the rate of cutting flux $d\Phi/dt$; the sense of the induced e.m.f. is such that any induced current opposes the change causing the induction, i.e. $E = -d\Phi/dt$. *See* FARADAY'S LAWS; LENZ'S LAW; NEUMANN'S LAW; INDUCTANCE.

electromagnetic interaction *See* FUNDAMENTAL INTERACTIONS.

electromagnetic pump A pump used for moving liquid metals, such as the liquid-sodium coolant in a fast nuclear reactor. The liquid is passed through a flattened pipe over two electrodes between which a direct current flows. A magnetic field at right angles to the current causes a force to be created directly on the liquid, along the axis of the tube. The pump has no moving parts and is therefore safe and trouble free.

electromagnetic radiation Energy resulting from the acceleration of electric charge and the associated electric fields and magnetic fields. The energy can be regarded as waves propagated through space (requiring no supporting medium) involving oscillating electric and magnetic fields at right angles to each other and to the direction of propagation. In a vacuum the waves travel with a constant speed (the speed of light) of 2.9979×10^8 metres per second; if material is present they are slower. Alternatively, the energy can be regarded as a stream of *photons travelling at the speed of light, each photon having an energy hc/λ, where h is the Planck constant, c is the speed of light, and λ is the wavelength of the associated wave. A fusion of these apparently conflicting concepts is possible using the methods of *quantum mechanics or *wave mechanics. The characteristics of the radiation depend on its wavelength. *See* ELECTROMAGNETIC SPECTRUM.

electromagnetic spectrum The range of wavelengths over which *electromagnetic radiation extends. The longest waves (10^5–10^{-3} metres) are radio waves, the next longest (10^{-3}–10^{-6} m) are infrared waves, then comes the narrow band (4–7×10^{-7} m) of visible light, followed by ultraviolet waves (10^{-7}–10^{-9} m), X-rays (10^{-9}–10^{-11} m), and gamma rays (10^{-11}–10^{-14} m).

electromagnetic units (e.m.u.) A system of electrical units formerly used in the c.g.s. system (*see* C.G.S. UNITS). The e.m.u. of electric current is the **abampere** (all e.m.u. have the prefix *ab-* attached to the names of practical units). The abampere is the current that, flowing in an arc of a circle (1 centimetre in diameter), exerts a force of 1 dyne on unit magnetic pole at the centre of the circle. In e.m.u. the magnetic constant is of unit magnitude. The system has now been replaced by *SI units for most purposes. *Compare* ELECTROSTATIC UNITS; GAUSSIAN UNITS; HEAVISIDE–LORENTZ UNITS.

electromagnetic wave *See* ELECTROMAGNETIC RADIATION; WAVE.

electrometallurgy The uses of electrical processes in the separation of metals from their ores, the refining of metals, or the forming or plating of metals.

electrometer A measuring instrument for determining a voltage difference without drawing an appreciable current from the source. Originally electrostatic instruments based on the electroscope, they are now usually based on operational amplifiers, solid-state devices with high input impedances. Electrometers are also used to measure low currents (nanoamperes), by passing the current through a high resistance.

electromotive force (e.m.f.) The greatest potential difference that can be generated by a particular source of electric current. In practice this may be observable only when the source is not supplying current, because of its *internal resistance.

electromotive series (electrochemical series) A series of chemical elements arranged in order of their *electrode potentials. The hydrogen electrode ($H^+ + e \rightarrow \frac{1}{2}H_2$) is taken as having zero electrode potential. Elements that have a greater tendency than hydrogen to lose electrons to their solution are taken as **electropositive**; those that gain electrons from their solution are below hydrogen in the series and are called **electronegative**. The series shows the order in which metals replace one another from their salts; electropositive metals will replace hydrogen from acids. The chief metals and hydrogen, placed in order in the series, are: potassium, calcium, sodium, magnesium, aluminium, zinc, cadmium, iron, nickel, tin, lead, hydrogen, copper, mercury, silver, platinum, gold. This type of series is sometimes referred to as an **activity series**.

electron An *elementary particle, classed as a *lepton, with a rest mass (symbol m_e) of 9.109 3897(54) $\times 10^{-31}$ kg and a negative charge of 1.602 177 33(49) $\times 10^{-19}$ coulomb. Electrons are present in all atoms in groupings called shells around the nucleus; when they are detached from the atom they are called **free electrons** (*see* FREE ELECTRON THEORY). The antiparticle of the electron is the **positron**.

The electron was discovered in 1897 by Joseph John *Thomson. The problem of the structure (if any) of the electron is unsolved. If the electron is taken to be a point charge, its *self-energy is infinite and difficulties arise for the *Dirac equation. It is possible to give the electron a nonzero size with a radius r_0, called the **classical electron radius**, given by $r_0 = e^2/(mc^2) = 2.82 \times 10^{-13}$ cm, where e and m are the charge and mass respectively of the electron and c is the speed of light. This model also causes difficulties, such as the necessity of postulating *Poincaré stresses. The view put forward by *superstring theory that particles such as electrons are strings in different excitation modes may solve all the difficulties associated with theories of the structure of the electron.

electron affinity Symbol A. The energy change occurring when an atom or molecule gains an electron to form a negative ion. For an atom or molecule X, it is the energy released for the electron-attachment reaction

$$X(g) + e \rightarrow X^-(g)$$

Often this is measured in electronvolts. Alternatively, the molar enthalpy change, ΔH, can be used.

electron biprism An arrangement of fields that splits a beam of electrons or other charged particles in an analogous way to an optical biprism.

electron capture 1. The formation of a negative ion by an atom or molecule when it acquires an extra free electron. **2.** A radioactive transformation in which a nucleus acquires an electron from an inner orbit of the atom, thereby transforming, initially, into a nucleus with the same mass number but an atomic number one less than that of the original nucleus (capture of the electron transforms a proton into a neutron). This type of capture is accompanied by emission of an X-ray photon as the vacancy in the inner orbit is filled by an outer electron.

electron diffraction *Diffraction of a beam of electrons by atoms or molecules. The fact that electrons can be diffracted in a similar way to light and X-rays shows that particles can act as waves (*see* DE BROGLIE WAVELENGTH). An electron (mass m, charge e) accelerated through a potential difference V acquires a kinetic energy $mv^2/2 = eV$, where v is the velocity of the electron. The (nonrelativistic) momentum (p) of the electron is $\sqrt{(2eVm)}$. The de Broglie wavelength (λ) of an electron is given by h/p, where h is the Planck constant, thus $\lambda = h/\sqrt{(2eVm)}$. For an accelerating voltage of 3600 V, the wavelength of the electron beam is 0.02 nanometre, some 3×10^4 times shorter than visible radiation.

Electrons then, like X-rays, show diffraction effects with molecules and crystals in which the interatomic spacing is comparable to the wavelength of the beam. They have the advantage that their wavelength can be set by adjusting the voltage. Unlike X-rays they have very low penetrating power. The first observation of electron diffraction was by George Thomson (1892–1975) in 1927, in an experiment in which he passed a beam of electrons in a vacuum through a very thin gold foil onto a photographic plate. Concentric circles were produced by diffraction of electrons by the lattice. The same year Clinton J. Davisson (1881–1958) and Lester Germer (1896–1971) performed a classic experiment in which they obtained diffraction patterns by glancing an electron beam off the surface of a nickel crystal. Both experiments were important verifications of de Broglie's theory and the new quantum theory.

Electron diffraction, because of the low penetration, cannot easily be used to investigate crystal structure. It is, however, employed to measure bond lengths and angles of molecules in gases. Moreover, it is extensively used in the study of solid surfaces and absorption. The main techniques are low-energy electron diffraction (**LEED**), in which the electron beam is reflected onto a fluorescent screen, and high-energy electron diffraction (**HEED**), used either with reflection or transmission in investigating thin films.

electronegative Describing elements that tend to gain electrons and form negative ions. The halogens are typical electronegative elements. For example, in hydrogen chloride, the chlorine atom is more electronegative than the hydrogen and the molecule is polar, with negative charge on the chlorine atom. There are various ways of assigning values for the **electronegativity** of an element. **Mulliken electronegativities** are calculated from $E = (I + A)/2$, where I is ionization potential and A is electron affinity. More commonly, **Pauling electronegativities** are used. These are based on bond dissociation energies using a scale in which fluorine, the most electronegative element, has a value 4. Some other values on this scale are B 2, C 2.5, N 3.0, O 3.5, Si 1.8, P 2.1, S 2.5, Cl 3.0, Br 2.8.

electron gun A device used in *cathode-ray tubes (including television tubes), electron microscopes, etc., to produce a steady narrow beam of electrons. It usually consists of a heated cathode, control grid, and two or more annular anodes inserted in an evacuated tube. The electrons emitted by the cathode are attracted to the final anode, through which they pass. The intensity of the beam is regulated by the control grid and potential differences between the anodes create electric fields that focus the diverging electrons into a narrow beam.

electronic mail (**e-mail**) Messages, documents, etc., sent between users of computer systems, the computer systems being used to transport and hold the e-mail. The service itself is also referred to as electronic mail. The sender and recipient(s) need not be at their computers at the same time to communicate, and the computer systems may be situated worldwide. The sender creates an e-mail by means of a mail-sending computer program, and a mail transport system then takes responsibility for delivering the e-mail to the indicated address(es).

electronics The study and design of control, communication, and computing devices that rely on the movement of electrons in circuits containing semiconductors, thermionic valves, resistors, capacitors, and inductors. See Chronology.

electron lens A device used to focus an electron beam. It is analogous to an optical lens but instead of using a refracting material, such as glass, it uses a coil or coils to produce a magnetic field or an arrangement of electrodes between which an electric field is created. Electron lenses are used in *electron microscopes and *cathode-ray tubes.

electron microscope A form of microscope that uses a beam of electrons instead of a beam of light (as in the optical microscope) to form a large image of a very small object. In optical microscopes the resolution

ELECTRONICS

1887	Radio waves are discovered by Heinrich Hertz.
1894	Oliver Lodge invents the 'coherer' for detecting radio waves. Marconi develops radio telegraphy.
1897	J. J. Thomson discovers the electron.
1902	US engineer Reginald Fessenden (1866–1932) develops radio telephony.
1903	Danish engineer Valdemar Poulsen (1869–1942) invents the arc transmitter for radio telegraphy.
1904	British engineer Ambrose Fleming (1849–1945) invents the diode thermionic valve.
1906	US engineer Lee De Forest (1873–1961) invents the triode thermionic valve. US electrical engineer Greenleaf Pickard (1877–1956) patents the crystal detector for radios. Fessenden introduces amplitude modulation in radio broadcasting.
1911	German physicist Karl Braun (1850–1918) invents cathode-ray tube scanning.
1912	Fessenden develops the heterodyne radio receiver.
1919	US electrical engineer Edwin Armstrong (1890–1954) develops the superheterodyne radio receiver.
1921	US physicist Albert Hull (1880–1966) invents the magnetron microwave-generating valve.
1923	Russian-born US engineer Vladimir Zworykin (1889–1982) invents the iconoscope television camera-tube.
1928	Scottish inventor John Logie Baird (1888–1946) and Vladimir Zworykin independently develop television.
1930	Swedish-born US electronics engineer Ernst Alexanderson (1878–1975) invents an all-electronic television system.
1933	US electrical engineer Edwin Armstrong (1890–1954) develops frequency modulation radio broadcasting.
1947	US physicists John Bardeen, Walter Brattain (1902–87), and William Shockley (1910–89) invent the point-contact transistor.
1950	US engineers develop the Videcon television camera tube.
1953	Chinese-born US computer engineer An Wang (1920–90) invents the magnetic core computer memory.
1954	US physicist Charles Townes (1915–) and Soviet physicists Nikolai Basov and Aleksandr Prokhorov (1916–) independently develop the maser.
1958	US electronics engineers Jack Kilby and Robert Noyce (1927–90) develop integrated circuits.
1960	US physicist Theodore Maiman (1927–) invents the ruby laser.
1961	US electronics engineer Steven Hofstein develops the field-effect transistor.
1971	US electronics engineer Marcian Edward Hoff (1937–) designs the first microprocessor (Intel 4004).
1977	US engineers transmit television signals along optical fibres.

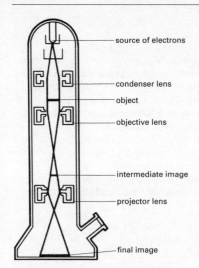

source of electrons

condenser lens

object

objective lens

intermediate image

projector lens

final image

Principle of transmission electron microscope

is limited by the wavelength of the light. High-energy electrons, however, can be associated with a considerably shorter wavelength than light; for example, electrons accelerated to an energy of 10^5 electronvolts have a wavelength of 0.004 nanometre (*see* DE BROGLIE WAVELENGTH) enabling a resolution of 0.2–0.5 nm to be achieved. The **transmission electron microscope** (see illustration) has an electron beam, sharply focused by *electron lenses, passing through a very thin metallized specimen (less than 50 nanometres thick) onto a fluorescent screen, where a visual image is formed. This image can be photographed. The **scanning electron microscope** can be used with thicker specimens and forms a perspective image, although the resolution and magnification are lower. In this type of instrument a beam of primary electrons scans the specimen and those that are reflected, together with any secondary electrons emitted, are collected. This current is used to modulate a separate electron beam in a TV monitor, which scans the screen at the same frequency, consequently building up a picture of the specimen. The resolution is limited to about 10–20 nm.

electron optics The study of the use of *electron lenses in the *electron microscope, *cathode-ray tubes, and other similar devices. The focusing of beams of positive or negative ions also relies on these methods.

electron probe microanalysis (EPM) A method of analysing a very small quantity of a substance (as little as 10^{-13} gram). The method consists of directing a very finely focused beam of electrons on to the sample to produce the characteristic X-ray spectrum of the elements present. It can be used quantitatively for elements with atomic numbers in excess of 11.

electron-spin resonance (ESR) A spectroscopic method of locating electrons within the molecules of a paramagnetic substance (*see* MAGNETISM) in order to provide information regarding its bonds and structure. The spin of an unpaired electron is associated with a *magnetic moment that is able to align itself in one of two ways with an applied external magnetic field. These two alignments correspond to different *energy levels, with a statistical probability, at normal temperatures, that there will be slightly more in the lower state than in the higher. By applying microwave radiation to the sample a transition to the higher state can be achieved. The precise energy difference between the two states of an electron depends on the surrounding electrons in the atom or molecule. In this way the position of unpaired electrons can be investigated. The technique is used particularly in studying free radicals and paramagnetic substances such as inorganic complexes. *See also* NUCLEAR MAGNETIC RESONANCE.

electronvolt Symbol eV. A unit of energy equal to the work done on an electron in moving it through a potential difference of one volt. It is used as a measure of particle energies although it is not an *SI unit. 1 eV = 1.602×10^{-19} joule.

electrophoresis (cataphoresis) A technique for the analysis and separation of colloids, based on the movement of charged colloidal particles in an electric field. There are various experimental methods. In one the sample is placed in a U-tube and a buffer solution added to each arm, so that there are sharp boundaries between buffer and sample. An electrode is placed in each arm, a voltage applied, and the motion of the boundaries under the influence of the field is observed. The rate of migration of the particles depends on the field, the charge on the particles, and on other factors, such as the size and shape of the particles. More simply, electrophoresis can be carried out using an adsorbent, such as a strip of filter paper, soaked in a buffer with two electrodes making contact. The sample is placed between the electrodes and a voltage applied. Different components of the mixture migrate at different rates, so the sample separates into zones. The components can be identified by the rate at which they move. This technique has also been known as **electrochromatography**.

Electrophoresis is used extensively in studying mixtures of proteins, nucleic acids, carbohydrates, enzymes, etc. In clinical medicine it is used for determining the protein content of body fluids.

electrophorus An early form of *electrostatic generator. It consists of a flat dielectric plate and a metal plate with an insulated handle. The dielectric plate is charged by friction and the metal plate is placed on it and momentarily earthed, which leaves the metal plate with an induced charge of opposite polarity to that of the dielectric plate. The process can be repeated until all of the original charge has leaked away.

electroplating A method of plating one metal with another by *electrodeposition. The articles to be plated are made the cathode of an electrolytic cell and a rod or bar of the plating metal is made the anode. Electroplating is used for covering metal with a decorative, more expensive, or corrosion-resistant layer of another metal.

electropositive Describing elements that tend to lose electrons and form positive ions. The alkali metals are typical electropositive elements.

electroscope A device for detecting electric charge and for identifying its polarity. In the **gold-leaf electroscope** two rectangular gold leaves are attached to the end of a conducting rod held in an insulated frame. When a charge is applied to a plate attached to the other end of the conducting rod, the leaves move apart owing to the mutual repulsion of the like charges they have received.

electrostatic field The *electric field that surrounds a stationary charged body.

electrostatic generator A device used to build up electric charge to an extreme potential usually for experimental purposes. The *electrophorus and the *Wimshurst machine were early examples; a more usual device now is the *Van de Graaff generator.

electrostatic precipitation A method of removing solid and liquid particles from suspension in a gas. The gas is exposed to an electric field so that the particles are attracted to and deposited on a suitably placed electrode. Electrostatic precipitation is widely used to remove dust and other pollutants from waste gases and from air. *See also* COTTRELL PRECIPITATOR.

electrostatics The study of electric charges at rest, the forces between them (*see* COULOMB'S LAW), and the electric fields associated with them. *Compare* ELECTRODYNAMICS.

electrostatic units (e.s.u.) A system of electrical units in the c.g.s. system (*see* C.G.S. UNITS). The e.s.u. of electric charge is the **statcoulomb** (all e.s.u. have the prefix *stat- attached to the names of practical units). The statcoulomb is the quantity of electric charge that will repel an equal quantity 1 centimetre distant with a force of 1 dyne. In e.s.u. the electric constant is of unit magnitude. The system has now been replaced for most purposes by *SI units. *Compare* ELECTROMAGNETIC UNITS; GAUSSIAN UNITS; HEAVISIDE–LORENTZ UNITS.

electrostriction A change in the dimensions of a body as a result of reorientation of its molecules when it is placed in an electric field. If the field is not homogeneous the body will tend to move; if its relative permittivity is higher than that of its surroundings it will tend to move into a region of higher field strength. *Compare* MAGNETOSTRICTION.

electroweak theory A *gauge theory (sometimes called **quantum flavourdynamics**, or **QFD**) that gives a unified description of the electromagnetic and weak interactions (*see* FUNDAMENTAL INTERACTIONS). A successful electroweak theory was proposed in 1967 by Steven Weinberg (1933–) and Abdus Salam (1926–96), known as the **Weinberg–Salam model** or **WS model**. Because early developments of these ideas were put forward by Sheldon Glashow (1926–), it is sometimes known as the **Glashow–Weinberg–Salam model** or **GWS model**. In this electroweak theory the gauge group is non-Abelian and the gauge symmetry is a *broken symmetry. The electroweak interaction is mediated by photons and by intermediate vector bosons, called the *W boson and the *Z boson. The observation of these particles in 1983–84, with their predicted energies, was a major success of the theory. The theory successfully accounts for existing data for electroweak processes and also predicts the existence of a heavy particle with spin 0, the *Higgs boson.

electrum **1.** An alloy of gold and silver containing 55–85% of gold. **2.** A *German silver alloy containing 52% copper, 26% nickel, and 22% zinc.

element A substance that cannot be decomposed into simpler substances. In an element, all the atoms have the same number of protons or electrons, although the number of neutrons may vary. There are 92 naturally occurring elements. See Appendix. *See also* PERIODIC LAW; TRANSURANIC ELEMENTS; TRANSACTINIDE ELEMENTS.

elementary particles The fundamental constituents of all the matter in the universe. By the beginning of the 20th century, the electron had been discovered. Subsequently, the proton was discovered. It was not until 1932 that the existence of the neutron was definitely established. Since 1932, it had been known that atomic nuclei consist of both protons and neutrons (except hydrogen, whose nucleus consists of a lone proton). Between 1900 and 1930, *quantum mechanics was also making progress in the understanding of physics on the atomic scale. Non-relativistic quantum theory was completed in an astonishingly brief period (1923–26), but it was the relativistic version that made the greatest impact on our understanding of elementary particles. Dirac's discovery in 1928 of the equation that bears his name led to the discovery of the positive electron or *positron. The mass of the positron is equal to that of the negative electron while its charge is equal in magnitude but opposite in

Name	Symbol	Charge (electron charges)	Rest mass (MeV/c^2)
electron	e^-	−1	0.511
electron neutrino	ν_e	0	*
muon	μ^-	−1	105.7
muon neutrino	ν_μ	0	*
tauon	τ^-	−1	1784
tau neutrino	ν_τ	0	*

Table of leptons

sign. Pairs of particles related to each other in this way are said to be antiparticles of each other. Positrons have only a transitory existence; that is, they do not form part of ordinary matter. Positrons and electrons are produced simultaneously in high-energy collisions of charged particles or gamma rays with matter in a process called *pair production.

The union of *relativity and quantum mechanics therefore led to speculation as early as 1932 that there might also be antiprotons and antineutrons, bearing a similar relationship to their respective ordinary particles as the positron does to the electron. However, it was not until 1955 that particle beams were made sufficiently energetic to enable these antimatter particles to be observed. It is now understood that all known particles have antimatter equivalents, which are predicted by relativistic quantum equations.

By the mid-1930s the list of known and theoretically postulated particles was still small but steadily growing. At this time the Japanese physicist Hideki Yukawa (1907–81) was studying the possible *fundamental interactions that could hold the nucleus together. Since the nucleus is a closely packed collection of positively charged protons and neutral neutrons, clearly it could not be held together by an electromagnetic force; there had to be a different and very large force capable of holding proton charges together at such close proximity. This force would necessarily be restricted to the short range of nuclear dimensions, because evidence of its existence only arose after the discovery of the constituents of the atomic nucleus. Guided by the properties required of this new force, Yukawa proposed the existence of a particle called the *meson, which was responsible for transmitting nuclear forces. He suggested that protons and neutrons in the nucleus could interact by emitting and absorbing mesons. For this reason this new type of force was called an *exchange force. Yukawa was even able to predict the mass of his meson (meaning 'middle weight'), which turned out to be intermediate between the proton and the electron.

Only a year after Yukawa had made this suggestion, a particle of intermediate mass was discovered in *cosmic radiation. This particle was named the **μ-meson** or **muon**. The μ^- has a charge equal to the electron, and its antiparticle μ^+ has a positive charge of equal magnitude. However, physicists soon discovered that muons do not interact with nuclear particles sufficiently strongly to be Yukawa's meson. It was not until 1947 that a family of mesons with the appropriate properties was discovered.

Quark symbol	Name	Charge
u	up	2/3
d	down	−1/3
c	charm	2/3
s	strange	−1/3
t	top	2/3
b	bottom	−1/3

Table of quarks (mass is not shown because quarks are never observed alone)

These were the **π-mesons** or **pions**, which occur in three types: positive, negative, and neutral. Pions, which interact strongly with nuclei, have in fact turned out to be the particles predicted by Yukawa in the 1930s. The nuclear force between protons and neutrons was given the name 'strong interaction' (*see* FUNDAMENTAL INTERACTIONS) and until the 1960s it was thought to be an exchange force as proposed by Yukawa.

A theory of the weak interaction was also in its infancy in the 1930s. The weak interaction is responsible for *beta decay, in which a radioactive nucleus is transformed into a slightly lighter nucleus with the emission of an electron. However, beta decays posed a problem because they appeared not to conserve energy and momentum. In 1931 *Pauli proposed the existence of a neutral particle that might be able to carry off the missing energy and momentum in a beta decay and escape undetected. Three years later, *Fermi included Pauli's particle in a comprehensive theory of beta decay, which seemed to explain many experimentally observed results. Fermi called this new particle the *neutrino, the existence of which was finally established in the 1950s.

A plethora of experiments involving the neutrino revealed some remarkable properties for this new particle. The neutrino was found to have an intimate connection with the electron and muon, and indeed never appeared without the simultaneous appearance of one or other of these particles. A conservation law was postulated to explain this observation. Numbers were assigned to the electron, muon, and neutrino, so that during interactions these numbers were conserved; i.e. their algebraic sums before and after these interactions were equal. Since these particles were among the lightest known at the time, these assigned numbers became known as **lepton numbers** (lepton: 'light ones'). In order to make the assignments of lepton number agree with experiment, it is necessary to postulate the existence of two types of neutrino. Each of these types is associated with either the electron or muon; there are thus muon neutrinos and electron neutrinos. In 1978 the **tau particle** or **tauon** was discovered and was added to the list of particles with assigned lepton numbers. The conservation of lepton number in the various interactions involving the tau requires the existence of an equivalent tau neutrino. The six particles with assigned lepton numbers are now known as *leptons.

Neutrinos have zero charge and were originally thought to have zero rest mass, but there has been some indirect experimental evidence to the contrary, beginning in the last twenty years of the 20th century. In 1985 a

Interaction	Mediator (exchange particle)	Rest mass (GeV/c^2)	Charge
strong	gluon	0	0
electromagnetic	photon	0	0
weak	W^+, W^-, Z°	81,81,93	+1,−1,0
gravitational	graviton	0	0

Table of mediators

Soviet team reported a measurement, for the first time, of a non-zero neutrino mass. The mass measured was extremely small (10 000 times less than the mass of the electron), but subsequent attempts independently to reproduce these results did not succeed. More recently (1998–99), Japanese and US groups have put forward theories and corroborating experimental evidence to suggest, indirectly, that neutrinos do have mass. In these experiments neutrinos are found to apparently 'disappear'. Since it is unlikely that momentum and energy are actually vanishing from the universe, a more plausible explanation is that the types of neutrinos detected are changing into types that cannot be detected. Present theoretical considerations imply that the masses of neutrinos involved cannot be equal to one another, and therefore they cannot all be zero. This speculative work has not yet yielded estimates of the neutrino masses, which is indicated by the use of asterisks in the table of leptons on p. 158.

In the 1960s, the development of high-energy accelerators and more sophisticated detection systems led to the discovery of many new and exotic particles. They were all unstable and existed for only small fractions of a second; nevertheless they set into motion a search for a theoretical description that could account for them all. The large number of these apparently fundamental particles suggested strongly that they do not, in fact, represent the most fundamental level of the structure of matter. Physicists found themselves in a position similar to Mendeleev when the periodic table was being developed. Mendeleev realized that there had to be a level of structure below the elements themselves, which explained the chemical properties and the interrelations between elements.

Murray Gell-Mann and his collaborators proposed the particle-physics equivalent of the periodic table between 1961 and 1964. In this structure, leptons were indeed regarded as fundamental particles, but the short-lived particles discovered in the 1960s were not. These particles were found to undergo strong interactions, which did not seem to affect the leptons. Gell-Mann called these strongly interacting particles the *hadrons and proposed that they occurred in two different types: baryons and mesons. These two different types corresponded to the two different ways of constructing hadrons from constituent particles, which Gell-Mann called **quarks**. These quarks came in three **flavours**, up (u), down (d), and strange (s). These three quarks were thought to be the fundamental constituents of hadrons, i.e. matter that undergoes strong interactions: baryons are composed of three quarks (u, d, or s) or three antiquarks ($\bar{u}$, $\bar{d}$, or $\bar{s}$); mesons are composed of (u, d, or s) quark–antiquark pairs.

No other combinations seemed to be necessary to describe the full variation of the observed hadrons. This scheme even led to the prediction of other particles that were not known to exist in 1961. For example, in 1961 Gell-Mann not only predicted the Ω^- (omega-minus) particle, but more importantly told experimentalists exactly how to produce it. The Ω^- particle was finally discovered in 1964.

Gell-Mann called his scheme 'the eight-fold way', after the similarly named Buddhist principle. The scheme requires that quarks have properties not previously allowed for fundamental particles. For example, quarks have fractional electric charges, i.e. charges of 1/3 and 2/3 of the electron charge. Quarks also have a strong affinity for each other through a new kind of charge known as **colour charge**. Colour charge is therefore responsible for strong interactions, and the force is known as the colour force. This is a revision of Yukawa's proposal in 1930. Yukawa's strong force was mediated by π-mesons. The strong force is now thought to be mediated by exchange of particles carrying colour charge known as **gluons**. The theory governing these colour charge combinations is modelled on *quantum electrodynamics and is known as *quantum chromodynamics.

In November 1974 the discovery of the ψ (psi) particle initiated what later came to be known as 'the November revolution'. Up to that time, any known hadron could be described as some combination of u, d, or s quarks. These hadrons were very short-lived with lifetimes of about 10^{-23} s. The ψ particle, however, had a lifetime of 10^{-20} s; i.e. a thousand times longer. This suggested a completely different species of particle. It is now universally accepted that the ψ represents a meson-bound state of a new fourth quark, the **charm** (c) quark and its antiquark. In 1977 the list of quarks once again increased with the discovery of a new even heavier meson, called the **Y** (**upsilon**) **meson**. This meson was found to have an even longer lifetime than the ψ, and was quickly identified as the carrier of a fifth quark, **bottom** (b).

Thus, by the end of 1977, five flavours of quark (u, d, s, c, b) were known to exist together with six flavours of lepton (e, μ, τ, ν_e, τ_μ, ν_τ). Assuming that quarks and leptons are the fundamental constituents of matter, many of the strong and weak interactions of hadrons and the weak interactions of leptons can be explained. However, anticipating a symmetry in nature's building blocks, it was expected that a sixth quark would eventually reveal itself. This quark, labelled **top** (t), would be the 2/3 electronic charge partner to the b quark (see table p. 159). In 1998 the top quark was found at CERN in Geneva and the symmetry of six quarks with six leptons was finally verified.

In 1978 the **standard model** was proposed as the definitive theory of the fundamental constituents of matter. In the current view, all matter consists of three kinds of particles: leptons, quarks, and mediators (see table p.160). The mediators are the particles by which the four fundamental interactions are mediated. In the standard model, each of these interactions has a particle mediator. For the electromagnetic reaction it is the *photon.

For weak interactions the force is mediated by three particles called W^+, W^-, and Z° *bosons; for the strong force it is the gluon. Current theories of quantum gravity propose the *graviton as the mediator for the gravitational interaction, but this work is highly speculative and the graviton has never been detected.

elements of an orbit Six parameters that can be used to define the path of a celestial body. The shape of the orbit is defined by its eccentricity (*see* CONIC) and semimajor axis. The orientation of the orbit is specified by the *inclination of the orbital plane to the reference plane (usually the *ecliptic) and by the longitude of the ascending *node (the angular distance from the vernal equinox to the ascending node). The position of the body in its orbit is defined by its eccentric *anomaly and the position as a function of time is calculated from the time of periapsis passage (*see* APSIDES).

elevation of boiling point An increase in the boiling point of a liquid when a solid is dissolved in it. The elevation is proportional to the number of particles dissolved (molecules or ions) and is given by $\Delta t = k_{B}C$, where C is the molal concentration of solute. The constant k_{B} is the **ebullioscopic constant** of the solvent and if this is known, the relative molecular mass of the solute can be calculated from the measured value of Δt. The elevation is measured by a Beckmann thermometer. *See also* COLLIGATIVE PROPERTIES.

Elinvar Trade name for a nickel–chromium steel containing about 36% nickel, 12% chromium, and smaller proportions of tungsten and manganese. Its elasticity does not vary with temperature and it is therefore used to make hairsprings for watches.

ellipse A *conic formed by the intersection of a plane with a right circular cone, so that the plane is inclined to the axis of the cone at an angle in excess of half the apex angle of the cone. The ellipse has two vertices, which are joined by a line called the **major axis**. The centre of the ellipse falls on this line, midway between the vertices. The **minor axis** is the line perpendicular to the major axis that passes through the centre and joins two points on the ellipse. The **foci** of an ellipse are two points on the major axis so placed that for any point on the ellipse the sum of the distances from that point to each focus is constant. (See illustration.) The area of an ellipse is πab, where a and b are half the major and minor axes, respectively. For an ellipse centred at the origin, the equation in Cartesian coordinates is $x^2/a^2 + y^2/b^2 = 1$. The foci are at $(ea, 0)$ and $(-ea, 0)$, where e is the eccentricity. Each of the two chords of the ellipse passing through a focus and parallel to the minor axis is called a **latus rectum** and has a length equal to $2b^2/a$.

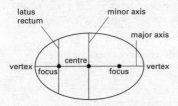

An ellipse

ellipsoid A solid body formed when an *ellipse is rotated about an axis. If it is rotated about its major axis it is a **prolate ellipsoid**; if it is rotated about its minor axis it is an **oblate ellipsoid**. For an ellipsoid centred at the origin the equation in Cartesian coordinates is:

$$x^2/a^2 + y^2/b^2 + z^2/c^2 = 1.$$

elliptical galaxy *See* GALAXY.

elliptical polarization *See* POLARIZATION OF LIGHT.

eluate *See* ELUTION.

eluent *See* ELUTION.

elution The process of removing an adsorbed material (**adsorbate**) from an adsorbent by washing it in a liquid (**eluent**). The solution consisting of the adsorbate dissolved in the eluent is the **eluate**.

elutriation The process of suspending finely divided particles in an upward flowing stream of air or water to wash and separate them into sized fractions.

e-mail *See* ELECTRONIC MAIL.

emanation The former name for the gas radon, of which there are three isotopes: Rn–222 (radium emanation), Rn–220 (thoron emanation), and Rn–219 (actinium emanation).

e.m.f. *See* ELECTROMOTIVE FORCE.

emission spectrum *See* SPECTRUM.

emissivity Symbol ε. The ratio of the power per unit area radiated by a surface to that radiated by a *black body at the same temperature. A black body therefore has an emissivity of 1 and a perfect reflector has an emissivity of 0. The emissivity of a surface is equal to its *absorptance.

emittance *See* EXITANCE.

emitter *See* TRANSISTOR.

emitter-coupled logic (ECL) A set of integrated *logic circuits. The input part of an ECL consists of an emitter-coupled *transistor pair which is a very good *differential amplifier. The output is through an *emitter follower. ECL circuits are very rapid logic circuits.

emitter follower An amplifying circuit using a bipolar junction *transistor with a *common-collector connection. The output is taken from the emitter.

empirical Denoting a result that is obtained by experiment or observation rather than from theory.

empirical formula *See* FORMULA.

emulsion A *colloid in which small particles of one liquid are dispersed in another liquid. Usually emulsions involve a dispersion of water in an oil or a dispersion of oil in water, and are stabilized by an **emulsifier**. Commonly emulsifiers are substances, such as detergents, that have lyophobic and lyophilic parts in their molecules.

enantiomers *See* OPTICAL ACTIVITY.

enantiomorphism *See* OPTICAL ACTIVITY.

endoergic Denoting a nuclear process that absorbs energy. *Compare* EXOERGIC.

endothermic Denoting a chemical reaction that takes heat from its surroundings. *Compare* EXOTHERMIC.

energy A measure of a system's ability to do work. Like work itself, it is measured in joules. Energy is conveniently classified into two forms: **potential energy** is the energy stored in a body or system as a consequence of its position, shape, or state (this includes gravitational energy, electrical energy, nuclear energy, and chemical energy); **kinetic energy** is energy of motion and is usually defined as the work that will be done by the body possessing the energy when it is brought to rest. For a body of mass m having a speed v, the kinetic energy is $mv^2/2$ (classical) or $(m - m_0)c^2$ (relativistic). The rotational kinetic energy of a body having an angular velocity ω is $I\omega^2/2$, where I is its moment of inertia.

The *internal energy of a body is the sum of the potential energy and the kinetic energy of its component atoms and molecules.

energy band A range of energies that electrons can have in a solid. In a single atom, electrons exist in discrete *energy levels. In a crystal, in which large numbers of atoms are held closely together in a lattice, electrons are influenced by a number of adjacent nuclei and the sharply defined levels of the atoms become bands of allowed energy (see illustration); this approach to energy levels in solids is often known as the *band theory. Each band represents a large number of allowed quantum states. Between the bands are **forbidden bands**. The outermost electrons of the atoms (i.e. the ones responsible for chemical bonding) form the **valence band** of the solid. This is the band, of those fully occupied, that has the highest energy.

The band structure of solids accounts for their electrical properties. In order to move through the solid, the electrons have to change from one quantum state to another. This can only occur if there are empty quantum states with the same energy. In general, if the valence band is full, electrons cannot change to new quantum states in the same band. For conduction to occur, the electrons have to be in an unfilled band – the **conduction band**. Metals are good conductors either because the conduction band is not filled or because the conduction band overlaps

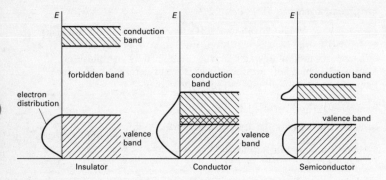

Energy bands

with the valence band; in either case vacant states are available. In insulators the conduction band and valence band are separated by a wide forbidden band and electrons do not have enough energy to 'jump' from one to the other.

In intrinsic *semiconductors the forbidden gap is narrow and, at normal temperatures, electrons at the top of the valence band can move by thermal agitation into the conduction band (at absolute zero, a semiconductor would act as an insulator). Doped semiconductors have extra bands in the forbidden gap.

energy level A definite fixed energy that a system described by *quantum mechanics, such as a molecule, atom, electron, or nucleus, can have. In an atom, for example, the atom has a fixed energy corresponding to the *orbitals in which its electrons move around the nucleus. The atom can accept a quantum of energy to become an excited atom (*see* EXCITATION) if that extra energy will raise an electron to a permitted orbital. Between the **ground state**, which is the lowest possible energy level for a particular system, and the first excited state there are no permissible energy levels. According to the *quantum theory, only certain energy levels are possible. An atom passes from one energy level to the next without passing through fractions of that energy transition. These levels are usually described by the energies associated with the individual electrons in the atoms, which are always lower than an arbitrary level for a free electron. The energy levels of molecules also involve quantized vibrational and rotational motion.

engine Any device for converting some forms of energy into mechanical work. *See* HEAT ENGINE; CARNOT CYCLE; INTERNAL-COMBUSTION ENGINE; STEAM ENGINE.

enrichment The process of increasing the abundance of a specified isotope in a mixture of isotopes. It is usually applied to an increase in the proportion of U–235, or the addition of Pu–239 to natural uranium for use in a nuclear reactor or weapon.

ensemble A set of systems of particles used in *statistical mechanics to describe a single system. The concept of an ensemble was put forward by the US scientist Josiah Willard Gibbs (1839–1903) in 1902 as a way of calculating the time average of the single system, by averaging over the systems in the ensemble at a fixed time. An ensemble of systems is constructed from knowledge of the single system and can be represented as a set of points in *phase space with each system of the ensemble represented by a point. Ensembles can be constructed both for isolated systems and for open systems.

enthalpy Symbol H. A thermodynamic property of a system defined by $H = U + pV$, where H is the enthalpy, U is the internal energy of the system, p its pressure, and V its volume. In a chemical reaction carried out in the atmosphere the pressure remains constant and the enthalpy of reaction, ΔH, is equal to $\Delta U + p\Delta V$. For an exothermic reaction ΔH is taken to be negative.

entropy Symbol S. A measure of the unavailability of a system's energy to do work; in a closed system an increase in entropy is accompanied by a decrease in energy availability. When a system undergoes a reversible change the entropy (S) changes by an amount equal to the energy (Q) transferred to the system by heat divided by the thermodynamic temperature (T) at which this occurs, i.e. $\Delta S = \Delta Q/T$. However, all real processes are to a certain extent irreversible changes and in any closed system an irreversible change is always accompanied by an increase in entropy.

In a wider sense entropy can be interpreted as a measure of disorder; the higher the entropy the greater the disorder. As any real change to a closed system tends towards higher entropy, and therefore higher disorder, it follows that the entropy of the universe (if it can be considered a closed system) is increasing and its available energy is decreasing (*see* HEAT DEATH OF THE UNIVERSE). This increase in the entropy of the universe is one way of stating the second law of *thermodynamics.

ephemeris A tabulation showing the calculated future positions of the sun, moon, and planets, together with other useful information for astronomers and navigators. It is published at regular intervals.

ephemeris time (ET) A time system that has a constant uniform rate as opposed to other systems that depend on the earth's rate of rotation, which has inherent irregularities. It is reckoned from an instant in 1900 (Jan 0d 12h) when the sun's mean longitude was 279.696 677 8°. The unit by which ephemeris time is measured is the tropical year, which contains 31 556 925.9747 **ephemeris seconds**. This fundamental definition of the *second was replaced in 1964 by the caesium second of atomic time.

epicentre The point on the surface of the earth directly above the focus of an earthquake or directly above or below a nuclear explosion.

epicycle A small circle whose centre rolls around the circumference of a larger fixed circle. The curve traced out by a point on the epicycle is called an **epicycloid**.

epidiascope An optical instrument used by lecturers, etc., for projecting an enlarged image of either a translucent object (such as a slide or transparency) or an opaque object (such as a diagram or printed page) onto a screen.

epitaxy (epitaxial growth) Growth of a layer of one substance on a single crystal of another, such that the crystal structure in the layer is the same as that in the substrate. It is used in making semiconductor devices.

epithermal neutron A neutron with an energy in excess of that associated with a thermal neutron (*see* MODERATOR) but less than that of a *fast neutron, i.e. a neutron having an energy in the range 0.1 to 100 eV.

EPM *See* ELECTRON PROBE MICROANALYSIS.

EPR experiment (Einstein–Podolsky–Rosen experiment) A thought experiment suggested by Einstein as an objection to the *Copenhagen interpretation of quantum mechanics. In a simplified form of this experiment one considers a particle with zero spin converted into two particles with spin. The particles fly apart and it is known that they must have opposite spins because spin is conserved. At a certain point one measures the spin direction of one of the particles, and immediately knows that the other particle's spin direction is opposite. However, according to Bohr, the spin is not defined until the measurement is made, at which point the wave function collapses. How does the second particle 'know' the result of the measurement on the first particle? According to Einstein, this implies that the spins of the particles were set at the time the particles were formed and that hidden variables are involved in quantum mechanics. In 1961 John Bell showed a way of testing this experimentally (*see* BELL'S THEOREM) and experiments were eventually done (*see* ASPECT EXPERIMENT). These seem to show that the Copenhagen interpretation is correct and that the particles are part of a single entangled state, even though they are far apart.

EPROM (erasable programmable read-only memory) A read-only memory device in a computer in which stored data can be erased by high-intensity ultraviolet light and which can be reprogrammed repeatedly using suitable voltage pulses.

equation of motion (kinematic equation) Any of four equations that apply to bodies moving linearly with uniform acceleration (a). The equations, which relate distance covered (s) to the time taken (t), are:

$v = u + at$

$s = (u + v)t/2$

$s = ut + at^2/2$

$v^2 = u^2 + 2as$

where u is the initial velocity of the body and v is its final velocity.

equation of state An equation that relates the pressure p, volume V, and thermodynamic temperature T of an amount of substance n. The simplest is the ideal *gas law:

$pV = nRT$,

where R is the universal gas constant. Applying only to ideal gases, this equation takes no account of the volume occupied by the gas molecules (according to this law if the pressure is infinitely great the volume becomes zero), nor does it take into account any forces between molecules. A more accurate equation of state would therefore be

$(p + k)(V - nb) = nRT$,

where k is a factor that reflects the decreased pressure on the walls of the container as a result of the attractive forces between particles, and nb is the volume occupied by the particles themselves when the pressure is infinitely high. In the **van der Waals equation of state**, proposed by the Dutch physicist J. D. van der Waals (1837–1923),

$k = n^2a/V^2$,

where a is a constant. This equation more accurately reflects the behaviour of real gases; several others have done better but are more complicated.

equation of time The length of time that must be added to the mean solar time, as shown on a clock, to give the apparent solar time, as shown by a sundial. The amount varies during the year, being a minimum of –14.2 minutes in February and a maximum of +16.4 minutes in October. It is zero on four days (April 15/16, June 14/15, Sept. 1/2, Dec. 25/26). The difference arises as a result of two factors: the eccentricity of the earth's orbit and the inclination of the ecliptic to the celestial equator.

equator 1. The great circle around the earth that lies in a plane perpendicular to the earth's axis. It is equidistant from the two geographical poles. **2.** The **magnetic equator** is a line of zero magnetic dip (*see* GEOMAGNETISM) that is close to the geographical equator but lies north of it in Africa and south of it in America. **3.** The **celestial equator** is the circle formed on the *celestial sphere by the extension of the earth's equatorial plane.

equilibrium A state in which a system has its energy distributed in the statistically most probable manner; a state of a system in which forces, influences, reactions, etc., balance each other out so that there is no net change.

A body is in **static equilibrium** if the resultants of all forces and all couples acting on it are both zero; it may be at rest and will certainly not be accelerated. Such a body at rest is in **stable equilibrium** if after a slight displacement it returns to its original position – for a body whose weight is the only downward force this will be the case if the vertical line through its centre of gravity always passes through its base. If a slight displacement causes the body to move to a new position, then the body is in **unstable equilibrium**.

A body is said to be in **thermal equilibrium** if no net heat exchange is taking place within it or between it and its surroundings. A system is in chemical equilibrium when a reaction and its reverse are proceeding at equal rates. These are examples of **dynamic equilibrium**, in which activity in one sense or direction is in aggregate balanced by comparable reverse activity.

equinox 1. Either of the two points on the *celestial sphere at which the *ecliptic intersects the celestial equator. The sun appears to cross the celestial equator from south to north at the **vernal equinox** and from north to south at the **autumnal equinox. 2.** Either of the two instants at which the centre of the sun appears to cross the celestial equator. In the northern hemisphere the vernal equinox occurs on or about March 21 and the autumnal equinox on or about Sept. 23. In the southern hemisphere the dates are reversed. *See* PRECESSION OF THE EQUINOXES.

equipartition of energy The theory, proposed by Ludwig *Boltzmann and given some theoretical support by James Clerk *Maxwell, that the energy of gas molecules in a large sample under thermal *equilibrium is equally divided among their available *degrees of freedom, the average energy for each degree of freedom being $kT/2$, where k is the *Boltzmann constant and T is the thermodynamic temperature. The proposition is not generally true if *quantum considerations are important, but is frequently a good approximation.

erecting prism A glass prism used in optical instruments to convert an inverted image into an erect image, as in prismatic binoculars.

erg A unit of work or energy used in the c.g.s. system and defined as the work done by a force of 1 dyne when it acts through a distance of 1 centimetre. $1 \text{ erg} = 10^{-7}$ joule.

ergodic hypothesis A hypothesis in *statistical mechanics concerning *phase space. If a system of N atoms or molecules is enclosed in a fixed volume, the state of this system is given by a point in $6N$-dimensional phase space with q_i representing coordinates and p_i representing momenta. Taking the energy E to be constant, a representative point in phase space describes an orbit on the surface $E(q_i, p_i) = c$, where c is a constant. The **ergodic hypothesis** states that the orbit of the representative point in phase space eventually goes through all points on the surface. The **quasi-ergodic hypothesis** states that the orbit of the representative point in phase space eventually comes close to all points on the surface. In general, it is very difficult to prove the ergodic or quasi-ergodic hypotheses for a given system. *See also* ERGODICITY; KAM THEOREM.

ergodicity A property of a system that obeys the *ergodic hypothesis. The ergodicity of systems has been discussed extensively in the foundations of *statistical mechanics, although it is now thought by many physicists to be irrelevant to the problem. Considerations of ergodicity occur in *dynamics, since the behaviour can be very complex even for simple *dynamical systems (*see* ATTRACTOR). Systems, such as *spin glasses, in which ergodicity is thought not to hold are described as having **broken ergodicity**. It is difficult to construct mathematically rigorous proofs of ergodicity in systems. *See also* KAM THEOREM.

ergonomics The study of the engineering aspects of the relationship between workers and their working environment.

Esaki diode *See* TUNNEL DIODE.

ESCA *See* PHOTOELECTRON SPECTROSCOPY.

escapement A device in a clock or watch that controls the transmission of power from the spring or falling weight to the hands. It is usually based on a balance wheel or pendulum. It thus allows energy to enter the mechanism in order to move the hands round the face, overcome friction in the gear trains, and maintain the balance wheel or pendulum in continuous motion.

escape velocity The minimum speed needed by a space vehicle, rocket, etc., to escape from the gravitational field of the earth, moon, or other celestial body. The gravitational force between a rocket of mass m and a celestial body of mass M and radius r is MmG/r^2 (*see* NEWTON'S LAW OF GRAVITATION). Therefore the gravitational potential energy of the rocket with respect to its possible position very far from the celestial body on which it is resting can be shown to be $-GmM/r$, assuming (by convention) that the potential energy is zero at an infinite distance from the celestial body. If the rocket is to escape from the gravitational field it must have a kinetic energy that exceeds this potential energy, i.e. the kinetic energy $mv^2/2$ must be greater than MmG/r, or $v > \sqrt{(2MG/r)}$. This is the value of the escape velocity. Inserting numerical values for the earth and moon into this relationship gives an escape velocity from the earth of 11 200 m s^{-1} and from the moon of 2370 m s^{-1}.

ESR *See* ELECTRON-SPIN RESONANCE.

ether (aether) A hypothetical medium once believed to be necessary to support the propagation of electromagnetic radiation. It is now regarded as unnecessary and in modern theory electromagnetic radiation can be propagated through empty space. The existence of the ether was first called into question as a result of the *Michelson–Morley experiment.

eudiometer An apparatus for measuring changes in volume of gases during chemical reactions. A simple example is a graduated glass tube sealed at one end and inverted in mercury. Wires passing into the tube allow the gas mixture to be sparked to initiate the reaction between gases in the tube.

eutectic mixture A solid solution consisting of two or more substances and having the lowest freezing point of any possible mixture of these components. The minimum freezing point for a set of components is called the **eutectic point**. Low melting-point alloys are usually eutectic mixtures.

evaporation The change of state of a liquid into a vapour at a temperature below the boiling point of the liquid. Evaporation occurs at the surface of a liquid, some of those molecules with the highest kinetic energies escaping into the gas phase. The result is a fall in the average kinetic energy of the molecules of the liquid and consequently a fall in its temperature.

evaporative cooling Cooling of a substance as a result of evaporation. *See also* LASER COOLING.

even–even nucleus An atomic nucleus containing an even number of protons and an even number of neutrons.

even–odd nucleus An atomic nucleus containing an even number of protons and an odd number of neutrons.

event horizon *See* BLACK HOLE; DEATH OF A STAR.

evolute The locus of the centres of curvature of all the points on a given curve (called the **involute**).

exa- Symbol E. A prefix used in the metric system to denote 10^{18} times. For example, 10^{18} metres = 1 exametre (Em).

exbi- *See* BINARY PREFIXES.

excess electron An electron in a *semiconductor that is not required in the bonding system of the crystal lattice and has been donated by an impurity atom. It is available for conduction (**excess conduction**).

exchange force 1. A force resulting from the continued interchange of particles in a manner that bonds their hosts together. Examples are the covalent bond involving electrons, and the strong interaction (*see* FUNDAMENTAL INTERACTIONS) in which mesons are exchanged between nucleons or gluons are exchanged between quarks (*see* ELEMENTARY PARTICLES). **2.** *See* MAGNETISM.

excimer *See* EXCIPLEX.

exciplex A combination of two different atoms that exists only in an excited state. When an exciplex emits a photon of electromagnetic radiation, it immediately dissociates into the atoms, rather than reverting to the ground state. A similar transient excited association of two atoms of the same kind is an **excimer**. An example of an exciplex is the species XeCl* (the asterisk indicates an excited state), which can be formed by an electric discharge in xenon and chlorine. This is used in the **exciplex laser**, in which a population inversion is produced by an electrical discharge.

excitation 1. A process in which a nucleus, electron, atom, ion, or molecule acquires energy that raises it to a quantum state (**excited state**) higher than that of its *ground state. The difference between the energy in the ground state and that in the excited state is called the **excitation energy**. *See* COLLECTIVE EXCITATION; ENERGY LEVEL; QUASIPARTICLE. **2.** The process of applying current to the winding of an electromagnet, as in an electric motor. **3.** The process of applying a signal to the base of a transistor or the control electrode of a thermionic valve.

exciton An electron–hole pair in a crystal that is bound in a manner analogous to the electron and proton of a hydrogen atom. It behaves like an atomic excitation that passes from one atom to another and may be long-lived. Exciton behaviour in *semiconductors is important.

exclusion principle *See* PAULI EXCLUSION PRINCIPLE.

exitance Symbol M. The radiant or luminous flux emitted per unit area of a surface. The **radiant exitance** (M_e) is measured in watts per square metre (W m^{-2}), while the **luminous exitance** (M_v) is measured in lumens per square metre (lm m^{-2}). Exitance was formerly called **emittance**.

exoergic Denoting a nuclear process that gives out energy. *Compare* ENDOERGIC.

exosphere *See* EARTH'S ATMOSPHERE.

exothermic Denoting a chemical reaction that releases heat into its surroundings. *Compare* ENDOTHERMIC.

exotic atom 1. An atom in which an electron has been replaced by another negatively charged particle, such as a muon (*see* LEPTON) or *meson. In this case the negative particle eventually collides with the nucleus with the emission of X-ray photons. **2.** A system in which the nucleus of an atom has been replaced by a positively charged meson. Such exotic atoms have to be created artifically and are unstable.

expansion The writing of a function or quantity as a *series of terms. The series may be finite or infinite. *See* BINOMIAL THEOREM; TAYLOR SERIES.

expansion of the universe The hypothesis, based on the evidence of the *redshift, that the distance between the galaxies is continuously increasing. The original theory, which was proposed in 1929 by Edwin *Hubble, assumes that the galaxies are flying apart as a consequence of the big bang with which the universe originated. Several variants have since been proposed. *See also* BIG-BANG THEORY; HUBBLE CONSTANT.

expansivity (thermal expansion) **1.** Linear expansivity is the fractional increase in length of a specimen of a solid, per unit rise in temperature. If a specimen increases in length from l_1 to l_2 when its temperature is raised $\theta°$, then the expansivity (α) is given by:

$$l_2 = l_1(1 + \alpha\theta).$$

This relationship assumes that α is independent of temperature. This is not, in general, the case and a more accurate relationship is:

$$l_2 = l_1(1 + a\theta + b\theta^2 + c\theta^3 \ldots),$$

where a, b, and c are constants.

2. Superficial expansivity is the fractional increase in area of a solid surface caused by unit rise in temperature, i.e.

$$A_2 = A_1(1 + \beta\theta),$$

where β is the superficial expansivity. To a good approximation $\beta = 2\alpha$.

3. Volume expansivity is the fractional increase in volume of a solid, liquid, or gas per unit rise in temperature, i.e.

$$V_2 = V_1(1 + \gamma\theta),$$

where γ is the cubic expansivity and $\gamma = 3\alpha$. For liquids, the expansivity observed directly is called the **apparent expansivity** as the container will also have expanded with the rise in temperature. The **absolute expansivity** is the apparent expansivity plus the volume expansivity of the container. For the expansion of gases, *see* CHARLES' LAW.

exponent A number or symbol indicating the power to which another number or expression is raised. For example, $(x + y)^n$ indicates that the expression $(x + y)$ is raised to the nth power; n is the exponent. Any number or expression in which the exponent is zero is equal to 1, i.e. $x^0 = 1$.

exponential A function that varies as the power of another quantity. If $y = a^x$, y varies exponentially with x. The function e^x, also written as $\exp(x)$, is called the **exponential function** (*see* E). It is equal to the sum of the **exponential series**, i.e.

$$e^x = 1 + x + x^2/2! + x^3/3! + \ldots + x^n/n! + \ldots$$

exponential growth A form of population growth in which the rate of growth is related to the number of individuals present. Increase is slow when numbers are low but rises sharply as numbers increase. If population number is plotted against time on a graph a characteristic J-shaped curve results (see graph). In animal and plant populations, such factors as overcrowding, lack of nutrients, and disease limit population increase beyond a certain point and the J-shaped exponential curve tails off giving an S-shaped (sigmoid) curve.

exposure meter A photocell that operates a meter to indicate the correct exposure for a specified film in photography. It enables the correct shutter speed and aperture to be chosen for any photographic circumstances. Some cameras have a built-in exposure meter that automatically sets the aperture according to the amount of light available and the chosen shutter speed.

extended ASCII A set of characters with *ASCII values between 128 and 255. These characters may include special symbols, graphics characters, and accented characters. The assignment of extended ASCII characters is not standard. It depends on the particular computer system and may also depend on the font being used.

extensive variable A quantity in a *macroscopic system that is proportional to the size of the system. Examples of extensive variables include the volume, mass, and total energy. If an extensive variable is divided by an arbitrary extensive variable, such as the volume, an *intensive variable results. A macroscopic system can be described by one extensive variable and a set of intensive variables.

extensometer Any device for measuring the extension of a specimen of a material under longitudinal stress. A common method is to make the specimen form part of a capacitor, the capacitance of which will change with a change in the specimen's dimensions.

extinction coefficient A measure of the extent by which the intensity of a beam of light is reduced by passing through a distance d of a solution having a molar concentration c of the dissolved substance. If the intensity of the light is reduced from I_1 to I_2, the extinction coefficient is $[\log(I_1/I_2)]/cd$.

extraction **1.** The process of obtaining a metal from its ore. **2.** The separation of a component from a mixture by selective solubility.

extraordinary ray *See* DOUBLE REFRACTION.

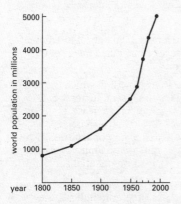

Graph showing exponential growth of the human population

extrapolation An *approximation technique for finding the value of a function or measurement beyond the values already known. If the values $f(x_0)$, $f(x_1)$,…,$f(x_n)$ of a function of a variable x are known in the interval $[x_0, x_n]$, the value of $f(x)$ for a value of x outside the interval $[x_0, x_n]$ can be found by extrapolation. The techniques used in extrapolation are usually not as good as those used in *interpolation.

extremely high frequency (EHF) A radio frequency between 30 000 megahertz and 300 gigahertz.

extrinsic semiconductor *See* SEMICONDUCTOR.

eye The organ of sight (see illustration). They normally occur in pairs, are nearly spherical, and filled with fluid. Light is refracted by the cornea through the pupil in the *iris and onto the *lens, which focuses images onto the retina. These images are received by light-sensitive cells in the retina (*see* CONE; ROD), which transmit impulses to the brain via the optic nerve.

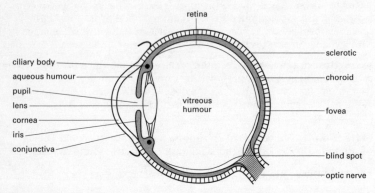

Structure of the vertebrate eye

eyepiece (ocular) The lens or system of lenses in an optical instrument that is nearest to the eye. It usually produces a magnified image of the previous image formed by the instrument.

Fabry–Pérot interferometer A type of *interferometer in which monochromatic light is passed through a pair of parallel half-silvered glass plates producing circular interference fringes. One of the glass plates is adjustable, enabling the separation of the plates to be varied. The wavelength of the light can be determined by observing the fringes while adjusting the separation. This type of instrument is used in spectroscopy.

face-centred cubic (f.c.c.) *See* CUBIC CRYSTAL.

facet A flat face of a crystal.

factorial The product of a given number and all the whole numbers below it. It is usually written $n!$, e.g. factorial $4 = 4! = 4 \times 3 \times 2 \times 1 = 24$. Factorial 0 is defined as 1.

Fahrenheit, Gabriel Daniel (1686–1736) German physicist, who became an instrument maker in Amsterdam. In 1714 he developed the mercury-in-glass thermometer, and devised a temperature scale to go with it (*see* FAHRENHEIT SCALE).

Fahrenheit scale A temperature scale in which (by modern definition) the temperature of boiling water is taken as 212 degrees and the temperature of melting ice as 32 degrees. It was invented in 1714 by G. D. Fahrenheit, who set the zero at the lowest temperature he knew how to obtain in the laboratory (by mixing ice and common salt) and took his own body temperature as 96°F. The scale is no longer in scientific use. To convert to the *Celsius scale the formula is $C = 5(F - 32)/9$.

fall-out (**radioactive fall-out**) Radioactive particles deposited from the atmosphere either from a nuclear explosion or from a nuclear accident. **Local fall-out**, within 250 km of an explosion, falls within a few hours of the explosion. **Tropospheric fall-out** consists of fine particles deposited all round the earth in the approximate latitude of the explosion within about one week. **Stratospheric fall-out** may fall anywhere on earth over a period of years. The most dangerous radioactive isotopes in fall-out are the fission fragments iodine–131 and strontium–90. Both can be taken up by grazing animals and passed on to human populations in milk, milk products, and meat. Iodine–131 accumulates in the thyroid gland and strontium–90 accumulates in bones.

false vacuum A state in *quantum field theory that is a local minimum but not the minimum energy state of the system overall (which is called the **true vacuum**). Tunnelling (*see* TUNNEL EFFECT) occurs from the false vacuum to the true vacuum, which can be calculated using instanton techniques. The false vacuum has never been observed but is predicted to exist in several quantum field theories of relevance to *elementary particles, including *grand unified theories. In particular, it has been suggested that the false vacuum could be responsible for the expansion in the inflationary universe, which could have been important in the *early universe.

farad Symbol F. The SI unit of capacitance, being the capacitance of a capacitor that, if charged with one coulomb, has a potential difference of one volt between its plates. $1 \text{ F} = 1 \text{ C V}^{-1}$. The farad itself is too large for most applications; the practical unit is the microfarad (10^{-6} F). The unit is named after Michael Faraday.

Faraday, Michael (1791–1867) British chemist and physicist, who received little formal education. He started to experiment on electricity and in 1812 attended lectures by Sir Humphry *Davy at the Royal Institution; a year later he became Davy's assistant. He remained at the Institution until 1861. Faraday's chemical discoveries include the liquefaction of chlorine (1823) and benzene (1825) as well as the laws of electrolysis (*see* FARADAY'S LAWS). He is probably best remembered for his work in physics: in 1821 he demonstrated electromagnetic rotation (the principle of the *electric motor) and in 1831 discovered *electromagnetic induction (the principle of the dynamo). In 1845 he discovered the *Faraday effect.

Faraday cage An earthed screen made of metal wire that surrounds an electric device in order to shield it from external electrical fields.

Faraday constant Symbol F. The electric charge carried by one mole of electrons or singly ionized ions, i.e. the product of the *Avogadro constant and the charge on an electron (disregarding sign). It has the value $9.648\ 5309(29) \times 10^{4}$ coulombs per mole. This number of coulombs is sometimes treated as a unit of electric charge called the **faraday**.

Faraday disc *See* HOMOPOLAR GENERATOR.

Faraday effect The rotation of the plane of polarization of electromagnetic radiation on passing through an isotropic medium exposed to a magnetic field. The angle of rotation is proportional to Bl, where l is the length of the path of the radiation in the medium and B is the magnetic flux density.

Faraday's laws Two laws describing electrolysis:
 (1) The amount of chemical change during electrolysis is proportional to the charge passed.
 (2) The charge required to deposit or liberate a mass m is given by $Q = Fmz/M$, where F is the Faraday constant, z the charge of the ion, and M the relative ionic mass.
These are the modern forms of the laws. Originally, they were stated by Faraday in a different form:
 (1) The amount of chemical change produced is proportional to the quantity of electricity passed.

(2) The amount of chemical change produced in different substances by a fixed quantity of electricity is proportional to the electrochemical equivalent of the substance.

Faraday's laws of electromagnetic induction (1) An e.m.f. is induced in a conductor when the magnetic field surrounding it changes. (2) The magnitude of the e.m.f. is proportional to the rate of change of the field. (3) The sense of the induced e.m.f. depends on the direction of the rate of change of the field.

fast neutron A neutron resulting from nuclear fission that has an energy in excess of 0.1 MeV (1.6×10^{-14} J), having lost little of its energy by collision. In some contexts **fast fission** is defined as fission brought about by fast neutrons, i.e. neutrons having energies in excess of 1.5 MeV (2.4×10^{-13} J), the fission threshold of uranium–238. *See also* NUCLEAR REACTOR; SLOW NEUTRON.

fast reactor *See* NUCLEAR REACTOR.

fatigue *See* METAL FATIGUE.

f.c.c. Face-centred cubic. *See* CUBIC CRYSTAL.

feedback The use of part of the output of a system to control its performance. In **positive feedback**, the output is used to enhance the input; an example is an electronic oscillator, or the howl produced by a loudspeaker that is placed too close to a microphone in the same circuit. A small random noise picked up by the microphone is amplified and reproduced by the loudspeaker. The microphone now picks it up again; it is further amplified, and fed from the speaker to microphone once again. This continues until the system is overloaded. In **negative feedback**, the output is used to reduce the input. In electronic amplifiers, stability is achieved, and distortion reduced, by using a system in which the input is decremented in proportion as the output increases. A similar negative feedback is used in *governors that reduce the fuel supply to an engine as its speed increases.

FEM *See* FIELD-EMISSION MICROSCOPE.

femto- Symbol f. A prefix used in the metric system to denote 10^{-15}. For example, 10^{-15} second = 1 femtosecond (fs).

Fermat's principle The path taken by a ray of light between any two points in a system is always the path that takes the least time. This principle leads to the law of the rectilinear propagation of light and the laws of reflection and refraction. It was discovered by the French mathematician, Pierre de Fermat (1601–65).

fermi A unit of length formerly used in nuclear physics. It is equal to 10^{-15} metre. In SI units this is equal to 1 femtometre (fm). It was named after Enrico Fermi.

Fermi, Enrico (1901–54) Italian-born US physicist. He became a professor at Rome University, where in 1934 he discovered how to produce slow (thermal) neutrons. He used these to create new radioisotopes, for which he was awarded the 1938 Nobel Prize. In 1938 he and his Jewish wife emigrated to the USA. In 1942 he led the team that built the first atomic pile (nuclear reactor) in Chicago.

Fermi constant Symbol G_W. The *coupling constant associated with the weak interaction (*see* FUNDAMENTAL INTERACTIONS), which gives rise to *beta decay. The Fermi constant has a value 1.435×10^{-36} joule metre^3. The Fermi constant characterizes the Fermi theory of weak interactions. The dimensional nature of G_W means that the Fermi theory is limited to low energies and is unrenormalizable (*see* RENORMALIZATION). This can be established by *dimensional analysis for *perturbation theory calculations of weak-interaction processes.

Fermi–Dirac statistics *See* QUANTUM STATISTICS.

Fermi level The energy in a solid at which the average number of particles per quantum state is ½; i.e. one half of the quantum states are occupied. The Fermi level in conductors lies in the conduction band (*see* ENERGY BANDS), in insulators and semiconductors it falls in the gap between the conduction band and the valence band.

Fermi pressure *See* DEATH OF A STAR.

fermion An *elementary particle (or bound state of an elementary particle, e.g. an atomic nucleus or an atom) with half-integral spin; i.e. a particle that conforms to Fermi–Dirac statistics (*see* QUANTUM STATISTICS). *Compare* BOSON.

fermium Symbol Fm. A radioactive metallic transuranic element belonging to the actinoids; a.n. 100; mass number of the most stable isotope 257 (half-life 10 days). Ten isotopes are known. The element was first identified by A. Ghiorso and associates in debris from the first hydrogen-bomb explosion in 1952.

ferrimagnetism *See* MAGNETISM.

ferrite **1.** A member of a class of mixed oxides $MO.Fe_2O_3$, where M is a metal such as cobalt, manganese, nickel, or zinc. The ferrites are ceramic materials that show either ferrimagnetism or ferromagnetism, but are not electrical conductors. For this reason they are used in high-frequency circuits as magnetic cores. **2.** *See* STEEL.

ferroalloys Alloys of iron with other elements made by smelting mixtures of iron ore and the metal ore; e.g. ferrochromium, ferrovanadium, ferromanganese, ferrosilicon, etc. They are used in making alloy *steels.

ferroelectric materials Ceramic dielectrics, such as Rochelle salt and barium titanate, that have a domain structure making them analogous to ferromagnetic materials (*see* MAGNETISM). They exhibit hysteresis and usually the *piezoelectric effect.

ferromagnetism *See* MAGNETISM.

fertile material A nuclide that can absorb a neutron to form a *fissile material. Uranium–238, for example, absorbs a neutron to form uranium–239, which decays to plutonium–239. This is the type of conversion that occurs in a breeder reactor (*see* NUCLEAR REACTOR).

FET *See* TRANSISTOR.

Feynman, Richard Phillips (1918–88) US theoretical physicist, who invented the *path integral formulation of quantum mechanics, was one of the inventors of *renormalization in quantum electrodynamics (QED), and made important contributions to the theories of strong and weak interactions, quantum gravity, and superfluidity. His teaching was very influential, as were his visionary articles on nanotechnology and quantum computers. He shared the 1965 Nobel Prize for physics for his work on QED.

Feynman diagram *See* QUANTUM ELECTRODYNAMICS.

fibre optics *See* OPTICAL FIBRES.

field A region in which a body experiences a *force as the result of the presence of some other body or bodies. A field is thus a method of representing the way in which bodies are able to influence each other. For example, a body that has mass is surrounded by a region in which another body that has mass experiences a force tending to draw the two bodies together. This is the gravitational field (*see* NEWTON'S LAW OF GRAVITATION). The other three *fundamental interactions can also be represented by means of fields of force. However in the case of the *magnetic field and *electric field that together create the electromagnetic interaction, the force can vary in direction according to the character of the field. For example, in the field surrounding a negatively charged body, a positively charged body will experience a force of attraction, while another negatively charged body is repelled.

   The strength of any field can be described as the ratio of the force experienced by a small appropriate specimen to the relevant property of that specimen, e.g. force/mass for the gravitational field. *See also* QUANTUM FIELD THEORY.

field coil The coil in an electrical machine that produces the magnetic field.

field-effect transistor (FET) *See* TRANSISTOR.

field emission The emission of electrons from cold metals by *electric fields. In order to build up sufficiently large electric fields, the metal is usually shaped to a sharp needle point. Field emission is an example of the *tunnel effect in *quantum mechanics, with an electron in the metal being in a *potential barrier. In field emission the probability of tunnelling, which can be calculated using the *semiclassical approximation, is related to the *work function of the metal. An important application of field emission is the *field-emission microscope.

field-emission microscope (FEM) A type of electron microscope in which a high negative voltage is applied to a metal tip placed in an evacuated vessel some distance from a glass screen with a fluorescent coating. The tip produces electrons by *field emission. The emitted electrons form an enlarged pattern on the fluorescent screen, related to the individual exposed planes of atoms. As the resolution of the instrument is limited by the vibrations of the metal atoms, it is helpful to cool the tip in liquid helium. Although the individual atoms forming the point are not displayed, individual adsorbed atoms of other substances can be, and their activity is observable. See illustration.

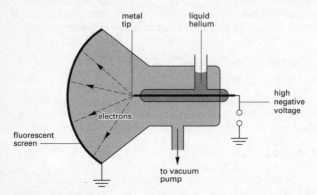

Field-emission microscope

field-ionization microscope (field-ion microscope; FIM) A type of electron microscope that is similar in principle to the *field-emission microscope, except that a high positive voltage is applied to the metal tip, which is surrounded by low-pressure gas (usually helium) rather than a vacuum. The image is formed in this case by **field ionization**: ionization at the surface of an unheated solid as a result of a strong electric field creating positive ions by electron transfer from surrounding atoms or molecules. The image is formed by ions striking the fluorescent screen. Individual atoms on the surface of the tip can be resolved and, in certain cases, adsorbed atoms may be detected.

field lens The lens in the compound eye-piece of an optical instrument that is furthest from the eye. Its function is to increase the field of view by refracting towards the main eye lens rays that would otherwise miss it.

field magnet The magnet that provides the magnetic field in an electrical machine. In some small dynamos and motors it is a permanent magnet but in most machines it is an electromagnet.

file A collection of data stored in a computer. It may consist of program instructions or numerical, textual, or graphical information. It usually consists of a set of similar or related records.

film badge A lapel badge containing masked photographic film worn by personnel who could be exposed to ionizing radiation. The film is developed to indicate the extent that the wearer has been exposed to harmful radiation.

filter 1. (in chemistry) A device for separating solid particles from a liquid or gas. The simplest laboratory filter for liquids is a funnel in which a cone of paper (**filter paper**) is placed. Special containers with a porous base of sintered glass are also used. **2.** (in physics) A device placed in the path of a beam of radiation to alter its frequency distribution. For example, a plane pigmented piece of glass may be placed over a camera lens to alter the relative intensity of the component wavelengths of the beam entering the camera. **3.** (in electronics) An electrical network that transmits signals within a certain frequency range but attenuates other frequencies.

filter pump A simple laboratory vacuum pump in which air is removed from a system by a jet of water forced through a narrow nozzle. The lowest pressure possible is the vapour pressure of water.

filtrate The clear liquid obtained by filtration.

filtration The process of separating solid particles using a filter. In **vacuum filtration**, the liquid is drawn through the filter by a vacuum pump. Ultrafiltration is filtration under pressure; for example, ultrafiltration of the blood occurs in the nephrons of the vertebrate kidney.

FIM *See* FIELD-IONIZATION MICROSCOPE.

finder A small low-powered astronomical telescope, with a wide field of view, that is fixed to a large astronomical telescope so that the large telescope can be pointed in the correct direction to observe a particular celestial body.

fine structure Closely spaced optical spectral lines arising from transitions between energy levels that are split by the vibrational or rotational motion of a molecule or by electron spin. They are visible only at high resolution. **Hyperfine structure**, visible only at very high resolution, results from the influence of the atomic nucleus on the allowed energy levels of the atom.

fine structure constant Symbol α. The dimensionless constant, with a value of about 1/137, that characterizes quantum electrodynamics. It is given by $\alpha = e^2/\hbar c$, where e is the charge on an electron, $\hbar$ is the Dirac constant, and c is the speed of light in free space.

finite series *See* SERIES.

fissile material A nuclide of an element that undergoes nuclear fission, either spontaneously or when irradiated by neutrons. Fissile nuclides, such as uranium–235 and plutonium–239, are used in *nuclear reactors and nuclear weapons. *Compare* FERTILE MATERIAL.

fission *See* NUCLEAR FISSION.

fission products *See* NUCLEAR FISSION.

fission-track dating A method of estimating the age of glass and other mineral objects by observing the tracks made in them by the fission fragments of the uranium nuclei that they contain. By irradiating the objects with neutrons to induce fission and comparing the density and number of the tracks before and after irradiation it is possible to estimate the time that has elapsed since the object solidified.

Fitzgerald contraction *See* LORENTZ–FITZGERALD CONTRACTION.

fixed point A temperature that can be accurately reproduced to enable it to be used as the basis of a *temperature scale.

fixed star One of very many heavenly bodies that does not appear to alter its position on the *celestial sphere. They were so called to distinguish them from the planets, which were once known as **wandering stars**. The discovery of the *proper motion of stars in the 18th century established that stars are not fixed in the sky although, because of their immense distances from the solar system, they may appear to be so.

fixed-target experiments *See* PARTICLE-BEAM EXPERIMENTS.

Fizeau, Armand Hippolyte Louis (1819–96) French physicist. In 1845 he and Léon Foucault (1819–68) took the first photographs of the sun. In 1849 he measured the *speed of light; he also analysed the *Doppler effect for light.

Fizeau's method A method of measuring the speed of light, invented by Armand Fizeau in 1849. A cogwheel rotating at high speed enables a series of flashes to be transmitted to a distant mirror. The light reflected back to the cogwheel is observed and the speed of light calculated from the rates of rotation of the wheel required to produce an eclipse of the returning light.

flame A hot luminous mixture of gases undergoing combustion. The chemical reactions in a flame are mainly free-radical chain reactions and the light comes from fluorescence of excited molecules or ions or from incandescence of small solid particles (e.g. carbon).

flash memory A form of semiconductor storage in which the data may

be altered electrically. The device does not need refreshing to maintain
the data, which is stored even when power is removed. Flash memory
finds application in computers, in digital cameras, and in portable storage
devices that emulate hard disks (e.g. *USB drives).

flash photolysis A technique for studying free-radical reactions in
gases. The apparatus used typically consists of a long glass or quartz tube
holding the gas, with a lamp outside the tube suitable for producing an
intense flash of light. This dissociates molecules in the sample creating
free radicals, which can be detected spectroscopically by a beam of light
passed down the axis of the tube. It is possible to focus the spectrometer
on an absorption line for a particular product and measure its change in
intensity with time using an oscilloscope. In this way the kinetics of very
fast free-radical gas reactions can be studied.

flash point The temperature at which the vapour above a volatile liquid
forms a combustible mixture with air. At the flash point the application of
a naked flame gives a momentary flash rather than sustained combustion,
for which the temperature is too low.

flavour *See* ELEMENTARY PARTICLES.

Fleming's rules Rules to assist in remembering the relative directions
of the field, current, and force in electrical machines. The left hand refers
to motors, the right hand to generators. If the forefinger, second finger,
and thumb of the left hand are extended at right angles to each other, the
forefinger indicates the direction of the field, the second finger the
direction of the current, and the thumb the direction of the force. If the
right hand is used the digits indicate these directions in a generator. The
mnemonic was invented by Sir John Ambrose Fleming (1849–1945).

flip-flop (bistable circuit) An electronic circuit that has two stable states.
It is switched from one stable state to the other by means of a triggering
pulse. They are extensively used as *logic circuits in computers.

floppy disk (diskette) A flexible plastic disk with a magnetic coating
encased in a stiff envelope. It is used to store information in a small
computer system. *See* MAGNETIC DISK.

fluctuation–dissipation theorem A theory relating quantities in
equilibrium and *nonequilibrium statistical mechanics and *microscopic
and *macroscopic quantities. The fluctuation–dissipation theorem was
first derived for electrical circuits with *noise in 1928 by H. Nyquist; a
general theorem in statistical mechanics was derived by H. B. Callen
and T. A. Welton in 1951. The underlying principle of the fluctuation–
dissipation theorem is that a nonequilibrium state may have been
reached either as a result of a random fluctuation or an external force
(such as an electric or magnetic field) and that the evolution towards
equilibrium is the same in both cases (for a sufficiently small fluctuation).
The fluctuation–dissipation theorem enables *transport coefficients to be
calculated in terms of response to external fields.

fluctuations Random deviations in the value of a quantity about some average value. In all systems described by *quantum mechanics fluctuations, called **quantum fluctuations**, occur – even at the *absolute zero of thermodynamic temperature as a result, ultimately, of the Heisenberg *uncertainty principle. In any system above absolute zero, fluctuations, called **thermal fluctuations**, occur. It is necessary to take fluctuations into account to obtain a quantitative theory of *phase transitions in three dimensions. The formation of structure in the *early universe is thought to be a result of quantum fluctuations.

fluidics The use of jets of fluid in pipes to perform many of the control functions usually performed by electronic devices. Being about one million times slower than electronic devices, fluidic systems are useful where delay lines are required. They are also less sensitive to high temperatures, strong magnetic fields, and ionizing radiation than electronic devices.

fluidization A technique used in some industrial processes in which solid particles suspended in a stream of gas are treated as if they were in the liquid state. Fluidization is useful for transporting powders, such as coal dust. **Fluidized beds**, in which solid particles are suspended in an upward stream, are extensively used in the chemical industry, particularly in catalytic reactions where the powdered catalyst has a high surface area. They are also used in furnaces, being formed by burning coal in a hot turbulent bed of sand or ash through which air is passed. The bed behaves like a fluid, enabling the combustion temperature to be reduced so that the production of polluting oxides of nitrogen is diminished. By adding limestone to the bed with the fuel, the emission of sulphur dioxide is reduced.

High-pressure fluidized beds are also used in power-station furnaces in a **combined cycle** in which the products of combustion from the fluidized bed are used to drive a gas turbine, while a steam-tube boiler in the fluid bed raises steam to drive a steam turbine. This system both increases the efficiency of the combustion process and reduces pollution.

fluid mechanics The study of fluids at rest and in motion. **Fluid statics** is concerned with the pressures and forces exerted on liquids and gases at rest. *Hydrostatics is specifically concerned with the behaviour of liquids at rest. In **fluid dynamics** the forces exerted on fluids, and the motion that results from these forces, are examined. It can be divided into *hydrodynamics: the motion of liquids (not only water); and aerodynamics: the motion of gases.

Fluid dynamics is an important science used to solve many of the problems arising in aeronautical, chemical, mechanical, and civil engineering. It also enables many natural phenomena, such as the flight of birds, the swimming of fish, and the development of weather conditions, to be studied scientifically.

fluorescence *See* LUMINESCENCE.

fluorescent light *See* ELECTRIC LIGHTING.

flux **1.** *See* LUMINOUS FLUX. **2.** *See* MAGNETIC FLUX. **3.** *See* ELECTRIC FLUX. **4.** The number of particles flowing per unit area of cross section in a beam of particles.

flux density **1.** *See* MAGNETIC FIELD. **2.** *See* ELECTRIC DISPLACEMENT.

fluxmeter An instrument used to measure *magnetic flux. It is used in conjunction with a coil (the **search coil**) and resembles a moving-coil galvanometer except that there are no restoring springs. A change in the magnetic flux induces a momentary current in the search coil and in the coil of the meter, which turns in proportion and stays in the deflected position. This type of instrument has been largely superseded by a type using the Hall probe (*see* HALL EFFECT).

flux quantum A quantum of magnetic flux that occurs in a superconducting ring such as a hollow cylinder. The magnetic flux is quantized in multiples of $h/(2e)$, where h is the Planck constant and e is the charge of an electron.

FM (frequency modulation) *See* MODULATION.

f-number *See* APERTURE.

focal length The distance between the *optical centre of a lens or pole of a spherical mirror and its *principal focus.

focal point *See* FOCUS.

focal ratio *See* APERTURE.

focus **1.** (in optics) Any point in an optical system through or towards which rays of light are converged. It is sometimes called the **focal point** and sometimes loosely used to mean *principal focus or (particularly by photographers) *focal length. **2.** (in mathematics) *See* CONIC; ELLIPSE.

Fokker–Planck equation An equation in *nonequilibrium statistical mechanics that describes a superposition of a dynamic friction (slowing-down) process and a diffusion process for the evolution of variables in a system. The Fokker–Planck equation, which can be used to analyse such problems as *Brownian movement, can be solved using statistical methods and the theory of probability. It is named after the Dutch physicist Adriaan Fokker (1887–1968) and Max *Planck. Results obtained from the Fokker–Planck equation are in agreement with results obtained from the *Langevin equation.

foot The unit of length in *f.p.s. units. It is equal to one-third of a yard and is now therefore defined as 0.3048 metre. Several units based on the foot were formerly used in science, including the units of work, the **foot-pound-force** and the **foot-poundal**, and the illumination units, the **foot-candle** and the **foot-lambert**. These have all been replaced by SI units.

forbidden band *See* ENERGY BANDS.

forbidden transitions Transitions between energy levels in a quantum-mechanical system that are not allowed to take place because of *selection rules. In practice, forbidden transitions can occur, but they do so with much lower probability than allowed transitions. There are three reasons why forbidden transitions may occur:

(1) the selection rule that is violated is only an approximate rule. An example is provided by those selection rules that are only exact in the absence of *spin–orbit coupling. When spin–orbit coupling is taken into account, the forbidden transitions become allowed – their strength increasing with the size of the spin–orbit coupling;

(2) the selection rule is valid for dipole radiation, i.e. in the interaction between a quantum-mechanical system, such as an atom, and an electromagnetic field, only the (variable) electric dipole moment is considered. Actual transitions may involve magnetic dipole radiation or quadrupole radiation;

(3) the selection rule only applies for an atom, molecule, etc., in isolation and does not necessarily apply if external fields, collisions, etc., are taken into account.

force Symbol F. The agency that tends to change the momentum of a massive body, defined as being proportional to the rate of increase of momentum. For a body of mass m travelling at a velocity v, the momentum is mv. In any coherent system of units the force is therefore given by $F = \mathrm{d}(mv)/\mathrm{d}t$. If the mass is constant $F = m\mathrm{d}v/\mathrm{d}t = ma$, where a is the acceleration (*see* NEWTON'S LAWS OF MOTION). The SI unit of force is the newton. Forces occur always in equal and opposite action–reaction pairs between bodies, though it is often convenient to think of one body being in a force *field.

forced convection *See* CONVECTION.

force ratio (mechanical advantage) The ratio of the output force (load) of a machine to the input force (effort).

formula **1.** (in chemistry) A way of representing a chemical compound using symbols for the atoms present. Subscripts are used for the numbers of atoms. The **molecular formula** simply gives the types and numbers of atoms present. For example, the molecular formula of ethanoic acid is $C_2H_4O_2$. The **empirical formula** gives the atoms in their simplest ratio; for ethanoic acid it is CH_2O. The **structural formula** gives an indication of the way the atoms are arranged. Commonly, this is done by dividing the formula into groups; ethanoic acid can be written $CH_3.CO.OH$ (or more usually simply CH_3COOH). Structural formulae can also show the arrangement of atoms or groups in space. **2.** (in mathematics and physics) A rule or law expressed in algebraic symbols.

Fortin barometer *See* BAROMETER.

fossil fuel Coal, oil, and natural gas, the fuels used by man as a source of energy. They are formed from the remains of living organisms and all have a high carbon or hydrogen content. Their value as fuels relies on the exothermic oxidation of carbon to form carbon dioxide ($C + O_2 \rightarrow CO_2$) and the oxidation of hydrogen to form water ($H_2 + \frac{1}{2}O_2 \rightarrow H_2O$).

Foucault pendulum A simple pendulum in which a heavy bob attached to a long wire is free to swing in any direction. As a result of the earth's rotation, the plane of the pendulum's swing slowly turns (at the poles of the earth it makes one complete revolution in 24 hours). It was devised by the French physicist Jean Bernard Léon Foucault (1819–68) in 1851, when it was used to demonstrate the earth's rotation.

Fourier analysis The representation of a function f(x), which is periodic in x, as an infinite series of sine and cosine functions,

$f(x) = a_0/2 + \sum_{n=1}^{\infty} (a_n \cos nx + b_n \sin nx)$.

A series of this type is called a *Fourier series. If the function is periodic with a period 2π, the coefficients a_0, a_n, b_n are:

$a_0 = \int_{-\pi}^{\pi} f(x)dx$,

$a_n = \int_{-\pi}^{\pi} f(x) \cos nx dx$, $(n = 1,2,3,\ldots)$,

$b_n = 1/\pi \int_{-\pi}^{\pi} f(x) \sin nx dx$, $(n = 1,2,3,\ldots)$.

Fourier analysis and Fourier series are named after the French mathematician and engineer Joseph Fourier (1768–1830). Fourier series have many important applications in mathematics, science, and engineering, having been invented by Fourier in the first quarter of the 19th century in his analysis of the problem of heat *conduction.

Fourier series An expansion of a periodic function as a series of trigonometric functions. Thus,

$f(x) = a_0 + (a_1 \cos x + b_1 \sin x) + (a_2 \cos 2x + b_2 \sin 2x) + \ldots$,

where a_0, a_1, b_1, b_2, etc., are constants, called **Fourier coefficients**. The series is used in *Fourier analysis.

fourth dimension *See* SPACE–TIME.

f.p.s. units The British system of units based on the foot, pound, and second. It has now been replaced for all scientific purposes by SI units.

fractal A curve or surface generated by a process involving successive subdivision. For example, a **snowflake curve** can be produced by starting with an equilateral triangle and dividing each side into three segments. The middle segments are then replaced by two equal segments, which form the sides of a smaller equilateral triangle. This gives a 12-sided star-shaped figure. The next stage is to subdivide each of the sides of this figure in the same way, and so on. The result is a developing figure that resembles a snowflake. In the limit, this figure has 'fractional dimension' – i.e. a dimension between that of a line (1) and a surface (2); the dimension of the snowflake curve is 1.26. The study of this type of **self-**

similarity in figures is used in certain branches of physics, such as crystal growth. Fractals are also important in *chaos theory and in computer graphics. *See also* MANDELBROT SET.

fractional quantum Hall effect *See* QUANTUM HALL EFFECT.

frame dragging *See* LENSE-THIRRING EFFECT.

frame of reference A set of axes, taken as being for practical purposes at rest, that enables the position of a point (or body) in space to be defined at any instant of time. In a four-dimensional continuum (*see* SPACE-TIME) a frame of reference consists of a set of four coordinate axes, three spatial and one of time.

Franklin, Benjamin (1706–90) American scientist and statesman who held various government posts. As an amateur scientist he experimented with electricity, introducing the concepts of 'positive' and 'negative'. In 1752 he carried out the extremely dangerous experiment of flying a kite during a thunderstorm and proved the electrical nature of lightning. He also invented the lightning conductor.

Frasch process A method of obtaining sulphur from underground deposits using a tube consisting of three concentric pipes. Superheated steam is passed down the outer pipe to melt the sulphur, which is forced up through the middle pipe by compressed air fed through the inner tube. The steam in the outer casing keeps the sulphur molten in the pipe. It is named after the German-born US chemist Hermann Frasch (1851–1914).

Fraunhofer, Josef von (1787–1826) German physicist, who trained as an optician. In 1814 he observed dark lines in the spectrum of the sun (*see* FRAUNHOFER LINES). He also studied *Fraunhofer diffraction.

Fraunhofer diffraction A form of *diffraction in which the light source and the receiving screen are in effect at infinite distances from the diffracting object, so that the wave fronts can be treated as planar rather than spherical. In practice it involves parallel beams of light. It can be regarded as an extreme case of *Fresnel diffraction but is of more practical use in explaining single and multiple slit patterns. It was studied by Joseph von Fraunhofer.

Fraunhofer lines Dark lines in the solar spectrum that result from the absorption by elements in the solar chromosphere of some of the wavelengths of the visible radiation emitted by the hot interior of the sun.

free electron theory The theory that treats *electrons as freely moving within the body of a metal crystal or fluid. See Feature (pp 192–193).

free energy A measure of a system's ability to do work. The **Gibbs free energy** (or **Gibbs function**), G, is defined by $G = H - TS$, where G is the energy liberated or absorbed in a reversible process at constant pressure and constant temperature (*T*), *H* is the *enthalpy, and *S* the *entropy of the system. Changes in Gibbs free energy, ΔG, are useful in indicating the conditions under which a chemical reaction will occur. If ΔG is positive the reaction will only occur if energy is supplied to force it away from the equilibrium position (i.e. when $\Delta G = 0$). If ΔG is negative the reaction will proceed spontaneously to equilibrium. It is named after Josiah Willard *Gibbs.

The **Helmholtz free energy** (or **Helmholtz function**), *F*, is defined by $F = U - TS$, where *U* is the *internal energy. For a reversible isothermal process, ΔF represents the useful work available. It is named after H. L. F. von *Helmholtz.

free fall Motion resulting from a gravitational field that is unimpeded by a medium that would provide a frictional retarding force or buoyancy. In the earth's gravitational field, free fall takes place at a constant acceleration, known as the *acceleration of free fall.

free space A region in which there is no matter and no electromagnetic or gravitational fields. It has a temperature of absolute zero, unit refractive index, and the speed of light is its maximum value. The electric constant (*see* PERMITTIVITY) and the magnetic constant (*see* PERMEABILITY) are defined for free space.

freeze drying A process used in dehydrating food, blood plasma, and other heat-sensitive substances. The product is deep-frozen and the ice trapped in it is removed by reducing the pressure and causing it to sublime. The water vapour is then removed, leaving an undamaged dry product.

freezing mixture A mixture of components that produces a low temperature. For example, a mixture of ice and sodium chloride gives a temperature of –20°C.

Frenkel defect *See* DEFECT.

Frenkel–Kontorowa model A one-dimensional model of atoms, such as xenon, adsorbed on a periodic substrate, such as graphite. This model, which can be used to investigate the nature of the lattice formed by the adsorbed gas, was invented in 1938 by Y. I. Frenkel and T. Kontorowa and independently in 1949 by F. C. Frank and J. H. Van der Merwe. A feature of the analysis is that it involves a *devil's staircase. The Frenkel–Kontorowa model can be used to investigate the *phase transition between a *commensurate lattice and an *incommensurate lattice.

frequency Symbol *f* or ν. The rate of repetition of a regular event. The number of cycles of a wave, or some other oscillation or vibration, per second is expressed in *hertz (cycles per second). The frequency (*f*) of a wave motion is given by $f = c/\lambda$, where *c* is the velocity of propagation and λ is the wavelength. The frequency associated with a quantum of electromagnetic energy is given by $f = E/h$, where *E* is the quantum's energy and *h* is the Planck constant.

FREE-ELECTRON THEORY

To a first approximation any *electron within the body of a metal crystal or fluid may be treated as freely moving .

A single electron within the ionic lattice can be analysed as a particle-in-a-box problem from *quantum mechanics. The possible *quantum states (ψ) and corresponding energies (E) are obtained on solving the *Schrödinger equation for the *wave function (ψ). The electron is assumed to be a free particle within the box; outside the box, however, the wave describing the properties of the electron is assumed to have zero amplitude. The Schrödinger equation for this problem is:

$$-(h/2\pi)^2(1/2m_e)\nabla^2\psi = E\psi \tag{1}$$

where m_e is the mass of the electron and ∇ is the vector operator del. The solution of this particle-in-a-box problem leads to an expression for the quantum states of the form:

$$\psi(n_1, n_2, n_3) = A.\sin(\pi.n_1.x/L).\sin(\pi.n_2.y/L).\sin(\pi.n_3.z/L) \tag{2}$$

where A is a constant, L is the linear dimension of the crystal, and n_1, n_2, n_3 are integers acting as labels for the different allowed quantum states of the electron in the crystal. The relationship between the energy of the system and the integers n_1, n_2, n_3 is obtained by substituting (2) into (1):

$$(h^2/8m_eL^2)(n_1^2 + n_2^2 + n_3^2)\psi = E\psi \tag{3}$$

therefore,

$$E = (h^2/8m_eL^2)n^2 \tag{4}$$

where $n^2 = (n_1^2 + n_2^2 + n_3^2)$.

The wave function $\psi(n_1, n_2, n_3)$ gives the possible ways of fitting the electron wave into the crystal box. However, for each of these ways, the electron also has two orientations of spin. Therefore, when a summation of the total number of states up to a particular n value is calculated, these two degenerate states of spin must be accounted for. The total number of electron states up to a particular n is:

$$G(n) = \pi n^3/3 \tag{5}$$

In terms of energy E, G becomes:

$$G(E) = 8\pi V(2m_eE)^{3/2}/3h^3 \tag{6}$$

where $V (=L^3)$ is the volume of the crystal.

The density of states

A useful quantity $g(E)$ is derived from $G(E)$ by considering how $G(E)$ varies with respect to the variable E. $g(E)$ is called the **density of states per unit energy interval** and is a measure of the number of states within a given arbitrarily narrow interval of energy that the electron may possess:

$$g(E) = 3G(E)/2E \tag{7}$$

It is obvious from (6) that the greater the volume the more states can be accommodated up to an energy E. This is a direct consequence of the Heisenberg *uncertainty principle – the greater the uncertainty in position, the less the uncertainty will be in the momentum and therefore the kinetic energy.

The density of states $g(E)$ is a useful quantity for analysing the behaviour of electrons in a metal at different temperatures. For example, at absolute zero

(0K), the first electron will be in the lowest energy state, and subsequent electrons will occupy the next higher states. Electrons cannot occupy already filled states because as *fermions they obey Fermi–Dirac statistics (see QUANTUM STATISTICS). If there are N electrons to be accommodated within the metal, the lowest N energy states will be filled up to an energy E_{max} given by:

$$N = 8\pi V(2m_e E_{max})^{3/2}/3h^3 \tag{8}$$

and

$$g(E_{max}) = 3G(E_{max})/2E_{max} \tag{9}$$

For a monovalent metal, such as sodium or copper, the electrons in the atom are arranged in closed shells with one extra electron in the outside shell. It is this single electron in each atom that is considered free. For a mole of the monovalent metal there will be 6×10^{23} of these free electrons, i.e. $N = 6 \times 10^{23}$. For a molar volume of about 10 cm^3, (8) gives a value of E_{max} at about 10^{-18} joules. An application of the theory of *equipartition of energy leads to an estimate of electron speeds of the order of 10^6 m s^{-1}. The free electron theory therefore predicts a very rapid random motion for conduction electrons inside a metal. Indeed, in this model free electrons are sometimes given the name 'free electron gas', because they behave like the atoms or molecules in an ideal gas.

At 0K the occupation of the lowest 6×10^{23} states in the mole of metal can be represented in a graph of $g(E)$ (see diagram 1). For temperatures above absolute zero, any variation of the electron energy is provided by the thermal energy of the crystal. The kinetic energies of the electrons are very large compared to these thermal excitations so the energy distribution is only changed very slightly to the form shown in diagram (2).

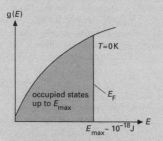

(1) $g(E)$ plotted as a function of energy (E) for the free electron model. At 0K all the states are occupied up to E_{max}.

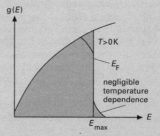

(2) At temperatures above 0K, the occupation of states in the region of E_{max} is slightly distorted.

The Fermi level

The line that delimits the occupied states at a given temperature is called the *Fermi level for that temperature and is given the symbol E_F. At 0K, $E_F = E_{max}$, but it is obvious that this is not the case at all other temperatures. In metals, however, it is sufficient to assume that E_F is equal to E_{max} as the temperature dependence is negligible (see diagram 2). In *semiconductors, in general, $E_F \neq E_{max}$.

frequency modulation (FM) *See* MODULATION; RADIO.

fresnel A unit of frequency equal to 10^{12} hertz. In SI units this is equal to 1 terahertz (THz). It was named after the French physicist A. J. Fresnel (1788–1827).

Fresnel diffraction A form of *diffraction in which the light source or the receiving screen, or both, are at finite distances from the diffracting object, so that the wavefronts are not plane, as in *Fraunhofer diffraction. It was studied by A. J. Fresnel.

Fresnel lens A lens with one face cut into a series of steps. It enables a relatively light and robust lens of short focal length though poor optical quality to be used in projectors (as condenser lenses), searchlights, spotlights, car headlights, and lighthouses. Such lenses have been made fine enough to serve as magnifiers for reading small print.

friction The force that resists the motion of one surface relative to another with which it is in contact. For a body resting on a horizontal surface there is a normal contact force, R, between the body and surface, acting perpendicularly to the surface. If a horizontal force B is applied to the body with the intention of moving it to the right, there will be an equal horizontal friction force, F, to the left, resisting the motion (see illustration). If B is increased until the body just moves, the value of F will also increase until it reaches the **limiting frictional force** (F_L), which is the maximum value of F. F_L is then equal to $\mu_s R$, where μ_s is the **coefficient of static friction**, the value of which depends on the nature of the surfaces. Once the body is moving with constant velocity, the value of F falls to a value F_k, which is equal to $\mu_k R$, where μ_k is the **coefficient of kinetic friction**. Both μ_s and μ_k are independent of the surface area of the body unless this is very small and μ_k is almost independent of the relative velocity of the body and surface.

The cause of friction is that surfaces, however smooth they may look to the eye, on the microscopic scale have many humps and crests. Therefore the actual area of contact is very small indeed, and the consequent very high pressure leads to local pressure welding of the surfaces. In motion the welds are broken and remade continually. *See also* ROLLING FRICTION.

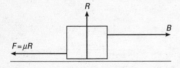

Friction

Friedmann–Lemaître–Robertson–Walker model A model of the universe corresponding to solutions of the Einstein field equations of general relativity theory associated with an expanding homogeneous isotropic universe. This model gives a good overall description of the evolution of the universe. It was proposed by Aleksandr Friedmann,

Georges Lemaître, Howard Robertson, and Arthur Walker in the 1920s and 1930s.

froth flotation A method of separating mixtures of solids, used industrially for separating ores from the unwanted gangue. The mixture is ground to a powder and water and a frothing agent added. Air is blown through the water. With a suitable frothing agent, the bubbles adhere only to particles of ore and carry them to the surface, leaving the gangue particles at the bottom.

frustration A situation in which there are competing interactions in a system. For example, in a *spin glass the magnetic atoms are subject to both ferromagnetic and antiferromagnetic interactions. Frustration also occurs in the folding of large polymer molecules due to parts of the molecule getting in each other's way. It is thought that proteins fold readily to specific shapes because they have evolved to minimize this type of steric frustration.

frustum A solid figure produced when two parallel planes cut a larger solid or when one plane parallel to the base cuts it.

fuel A substance that is oxidized or otherwise changed in a furnace or heat engine to release useful heat or energy. For this purpose wood, vegetable oil, and animal products have largely been replaced by *fossil fuels since the 18th century.

The limited supply of fossil fuels and the expense of extracting them from the earth has encouraged the development of nuclear fuels to produce electricity (*see* NUCLEAR ENERGY).

fuel cell A cell in which the chemical energy of a fuel is converted directly into electrical energy. The simplest fuel cell is one in which hydrogen is oxidized to form water over porous sintered nickel electrodes. A supply of gaseous hydrogen is fed to a compartment containing the porous cathode and a supply of oxygen is fed to a compartment containing the porous anode; the electrodes are separated by a third compartment containing a hot alkaline electrolyte, such as potassium hydroxide. The electrodes are porous to enable the gases to react with the electrolyte, with the nickel in the electrodes acting as a catalyst. At the cathode the hydrogen reacts with the hydroxide ions in the electrolyte to form water, with the release of two electrons per hydrogen molecule:

$$H_2 + 2OH^- \rightarrow 2H_2O + 2e^-$$

At the anode, the oxygen reacts with the water, taking up electrons, to form hydroxide ions:

$$\tfrac{1}{2}O_2 + H_2O + 2e^- \rightarrow 2OH^-$$

The electrons flow from the cathode to the anode through an external circuit as an electric current. The device is a more efficient converter of electric energy than a heat engine, but it is bulky and requires a continuous supply of gaseous fuels. Their use to power electric vehicles is being actively explored.

The second generation of fuel cells uses molten salts, especially carbonates of metals such as lithium and potassium, as electrolytes. The third generation of fuel cells uses conducting solid ionic oxides as electrolytes.

fuel element *See* NUCLEAR REACTOR.

fugacity Symbol f. A thermodynamic function used in place of partial pressure in reactions involving real gases and mixtures. For a component of a mixture, $d(\ln f) = d\mu/RT$, where μ is the chemical potential. It has the same units as pressure and the fugacity of a gas is equal to the pressure if the gas is ideal. The fugacity of a liquid or solid is the fugacity of the vapour with which it is in equilibrium. The ratio of the fugacity to the fugacity in some standard state is the *activity. For a gas, the standard state is chosen to be the state at which the fugacity is 1. The activity then equals the fugacity.

fullerene *See* BUCKMINSTERFULLERENE.

fullerite *See* BUCKMINSTERFULLERENE.

full-wave rectifier *See* RECTIFIER.

function Any operation or procedure that relates one variable to one or more other variables. If y is a function of x, written $y = f(x)$, a change in x produces a change in y, and if x is known, y can be determined. x is known as the **independent variable** and y is the **dependent variable**.

functional A function of a function. Functionals are used very extensively in the quantum-mechanical many-body problem, statistical mechanics, and quantum field theory.

fundamental *See* HARMONIC.

fundamental constants (universal constants) Those parameters that do not change throughout the universe. The charge on an electron, the speed of light in free space, the Planck constant, the gravitational constant, the electric constant, and the magnetic constant are all thought to be examples. It has been suggested that some fundamental constants might change with time but there is no conclusive evidence that this occurs.

fundamental interactions The four different types of interaction that can occur between bodies. These interactions can take place even when the bodies are not in physical contact and together they account for all the observed forces that occur in the universe. While the unification of these four types of interaction into one model, theory, or set of equations has long been the aim of physicists, this has not yet been achieved, although progress has been made in the unification of the electromagnetic and weak interactions. *See also* ELEMENTARY PARTICLES; GAUGE THEORY; UNIFIED-FIELD THEORY.

The **gravitational interaction**, some 10^{40} times weaker than the electromagnetic interaction, is the weakest of all. The force that it generates acts between all bodies that have mass and the force is always attractive. The interaction can be visualized in terms of a classical *field of force in which the strength of the force falls off with the square of the distance between the interacting bodies (*see* NEWTON'S LAW OF GRAVITATION). The hypothetical gravitational quantum, the **graviton**, is also a useful concept in some contexts. On the atomic scale the gravitational force is negligibly weak, but on the cosmological scale, where masses are enormous, it is immensely important in holding the components of the universe together. Because gravitational interactions are long-ranged, there is a well-defined macroscopic theory in general relativity. At present, there is no satisfactory quantum theory of gravitational interaction. It is possible that *superstring theory may give a consistent quantum theory of gravity as well as unifying gravity with the other fundamental interactions.

The **weak interaction**, some 10^{10} times weaker than the electromagnetic interaction, occurs between *leptons and in the decay of hadrons. It is responsible for the *beta decay of particles and nuclei. In the current model, the weak interaction is visualized as a force mediated by the exchange of virtual particles, called intermediate vector bosons. The weak interactions are described by *electroweak theory, which unifies them with the electromagnetic interactions.

The **electromagnetic interaction** is responsible for the forces that control atomic structure, chemical reactions, and all electromagnetic phenomena. It accounts for the forces between charged particles, but unlike the gravitational interaction, can be either attractive or repulsive. Some neutral particles decay by electromagnetic interaction. The interaction is either visualized as a classical field of force (*see* COULOMB'S LAW) or as an exchange of virtual *photons. As with gravitational interactions, the fact that electromagnetic interactions are long-ranged means that they have a well-defined classical theory given by *Maxwell's equations. The quantum theory of electromagnetic interactions is described by *quantum electrodynamics, which is a simple form of gauge theory.

The **strong interaction**, some 10^2 times stronger than the electromagnetic interaction, functions only between *hadrons and is responsible for the force between nucleons that gives the atomic nucleus its great stability. It operates at very short range inside the nucleus (as little as 10^{-15} metre) and is visualized as an exchange of virtual mesons. The strong interactions are described by a gauge theory called *quantum chromodynamics.

fundamental units A set of independently defined *units of measurement that forms the basis of a system of units. Such a set requires three mechanical units (usually of length, mass, and time) and, in some systems, one electrical unit; it has also been found convenient to treat certain other quantities as fundamental, even though they are not strictly independent. In the metric system the centimetre–gram–second (c.g.s.) system was replaced by the metre–kilogram–second (m.k.s.) system; the latter has now been adapted to provide the basis for *SI units. In British Imperial units the foot–pound–second (f.p.s.) system was formerly used.

fuse A length of thin wire made of tinned copper or a metal alloy of low melting point that is designed to melt at a specified current loading in order to protect an electrical device or circuit from overloading. The wire is often enclosed in a small glass or ceramic cartridge with metal ends.

fusible alloys Alloys that melt at low temperature (around 100°C). They have a number of uses, including constant-temperature baths, pipe bending, and automatic sprinklers to provide a spray of water to prevent fires from spreading. Fusible alloys are usually *eutectic mixtures of bismuth, lead, tin, and cadmium. Wood's metal and Lipowitz's alloy are examples of alloys that melt at about 70°C.

fusion 1. Melting. **2.** *See* NUCLEAR FUSION.

fusion reactor *See* THERMONUCLEAR REACTOR.

fuzzy logic A form of logic that allows for degrees of imprecision, used in *artificial intelligence studies. More traditional logics deal with two truth values: 'true' and 'false'. Fuzzy logics are multivalued dealing with such concepts as 'fairly true' and 'more or less true'. These can be represented by numbers within a range [0,1], with the number representing the degree of truth. **Fuzzy control** is the application of fuzzy logic to the computer-control of processes.

Gabor, Dennis (1900–79) Hungarian-born British physicist, who worked as a research engineer from 1927 until 1933, when he joined the British Thomson-Houston company. In 1948 he joined the staff of Imperial College, London. In that same year, while working on electron microscopes, he invented *holography, for which he was awarded the 1971 Nobel Prize.

gain *See* AMPLIFIER.

galaxy A vast collection of stars, dust, and gas held together by the gravitational attraction between its components. Galaxies are usually classified as elliptical, spiral, or irregular in shape. **Elliptical galaxies** appear like ellipsoidal clouds of stars, with very little internal structure apart from (in some cases) a denser nucleus. **Spiral galaxies** are flat disc-shaped collections of stars with prominent spiral arms. **Irregular galaxies** have no apparent structure or shape.

The sun belongs to a spiral galaxy known as the **Galaxy** (with a capital G) or the **Milky Way System**. There are some 10^{11} stars in the system, which is about 30 000 parsecs across with a maximum thickness at the centre of about 4000 parsecs. The sun is about 10 000 parsecs from the centre of the Galaxy.

The galaxies are separated from each other by enormous distances, the nearest large galaxy to our own (the Andromeda galaxy) being about 6.7×10^5 parsecs away.

galaxy cluster A group of *galaxies containing many hundreds of members extending over a radius of up to a few megaparsecs (there also exist small groups of galaxies, such as the *Local Group, with a few tens of members). The richest and most regular clusters, such as the **Coma cluster**, with thousands of members, are gravitationally bound systems; it is not certain whether other less regular and less concentrated clusters are also bound. As well as galaxies, the clusters contain hot **intracluster gas**, at temperatures between 10^7 and 10^8 K; this can be detected by its X-ray emission. On a scale larger than clusters there are also **superclusters**, with extents of the order of a hundred megaparsecs, containing about a hundred galaxies. It is not known whether superclusters are gravitationally bound. *See also* MISSING MASS.

galaxy formation *See* STRUCTURE FORMATION.

Galilean telescope *See* TELESCOPE.

Galilean transformations A set of equations for transforming the position and motion parameters from a frame of reference with origin at O and coordinates (x,y,z) to a frame with origin at O′ and coordinates at (x',y',z'). They are:

$x' = x - vt$

$y' = y$

$z' = z$

$t' = t$

The equations, which conform to Newtonian mechanics, were named after *Galileo Galilei. *Compare* LORENTZ TRANSFORMATIONS.

Galileo Galilei (1564–1642) Italian astronomer and physicist. In 1583 he noticed that the time of swing of a *pendulum is independent of its amplitude, and three years later invented a hydrostatic balance for measuring *relative densities. He became a professor in Padua in 1592 and it was there (in 1610) that he made his first astronomical telescope. With it he discovered four satellites of Jupiter, mountains on the moon, and sunspots. Returning to Pisa, his birthplace, he studied motion, demonstrating that the speed of a falling body is independent of its weight. He also gave open support to the sun-centred theory of the universe advocated by *Copernicus, a stand that brought him into conflict with the church. He was summoned to Rome, forced to retract before the Inquisition, and banished under house arrest.

gallon 1. (Imperial gallon) The volume occupied by exactly ten pounds of distilled water of density 0.998 859 gram per millilitre in air of density 0.001 217 gram per millilitre. 1 gallon = 4.546 09 litres (cubic decimetres). **2.** A unit of volume in the US Customary system equal to 0.832 68 Imperial gallon, i.e. 3.785 44 litres.

Galvani, Luigi (1737–98) Italian physiologist. In the late 1770s he observed that the muscles of a dead frog twitched when touched by two different metals. He concluded that the muscle was producing electricity, later disproved by *Volta (who showed that the two metals and body fluids formed a battery). Galvani invented *galvanized iron and the *galvanometer.

galvanic cell *See* VOLTAIC CELL.

galvanized iron Iron or steel that has been coated with a layer of zinc to protect it from corrosion. The process was invented by Luigi *Galvani. Corrugated mild-steel sheets for roofing and mild-steel sheets for dustbins, etc., are usually galvanized by dipping them in molten zinc. The formation of a brittle zinc–iron alloy is prevented by the addition of small quantities of aluminium or magnesium. Wire is often galvanized by a cold electrolytic process as no alloy forms in this process. Galvanizing is an effective method of protecting steel because even if the surface is scratched, the zinc still protects the underlying metal. *See* SACRIFICIAL PROTECTION.

galvanometer An instrument for detecting and measuring small electric currents, invented by Luigi *Galvani. In the moving-coil instrument a pivoted coil of fine insulated copper wire surrounds a fixed soft-iron core between the poles of a permanent magnet. The interaction between the field of the permanent magnet and the sides of the coil, produced when a current flows through it, causes a torque on the coil. The moving coil carries either a pointer or a mirror that deflects a light beam when it moves; the extent of the deflection is a measure of the strength of the current. The galvanometer can be converted into an *ammeter or a *voltmeter. Digital electronic instruments are increasingly replacing the moving-coil type. *See also* BALLISTIC GALVANOMETER.

gamma camera (scintillation camera) A device that detects and records the spatial distribution of radioactive compounds in the human body. It consists of a large scintillation crystal (or multiple crystals) of sodium iodide and a collection of *photomultiplier tubes above the crystal, connected to it by a transparent material.

Radiation in the form of low-energy *gamma radiation from an emitter causes a scintillation at a certain point P in the photomultipliers. This causes pulses in the photomultipliers, the size of the pulses depending on the position of both P and the tubes. All the pulses produced in the photomultiplier are analysed to enable the sizes and positions of the sources of radiation to be found. This information can be converted into spots on a *cathode-ray tube.

gamma radiation Electromagnetic radiation emitted by excited atomic nuclei during the process of passing to a lower excitation state. Gamma radiation ranges in energy from about 10^{-15} to 10^{-10} joule (10 keV to 10 MeV) corresponding to a wavelength range of about 10^{-10} to 10^{-14} metre. A common source of gamma radiation is cobalt–60:

$$^{60}_{27}\text{Co} \xrightarrow{\beta} {}^{60}_{28}\text{Ni} \xrightarrow{\gamma} {}^{60}_{28}\text{Ni}$$

The de-excitation of nickel–60 is accompanied by the emission of gamma-ray photons having energies 1.17 MeV and 1.33 MeV.

gamma-ray astronomy *Astronomy involving gamma-ray photons (with energies in excess of 100 MeV). The cosmic radiation with the highest energy can be detected by electron–photon cascades, which take place in the atmosphere. Gamma rays having lower energies can only be detected above the atmosphere. Many high-energy processes in *astrophysics are responsible for the production of gamma rays; one example is the decay of neutral *pions.

An interesting phenomenon is the **gamma-ray burst**. These events last for a few seconds, during which they are the strongest source of gamma rays in the sky. Many theories have been put forward to explain gamma-ray bursts. It is thought that a gamma-ray burst occurs when a black hole is born, either when a very large star collapses or when two neutron stars collide.

gamma-ray burst *See* GAMMA-RAY ASTRONOMY.

gas A state of matter in which the matter concerned occupies the whole of its container irrespective of its quantity. In an *ideal gas, which obeys the *gas laws exactly, the molecules themselves would have a negligible volume and negligible forces between them, and collisions between molecules would be perfectly elastic. In practice, however, the behaviour of real gases deviates from the gas laws because their molecules occupy a finite volume, there are small forces between molecules, and in polyatomic gases collisions are to a certain extent inelastic (*see* EQUATION OF STATE).

gas chromatography A technique for separating or analysing mixtures of gases by chromatography. The apparatus consists of a very long tube containing the stationary phase. This may be a solid, such as kieselguhr (**gas–solid chromatography**, or **GSC**), or a nonvolatile liquid, such as a hydrocarbon oil coated on a solid support (**gas–liquid chromatography**, or **GLC**). The sample is often a volatile liquid mixture, which is vaporized and swept through the column by a carrier gas (e.g. hydrogen). The components of the mixture pass through the column at different rates because they adsorb to different extents on the stationary phase. They are detected as they leave, either by measuring the thermal conductivity of the gas or by a flame detector.

Gas chromatography is usually used for analysis; components can be identified by the time they take to pass through the column. It is sometimes also used for separating mixtures.

Gas chromatography is often used to separate a mixture into its components, which are then directly injected into a mass spectrometer. This technique is known as **gas chromatography–mass spectroscopy** or **GCMS**.

gas constant (**universal molar gas constant**) Symbol R. The constant that appears in the **universal gas equation** (*see* GAS LAWS). It has the value $8.314\,510(70)\ \mathrm{J\,K^{-1}\,mol^{-1}}$.

gas-cooled reactor *See* NUCLEAR REACTOR.

gas equation *See* GAS LAWS.

gas laws Laws relating the temperature, pressure, and volume of an *ideal gas. *Boyle's law states that the pressure (p) of a specimen is inversely proportional to the volume (V) at constant temperature ($pV =$ constant). The modern equivalent of *Charles' law states that the volume is directly proportional to the thermodynamic temperature (T) at constant pressure ($V/T =$ constant); originally this law stated the constant expansivity of a gas kept at constant pressure. The pressure law states that the pressure is directly proportional to the thermodynamic temperature for a specimen kept at constant volume. The three laws can be combined in the **universal gas equation**, $pV = nRT$, where n is the amount of gas in the specimen and R is the *gas constant. The gas laws were first established experimentally for real gases, although they are obeyed by real gases to only a limited extent; they are obeyed best at high temperatures and low pressures. *See also* EQUATION OF STATE.

gas thermometer A device for measuring temperature in which the working fluid is a gas. It provides the most accurate method of measuring temperatures in the range 2.5 to 1337 K. Using a fixed mass of gas a **constant-volume thermometer** measures the pressure of a fixed volume of gas at relevant temperatures, usually by means of a mercury *manometer and a *barometer.

gas turbine An internal-combustion engine in which the products of combustion of a fuel burnt in compressed air are expanded through a turbine. Atmospheric air is compressed by a rotary compressor driven by the turbine, fed into a combustion chamber, and mixed with the fuel (kerosene, natural gas, etc.); the expanding gases drive the turbine and power is taken from the unit by means of rotation of the turbine shaft (as in locomotives) or thrust from a jet (as in aircraft).

gate 1. An electronic circuit with a single output and one or more inputs; the output is a function of the input or inputs. In the **transmission gate** the output waveform is a replica of a selected input during a specific interval. In the **switching gate** a constant output is obtained for a specified combination of inputs. These gates are the basic components of digital computers. *See* LOGIC CIRCUITS. **2.** The electrode in a field-effect *transistor that controls the current through the channel.

gauge boson A spin-one vector boson that mediates interactions governed by gauge theories. Examples of gauge bosons are photons in *quantum electrodynamics, gluons in *quantum chromodynamics, and W and Z bosons that mediate the interactions in the Weinberg–Salam model (*see* ELECTROWEAK THEORY) unifying electromagnetic and weak interactions. If the gauge symmetry of the theory is unbroken, the gauge boson is massless. Examples of massless gauge bosons include the photon and gluon. If the gauge symmetry of the theory is a *broken symmetry, the gauge boson has a nonzero mass, examples being the W and Z bosons. Treating gravity, as described by the general theory of relativity, as a gauge theory, the gauge boson is the massless spin-two *graviton.

gauge theory Any of a number of *quantum field theories put forward to explain fundamental interactions. A gauge theory involves a symmetry group (*see* GROUP THEORY) for the fields and potentials (the **gauge group**). In the case of electrodynamics, the group is Abelian whereas the gauge theories for strong and weak interactions use non-Abelian groups. Non-Abelian gauge theories are known as **Yang–Mills theories**. This difference explains why *quantum electrodynamics is a much simpler theory than *quantum chromodynamics, which describes the strong interactions, and *electroweak theory, which is the unified theory of the weak and electromagnetic interactions. In the case of quantum gravity, the gauge group is even more complicated than the gauge groups for either the strong or weak interactions.

In gauge theories the interactions between particles can be explained by the exchange of particles (intermediate vector bosons, or *gauge bosons), such as gluons, photons, and W and Z bosons.

gauss Symbol G. The c.g.s. unit of magnetic flux density. It is equal to 10^{-4} tesla. It is named after Karl Gauss.

Gauss, Karl Friedrich (1777–1855) German mathematician and physicist, who became director of Göttingen Observatory in 1806. One of the greatest mathematicians of all time, he contributed to the theory of numbers and proved the fundamental theorem of algebra. He also initiated the study of curved surfaces, thus paving the way for the geometry used in general relativity theory. His collaboration with Wilhelm *Weber on electromagnetism led to the invention of an electric telegraph, and he worked out the relationship between electric flux and electric field (*see* GAUSS' LAW). He also investigated the magnetism of the earth.

Gaussian units A system of units for electric and magnetic quantities based upon c.g.s. electrostatic and electromagnetic units. Although replaced by *SI units in most branches of science, they are, like Heaviside–Lorentz units, still used in relativity theory and in particle physics. In Gaussian units, the electric and magnetic constants are both equal to unity.

Gauss' law The total electric flux normal to a closed surface in an electric field is proportional to the algebraic sum of the electric charges within the surface. A similar law applies to surfaces drawn in a magnetic field and the law can be generalized for any vector field through a closed surface. It was first stated by Karl Gauss.

gaussmeter A *magnetometer, especially one calibrated in gauss.

Gauss' theorem *See* DIVERGENCE THEOREM.

Gay-Lussac, Joseph Louis (1778–1850) French scientist whose discovery of the laws of chemical combination in gases helped to establish the atomic theory. It also led to *Avogadro's law. *See also* CHARLES' LAW.

Gay-Lussac's law 1. When gases combine chemically the volumes of the reactants and the volume of the product, if it is gaseous, bear simple relationships to each other when measured under the same conditions of temperature and pressure. The law was first stated in 1808 by J. L. Gay-Lussac and led to *Avogadro's law. **2.** *See* CHARLES' LAW.

gebi- *See* BINARY PREFIXES.

Gegenschein (German: counterglow) A faint elliptical patch of light visible on a moonless night on the ecliptic at a point 180° from the position of the sun. It is caused by the reflection of sunlight by meteoric particles (*see also* ZODIACAL LIGHT).

Geiger, Hans Wilhelm (1882–1945) German physicist, who carried out research with *Rutherford at Manchester University before returning to Germany in 1912. In 1908 he and Rutherford produced the *Geiger counter, improved in 1928 as the Geiger-Muller counter. In 1909 his scattering experiments with alpha particles led to Rutherford's nuclear theory of the atom.

Geiger counter (Geiger–Müller counter) A device used to detect and measure *ionizing radiation. It consists of a tube containing a low-pressure gas (usually a mixture of methane with argon or neon) and a cylindrical hollow cathode through the centre of which runs a fine-wire anode. A potential difference of about 1000 volts is maintained between the electrodes. An ionizing particle or photon passing through a window into the tube will cause an ion to be produced and the high p.d. will accelerate it towards its appropriate electrode, causing an avalanche of further ionizations by collision. The consequent current pulses can be counted in electronic circuits or simply amplified to work a small loudspeaker in the instrument. It was first devised in 1908 by Hans Geiger. Geiger and W. Müller produced an improved design in 1928.

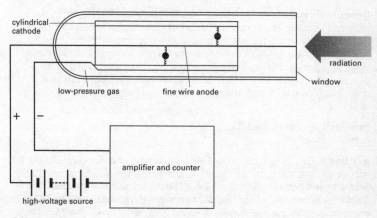

Geiger counter

Geissler tube An early form of gas-discharge tube designed to demonstrate the luminous effects of an electric discharge passing through a low-pressure gas between two electrodes. Modified forms are used in spectroscopy as a source of light. It was invented in 1858 by Heinrich Geissler (1814–79).

gel A lyophilic *colloid that has coagulated to a rigid or jelly-like solid. In a gel, the disperse medium has formed a loosely-held network of linked molecules through the dispersion medium. Examples of gels are silica gel and gelatin.

Gell-Mann, Murray (1929–) US theoretical physicist, who held a professorship at the California Institute of Technology. In 1955 he proposed the property of *strangeness for certain fundamental particles. In 1961 he and Yuval Ne'eman (1925–) proposed the eightfold way to define the structure of particles. This led to Gell-Mann's postulate of the quark (*see* ELEMENTARY PARTICLES). In 1969 he was awarded the Nobel Prize. ·

general theory of relativity *See* RELATIVITY.

generation time The average time that elapses between the creation of a neutron by fission in a nuclear reactor and a fission produced by that neutron.

generator Any machine that converts mechanical power into electrical power. Electromagnetic generators are the main source of electricity and may be driven by steam turbines, water turbines, internal-combustion engines, windmills, or by some moving part of any other machine. In power stations, generators produce alternating current and are often called **alternators**.

geocentric universe A view of the universe in which the earth is regarded as being at its centre. *Galileo Galilei finally established that the earth revolves round the sun (not the other way round, as the church believed); during the 20th century it has become clear from advances in observational astronomy that the earth is no more than one of nine planets orbiting the sun, which is one of countless millions of similar stars, many of which undoubtedly possess planetary bodies on which life could have evolved.

geodesic (geodesic line) The shortest distance between two points on a curved surface.

geodesy The science concerned with surveying and mapping the earth's surface to determine, for example, its exact size, shape, and gravitational field. The information supplied by geodesy in the form of locations, distances, directions, elevations, and gravity information is of use in civil engineering, navigation, geophysics, and geography.

geodynamics The study of the motions of the earth; it includes those of the crust, mantle, and core and the earth's rotation.

geomagnetism The science concerned with the earth's magnetic field. If a bar magnet is suspended at any point on the earth's surface so that it can move freely in all planes, the north-seeking end of the magnet (N-pole) will point in a broadly northerly direction. The angle (D) between the horizontal direction in which it points and the geographic meridian at that point is called the **magnetic declination**. This is taken to be positive to the east of geographic north and negative to the west. The needle will not, however, be horizontal except on the **magnetic equator**. In all other positions it will make an angle (I) with the horizontal, called the **inclination** (or **magnetic dip**). At the **magnetic poles** $I = 90°$ ($+90°$ at the N-pole, $-90°$ at the S-pole) and the needle will be vertical. The positions of the poles, which vary with time, were in the 1970s approximately 76.1°N, 100°W (N) and 65.8°S, 139°E (S). The vector intensity $\mathbf{F}$ of the geomagnetic field is specified by I, D, and F, where F is the local magnetic intensity of the field measured in gauss or tesla (1 gauss = 10^{-4} tesla). F, I, and D, together with the horizontal and vertical components of F, and its north and east components, are called the **magnetic elements**. The value of F varies from about 0.2 gauss to 0.6 gauss, in general being higher in the region of the poles than at the equator, but values vary irregularly over the earth's surface with no correlation with surface features. There is also a slow unpredictable change in the local values of the magnetic elements called the **secular magnetic variation**. For example, in London between 1576 and 1800 D changed from $+11°$ to $-24°$ and I varied between 74° and 67°. The study of *palaeomagnetism has extended knowledge of the secular magnetic variation into the geological past and it is clear that the direction of the geomagnetic field has reversed many times. The source of the field and the cause of the variations are not known with any certainty but the source is believed to be associated with *dynamo action in the earth's liquid core.

geometrical optics *See* OPTICS.

geometric average (**geometric mean**) *See* AVERAGE.

geometric series A series of numbers or terms in which the ratio of any term to the subsequent term is constant. For example, 1, 4, 16, 64, 256,... has a **common ratio** of 4. In general, a geometric series can be written:

$$a + ar + ar^2 \ldots + ar^{n-1}$$

and the sum of n terms is:

$$a(r^n - 1)/(r - 1).$$

geometrized units A system of units, used principally in general relativity, in which all quantities that have dimensions involving length, mass, and time are given dimensions of a power of length only. This is equivalent to setting the gravitational constant and the speed of light both equal to unity. *See also* GAUSSIAN UNITS; HEAVISIDE–LORENTZ UNITS; NATURAL UNITS; PLANCK UNITS.

geophysics The branch of science in which the principles of mathematics and physics are applied to the study of the earth's crust and interior. It includes the study of earthquake waves, geomagnetism, gravitational fields, and electrical conductivity using precise quantitative principles. In applied geophysics the techniques are applied to the discovery and location of economic minerals (e.g. petroleum). Meteorology and physical oceanography can also be considered as geophysical sciences.

geostationary orbit *See* SYNCHRONOUS ORBIT.

geosynchronous orbit *See* SYNCHRONOUS ORBIT.

geothermal energy Heat within the earth's interior that is a potential source of energy. Volcanoes, geysers, hot springs, and fumaroles are all sources of geothermal energy. The main areas of the world in which these energy sources are used to generate power include Larderello (Italy), Wairakei (New Zealand), Geysers (California, USA), and Reykjavik (Iceland). High-temperature porous rock also occurs in the top few kilometres of the earth's crust. Thermal energy from these reservoirs can be tapped by drilling into them and extracting their thermal energy by conduction to a fluid. The hot fluid can then be used for direct heating or to raise steam to drive a turbogenerator.

German silver (nickel silver) An alloy of copper, zinc, and nickel, often in the proportions 5:2:2. It resembles silver in appearance and is used in cheap jewellery and cutlery and as a base for silver-plated wire. *See also* ELECTRUM.

getter A substance used to remove small amounts of other substances from a system by chemical combination. For example, a metal such as magnesium may be used to remove the last traces of air when achieving a high vacuum. Various getters are also employed to remove impurities from semiconductors.

GeV Gigaelectronvolt, i.e. 10^9 eV. In the USA this is often written BeV, billion-electronvolt.

giant star A very large star that is highly luminous. Lying above the main sequence on a *Hertzsprung–Russell diagram, giant stars represent a late stage in *stellar evolution. *See also* RED GIANT; SUPERGIANT.

gibbous *See* PHASES OF THE MOON.

Gibbs, Josiah Willard (1839–1903) US mathematician and physicist, who spent his entire academic career at Yale University. During the 1870s he developed the theory of chemical thermodynamics, devising functions such as Gibbs *free energy; he also derived the *phase rule and was one of the founders of *statistical mechanics. In mathematics he introduced *vector notation.

Gibbs free energy (Gibbs function) *See* FREE ENERGY. It is named after J. W. Gibbs.

gibi- *See* BINARY PREFIXES.

giga- Symbol G. A prefix used in the metric system to denote one thousand million times. For example, 10^9 joules = 1 gigajoule (GJ).

gilbert Symbol Gb. The c.g.s. unit of *magnetomotive force equal to $10/4\pi$ (= 0.795 77) ampere-turn. It is named after William Gilbert.

Gilbert, William (1544–1603) English physician and physicist. He was physician to Queen Elizabeth I, and in 1600 published his famous book about magnetism, in which he likened the earth to a huge bar magnet. He was the first to use the terms 'magnetic pole' and 'electricity'.

gimbal A type of mount for an instrument (such as a *gyroscope or compass) in which the instrument is free to rotate about two perpendicular axes.

Giorgi units *See* M.K.S. UNITS.

Glaser, Donald Arthur (1926–) US physicist. In 1952, at the University of Michigan, he devised the *bubble chamber for detecting ionizing radiation. For this work he was awarded the 1960 Nobel Prize.

Glashow–Weinberg–Salam model (GWS model) *See* ELECTROWEAK THEORY.

glass An *amorphous solid in which the atoms form a random network. Glasses do not have the rigidity of *crystals and have a viscosity that increases as the temperature is lowered. At very low temperatures, when the viscosity is very large, glasses can become elastic and brittle. Because the simplifications that can be applied to a periodic system are not applicable it is more difficult to construct theories relating to the properties of glasses than it is for regular crystals and crystals with isolated *defects. At the present, a fully quantitative theory of the properties of glasses, particularly those properties that are distinct from the properties of crystals, does not exist. Because glasses consist of a random network it is usually necessary to use statistical methods to discuss their properties. Experimentally, a glass is a *disordered solid with nonzero shear (rigidity) modulus and a viscosity greater than 10^{13} poise.

glass fibres Melted glass drawn into thin fibres some 0.005 mm–0.01 mm in diameter. The fibres may be spun into threads and woven into fabrics, which are then impregnated with resins to give a material that is both strong and corrosion resistant for use in car bodies and boat building.

glass transition The transition that occurs when a liquid is cooled very rapidly to become a *glass. The glass transition temperature of a material is not a clear-cut temperature like the melting point. This is the case because the glassy state is a metastable state, with the transition temperature becoming lower as the rate of cooling decreases. The freezing temperature for a material is always higher than the glass transition temperature for that material.

Glauber model A model describing critical dynamics in which the dynamics is dissipative and the *order parameter is not conserved. The Glauber model was introduced by the US physicist R. Glauber in 1963. The Glauber model is also called the time-dependent Ginzburg–Landau model. *See also* CAHN–HILLIARD MODEL.

glide A symmetry element in a crystal lattice that consists of a combination of a translation with a reflection about a plane. *See also* SCREW.

global positioning system (GPS) A satellite-based navigational system that, with the use of a GPS receiver, can determine any point on or above the earth's surface with a high degree of accuracy. The system uses a network of 24 satellites, designed and controlled by the US Department of Defense, originally for military use. Uses include marine and terrestrial navigation systems (e.g. satellite navigation systems in vehicles), surveying, and mapping.

globular cluster *See* STAR CLUSTER.

glove box A metal box that has gloves fitted to ports in its walls. It is used to manipulate mildly radioactive materials and in laboratory techniques in which an inert, sterile, dry, or dust-free atmosphere has to be maintained.

glow discharge An electrical discharge that passes through a gas at low pressure and causes the gas to become luminous. The glow is produced by the decay of excited atoms and molecules.

glueball A hypothetical bound state consisting of two or more gluons (*see* ELEMENTARY PARTICLES). Glueballs are thought to be unstable and decay rapidly into *hadrons. There is some indirect experimental evidence for glueballs.

gluino *See* SUPERSYMMETRY.

gluon *See* ELEMENTARY PARTICLES.

Golay cell A transparent cell containing gas, used to detect *infrared radiation. Incident radiation is absorbed by a thin film within the cell, causing a rise in the gas temperature and pressure. The amount of incident radiation can be measured from the pressure rise in the tube.

Gold, Thomas *See* HOYLE, SIR FRED.

Goldstone's theorem The theorem in relativistic quantum field theory that if there is an exact continuous symmetry of the *Hamiltonian or *Lagrangian defining the system, and this is not a symmetry of the *vacuum state (i.e. there is *broken symmetry), then there must be at least one spin-zero massless particle called a **Goldstone boson**. In the quantum theory of many-body systems Goldstone bosons are *collective excitations such as spin waves. An important exception to Goldstone's theorem is provided in *gauge theories with the Higgs mechanism, whereby the Goldstone bosons gain mass and become *Higgs bosons.

Gopher A computer program used on a computer connected to the *Internet that carries out routine tasks of collecting information for the user from services attached to the Internet. Gopher presents the user with a directory of material accessible at a particular point, a set of documents that can be searched using keywords, or a document containing text or other forms of material that the computer can display. Gopher is simpler but rather less flexible than the *World Wide Web.

governor A device that maintains a motor or engine at a constant speed despite variations in the load, using the principle of negative feedback. A common method uses a set of flying balls that reduce the fuel intake as the speed increases. The balls, attached by flexible steel strips to a collar capable of moving vertically up and down a rotating shaft, move outwards as the speed increases. The collar rises as the balls fly out and is coupled to a lever that controls the fuel intake.

GPS *See* GLOBAL POSITIONING SYSTEM.

grad *See* GRADIENT OPERATOR.

gradient **1.** The slope of a line. In Cartesian coordinates, a straight line $y = mx + c$, has a gradient m. For a curve, $y = f(x)$, the gradient at a point is the derivative dy/dx at that point, i.e. the slope of the tangent to the curve at that point. **2.** *See* GRADIENT OPERATOR.

gradient operator (grad) The *operator

$$\nabla = \boldsymbol{i}\, \partial/\partial x + \boldsymbol{j}\, \partial/\partial y + \boldsymbol{k}\, \partial/\partial z,$$

where $\boldsymbol{i}$, $\boldsymbol{j}$, and $\boldsymbol{k}$ are unit vectors in the x, y, and z directions. Given a scalar function f and a unit vector $\boldsymbol{n}$, the *scalar product $\boldsymbol{n}.\nabla$f is the rate of change of f in the direction of $\boldsymbol{n}$. *See also* CURL; DIVERGENCE.

Graham, Thomas (1805–69) Scottish chemist, who became professor of chemistry at Glasgow University in 1830, moving to University College, London, in 1837. His 1829 paper on gaseous diffusion introduced *Graham's law. He went on to study diffusion in liquids, leading in 1861 to the definition of *colloids.

Graham's law The rates at which gases diffuse is inversely proportional to the square roots of their densities. This principle is made use of in the diffusion method of separating isotopes. The law was formulated in 1829 by Thomas Graham.

gram Symbol g. One thousandth of a kilogram. The gram is the fundamental unit of mass in *c.g.s. units and was formerly used in such units as the **gram-atom**, **gram-molecule**, and **gram-equivalent**, which have now been replaced by the *mole.

grand unified theory (GUT) A theory that attempts to combine the strong, weak, and electromagnetic interactions into a single *gauge theory with a single symmetry group. There are a number of different theories, most of which postulate that the interactions merge at high energies into a single interaction (the standard model (*see* ELEMENTARY PARTICLES) emerges from the GUT as a result of *broken symmetry). The energy above which the interactions are the same is around 10^{15} GeV, which is much higher than those obtainable with existing accelerators.

One prediction of GUTs is the occurrence of *proton decay. Some also predict that the neutrino has nonzero mass. There is no evidence for proton decay at the moment. There is evidence that neutrinos have nonzero masses. It seems likely that GUTs need to be combined with supersymmetry. *See also* SUPERSTRING THEORY.

graph A diagram that illustrates the relationship between two variables. It usually consists of two perpendicular axes, calibrated in the units of the variables and crossing at a point called the **origin**. Points are plotted in the spaces between the axes and the points are joined to form a curve. *See also* CARTESIAN COORDINATES; POLAR COORDINATES.

graphite-moderated reactor *See* NUCLEAR REACTOR.

graticule (in optics) A network of fine wires or a scale in the eyepiece of a telescope or microscope or on the stage of a microscope, or on the screen of a cathode-ray oscilloscope for measuring purposes.

grating *See* DIFFRACTION GRATING.

gravitation *See* NEWTON'S LAW OF GRAVITATION.

gravitational collapse A phenomenon predicted by the general theory of *relativity in which matter collapses as a consequence of gravitational attraction until it becomes a compact object such as a *white dwarf, a *neutron star, or a *black hole. The process of gravitational collapse is thought to be important in *astrophysics since it gives rise to phenomena such as *supernova explosions and gamma-ray bursts. The compact object formed depends on the mass of the initial star.

gravitational constant Symbol G. The constant that appears in *Newton's law of gravitation; it has the value $6.672\ 59(85) \times 10^{-11}$ N m^2 kg^{-2}. G is usually regarded as a universal constant although, in some models of the universe, it is proposed that it decreases with time as the universe expands.

gravitational field The region of space surrounding a body that has the property of *mass. In this region any other body that has mass will experience a force of attraction. The ratio of the force to the mass of the second body is the **gravitational field strength**.

gravitational interaction *See* FUNDAMENTAL INTERACTIONS.

gravitational lens An object that deflects light by gravitation as described by the general theory of *relativity; it is analogous to a lens in *optics. The prediction of a gravitational lensing effect in general relativity theory has been confirmed in observations on *quasars. In 1979 a 'double' quasar was discovered, due to the multiple image of a single quasar caused by gravitational lensing by a galaxy, or cluster of galaxies, along the line of sight between the observer and the quasar. The images obtained by gravitational lensing can be used to obtain information about the mass distribution of the galaxy or cluster of galaxies.

gravitational mass *See* MASS.

gravitational potential The gravitational potential at a point due to the gravitational field of a body of mass M is equal to the work done in taking a body of unit mass from infinity to that point. It is customary to take the gravitational potential at infinity to have the value zero. If the body of mass M is taken to be a sphere, then the gravitational potential at a distance R from the centre of the sphere is given by $-GM/R$, where G is the *gravitational constant. The significance of the negative sign is that the potential at infinity has a higher value than it does near the body.

gravitational shift *See* REDSHIFT.

gravitational waves Waves propagated through a *gravitational field. The prediction that an accelerating mass will radiate gravitational waves (and lose energy) comes from the general theory of *relativity. Many attempts have been made to detect waves from space directly using large metal detectors. The theory suggests that a pulse of gravitational radiation (as from a supernova explosion or *black hole) causes the detector to vibrate, and the disturbance is detected by a transducer. The interaction is very weak and extreme care is required to avoid external disturbances and the effects of thermal noise in the detecting system. So far, no accepted direct observations have been made. However, indirect evidence of gravitational waves has come from observations of a pulsar in a binary system with another star.

graviton A hypothetical particle or quantum of energy exchanged in a gravitational interaction (*see* FUNDAMENTAL INTERACTIONS). Such a particle has not been observed but is postulated to make the gravitational interaction consistent with quantum mechanics. It would be expected to travel at the speed of light and have zero rest mass and charge, and spin 2.

gravity The phenomenon associated with the gravitational force acting on any object that has mass and is situated within the earth's *gravitational field. The weight of a body (*see* MASS) is equal to the force of gravity acting on the body. According to Newton's second law of motion $F = ma$, where F is the force producing an acceleration a on a body of mass m. The weight of a body is therefore equal to the product of its mass and the acceleration due to gravity (g), which is now called the *acceleration of free fall. By combining the second law of motion with *Newton's law of gravitation ($F = GM_1M_2/d^2$) it follows that: $g = GM/d^2$, where G is the *gravitational constant, M is the mass of the earth, and d is the distance of the body from the centre of the earth. For a body on the earth's surface g = 9.806 65 m s^{-2}.

A force of gravity also exists on other planets, moons, etc., but because it depends on the mass of the planet and its diameter, the strength of the force is not the same as it is on earth. If F_e is the force acting on a given mass on earth, the force F_p acting on the same mass on another planet will be given by:

$$F_p = F_e d_e^2 M_p / M_e d_p^2,$$

where M_p and d_p are the mass and diameter of the planet, respectively. Substituting values of M_p and d_p for the moon shows that the force of gravity on the moon is only 1/6 of the value on earth.

gray Symbol Gy. The derived SI unit of absorbed *dose of ionizing radiation (*see* RADIATION UNITS). It is named after the British radiobiologist L. H. Gray (1905–65).

Great Attractor A huge concentration of mass, equivalent to about a million galaxies, beyond the Hydra and Centaurus constellations. Our own Galaxy and others near to it are heading towards the Great Attractor at a rate of about 600 kilometres per second.

great circle Any circle on a sphere formed by a plane that passes through the centre of the sphere. The equator and the meridians of longitude are all great circles on the earth's surface.

greenhouse effect 1. The effect within a greenhouse in which solar radiation mainly in the visible range of the spectrum passes through the glass roof and walls and is absorbed by the floor, earth, and contents, which re-emit the energy as infrared radiation. Because the infrared radiation cannot escape through the glass, the temperature inside the greenhouse rises. **2.** A similar effect in which the earth's atmosphere behaves like the greenhouse and the surface of the earth absorbs most of the solar radiation, re-emitting it as infrared radiation. This is absorbed by carbon dioxide, water, and ozone in the atmosphere as well as by clouds and reradiated back to earth. At night this absorption prevents the temperature falling rapidly after a hot day, especially in regions with a high atmospheric water content.

Gregorian telescope *See* TELESCOPE.

grey body A body that emits radiation of all wavelengths at a given temperature, which is a constant fraction of the energy of *black-body radiation of the same wavelength at the same temperature.

grid 1. (in electricity) The system of overhead wires or underground cables by which electrical power is distributed from power stations to users. The grid is at a high voltage, up to 750 kV in some countries. **2.** (in electronics) *See* CONTROL GRID.

ground state The lowest stable energy state of a system, such as a molecule, atom, or nucleus. *See* ENERGY LEVEL.

ground wave A radio wave that travels in approximately a straight line between points on the earth's surface. For transmission over longer distances sky waves have to be involved. *See* RADIO TRANSMISSION.

group theory The study of the symmetries that define the properties of a system. Invariance under symmetry operations enables much about a system to be deduced without knowing explicitly the solutions to the equations of motion. *Newton's law of gravitation, for instance, exhibits spherical symmetry. The force of gravity due to the attraction of a planet to a star is the same for all positions that are equidistant from the centre of mass of the star. However, the possible trajectories of the planet include non-symmetric elliptical orbits. These elliptical orbits are solutions to the Newtonian equations. However, one discovers on solving them that the planet does not move at a constant speed around the ellipse: it speeds up when it approaches perihelion and slows down approaching aphelion, which is consistent with what one might expect from a spherically symmetric force law. This behaviour, first formulated as one of *Kepler's laws of planetary motion, is now accepted as a result of the conservation of angular momentum.

This association of a dynamical symmetry with a conservation law was first suggested by A. E. Noether in 1918 (*see* NOETHER'S THEOREM). For example, the laws of physics are invariant under translations in time; they are the same today as they were yesterday. Noether's theorem relates this invariance to the conservation of energy. If a system is invariant under translations in space, linear momentum is conserved.

The symmetry operations on any physical system must possess the following properties:
1. **Closure**. If R_i and R_j are in the set of all symmetry operations, then the combination, $R_i R_j$ – meaning: first perform R_i, then perform R_j – is also a member of the set; that is, there exists an R_k, which is a member of the set such that $R_k = R_i R_j$.
2. **Identity**. There is an element I, which is also a member of the set of symmetry operations, such that $I R_i = R_i I = R_i$, for all elements R_i in the set.
3. **Inverse**. For every element R_i there is an inverse, R_i^{-1}, such that $R_i R_i^{-1} = R_i^{-1} R_i = I$.
4. **Associativity**. $R_i (R_j R_k) = (R_i R_j) R_k$.

These are the defining properties of a group in group theory. Group elements need not commute; i.e., $R_i R_j \neq R_j R_i$, in general; if all the elements do commute then the group is said to be **Abelian**. Though translations in space and time are Abelian, it is easily verified that rotations about axes in 3D space are not. Groups can be **finite** (as the group of rotations of an equilateral triangle) or **infinite** (for example, the set of all integers, with addition used to combine the members). Groups can also be classed as **continuous** or **discrete**. An example of a continuous group is the group of all continuous translations of a point on a spherical surface. The symmetries of the star-planet system are the elements of this spherical group. Discrete groups have elements that may be labelled by an index that takes on only integer values. All finite groups and some infinite groups, such as the group of integers described above, are discrete.

Group elements are conveniently represented in *matrix form. The *Lorentz group (transformations between inertial frames of reference in special relativity), for instance, consists of a set of 4×4 matrices, which act on space-time coordinates. In elementary particle physics, the most common groups are called $U(n)$: the set of all $n \times n$ matrices that describe *unitary transformations. If the unitary matrices are further restricted to have a determinant of 1, the group is called a 'special unitary group' or $SU(n)$. When the elements of the unitary matrices are real numbers the group is called $O(n)$. When the elements of the unitary matrices are real numbers the group is called $O(o)$ or the group of **orthogonal** matrices in n dimensions. Finally, the group of real, orthogonal, $n \times n$ matrices of determinant 1 is $SO(n)$. $SO(n)$ may be thought of as the group of all rotations in n-dimensional space. Thus $SO(3)$ is the name of the group of symmetries of the aforementioned star-planet system, and the symmetry that is related to the conservation of angular momentum by Noether's theorem.

In discussing the particles of high-energy physics and the special unitary groups $SU(2)$ and $SU(3)$, it is helpful to consider an analogous example with $SO(3)$ symmetry, such as the hydrogen atom. The hydrogen atom is composed of an electron trapped within the spherically symmetric potential of the atomic nucleus (a proton). The quantum mechanical treatment of this problem leads to a description of the electron states in terms of standing waves on a spherical surface. These standing waves themselves are in general less than spherically symmetric and are labelled by three integers or quantum numbers n, l, and m (see ATOM). The quantum numbers l and m essentially describe the *nodes and *antinodes on the spherical surface undergoing standing wave oscillations. The energy associated with these standing waves, however, is $(2l + 1)$-fold degenerate; that is, for a given standing wave characterized by the quantum numbers n and l, there are $(2l + 1)$ values of m characterizing states of the same energy. The reason for this degeneracy lies in the fact that the potential is spherically symmetric and independent of the angular position of the electron with respect to the nucleus. As a consequence of the spherical symmetry of the potential, the angular momentum L is conserved.

The degeneracy of the hydrogen atom may be eliminated by the application of a magnetic field. The directional nature of this field destroys the spherical symmetry of the problem and leads to the *Zeeman effect. The degeneracy is lifted by this symmetry-breaking term.

In the 1930s, Heisenberg proposed a model for nuclear forces that was analogous to the group theory of the hydrogen atom. The only known nuclear particles at the time were the proton and the neutron, which were very similar in mass. Heisenberg proposed that nuclear interactions were independent of electric charge and that the proton and neutron were the same particle in the absence of electromagnetic interactions. This meant that the proton and neutron were members of a degenerate state and that the electromagnetic interaction broke the symmetry of nuclear interactions lifting this degeneracy. Just as L is the conserved quantity for the hydrogen atom analogy, the conserved quantity for nuclear interactions between protons and neutrons is a quantity labelled I that was later called isospin (*see* ISOTOPIC SPIN).

In the absence of the electromagnetic interaction, isospin is conserved (the proton and neutron have the same mass) and there is a two-fold degeneracy. Equivalently, the nuclear interaction (strong interaction) must be invariant under the group that has matrices that are described by isospin matrices. This group is $SU(2)$.

By the 1960s many more particles had been discovered and the ideas of isospin were applied to them. It turns out that it is convenient to describe these particles by characteristic quantum numbers, I for isospin, and Y for *hypercharge. The particles may be grouped into charge or isospin degenerate multiplets. The hypercharge Y may be taken as twice the average charge of the multiplet. For example, the proton-neutron multiplet has a hypercharge given by:

$Y = 2 \cdot \frac{1}{2} (0+1) = 1$.

The hypercharge and isospin values of different particles are tabulated below:

Multiplet	Particle	Mass (MeV)	Y	I	I_3
Ξ	Ξ^-	1321.300	-1	$1/2$	$-1/2$
	Ξ^0	1314.900			$+1/2$
Σ	Σ^-	1197.410			-1
	Σ^0	1192.540	0	1	0
	Σ^+	1189.470			$+1$
Λ	Λ	1115.500	0	0	0
N	n	939.550	1	$1/2$	$-1/2$
	p	938.256			$+1/2$

The value of I_3 given in the last column is similar to the quantum number m in the hydrogen atom analogy. It labels the degenerate states in the absence of the electromagnetic interaction. Experiments involving these particles show that hypercharge Y and isospin I are conserved under strong interactions.

In 1961 *Gell-Mann, and independently Ne'eman, suggested that the strong interaction should be invariant under $SU(3)$; that is, should have $SU(3)$ symmetry. Matrices for Y and I may be associated with matrices which define transformations of $SU(3)$. Gell-Mann called this scheme the **eight-fold way**. In the absence of any other interaction, the strong interaction sees all eight particles in the table above as the same particle; i.e., the same state. It is the effect of the weak and electromagnetic interactions that break the $SU(3)$ symmetry and lead to the lifting of the degeneracy and the splitting of the masses of these particles.

gun metal A type of bronze usually containing 88–90% copper, 8–10% tin, and 2–4% zinc. Formerly used for cannons, it is still used for bearings and other parts that require high resistance to wear and corrosion.

GUT *See* GRAND UNIFIED THEORY.

GWS model Glashow–Weinberg–Salam model. *See* ELECTROWEAK THEORY.

gyrocompass A *gyroscope that is driven continuously so that it can be used as a nonmagnetic compass. When the earth rotates the gyroscope experiences no torque if its spin axis is parallel to the earth's axis; if these axes are not parallel, however, the gyroscope experiences a sequence of restoring torques that tend to make it align itself with the earth's axis. The gyrocompass is therefore an accurate north-seeking device that is uninfluenced by metallic or magnetic objects and it is more consistent than the magnetic compass. It is widely used on ships, aircraft, missiles, etc.

gyromagnetic ratio Symbol γ. The ratio of the angular momentum of an atomic system to its magnetic moment. The inverse of the gyromagnetic ratio is called the **magnetomechanical ratio**.

gyroscope A disc with a heavy rim mounted in a double *gimbal so that its axis can adopt any orientation in space. When the disc is set spinning the whole contrivance has two useful properties: (1) Gyroscopic inertia, i.e. the direction of the axis of spin resists change so that if the gimbals are turned the spinning disc maintains the same orientation in space. This property forms the basis of the *gyrocompass and other navigational devices. (2) Precession, i.e. when a gyroscope is subjected to a torque that tends to alter the direction of its axis, the gyroscope turns about an axis at right angles both to the axis about which the torque was applied and to its main axis of spin. This is a consequence of the need to conserve *angular momentum.

In the gyrostabilizer for stabilizing a ship, aircraft, or platform, three gyroscopes are kept spinning about mutually perpendicular axes so that any torque tending to alter the orientation of the whole device affects one of the gyroscopes and thereby activates a servomechanism that restores the original orientation.

habit *See* CRYSTAL.

hadron Any of a class of subatomic particles that interact by the strong interaction (*see* FUNDAMENTAL INTERACTIONS). The class includes protons, neutrons, and pions. Hadrons are believed to have an internal structure and to consist of quarks; they are therefore not truly elementary. Hadrons are either *baryons, including protons, and are believed to consist of three quarks, or *mesons, which decay into *leptons and photons or into proton pairs and are believed to consist of a quark and an antiquark. *See* ELEMENTARY PARTICLES.

Hahn, Otto (1879–1968) German chemist, who studied in London (with William *Ramsay) and Canada (with Ernest *Rutherford) before returning to Germany in 1907. In 1917, together with Lise *Meitner, he discovered protactinium. In the late 1930s he collaborated with Fritz Strassmann (1902–) and in 1938 bombarded uranium with slow neutrons. Among the products was barium, but it was Meitner (now in Sweden) who the next year interpreted the process as *nuclear fission. In 1944 Hahn received the Nobel Prize for chemistry.

half cell An electrode in contact with a solution of ions, forming part of a *cell. Various types of half cell exist, the simplest consisting of a metal electrode immersed in a solution of metal ions. Gas half cells have a gold or platinum plate in a solution with gas bubbled over the metal plate. The commonest is the *hydrogen half cell. Half cells can also be formed by a metal in contact with an insoluble salt or oxide and a solution. The calomel half cell is an example of this. Half cells are commonly referred to as **electrodes**.

half-life *See* DECAY.

half-thickness The thickness of a specified material that reduces the intensity of a beam of radiation to half its original value.

half-wave plate *See* RETARDATION PLATE.

half-wave rectifier *See* RECTIFIER.

half-width Half the width of a spectrum line (or in some cases the full width) measured at half its height.

Hall effect The production of an e.m.f. within a conductor or semiconductor through which a current is flowing when there is a strong transverse magnetic field. The potential difference develops at right angles to both the current and the field. It is caused by the deflection of charge carriers by the field and was first discovered by Edwin Hall (1855–1938). The strength of the electric field E_H produced is given by the relationship $E_H = R_H jB$, where j is the current density, B is the magnetic flux density, and R_H is a constant called the **Hall coefficient**. The value of R_H can be shown to be $1/ne$, where n is the number of charge carriers per unit volume and e is the electronic charge. The effect is used to investigate the nature of charge carriers in metals and semiconductors, in the **Hall probe** for the measurement of magnetic fields, and in magnetically operated switching devices. *See also* QUANTUM HALL EFFECT.

Halley, Edmund (1656–1742) British astronomer and mathematician, who published a catalogue of southern stars in 1679, made improvements to barometers, and investigated the optics of *rainbows. In 1705 he calculated the orbit of *Halley's comet and in 1718 discovered the *proper motion of the stars.

Halley's comet A bright *comet with a period of 76 years. Its last visit was in 1986. The comet moves around the sun in the opposite direction to the planets. Its orbit was first calculated in 1705 by Edmund Halley, after whom it is named.

halo 1. A luminous ring that sometimes can be observed around the sun or the moon. It is caused by diffraction of their light by particles in the earth's atmosphere; the radius of the ring is inversely proportional to the predominant particle radius. **2.** Broad rings appearing in the *electron diffraction, neutron diffraction, or X-ray diffraction patterns of materials that are not crystals. For example, the halo associated with neutron diffraction is called a **neutron halo**. Haloes of this type occur in gases and liquids as well as in noncrystalline solids. **3.** The glow in a *cathode-ray tube that briefly remains after the beam has passed.

halo nucleus A type of nucleus in which there are many more neutrons (or, more rarely, more protons) than are present in stable isotopes of that element. Sometimes a few of the extra neutrons are only weakly bound to the rest of the nucleus and are relatively far from the centre of the nucleus. Halo nuclei are highly unstable; examples include beryllium–11 and carbon–19.

Hamiltonian Symbol H. A function used to express the energy of a system in terms of its momentum and positional coordinates. In simple cases this is the sum of its kinetic and potential energies. In **Hamiltonian equations**, the usual equations used in mechanics (based on forces) are replaced by equations expressed in terms of momenta. These concepts, often called **Hamiltonian mechanics**, were formulated by Sir William Rowan Hamilton (1805–65).

hard ferromagnetic materials *See* SOFT IRON.

hard radiation Ionizing radiation of high penetrating power, usually gamma rays or short-wavelength X-rays. *Compare* SOFT RADIATION.

hardware *See* COMPUTER.

harmonic An oscillation having a frequency that is a simple multiple of a **fundamental** sinusoidal oscillation. The fundamental frequency of a sinusoidal oscillation is usually called the **first harmonic**. The **second harmonic** has a frequency twice that of the fundamental and so on (see illustration). A taut string or column of air, as in a violin or organ, will sound upper harmonics at the same time as the fundamental sounds. This is because the string or column of air divides itself into sections, each section then vibrating as if it were a whole. The upper harmonics are also called **overtones**, but the second harmonic is the first overtone, and so on. Musicians, however, often regard harmonic and overtone as synonymous, not counting the fundamental as a harmonic.

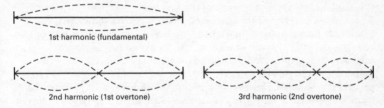

1st harmonic (fundamental)

2nd harmonic (1st overtone)   3rd harmonic (2nd overtone)

Harmonics

harmonic motion *See* SIMPLE HARMONIC MOTION.

harmonic oscillator A system (in either *classical physics or *quantum mechanics) that oscillates with *simple harmonic motion. The harmonic oscillator is exactly soluble in both classical mechanics and quantum mechanics. Many systems exist for which harmonic oscillators provide very good approximations. An example in classical mechanics is a simple *pendulum, while at low temperatures atoms vibrating about their mean positions in molecules or crystal lattices can be regarded as good approximations to harmonic oscillators in quantum mechanics. Even if a system is not exactly a harmonic oscillator the solution of the harmonic oscillator is frequently a useful starting point for solving such systems using *perturbation theory. *Compare* ANHARMONIC OSCILLATOR.

harmonic series (harmonic progression) A series or progression in which the reciprocals of the terms have a constant difference between them, e.g.

$1 + 1/2 + 1/3 + 1/4 \dots + 1/n$.

Harvard classification *See* SPECTRAL CLASS.

Hawking, Stephen William (1942–) British cosmologist and physicist, who in 1979 became a professor of mathematics at Oxford University. Working with Roger Penrose (1931–), who had shown how a singularity results from a *black hole, he postulated that the original big bang must have come from a singularity (*see* BIG-BANG THEORY). He also showed how black holes can emit particles by the *Hawking process.

Hawking process Emission of particles by a *black hole as a result of quantum-mechanical effects. It was discovered by Stephen Hawking. The gravitational field of the black hole causes production of particle–antiparticle pairs in the vicinity of the event horizon (the process is analogous to that of pair production). One member of each pair (either the particle or the antiparticle) falls into the black hole, while the other escapes. To an external observer, it appears that the black hole is emitting radiation (**Hawking radiation**). Furthermore, it turns out that the energy of the particles that fall in is negative and exactly balances the (positive) energy of the escaping particles. This negative energy reduces the mass of the black hole and the net result of the process is that the emitted particle flux appears to carry off the black-hole mass. It can be shown that the black hole radiates like a *black body, with the energy distribution of the particles obeying *Planck's radiation law for a temperature that is inversely proportional to the mass of the hole. For a black hole of the mass of the sun, this temperature turns out to be only about 10^{-7} K, so the process is negligible. However, for a 'mini' black hole, such as might be formed in the early universe, with a mass of order 10^{12} kg (and a radius of order 10^{-15} m), the temperature would be of order 10^{11} K and the hole would radiate copiously (at a rate of about 6×10^{9} W) a flux of gamma rays, neutrinos, and electron–positron pairs. (The observed levels of cosmic gamma rays put strong constraints on the number of such 'mini' black holes, suggesting that there are too few of them to solve the *missing-mass problem.) Further insight into Hawking radiation has been obtained using *superstring theory.

health physics The branch of medical physics concerned with the protection of medical, scientific, and industrial workers from the hazards of ionizing radiation and other dangers associated with atomic physics. Establishing the maximum permissible *dose of radiation, the disposal of radioactive waste, and the shielding of dangerous equipment are the principal activities in this field.

heat The process of energy transfer from one body or system to another as a result of a difference in temperature. The energy in the body or system before or after transfer is also sometimes called heat but this leads to confusion, especially in thermodynamics.

A body in equilibrium with its surroundings contains energy (the kinetic and potential energies of its atoms and molecules) but this is called **internal energy**, U, rather than heat. When such a body changes its temperature or phase there is a change in internal energy, ΔU, which (according to the first law of *thermodynamics) is given by $\Delta U = Q - W$, where Q is the heat absorbed by the body from the surroundings and W is the work done simultaneously on the surroundings. To use the word 'heat' for both U and Q is clearly confusing. Note also that certain physical quantities are described as *heat of atomization, *heat of combustion, etc. What is usually used is a standard molar *enthalpy change for the process under consideration. The units are kJ mol^{-1}; a negative value indicates that energy is liberated. *See also* HEAT CAPACITY; HEAT TRANSFER; LATENT HEAT.

heat balance 1. A balance sheet showing all the heat inputs to a system (such as a chemical process, furnace, etc.) and all the heat outputs. **2.** The equilibrium that exists on the average between the radiation received from the sun by the earth and its atmosphere and that reradiated or reflected by the earth and the atmosphere. In general, the regions of the earth nearer the equator than about 35°N or S receive more energy from the sun than they are able to reradiate, whereas those regions polewards of 35°N or S receive less energy than they lose. The excess of heat received by the low latitudes is carried to the higher latitudes by atmospheric and oceanic circulations.

heat capacity (thermal capacity) The ratio of the heat supplied to an object or specimen to its consequent rise in temperature. The **specific heat capacity** is the ratio of the heat supplied to unit mass of a substance to its consequent rise in temperature. The **molar heat capacity** is the ratio of the heat supplied to unit amount of a substance to its consequent rise in temperature. In practice, heat capacity (C) is measured in joules per kelvin, specific heat capacity (c) in J K^{-1} kg^{-1}, and molar heat capacity (C_{m}) in J K^{-1} mol^{-1}. For a gas, the values of c and C_{m} are commonly given either at **constant volume**, when only its *internal energy is increased, or at **constant pressure**, which requires a greater input of heat as the gas is allowed to expand and do work against the surroundings. The symbols for the specific and molar heat capacities at constant volume are c_{v} and C_{v}, respectively; those for the specific and molar heat capacities at constant pressure are c_{p} and C_{p}.

heat death of the universe The condition of the universe when *entropy is maximized and all large-scale samples of matter are at a uniform temperature. In this condition no energy is available for doing work and the universe is finally unwound. The condition was predicted by Rudolph *Clausius, who introduced the concept of entropy. Clausius's dictum that "the energy of the universe is constant, its entropy tends to a maximum" is a statement of the first two laws of thermodynamics. These laws apply in this sense only to closed systems and, for the predicted heat death to occur, the universe must be a closed system. Whether the heat death of the universe will occur is a contentious issue that has been extensively discussed.

heat engine A device for converting heat into work. The heat is derived from the combustion of a fuel. In an *internal-combustion engine the fuel is burnt inside the engine, whereas in a *steam engine or steam turbine, examples of external-combustion engines, the fuel is used to raise steam outside the engine and then some of the steam's internal energy is used to do work inside the engine. Engines usually work on cycles of operation, the most efficient of which would be the *Carnot cycle. This cannot be realized in practice, but the *Rankine cycle is approximated by some engines.

heat exchanger A device for transferring heat from one fluid to another without permitting the two fluids to contact each other. A simple industrial heat exchanger consists of a bundle of parallel tubes, through which one fluid flows, enclosed in a container, through which the other fluid flows in the opposite direction (a **counter-current heat exchanger**).

heat of atomization The energy required to dissociate one mole of a given substance into atoms. *See* HEAT.

heat of combustion The energy liberated when one mole of a given substance is completely oxidized. *See* HEAT.

heat of formation The energy liberated or absorbed when one mole of a compound is formed in its *standard state from its constituent elements.

heat of neutralization The energy liberated in neutralizing one mole of an acid or base.

heat of reaction The energy liberated or absorbed as a result of the complete chemical reaction of molar amounts of the reactants.

heat of solution The energy liberated or absorbed when one mole of a given substance is completely dissolved in a large volume of solvent (strictly, to infinite dilution).

heat pump A device for transferring heat from a low temperature source to a high temperature region by doing work. It is essentially a refrigerator with a different emphasis. The working fluid is at one stage a vapour, which is compressed by a pump adiabatically so that its temperature rises. It is then passed to the radiator, where heat is given out to the surroundings (the space to be heated) and the fluid condenses to a liquid. It is then expanded into an evaporator where it takes up heat from its surroundings and becomes a vapour again. The cycle is completed by returning the vapour to the compressor. Heat pumps are sometimes adapted to be used as dual-purpose space-heating-in-winter and air-conditioning-in-summer devices.

heat radiation (radiant heat) Energy in the form of electromagnetic waves emitted by a solid, liquid, or gas as a result of its temperature. It can be transmitted through space; if there is a material medium this is not warmed by the radiation except to the extent that it is absorbed. Although it covers the whole electromagnetic spectrum, the highest proportion of this radiation lies in the infrared portion of the spectrum at normal temperatures. *See* BLACK BODY; PLANCK'S RADIATION LAW; STEFAN'S LAW; WIEN'S DISPLACEMENT LAW.

heat shield A specially prepared surface that prevents a spacecraft or capsule from overheating as it re-enters the earth's atmosphere. The surface is coated with a plastic impregnated with quartz fibres, which is heated by friction with air molecules as the craft enters the atmosphere, causing the outer layer to vaporize. In this way about 80% of the energy is reradiated and the craft is safeguarded from an excessive rise in temperature.

heat transfer The transfer of energy from one body or system to another as a result of a difference in temperature. The heat is transferred by *conduction (*see also* CONDUCTIVITY), *convection, and radiation (*see* HEAT RADIATION).

Heaviside–Kennelly layer *See* EARTH'S ATMOSPHERE.

Heaviside–Lorentz units A system of units for electric and magnetic quantities based upon c.g.s. electrostatic and electromagnetic units. They are the rationalized forms of *Gaussian units and, like the latter, are widely used in particle physics and relativity in preference to the *SI units now employed for general purposes in physics. This system of units was devised by Oliver Heaviside (1850–1925) and Hendrik *Lorentz.

heavy-fermion system A solid in which the electrons have a very high effective mass; i.e. they act as if they had masses several hundred times the normal mass of the electron. An example of a heavy-fermion system is the cerium–copper–silicon compound $CeCuSi_2$. The electrons with high effective mass are f-electrons in narrow energy bands associated with strong many-body effects. Substances containing such electrons have unusual thermodynamic, magnetic, and superconducting properties, which are still not completely understood. The *superconductivity of such materials has a more complicated mechanism than that for metals described by the BCS theory, since the Cooper pairs are formed from *quasiparticles with very high effective masses rather than from electrons.

heavy hydrogen *See* DEUTERIUM.

heavy water (deuterium oxide) Water in which hydrogen atoms, ^1H, are replaced by the heavier isotope deuterium, ^2H (symbol D). It is a colourless liquid, which forms hexagonal crystals on freezing. Its physical properties differ from those of 'normal' water; r.d. 1.105; m.p. 3.8°C; b.p. 101.4°C. Deuterium oxide, D_2O, occurs to a small extent (about 0.003% by weight) in natural water, from which it can be separated by fractional distillation or by electrolysis. It is useful in the nuclear industry because of its ability to reduce the energies of fast neutrons to thermal energies (*see* MODERATOR) and because its absorption cross-section is lower than that of hydrogen and consequently it does not appreciably reduce the neutron flux. Water also contains the compound HDO.

hecto- Symbol h. A prefix used in the metric system to denote 100 times. For example, 100 coulombs = 1 hectocoulomb (hC).

Heisenberg, Werner Karl (1901–76) German physicist, who became a professor at the University of Leipzig and, after World War II, at the Kaiser Wilhelm Institute in Göttingen. In 1923 he was awarded the Nobel Prize for his work on *matrix mechanics, but he is best known for his 1927 discovery of the *uncertainty principle.

Heisenberg uncertainty principle *See* UNCERTAINTY PRINCIPLE.

heliocentric universe A view of the universe in which the sun is taken to be at its centre. The model was first proposed by Aristarchus of Samos (310–230 BC) but dropped in favour of the *geocentric universe proposed by Ptolemy (c. 90–168 AD). *Copernicus revived an essentially heliocentric view, which was upheld by *Galileo Galilei against strong opposition from the church on the grounds that if the earth was not at the centre of the universe man's position in it was diminished. In the modern view the sun is at the centre of the *solar system, but the solar system is one of an enormous number of stars in the Galaxy, which is itself one of an enormous number of *galaxies.

Helmholtz, Hermann Ludwig Ferdinand von (1821–94) German physiologist, physicist, and mathematician. In 1850 he measured the speed of a nerve impulse and in 1851 invented the ophthalmoscope. In physics, he discovered the conservation of energy (1847) and introduced the concept of *free energy.

Helmholtz coils Two coaxial parallel flat coils with the same radius placed a distance apart that is equal to the radius. If the same current is flowing in both coils then the value of the magnetic field strength is approximately uniform between the coils. Coils of this type are used to create fields and, in some cases, are employed to counter the effect of the earth's magnetic field. Helmholtz coils are also used in magnetic measurement, with the coils connected to a *fluxmeter. If a small magnet is placed between the coils and then removed, the integrated signal on the fluxmeter is proportional to the magnet's magnetic moment.

Helmholtz free energy *See* FREE ENERGY.

henry Symbol H. The *SI unit of inductance equal to the inductance of a closed circuit in which an e.m.f. of one volt is produced when the electric current in the circuit varies uniformly at a rate of one ampere per second. It is named after Joseph Henry.

Henry, Joseph (1797–1878) US physicist, who became professor of natural philosophy at Princeton in 1832. In 1829 he made an *electric motor, and used insulated windings to produce a powerful *electromagnet. A year later he discovered *electromagnetic induction (independently of *Faraday), and in 1832 he discovered self-induction (*see* INDUCTANCE). In 1835 he invented the electric *relay.

Henry's law At a constant temperature the mass of gas dissolved in a liquid at equilibrium is proportional to the partial pressure of the gas. The law, discovered in 1801 by the British chemist and physician William Henry (1775–1836), is a special case of the partition law. It applies only to gases that do not react with the solvent.

hertz Symbol Hz. The *SI unit of frequency equal to one cycle per second. It is named after Heinrich Hertz.

Hertz, Heinrich Rudolph (1857–94) German physicist, who worked as an engineer before attending Berlin University. He is best known for his 1888 discovery of *radio waves, as predicted by James Clerk *Maxwell. The SI unit of frequency is named after him.

Hertzsprung–Russell diagram (H–R diagram) A graphical representation of the absolute magnitude of stars (usually along the y-axis) plotted against the spectral class or colour index (x-axis): see illustration. The y-axis then represents the energy output of the star and the x-axis its surface temperature. The majority of stars on such a diagram fall on a band running from the top left to the bottom right of the graph. These are called **main-sequence stars** (the sun falls into this class). The few stars falling in the lower left portion are called *white dwarfs. The *giant stars fall in a cluster above the main sequence and the *supergiants are above them. The diagram, which was first devised in 1911 by Ejnar Hertzsprung

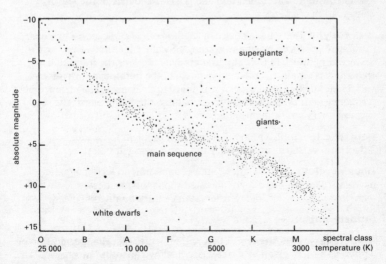

Hertzsprung–Russell diagram

(1873–1969) and in 1913 by H. N. Russell (1897–1957), forms the basis of the theory of *stellar evolution.

Hess's law If reactants can be converted into products by a series of reactions, the sum of the heats of these reactions (with due regard to their sign) is equal to the heat of reaction for direct conversion from reactants to products. More generally, the overall energy change in going from reactants to products does not depend on the route taken. The law can be used to obtain thermodynamic data that cannot be measured directly. For example, the heat of formation of ethane can be found by considering the reaction:

$$2C(s) + 3H_2(g) + 3\tfrac{1}{2}O_2(g) \rightarrow 2CO_2(g) + 3H_2O(l)$$

The heat of this reaction is $2\Delta H_C + 3\Delta H_H$, where ΔH_C and ΔH_H are the heats of combustion of carbon and hydrogen respectively, which can be measured. By Hess's law, this is equal to the sum of the energies for two stages:

$$2C(s) + 3H_2(g) \rightarrow C_2H_6(g)$$

(the heat of formation of ethane, ΔH_f) and

$$C_2H_6(g) + 3\tfrac{1}{2}O_2 \rightarrow 2CO_2(g) + 3H_2O(l)$$

(the heat of combustion of ethane, ΔH_E). As ΔH_E can be measured and as

$$\Delta H_f + \Delta H_E = 2\Delta H_c + 3\Delta H_H$$

ΔH_f can be found. Another example is the use of the *Born–Haber cycle to obtain lattice energies. The law was first put forward in 1840 by the Russian chemist Germain Henri Hess (1802–50). It is sometimes called the **law of constant heat summation** and is a consequence of the law of conservation of energy.

heterodyne Denoting a device or method of radio reception in which *beats are produced by superimposing a locally generated radio wave on an incoming wave. In the *superheterodyne receiver the intermediate frequency is amplified and demodulated. In the **heterodyne wavemeter**, a variable-frequency local oscillator is adjusted to give a predetermined beat frequency with the incoming wave, enabling the frequency of the incoming wave to be determined.

heuristic Denoting a method of solving a problem for which no *algorithm exists. It involves trial and error, as in *iteration.

Heusler alloys Ferromagnetic alloys containing no ferromagnetic elements. The original alloys contained copper, manganese, and tin and were first made by Conrad Heusler (19th century mining engineer).

hidden matter *See* MISSING MASS.

hidden-variables theory A theory that denies that the specification of a physical system given by a state described by *quantum mechanics is a complete specification. A successful hidden-variables theory has never been constructed. There are important considerations preventing the construction of a simple hidden-variables theory, notably *Bell's theorem.

The only type of hidden-variables theories that appears not to have been ruled out are **nonlocal hidden-variables theories**, i.e. theories in which hidden parameters can affect parts of the system in arbitrarily distant regions simultaneously. A hidden-variables theory that does not satisfy this definition is called a **local hidden-variables theory**.

Higgs boson A spin-zero particle with a nonzero mass, predicted by Peter Higgs (1929–) to exist in certain *gauge theories, in particular in *electroweak theory. The Higgs boson has not yet been found but it is thought likely that it will be found by larger *accelerators in the next few years, especially since other associated features of the theory, including W and Z bosons, have been found. *See also* GOLDSTONE'S THEOREM.

Higgs field The symmetry-breaking field associated with a *Higgs boson. The Higgs field can either be an elementary *scalar quantity or the field associated with a *bound state of two *fermions. In the *Weinberg–Salam model, the Higgs field is taken to be a scalar field. It is not known whether this assumption is correct, although attempts to construct *electroweak theory involving bound states for the Higgs field, known as **technicolour theory**, have not been successful. Higgs fields also occur in many-body systems, which can be expressed in terms of *quantum field theory with Higgs bosons; an example is the BCS theory of *superconductivity, in which the Higgs field is associated with a Cooper pair, rather than an elementary scalar field.

high frequency (HF) A radio frequency in the range 3–30 megahertz; i.e. having a wavelength in the range 10–100 metres.

high-speed steel A steel that will remain hard at dull red heat and can therefore be used in cutting tools for high-speed lathes. It usually contains 12–22% tungsten, up to 5% chromium, and 0.4–0.7% carbon. It may also contain small amounts of vanadium, molybdenum, and other metals.

high-temperature superconductivity *See* SUPERCONDUCTIVITY.

high tension (HT) A high potential difference, usually one of several hundred volts or more. Batteries formerly used in the anode circuits of radio devices using valves were usually called **high-tension batteries** to distinguish them from the batteries supplying the heating filaments.

Hilbert space A linear *vector space that can have an infinite number of dimensions. The concept is of interest in physics because the state of a system in *quantum mechanics is represented by a vector in Hilbert space. The dimension of the Hilbert space has nothing to do with the physical dimension of the system. The Hilbert space formulation of quantum mechanics was put forward by the Hungarian-born US mathematician John von Neumann (1903–57) in 1927. Other formulations of quantum mechanics, such as *matrix mechanics and *wave mechanics, can be deduced from the Hilbert space formulation. Hilbert space is named after the German mathematician David Hilbert (1862–1943), who invented the concept early in the 20th century.

Hildebrand rule A rule stating that if the density of the vapour phase above a liquid is constant, the entropy of vaporization of a mole of liquid is constant. This law does not hold if molecular association takes place in the liquid or if the liquid is subject to quantum-mechanical effects, e.g. as in *superfluidity. The Hildebrand rule is named after the US chemist Joel Henry Hildebrand (1881–1983).

hoarfrost *See* DEW.

HOE *See* HOLOGRAPHIC OPTICAL ELEMENT.

hole A vacant electron position in the lattice structure of a solid that behaves like a mobile positive *charge carrier with a negative *rest energy. *See* SEMICONDUCTOR.

holographic hypothesis A key principle of *quantum gravity postulating that it is possible to obtain information about the bulk of a system, such as a black hole, from the surface of that system. This hypothesis was suggested as a general feature of quantum gravity in the mid-1990s and found to hold in superstring theory a few years later.

holographic optical element (HOE) An optical device consisting of a hologram, used to focus or deflect electromagnetic radiation. Holographic optical elements are relatively easy to make and can function as lenses, mirrors, diffraction gratings, beam splitters, etc., in place of more traditional optical components.

holography A method of recording and displaying a three-dimensional image of an object, usually using *coherent radiation from a *laser and photographic plates (see illustration). The light from a laser is divided so that some of it (the reference beam) falls directly on a photographic plate. The other part illuminates the object, which reflects it back onto the photographic plate. The two beams form interference patterns on the plate, which when developed is called the **hologram.** To reproduce the image of the object, the hologram is illuminated by coherent light, ideally the original reference beam. The hologram produces two sets of diffracted waves; one set forms a virtual image coinciding with the original object

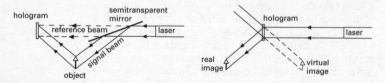

Holography

position and the other forms a real image on the other side of the plate. Both are three-dimensional. The method was invented by Dennis *Gabor in 1948. More recent techniques can produce holograms visible in white light.

homopolar generator A simple electric generator consisting of a metal disc rotating between two poles of a magnet. Contacts are made to the axle and the rim of the disc. A radial e.m.f. is produced. At constant rotational speed, the device produces a steady direct current and generators of this type are used in certain specialized applications. It can also be used as a simple motor if a direct current is supplied. A device of this type (known as the **Barlow wheel**) was invented in 1822 by the physicist Peter Barlow (1776–1862). This had a star-shaped wheel with the points of the star dipping into a pool of mercury to give the electrical contact. A generator with a disc was used by Michael Faraday in his experiments, and the device is sometimes known as the **Faraday disc**.

Hooke, Robert (1635–1703) English physicist, who worked at Oxford University, where he assisted Robert *Boyle. Among his many achievements were the law of elasticity (*see* HOOKE'S LAW), the watch balance wheel, and the compound *microscope. In 1665, using his microscope to study vegetable tissues, he saw 'little boxes' which he named 'cells'.

Hooke's law The *stress applied to any solid is proportional to the *strain it produces within the elastic limit for that solid. The ratio of longitudinal stress to strain is equal to the Young modulus of elasticity (*see* ELASTIC MODULUS). The law was first stated by Robert Hooke in the form "Ut tensio, sic vis."

horizon A boundary between regions of space–time in general relativity theory that signals cannot cross. For example, in an *event horizon around a *black hole any light that is inside the black hole cannot escape across the event horizon (in classical general relativity theory). Horizons also occur in cosmology in models of the expanding universe.

horsepower (hp) An imperial unit of power originally defined as 550 foot-pound force per second; it is equal to 745.7 watts.

hot-wire instrument An electrical measuring instrument (basically an ammeter) in which the current to be measured is passed through a thin wire and causes its temperature to rise. The temperature rise, which is proportional to the square of the current, is measured by the expansion of the wire. Such instruments can be used for either direct current or alternating current.

howl 1. A high-pitched audiofrequency tone caused by *feedback.
2. (screeching) A harsh-sounding unstable form of combustion that can occur in rockets and sometimes in gas-turbine engines, causing great damage very quickly.

Hoyle, Sir Fred (1915–2001) British astronomer, who in 1958 became a professor of astronomy at Cambridge University. He is best known for his proposal in 1948, with Hermann Bondi (1919–) and Thomas Gold (1920–2004), of the *steady-state theory of the universe. He also carried out theoretical work on the formation of elements in stars.

Hubble, Edwin Powell (1889–1953) US astronomer, who worked at both the Yerkes Observatory and the Mount Wilson Observatory. Most of his studies involved nebulae and galaxies, which he classified in 1926. In 1929he established the *Hubble constant, which enabled him to estimate the size and age of the universe. The *Hubble space telescope is named after him.

Hubble constant The rate at which the velocity of recession of the galaxies increases with distance as determined by the *redshift. The value is not agreed upon but current measurements indicate that it lies between 49 and 95 km s^{-1} per megaparsec. The reciprocal of the Hubble constant, the **Hubble time**, is a measure of the age of the universe, assuming that the expansion rate has remained constant. In fact, it is necessary to take into account the observation that the expansion of the universe is accelerating to get an accurate determination of the true age of the universe. The constant is named after Edwin Hubble.

Hubble's law An empirical law in astronomy stating that the velocity v at which a distant galaxy is receding from the Earth is proportional to its distance d from the Earth. This law is usually expressed in the form $v = Hd$, where H is the *Hubble constant. Hubble's law was stated by Edwin Hubble in 1929. It can be derived from models of the expanding universe in general relativity theory.

Hubble Space Telescope A *telescope used to take pictures from space, avoiding the difficulties arising from the earth's atmosphere. These pictures give important information on many aspects of *astronomy and *cosmology. The Hubble Space Telescope (named after Edwin Hubble) was launched in 1990 by the National Aeronautics and Space Administration (NASA) of the USA. Since a defective mirror was corrected in 1993 the scientific results obtained have included: (1) measurements of the *Hubble constant and hence the *age of the universe; these measurements indicate that the *big-bang theory may be in need of modification; (2) evidence for a *black hole at the centre of a galaxy; (3) evidence that in several cases a *quasar is not surrounded by a galaxy; (4) pictures of fragments of the comet Shoemaker–Levy 9 crashing into Jupiter; (5) pictures of some of the most distant galaxies known, which indicate the need for changes in theories of the evolution of galaxies.

hue *See* COLOUR.

hum An undesirable contribution to sound reproduction caused by a nearby electric power circuit. In countries supplying alternating current at a frequency of 50 hertz, the hum usually has this frequency.

humidity The concentration of water vapour in the atmosphere. The **absolute humidity** is the mass of water vapour per unit volume of air, usually expressed in kg m^{-3}. A useful measure is the **relative humidity**, the ratio, expressed as a percentage, of the moisture in the air to the moisture it would contain if it were saturated at the same temperature and pressure. The **specific humidity** is also sometimes used: this is the mass of water vapour in the atmosphere per unit mass of air.

Humphreys series *See* HYDROGEN SPECTRUM.

Hund coupling cases *See* COUPLING.

Hund's rules Empirical rules in atomic *spectra that determine the lowest energy level for a configuration of two equivalent electrons (i.e. electrons with the same n and l quantum numbers) in a many-electron *atom. (1) The lowest energy state has the maximum *multiplicity consistent with the *Pauli exclusion principle. (2) The lowest energy state has the maximum total electron *orbital angular momentum quantum number, consistent with rule (1). These rules were put forward by the German physicist Friedrich Hund (1896–1997) in 1925. Hund's rules are explained by the quantum theory of atoms by calculations involving the repulsion between two electrons and the attraction between these electrons and nuclei.

hunting The oscillation of a gauge needle or engine speed about a mean value. In a rotating mechanism, set to operate at a constant speed, pulsation above and below the set speed can occur, especially if the speed is controlled by a governor. It can be corrected using a damping device.

Huygens, Christiaan (1629–95) Dutch astronomer and physicist, who worked at the Academy of Science in Paris from 1666 to 1681. Returning to The Hague, his birthplace, in 1657, he designed the first pendulum clock and made improvements to astronomical telescopes (with which he discovered Saturn's satellite Titan in 1655). His greatest achievement, however, was the wave theory of *light, announced in 1690 (*see* HUYGENS' CONSTRUCTION).

Huygens' construction (Huygens' principle) Every point on a wavefront may itself be regarded as a source of secondary waves. Thus, if the position of a wavefront at any instant is known, a simple construction enables its position to be drawn at any subsequent time. The construction was first used by Christiaan Huygens.

hydraulic press A device in which a force (F_1) applied to a small piston (A_1) creates a pressure (p), which is transmitted through a fluid to a larger piston (A_2), where it gives rise to a larger force (F_2): see illustration. This depends on Pascal's principle that the pressure applied anywhere in an

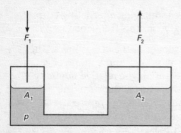

Hydraulic press

enclosed fluid is transmitted equally in all directions. The principle of the hydraulic press is widely used in jacks, vehicle brakes, presses, and earth-moving machinery, usually with oil as the working fluid.

hydraulics The study of water or other fluids at rest or in motion, particularly with respect to their engineering uses. The study is based on the principles of *hydrostatics and *hydrodynamics.

hydrodynamics The study of the motion of incompressible fluids and the interaction of such fluids with their boundaries.

hydroelectric power Electric power generated by a flow of water. A natural waterfall provides a source of energy, in the form of falling water, which can be used to drive a water *turbine. This turbine can be coupled to a generator to provide electrical energy. Hydroelectric generators can be arranged to work in reverse so that during periods of low power demand current can be fed to the generator, which acts as a motor. This motor drives the turbine, which then acts as a pump. The pump then raises water to an elevated reservoir so that it can be used to provide extra power at peak-load periods.

hydrogen bomb *See* NUCLEAR WEAPONS.

hydrogen electrode *See* HYDROGEN HALF CELL.

hydrogen half cell (**hydrogen electrode**) A type of half cell in which a metal foil is immersed in a solution of hydrogen ions and hydrogen gas is bubbled over the foil. The standard hydrogen electrode, used in measuring standard *electrode potentials, uses a platinum foil with a 1.0 M solution of hydrogen ions, the gas at 1 atmosphere pressure, and a temperature of 25°C. It is written $Pt(s)|H_2(g)$, $H^+(aq)$, the effective reaction being summarized in the equation $H_2 \rightarrow 2H^+ + 2e$.

hydrogen ion *See* pH SCALE.

hydrogen spectrum The atomic spectrum of hydrogen is characterized by lines corresponding to radiation quanta of sharply defined energy. A graph of the frequencies at which these lines occur against the ordinal number that characterizes their position in the series of lines, produces a smooth curve indicating that they obey a formal law. In 1885 J. J. Balmer (1825–98) discovered the law having the form:

$1/\lambda = R(1/n_1^2 - 1/n_2^2)$

This law gives the so-called **Balmer series** of lines in the visible spectrum in which $n_1 = 2$ and $n_2 = 3,4,5…$, λ is the wavelength associated with the lines, and R is the *Rydberg constant.

In the **Lyman series**, discovered by Theodore Lyman (1874–1954), $n_1 = 1$ and the lines fall in the ultraviolet. The Lyman series is the strongest feature of the solar spectrum as observed by rockets and satellites above the earth's atmosphere. In the **Paschen series**, discovered by F. Paschen (1865–1947), $n_1 = 3$ and the lines occur in the far infrared. The **Brackett series** ($n_1 = 4$), **Pfund series** ($n_1 = 5$), and **Humphreys series** ($n_1 = 6$) also occur.

hydrometeor Any product of the *condensation of the water vapour in the atmosphere, which can occur either in the atmosphere or on the surface of the earth. Hydrometeors can consist of solid or liquid particles that form and remain in suspension in the atmosphere, e.g. **cloud**, **fog**, and **mist**. Liquid *precipitation can occur as **rain**, while solid precipitation can take the form of **hail** and **snow**. Solid or liquid particles rising from the surface of the earth include blowing snow and blowing spray from the sea. Solid or liquid deposits on exposed objects include *dew and hoarfrost.

hydrometer An instrument for measuring the density or relative density of liquids. It usually consists of a glass tube with a long bulb at one end. The bulb is weighted so that the device floats vertically in the liquid, the relative density being read off its calibrated stem by the depth of immersion.

hydrostatics The study of liquids at rest, with special reference to storage tanks, dams, bulkheads, and hydraulic machinery.

hygrometer An instrument for measuring *humidity in the atmosphere. The mechanical type uses an organic material, such as human hair, which expands and contracts with changes in atmospheric humidity. The expansion and contraction is used to operate a needle. In the electric type, the change in resistance of a hygroscopic substance is used as an indication of humidity. In **dew-point hygrometers** a polished surface is reduced in temperature until water vapour from the atmosphere forms on it. The temperature of this dew point enables the relative humidity of the atmosphere to be calculated. In the **wet-and-dry bulb hygrometer**, two thermometers are mounted side by side, the bulb of one being surrounded by moistened muslin. The thermometer with the wet bulb will register a lower temperature than that with a dry bulb owing to the cooling effect of the evaporating water. The temperature difference enables the relative humidity to be calculated. Only the dew-point hygrometer can be operated as an absolute instrument; all the others must ultimately be calibrated against this.

hyper- A prefix denoting over, above, high; e.g. hypersonic.

hyperbola A *conic with eccentricity $e > 1$. It has two branches (see graph). For a hyperbola centred at the origin, the **transverse axis** runs along the x-axis between the vertices and has length $2a$. The **conjugate axis** runs along the y-axis and has length $2b$. There are two **foci** on the x-axis at $(ae, 0)$ and $(-ae, 0)$. The **latus rectum**, the chords through the foci perpendicular to the transverse axis, have length $2b^2/a$. The equation of the hyperbola is:

$$x^2/a^2 - y^2/b^2 = 1,$$

and the asymptotes are $y = \pm bx/a$.

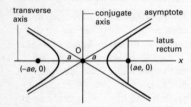

A hyperbola

hyperbolic functions A set of functions, **sinh**, **cosh**, and **tanh**, that have similar properties to *trigonometric functions but are related to the hyperbola in the manner that trigonometric functions are related to the circle. The hyperbolic sine (sinh) of the angle x is defined by:

$$\sinh x = \tfrac{1}{2}(e^x - e^{-x}).$$

Similarly,

$$\cosh x = \tfrac{1}{2}(e^x + e^{-x})$$
$$\tanh x = (e^x - e^{-x})/(e^x + e^{-x})$$

Hyperbolic secant (sech), cosecant (cosech), and cotangent (coth) are the reciprocals of cosh, sinh, and tanh, respectively.

hypercharge A quantized property of *baryons (see ELEMENTARY PARTICLES) that provides a formal method of accounting for the nonoccurrence of certain expected decays by means of the strong interaction (see FUNDAMENTAL INTERACTIONS). Hypercharge is in some respects analogous to electric charge but it is not conserved in weak interactions. Nucleons have a hypercharge of +1, and the *pion has a value of 0. Quarks would be expected to have fractional hypercharges.

hyperfine structure *See* FINE STRUCTURE.

hypermetropia (hyperopia) Long-sightedness. A vision defect in which the lens of the eye is unable to accommodate sufficiently to throw the image of near objects onto the retina. It is caused usually by shortness of

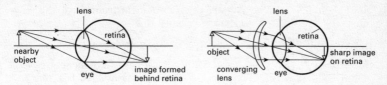

Hypermetropia

the eyeball rather than any fault in the lens system. The subject requires spectacles with converging lenses to bring the image from behind the retina back on to its surface. See illustration.

hypernova An explosive event in which a large star collapses with the formation of a *black hole. An event of this type would be even more violent than a *supernova (in which a star collapses to form a neutron star). Hypernovae may be the cause of gamma-ray bursts.

hyperon A shortlived *elementary particle; it is classified as a *baryon and has a nonzero *strangeness.

hypersonic Denoting a velocity in excess of Mach 5 (*see* MACH NUMBER). **Hypersonic flight** is flight at hypersonic speeds in the earth's atmosphere. **Hypersonics** is the study of phenomena, including air flow, involving bodies moving at these speeds.

hypertext A technique by which textual documents can be created and viewed on a computer screen so that one or more documents can be browsed in any order by the selection of key words or phrases by the user. The selected text leads (by underlying searches through associated files, indexes, etc.) to the display of another part of the document, or of some other document. **Hypermedia** is an extension of this technique enabling links to be made between text, images, sounds, etc. *See also* WORLD WIDE WEB.

hypertonic solution A solution that has a higher osmotic pressure than some other solution.

hypo- A prefix denoting under, below, low; e.g. hypotonic.

hypothesis *See* LAWS, THEORIES, AND HYPOTHESES.

hypotonic solution A solution that has a lower osmotic pressure than some other solution.

hypsometer A device for calibrating thermometers at the boiling point of water. As the boiling point depends on the atmospheric pressure, which in turn depends on the height above sea level, the apparatus can be used to measure height above sea level.

hysteresis A phenomenon in which two physical quantities are related

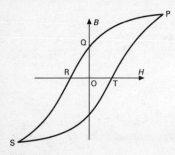

Hysteresis

in a manner that depends on whether one is increasing or decreasing in relation to the other.

The repeated measurement of *stress against *strain, with the stress first increasing and then decreasing, will produce for some specimens a graph that has the shape of a closed loop. This is known as a **hysteresis cycle**. The most familiar hysteresis cycle, however, is produced by plotting the magnetic flux density (B) within a ferromagnetic material against the applied magnetic field strength (H).

If the material is initially unmagnetized at O it will reach saturation at P as H is increased. As the field is reduced and again increased the loop PQRSTP is formed (see graph). The area of this loop is proportional to the energy loss (**hysteresis loss**) occurring during the cycle. The value of B equal to OQ is called the **remanence** (or retentivity) and is the magnetic flux density remaining in the material after the saturating field has been reduced to zero. This is a measure of the tendency of the magnetic domain patterns (*see* MAGNETISM) to remain distorted even after the distorting field has been removed. The value of H equal to OR is called the **coercive force** (or coercivity) and is the field strength required to reduce the remaining flux density to zero. It is a measure of the difficulty of restoring the symmetry of the domain patterns.

IBFM *See* INTERACTING-BOSON MODEL.

IBM *See* INTERACTING-BOSON MODEL.

ice point The temperature at which there is equilibrium between ice and water at standard atmospheric pressure (i.e. the freezing or melting point under standard conditions). It was used as a fixed point (0°) on the Celsius scale, but the kelvin and the International Practical Temperature Scale are based on the *triple point of water.

iconoscope A form of television camera tube (*see* CAMERA) in which the beam of light from the scene is focused on to a thin mica plate. One side of the plate is faced with a thin metallic electrode, the other side being coated with a mosaic of small globules of a photoemissive material. The light beam falling on the mosaic causes photoemission of electrons, creating a pattern of positive charges in what is effectively an array of tiny capacitors. A high-velocity electron beam scans the mosaic, discharging each capacitor in turn through the metallic electrode. The resulting current is fed to amplification circuits, the current from a particular section of the mosaic depending on the illumination it has received. In this way the optical information in the light beam is converted into an electrical signal.

icosahedron A polyhedron having 20 triangular faces with five edges meeting at each vertex. The symmetry of an icosahedron (known as **icosahedral symmetry**) has fivefold rotation axes. It is impossible in *crystallography to have a periodic *crystal with the point group symmetry of an icosahedron (icosahedral packing). However, it is possible for short-range order to occur with icosahedral symmetry in certain *liquids and *glasses because of the dense packing of icosahedra. Icosahedral symmetry also occurs in certain *quasicrystals, such as alloys of aluminium and manganese.

ideal crystal A single crystal with a perfectly regular lattice that contains no impurities, imperfections, or other defects.

ideal gas (perfect gas) A hypothetical gas that obeys the *gas laws exactly. An ideal gas would consist of molecules that occupy negligible space and have negligible forces between them. All collisions made between molecules and the walls of the container or between molecules and other molecules would be perfectly elastic, because the molecules would have no means of storing energy except as translational kinetic energy.

ideal solution *See* RAOULT'S LAW.

identity Symbol ≡. A statement of equality that applies for all values of the unknown quantity. For example, $5y \equiv 2y + 3y$.

ignition temperature 1. The temperature to which a substance must be heated before it will burn in air. **2.** The temperature to which a *plasma has to be raised in order that nuclear fusion will occur.

illuminance (illumination) Symbol E. The energy in the form of visible radiation reaching a surface per unit area in unit time; i.e. the luminous flux per unit time. It is measured in *lux (lumens per square metre).

image A representation of a physical object formed by a lens, mirror, or other optical instrument. If the rays of light actually pass through the image, it is called a **real image**. If a screen is placed in the plane of a real image it will generally become visible. If the image is seen at a point from which the rays appear to come to the observer, but do not actually do so, the image is called a **virtual image**. No image will be formed on a screen placed at this point. Images may be **upright** or **inverted** and they may be **magnified** or **diminished**.

image converter An electronic device in which an image formed by invisible radiation (usually gamma rays, X-rays, ultraviolet, or infrared) is converted into a visible image. Commonly the invisible radiation is focused on to a photocathode, which emits electrons when it is exposed to the radiation. These electrons fall on a fluorescent anode screen, after acceleration and focusing by a system of electron lenses. The fluorescent screen produces a visible image. The device is used in fluoroscopes, infrared telescopes, ultraviolet microscopes, and other devices.

image intensifier A device used to produce a brighter optical image than an initial image. An image intensifier can be used at very low levels of brightness without introducing *noise into the brightness levels of the intensified image. The principle is similar to that of the *image converter.

imaginary number A number that is a multiple of $\sqrt{-1}$, which is denoted by i; for example $\sqrt{-3} = i\sqrt{3}$. *See also* COMPLEX NUMBER.

immersion objective An optical microscope objective in which the front surface of the lens is immersed in a liquid on the cover glass of the microscope specimen slide. Cedar-wood oil or sugar solution is frequently used. It has the same refractive index as the glass of the slide, so that the object is effectively immersed in it. The presence of the liquid increases the effective aperture of the objective, thus increasing the resolution.

impact printer *See* PRINTER.

impedance Symbol Z. The quantity that measures the opposition of a circuit to the passage of a current and therefore determines the amplitude of the current. In a d.c. circuit this is the resistance (R) alone. In an a.c.

circuit, however, the *reactance (X) also has to be taken into account, according to the equation: $Z^2 = R^2 + X^2$, where Z is the impedance. The **complex impedance** is given by $Z = R + iX$, where $i = \sqrt{-1}$. The real part of the complex impedance, the resistance, represents the loss of power according to *Joule's law. The ratio of the imaginary part, the reactance, to the real part is an indication of the difference in phase between the voltage and the current.

Imperial units The British system of units based on the pound and the yard. The former f.p.s. system was used in engineering and was loosely based on Imperial units; for all scientific purposes *SI units are now used. Imperial units are also being replaced for general purposes by metric units.

implosion An inward collapse of a vessel, especially as a result of evacuation.

impulse Symbol J. The product of a force F and the time t for which it acts. If the force is variable, the impulse is the integral of Fdt from t_0 to t_1. The impulse of a force acting for a given time interval is equal to the change in momentum produced over that interval, i.e. $J = m(v_1 - v_0)$, assuming that the mass (m) remains constant while the velocity changes from v_0 to v_1.

incandescence The emission of light by a substance as a result of raising it to a high temperature. An **incandescent lamp** is one in which light is emitted by an electrically heated filament. *See* ELECTRIC LIGHTING.

inclination 1. *See* GEOMAGNETISM. **2.** The angle between the orbital plane of a planet, satellite, or comet and the plane of the earth's *ecliptic.

incoherent scattering *See* COHERENT SCATTERING.

incommensurate lattice A lattice with long-range periodic order that has two or more periodicities in which an irrational number gives the ratio between the periodicities. An example of an incommensurate lattice occurs in certain magnetic systems in which the ratio of the magnetic period to the atomic lattice is an irrational number. The *phase transition between a *commensurate lattice and an incommensurate lattice can be analysed using the *Frenkel–Kontorowa model.

indefinite integral *See* INTEGRATION.

indeterminacy *See* UNCERTAINTY PRINCIPLE.

index theorems Theories in mathematics that relate the solutions of an equation to *topology associated with the equation. There are many applications of index theorems to *gauge theories, *quantum gravity, *supersymmetry, *string theory, and certain problems in the theory of condensed matter. Examples of index theorems in the case of gauge fields coupled to *fermion fields (of the type relevant to the description of elementary particles) are that solutions of the Dirac equation are related to topological invariants of gauge theories. This application of index theorems is closely related to chiral anomalies. Index theorems also give the number of parameters associated with instantons and *magnetic monopoles.

indirect transition *See* DIRECT TRANSITION.

induced emission (**stimulated emission**) The emission of a photon by an atom in the presence of *electromagnetic radiation. The atom can become excited by the absorption of a photon of the right energy and, having become excited, the atom can emit a photon. The rate of absorption is equal to the rate of induced emission, both rates being proportional to the density of photons of the electromagnetic radiation. The relation between induced emission and *spontaneous emission is given by the *Einstein coefficients. The process of induced emission is essential for the operation of *lasers and *masers. *See also* QUANTUM THEORY OF RADIATION.

inductance The property of an electric circuit or component that causes an e.m.f. to be generated in it as a result of a change in the current flowing through the circuit (**self inductance**) or of a change in the current flowing through a neighbouring circuit with which it is magnetically linked (**mutual inductance**). In both cases the changing current is associated with a changing magnetic field, the linkage with which in turn induces the e.m.f. In the case of self inductance, L, the e.m.f., E, generated is given by $E = -L.dI/dt$, where I is the instantaneous current and the minus sign indicates that the e.m.f. induced is in opposition to the change of current. In the case of mutual inductance, M, the e.m.f., E_1, induced in one circuit is given by $E_1 = -M.dI_2/dt$, where I_2 is the instantaneous current in the other circuit.

induction A change in the state of a body produced by a field. *See* ELECTROMAGNETIC INDUCTION; INDUCTANCE.

induction coil A type of *transformer used to produce a high-voltage alternating current or pulses of high-voltage current from a low-voltage direct-current source. The induction coil is widely used in spark-ignition *internal-combustion engines to produce the spark in the sparking plugs. In such an engine the battery is connected to the primary winding of the coil through a circuit-breaking device driven by the engine and the e.m.f. generated in the secondary winding of the coil is led to the sparking plugs through the distributor. The primary coil consists of relatively few turns, whereas the secondary consists of many turns of fine wire.

induction heating The heating of an electrically conducting material by *eddy currents induced by a varying electromagnetic field. Induction heating may be an undesirable effect leading to power loss in transformers and other electrical devices. It is, however, useful for melting and heat-treating and in forging and rolling metals, as well as for welding, brazing, and soldering. The material to be heated is inserted into a coil through which an alternating current flows and acts as the short-circuited secondary of a *transformer. Eddy currents induced in the material within the coil cause the temperature of the material to rise.

induction motor *See* ELECTRIC MOTOR.

inelastic collision A collision in which some of the kinetic energy of the colliding bodies is converted into internal energy in one body so that kinetic energy is not conserved. In collisions of macroscopic bodies some kinetic energy is turned into vibrational energy of the atoms, causing a heating effect. Collisions between molecules of a gas or liquid may also be inelastic as they cause changes in vibrational and rotational *energy levels. In nuclear physics, an inelastic collision is one in which the incoming particle causes the nucleus it strikes to become excited or to break up. **Deep inelastic scattering** is a method of probing the structure of sub-atomic particles in much the same way as Rutherford probed the inside of the atom (*see* RUTHERFORD SCATTERING). Such experiments were performed on protons in the late 1960s using high-energy electrons at the Stanford Linear Accelerator Center (SLAC). As in Rutherford scattering, deep inelastic scattering of electrons by proton targets revealed that most of the incident electrons interacted very little because they pass straight through the target; only a small number bounced back very sharply. This means that the charge in the proton must be concentrated in small lumps, reminiscent of Rutherford's discovery that charge in an atom is concentrated at the nucleus. However, in the case of the proton the evidence suggested that three distinct concentrations of charge existed rather than one (*see* ELEMENTARY PARTICLES; PARTON; QUANTUM CHROMODYNAMICS).

inequality A relationship between two quantities in which one of the quantities is not equal to (or not necessarily equal to) the other quantity. If the quantities are a and b, two inequalities exist: a is greater than b, written $a > b$, and a is less than b, i.e. $a < b$. Similar statements can take the form: a is greater than or equal to b, written $a \geq b$, and a is less than or equal to b, which is denoted $a \leq b$. There are many applications of inequalities in physical science, an example being the *Heisenberg uncertainty principle.

inertia The property of matter that causes it to resist any change in its motion. Thus, a body at rest remains at rest unless it is acted upon by an external force and a body in motion continues to move at constant speed in a straight line unless acted upon by an external force. This is a statement of Newton's first law of motion. The *mass of a body is a measure of its inertia. *See* MACH'S PRINCIPLE; INERTIAL FRAME.

inertial confinement *See* THERMONUCLEAR REACTOR.

inertial frame A *frame of reference in which bodies move in straight lines with constant speeds unless acted upon by external forces, i.e. a frame of reference in which free bodies are not accelerated. Newton's laws of motion are valid in an inertial system but not in a system that is itself accelerated with respect to such a frame.

inertial mass *See* MASS.

infinite series *See* SERIES.

infinitesimal Vanishingly small but not zero. Infinitesimal changes are notionally made in the *calculus, which is sometimes called the **infinitesimal calculus**.

infinity Symbol ∞. A quantity having a value that is greater than any assignable value. Minus infinity, $-\infty$, is a quantity having a value that is less than any assignable value.

inflation *See* EARLY UNIVERSE.

inflationary universe *See* EARLY UNIVERSE.

inflection A point on a curve at which the tangent changes from rotation in one direction to rotation in the opposite direction. If the curve $y = f(x)$ has a stationary point $dy/dx = 0$, there is either a maximum, minimum, or inflection at this point. If $d^2y/dx^2 = 0$, the stationary point is a point of inflection.

information technology *See* IT.

information theory The branch of mathematics that analyses information mathematically. Several branches of physics have been related to information theory. For example, an increase in *entropy has been expressed as a decrease in information. It has been suggested that it may be possible to express the basic laws of physics using information theory. *See also* LANDAUER'S PRINCIPLE; ZEILINGER'S PRINCIPLE.

infrared astronomy The study of radiation from space in the infrared region of the spectrum (*see* INFRARED RADIATION). Some infrared radiation is absorbed by water and carbon dioxide molecules in the atmosphere but there are several narrow atmospheric *windows in the near-infrared (1.15–1.3 μm, 1.5–1.75 μm, 2–2.4 μm, 3.4–4.2 μm, 4.6–4.8 μm, 8–13 μm, and 16–18 μm). Longer wavelength observations must be made from balloons, rockets, or satellites. Infrared sources are either thermal, i.e. emitted by the atoms or molecules of gases or dust particles in the temperature range 100–3000 K, or electronic, i.e. emitted by high-energy electrons interacting with magnetic fields as in *synchrotron radiation. Detectors are either modified reflecting *telescopes or solid-state photon detectors, usually incorporating photovoltaic devices.

infrared radiation (IR) Electromagnetic radiation with wavelengths longer than that of red light but shorter than radiowaves, i.e. radiation in the wavelength range 0.7 micrometre to 1 millimetre. It was discovered in 1800 by William Herschel (1738–1822) in the sun's spectrum. The natural vibrational frequencies of atoms and molecules and the rotational frequencies of some gaseous molecules fall in the infrared region of the electromagnetic spectrum. The infrared absorption spectrum of a molecule is highly characteristic of it and the spectrum can therefore be used for molecular identification. Glass is opaque to infrared radiation of wavelength greater than 2 micrometres and other materials, such as germanium, quartz, and polyethylene, have to be used to make lenses and prisms. Photographic film can be made sensitive to infrared up to about 1.2 μm.

infrasound Soundlike waves with frequencies below the audible limit of about 20 hertz.

insolation (from *in*coming *so*lar radi*ation*) The solar radiation that is received at the earth's surface per unit area. It is related to the *solar constant, the duration of daylight, the altitude of the sun, and the latitude of the receiving surface. It is measured in MJ m^{-2}.

instantaneous value The value of any varying quantity at a specified instant.

insulator A substance that is a poor conductor of heat and electricity. Both properties usually occur as a consequence of a lack of mobile electrons. *See* ENERGY BANDS.

integer quantum Hall effect *See* QUANTUM HALL EFFECT.

integral calculus *See* CALCULUS.

integrand *See* INTEGRATION.

integrated circuit A miniature electronic circuit produced within a single crystal of a *semiconductor, such as silicon. They range from simple logic circuits, little more than 1 mm square, to large-scale circuits measuring up to 8 mm square and containing a million or so transistors (active components) and resistors or capacitors (passive components). They are widely used in memory circuits, microcomputers, pocket calculators, and electronic watches on account of their low cost and bulk, reliability, and high speed. They are made by introducing impurities into specific regions of the semiconductor crystal by a variety of techniques.

integration The process of continuously summing changes in a function f(x). It is the basis of the integral *calculus and the opposite process to *differentiation. The function to be integrated is called the **integrand** and the result of integration on the integrand is called the **integral**. For example, the integration of f(x) is written ∫f(x)dx, the differential dx being added to indicate that f(x) must be integrated with respect to x. To complete the integration, a **constant of integration**, C, must be added where no interval over which the integration takes place is given. This is called an **indefinite integral**. If the interval is specified, e.g.

$$\int_a^b f(x)dx,$$

no constant of integration is required and the result is called a **definite integral**. This means that f(x) is to be integrated between the values $x = a$ and $x = b$ to give a definite value.

intensity 1. The rate at which radiant energy is transferred per unit area. *See* RADIANT INTENSITY. **2.** The rate at which sound energy is transferred as measured relative to some reference value. *See* DECIBEL. **3.** Magnetic intensity. *See* MAGNETIC FIELD. **4.** Electric intensity. *See* ELECTRIC FIELD. **5.** *See* LUMINOUS INTENSITY.

intensive variable A quantity in a *macroscopic system that has a well defined value at every point inside the system and that remains (nearly) constant when the size of the system is increased. Examples of intensive variables are the pressure, temperature, density, specific heat capacity at constant volume, and viscosity. An intensive variable results when any *extensive variable is divided by an arbitrary extensive variable such as the volume. A macroscopic system can be described by one extensive variable and a set of intensive variables.

interaction An effect involving a number of bodies, particles, or systems as a result of which some physical or chemical change takes place to one or more of them. *See also* FUNDAMENTAL INTERACTIONS.

interacting-boson model (IBM) A model of nuclear structure in which nucleons pair up and form bosons from which collective states are formed. The IBM has had considerable success in describing the spectra of nuclei with even numbers of protons and neutrons. An extension of the model to nuclei in which there are unpaired protons is called the **interacting boson–fermion model** (IBFM). Here the nuclei are analysed in terms of interaction between the unpaired fermions and the bosons made up of paired fermions.

intercalation cell A type of secondary cell in which layered electrodes, usually made of metal oxides or graphite, store positive ions between the crystal layers of an electrode. In one type, lithium ions form an intercalation compound with a graphite electrode when the cell is charged. During discharge, the ions move through an electrolyte to the other electrode, made of manganese oxide, where they are more tightly bound. When the cell is being charged, the ions move back to their positions in the graphite. This backwards and forwards motion of the ions has led to the name **rocking-chair cell** for this type of system. Such cells have the advantage that only minor physical changes occur to the electrodes during the charging and discharging processes and the electrolyte is not decomposed but simply serves as a conductor of ions. Consequently, such cells can be recharged many more times than, say, a lead-acid accumulator, which eventually suffers from degeneration of the electrodes. **Lithium cells**, based on this principle, have been used in portable electronic equipment, such as camcorders. They have also been considered for use in electric vehicles.

interference The interaction of two or more wave motions affecting the same part of a medium so that the instantaneous disturbances in the resultant wave are the vector sum of the instantaneous disturbances in the interfering waves.

The phenomenon was first described by Thomas *Young in 1801 in light waves; it provided strong evidence for the wave theory of light. In the apparatus known as **Young's slits**, light is passed from a small source through a slit in a screen and the light emerging from this slit is used to illuminate two adjacent slits on a second screen. By allowing the light from these two slits to fall on a third screen, a series of parallel interference fringes is formed. Where the maximum values of the two waves from the slits coincide a bright fringe occurs (**constructive interference**) and where the maxima of one wave coincide with the minima of the other dark fringes are produced (**destructive interference**). *Newton's rings are also an interference effect. Because *lasers produce *coherent radiation they are also used to produce interference effects, one application of their use being *holography. *See also* INTERFEROMETER.

interferometer An instrument designed to produce optical *interference fringes for measuring wavelengths, testing flat surfaces, measuring small distances, etc. (*see also* ECHELON; FABRY–PÉROT INTERFEROMETER; MICHELSON–MORLEY EXPERIMENT). In astronomy, radio interferometers are one of the two basic types of *radio telescopes.

intermediate coupling *See* J-J COUPLING.

intermediate frequency *See* HETERODYNE; SUPERHETERODYNE RECEIVER.

intermediate neutron A *neutron with kinetic energy in the range 10^2–10^5 electronvolts (1.6×10^{-17}–1.6×10^{-14} joule).

intermediate vector boson *See* W BOSON; Z BOSON.

intermetallic compound A compound consisting of two or more metallic elements present in definite proportions in an alloy.

intermolecular forces Weak forces occurring between molecules. *See* VAN DER WAALS' FORCES.

internal-combustion engine A *heat engine in which fuel is burned in combustion chambers within the engine rather than in a separate furnace (as with the steam engine). The first working engine was the four-stroke **Otto engine** produced in 1876 by Nikolaus Otto (1832–91). In this type of engine a piston descends in a cylinder, drawing in a charge of fuel and air through an inlet valve; after reaching the bottom of its stroke the piston rises in the cylinder with the valves closed and compresses the

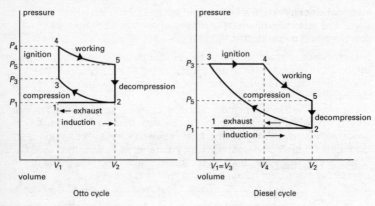

Otto cycle   Diesel cycle

Internal-combustion engines

charge; at or near the top of its stroke the charge is ignited by a spark and the resulting increase in pressure from the explosion forces the piston down again; on the subsequent upstroke the exhaust valve opens and the burnt gases are pushed out of the combustion chamber. The cycle is then repeated. Otto's engine used gas as a fuel; however, the invention of the carburettor and the development of the oil industry at the end of the 19th century enabled the Otto engine to become the source of power for the emerging motor car. A variation of the Otto four-stroke engine is the two-stroke engine that has no complicated valve system, the explosive charge entering and leaving the cylinder through ports in the cylinder that are covered and uncovered by the moving piston.

An alternative to the Otto engine, especially for heavy vehicles where weight is not a problem, is the compression-ignition **Diesel engine** invented by Rudolf Diesel (1858–1913) in about 1896. In this type of engine there are no sparking plugs; instead air is compressed in the cylinder, causing its temperature to rise to about 550°C. Oil is then sprayed into the combustion chamber and ignites on contact with the hot air. While the spark-ignition petrol engine typically works on a *compression ratio of 8 or 9 to 1, the Diesel engine has to have a compression ratio of between 15 and 25 to 1. This requires a much heavier, and therefore more expensive, engine. *See also* GAS TURBINE.

internal conversion A process in which an excited atomic nucleus (*see* EXCITATION) decays to the *ground state and the energy released is transferred by electromagnetic coupling to one of the bound electrons of that atom rather than being released as a photon. The coupling is usually with an electron in the K-, L-, or M-shell of the atom, and this **conversion electron** is ejected from the atom with a kinetic energy equal to the difference between the nuclear transition energy and the binding energy of the electron. The resulting ion is itself in an excited state and usually subsequently emits an Auger electron or an X-ray photon.

internal energy Symbol U. The total of the kinetic energies of the atoms and molecules of which a system consists and the potential energies associated with their mutual interactions. It does not include the kinetic and potential energies of the system as a whole nor their nuclear energies or other intra-atomic energies. The value of the absolute internal energy of a system in any particular state cannot be measured; the significant quantity is the change in internal energy, ΔU. For a closed system (i.e. one that is not being replenished from outside its boundaries) the change in internal energy is equal to the heat absorbed by the system (Q) from its surroundings, less the work done (W) by the system on its surroundings, i.e. $\Delta U = Q - W$. *See also* ENERGY; HEAT.

internal resistance The resistance within a source of electric current, such as a cell or generator. It can be calculated as the difference between the e.m.f. (E) and the potential difference (V) between the terminals divided by the current being supplied (I), i.e. $r = (E - V)/I$, where r is the internal resistance.

international candle A former unit of *luminous intensity. It has now been replaced by the *candela, to which it is approximately equal.

international date line An imaginary line on the earth's surface that joins the north and south poles and approximately follows the 180° meridian through the Pacific Ocean. This line has been agreed internationally to mark the beginning and end of a day. A traveller moving towards the east, against the sun's apparent movement, gains 1 hour for every 15° of longitude; westward he loses time at the same rate. In crossing the dateline therefore he is deemed to compensate for this by losing or gaining (respectively) one day. The 180° meridian was chosen as the date line by the International Meridian Conference in 1884.

International Practical Temperature Scale *See* TEMPERATURE SCALES.

Internet (Net) The global network that links most of the world's computer networks. It does not offer services to users, but serves primarily to interconnect other networks on which services are located. These include basic services for *electronic mail, the transfer of computer files, and remote log-in, and high-level services including the *World Wide Web. The Internet is informal, with a minimal level of administration by governing bodies.

interplanetary space The space between the sun and the planets within the *solar system. The **interplanetary matter** that occupies this region of space mostly originates in the sun, from which it flows as the *solar wind. The solar wind consists primarily of protons emerging from the sun at a rate of about 10^9 kilograms per second. At the earth's distance from the sun the particle density has fallen to a few particles per cm^3. Apart from this very tenuous gas, there are also dust particles in interplanetary space, largely believed to originate in the belt of asteroids.

Particles weighing about 1 g produce visible meteors in the earth's atmosphere; micrometeorites as small as 1 nanogram can be detected by their impact on spacecraft.

interpolation An *approximation technique for finding the value of a function or a measurement that lies within known values. If the values $f(x_0)$, $f(x_1)$, ..., $f(x_n)$ of a function f of a variable x are known in the interval $[x_0, x_n]$, the value of $f(x)$ for a value of x inside the interval $[x_0, x_n]$ can be found by interpolation. One method of interpolation, called **linear interpolation** for $x_0 < x < x_1$, gives:

$$f(x) \approx f(x_0) + [f(x_1) - f(x_0)] \, (x - x_0)/(x_1 - x_0),$$

which is derived using the assumption that between the points x_0 and x_1, the graph of the function $f(x)$ can be regarded as a straight line. More complicated methods of interpolation exist, using more than two values for the function. The techniques used for interpolation are usually much better than the techniques used in *extrapolation.

interstellar space The space between the stars. The **interstellar matter** that occupies this space constitutes several percent of the Galaxy's total mass and it is from this matter that new stars are formed. The matter is primarily hydrogen, in which a number of other molecules and radicals have been detected, together with small solid dust grains. On average the density of matter in interstellar space is about 10^6 hydrogen atoms per cubic metre, but the gas is not uniformly distributed, being clumped into **interstellar clouds** of various sizes and densities.

interstitial *See* DEFECT.

interstitial compound A compound in which ions or atoms of a nonmetal occupy interstitial positions in a metal lattice. Such compounds often have metallic properties. Examples are found in the carbides, borides, and silicides.

intrinsic semiconductor *See* SEMICONDUCTOR.

Invar A tradename for an alloy of iron (63.8%), nickel (36%), and carbon (0.2%) that has a very low *expansivity over a a restricted temperature range. It is used in watches and other instruments to reduce their sensitivity to changes in temperature.

inverse Compton effect The gain in energy of low-energy photons when they are scattered by free electrons of much higher energy. As a consequence, the electrons lose energy. The inverse Compton effect is thought to be important in some processes in astrophysics. *See also* COMPTON EFFECT.

inverse functions If $y = f(x)$ and a function can be found so that $x = g(y)$, then $g(y)$ is said to be the inverse function of $f(x)$. If y is a trigonometrical function of the angle x, say $y = \sin x$, then x is the **inverse trigonometrical function** of y, written $x = \arcsin y$ or $\sin^{-1} y$. Similarly, the other trigonometrical functions form the inverse trigonometrical functions $\cos^{-1}y$, $\tan^{-1}y$, $\cot^{-1}y$, $\sec^{-1}y$, and $\mathrm{cosec}^{-1}y$. **Inverse hyperbolic functions** are also formed in this way, e.g. arcsinh$y$ or $\sinh^{-1}y$, $\cosh^{-1}y$, and $\tanh^{-1}y$.

inverse-square law A law in which the magnitude of a physical quantity is proportional to the reciprocal of the square of the distance from the source of that property. *Newton's law of gravitation and *Coulomb's law are both examples.

inversion layer *See* TRANSISTOR.

inversion temperature *See* JOULE–THOMSON EFFECT.

involute *See* EVOLUTE.

ion An atom or group of atoms that has either lost one or more electrons, making it positively charged (a cation), or gained one or more electrons, making it negatively charged (an anion). *See also* IONIZATION.

ion engine A type of jet-propulsion engine that may become important for propelling or controlling spacecraft. It consists of a unit producing a beam of ions, which are accelerated by an electric or electromagnetic field. Reaction forces from the high-speed ions causes propulsion in much the same way as that caused by exhaust gas of a rocket. However, a separate beam of electrons or ions of opposite polarity to the propelling beam must also be ejected from the engine to enable recombination to take place behind the vehicle (to avoid the vehicle becoming charged). Ion engines provide high *specific impulse and therefore low propellant consumption. The three main components of an ion engine are the power generator, the propellant feed, and the thruster. The power generator may be a nuclear reactor or a solar-energy collector. If it is the former, a gas turbine is coupled to the reactor and the turbine drives an electric generator. A solar-energy unit provides electricity direct. The propellant chosen needs to have an ion of medium mass (low mass for high specific impulse, high mass for high thrust) and a low first *ionization potential. Caesium and mercury are materials currently envisaged as suitable propellants. The thruster consists of an ionizer to produce the ions, an accelerator to provide and shape the accelerating field, and a neutralizer (usually an electron emitter) to neutralize the fast-moving ion beam after ejection.

ion exchange The exchange of ions of the same charge between a solution (usually aqueous) and a solid in contact with it. The process occurs widely in nature, especially in the absorption and retention of water-soluble fertilizers by soil. For example, if a potassium salt is dissolved in water and applied to soil, potassium ions are absorbed by the soil and sodium and calcium ions are released from it.

The soil, in this case, is acting as an ion exchanger. Synthetic **ion-exchange resins** consist of various copolymers having a cross-linked three-dimensional structure to which ionic groups have been attached. An **anionic resin** has negative ions built into its structure and therefore exchanges positive ions. A **cationic resin** has positive ions built in and exchanges negative ions. Ion-exchange resins, which are used in sugar refining to remove salts, are synthetic organic polymers containing side groups that can be ionized. In anion exchange, the side groups are ionized basic groups, such as $-NH_3^+$ to which anions X^- are attached. The exchange reaction is one in which different anions in the solution displace the X^- from the solid. Similarly, cation exchange occurs with resins that have ionized acidic side groups such as $-COO^-$ or $-SO_2O^-$, with positive ions M^+ attached.

Ion exchange also occurs with inorganic polymers such as zeolites, in which positive ions are held at sites in the silicate lattice. These are used for water-softening, in which Ca^{2+} ions in solution displace Na^+ ions in the zeolite. The zeolite can be regenerated with sodium chloride solution. **Ion-exchange membranes** are used as separators in electrolytic cells to remove salts from sea water and in producing deionized water. Ion-exchange resins are also used as the stationary phase in **ion-exchange chromatography**.

ionic radius A value assigned to the radius of an ion in a crystalline solid, based on the assumption that the ions are spherical with a definite size. X-ray diffraction can be used to measure the internuclear distance in crystalline solids. For example, in NaF the Na – F distance is 23.1 nm, and this is assumed to be the sum of the Na^+ and F^- radii. By making certain assumptions about the shielding effect that the inner electrons have on the outer electrons, it is possible to assign individual values to the ionic radii – Na^+ 9.6 nm; F^- 13.5 nm. In general, negative ions have larger ionic radii than positive ions. The larger the negative charge, the larger the ion; the larger the positive charge, the smaller the ion.

ionic strength Symbol I. A function expressing the effect of the charge of the ions in a solution, equal to the sum of the molality of each type of ion present multiplied by the square of its charge. $I = \Sigma m_i z_i^2$.

ion implantation The technique of implanting ions in the lattice of a semiconductor crystal in order to modify its electronic properties. It is used as an alternative to diffusion, or in conjunction with it, in the manufacture of integrated circuits and solid-state components.

ionization The process of producing *ions. Certain molecules (*see* ELECTROLYTE) ionize in solution; for example, acids ionize when dissolved in water:

$$HCl \rightarrow H^+ + Cl^-$$

Electron transfer also causes ionization in certain reactions; for example, sodium and chlorine react by the transfer of a valence electron from the sodium atom to the chlorine atom to form the ions that constitute a sodium chloride crystal:

$$Na + Cl \rightarrow Na^+Cl^-$$

Ions may also be formed when an atom or molecule loses one or more electrons as a result of energy gained in a collision with another particle or a quantum of radiation (*see* PHOTOIONIZATION). This may occur as a result of the impact of *ionizing radiation or of thermal ionization and the reaction takes the form

$$A \rightarrow A^+ + e$$

Alternatively, ions can be formed by electron capture, i.e.

$$A + e \rightarrow A^-$$

ionization chamber An instrument for detecting *ionizing radiation. It consists of two electrodes contained in a gas-filled chamber with a potential difference maintained between them. Ionizing radiation entering the chamber ionizes gas atoms, creating electrons and positive ions. The electric field between the electrodes drives the electrons to the anode and the positive ions to the cathode. This current is, in suitable conditions, proportional to the intensity of the radiation. *See also* GEIGER COUNTER.

ionization gauge A vacuum gauge consisting of a three-electrode system inserted into the container in which the pressure is to be measured. Electrons from the cathode are attracted to the grid, which is positively biased. Some pass through the grid but do not reach the anode, as it is maintained at a negative potential. Some of these electrons do, however, collide with gas molecules, ionizing them and converting them to positive ions. These ions are attracted to the anode; the resulting anode current can be used as a measure of the number of gas molecules present. Pressure as low as 10^{-6} pascal can be measured in this way.

ionization potential (IP) Symbol I. The minimum energy required to remove an electron from a specified atom or molecule to such a distance that there is no electrostatic interaction between ion and electron. Originally defined as the minimum potential through which an electron would have to fall to ionize an atom, the ionization potential was measured in volts. It is now, however, defined as the energy to effect an ionization and is conveniently measured in electronvolts (although this is not an SI unit).

The energy to remove the least strongly bound electrons is the **first ionization potential.** Second, third, and higher ionization potentials can also be measured, although there is some ambiguity in terminology. Thus, in chemistry the second ionization potential is often taken to be the minimum energy required to remove an electron from the singly charged ion; the second IP of lithium would be the energy for the process

$$Li^+ \rightarrow Li^{2+} + e$$

In physics, the second ionization potential is the energy required to remove an electron from the next to highest energy level in the neutral atom or molecule; e.g.

$Li \rightarrow Li^{*+} + e,$

where Li^{*+} is an excited singly charged ion produced by removing an electron from the K-shell.

ionizing radiation Radiation of sufficiently high energy to cause *ionization in the medium through which it passes. It may consist of a stream of high-energy particles (e.g. electrons, protons, alpha-particles) or short-wavelength electromagnetic radiation (ultraviolet, X-rays, gamma-rays). This type of radiation can cause extensive damage to the molecular structure of a substance either as a result of the direct transfer of energy to its atoms or molecules or as a result of the secondary electrons released by ionization (*see* SECONDARY EMISSION). In biological tissue the effect of ionizing radiation can be very serious, usually as a consequence of the ejection of an electron from a water molecule and the oxidizing or reducing effects of the resulting highly reactive species:

$2H_2O \rightarrow e^- + H_2O^* + H_2O^+$

$H_2O^* \rightarrow .OH + .H$

$H_2O^+ + H_2O \rightarrow .OH + H_3O^+$

where the dot before a radical indicates an unpaired electron and * denotes an excited species.

ion-microprobe analysis A technique for analysing the surface composition of solids. The sample is bombarded with a narrow beam (as small as 2 μm diameter) of high-energy ions. Ions ejected from the surface by sputtering are detected by mass spectrometry. The technique allows quantitative analysis of both chemical and isotopic composition for concentrations as low as a few parts per million.

ionosphere *See* EARTH'S ATMOSPHERE; RADIO TRANSMISSION.

ionospheric wave *See* RADIO TRANSMISSION.

ion pair A pair of oppositely charged ions produced as a result of a single ionization; e.g.

$HCl \rightarrow H^+ + Cl^-.$

Sometimes a positive ion and an electron are referred to as an ion pair, as in

$A \rightarrow A^+ + e^-.$

ion pump A type of *vacuum pump that can reduce the pressure in a container to about 1 nanopascal by passing a beam of electrons through the residual gas. The gas is ionized and the positive ions formed are attracted to a cathode within the container where they remain trapped. The pump is only useful at very low pressures, i.e. below about 1 micropascal. The pump has a limited capacity because the absorbed ions eventually saturate the surface of the cathode. A more effective pump can be made by simultaneously *sputtering a film of metal, so that fresh surface is continuously produced. The device is then known as a **sputter-ion pump**.

IP *See* IONIZATION POTENTIAL.

IR *See* INFRARED RADIATION.

iris **1.** (in anatomy) The pigmented ring of muscular tissue, lying between the cornea and the lens, in the eyes of vertebrates and some cephalopod molluscs. It has a central hole (the **pupil**) through which light enters the eye and it contains both circular and radial muscles. Reflex contraction of the former occurs in bright light to reduce the diameter of the pupil; contraction of the radial muscles in dim light increases the pupil diameter and therefore the amount of light entering the eye. Colour is determined by the amount of the pigment melanin in the iris. Blue eyes result from relatively little melanin; grey and brown eyes from increasingly larger amounts. **2.** (in physics) *See* DIAPHRAGM.

irradiance Symbol E. The *radiant flux per unit area reaching a surface; in SI units it is measured in watts per square metre ($W\,m^{-2}$). Irradiance refers to electromagnetic radiation of all kinds, whereas *illuminance refers only to visible radiation.

irradiation Exposure to any form of radiation, often exposure to *ionizing radiation is implied.

irrational number A number that cannot be expressed as the ratio of two integers. An irrational number may be a *surd, such as $\sqrt{2}$ or $\sqrt{3}$, which can be expressed to any desired degree of accuracy but cannot be assigned an exact value. Alternatively, it may be a *transcendental number, such as π or e. *Compare* RATIONAL NUMBER.

irreversibility The property of a system that precludes a change to the system from being a *reversible process. The paradox that although the equations describing the bodies in a system, such as Newton's laws of motion, Maxwell's equation, or Schrödinger's equation are invariant under *time reversal, events involving systems made up from large numbers of these bodies are not reversible. The process of scrambling an egg is an example. The resolution of this paradox requires the concept of *entropy using *statistical mechanics. Irreversibility occurs in the transition from an ordered arrangement to a disordered arrangement, which is a natural trend, since changes in a closed system occur in the direction of increasing entropy. Irreversibility also occurs in processes that violate T symmetry. According to the *CPT theorem, processes that violate CP also violate T and hence are irreversible. This has been observed in some weak-interaction processes.

irreversible process *See* REVERSIBLE PROCESS.

isentropic process Any process that takes place without a change of *entropy. The quantity of heat transferred, δQ, in a reversible process is proportional to the change in entropy, δS, i.e. $\delta Q = T\delta S$, where T is the thermodynamic temperature. Therefore, a reversible *adiabatic process is isentropic, i.e. when $\delta Q = 0$, δS also equals 0.

Ising model A simplified model of a magnetic system consisting of an array of magnetic spins. Spins may have one of two values and interactions occur with nearest neighbours. There are also random thermal fluctuations depending on the temperature of the system. At low temperatures there is a net magnetization as a result of alignment of spins. At high temperature there is no net magnetization. The model was first proposed by the German physicist Ernst Ising (1900–98), who studied the one-dimensional case in 1924. The two-dimensional case for a square lattice was solved exactly by Lars Onsager in 1944. Only approximate solutions have been found for three-dimensional models. The Ising model is very important in statistical mechanics and can be used to investigate other types of phase transition.

isobar 1. A line on a map or chart that joints points or places that have the same atmospheric pressure. **2.** A curve on a graph representing readings taken at constant pressure. **3.** One of two or more nuclides that have the same number of nucleons but different *atomic numbers. Radium–88, actinium–89, and thorium–90 are isobars as each has a *nucleon number of 228.

isobaric spin *See* ISOTOPIC SPIN.

isocline A line on a map or chart joining points or places of equal magnetic dip (*see* GEOMAGNETISM).

isodiaphere One of two or more nuclides in which the difference between the number of neutrons and the number of protons is the same. A nuclide and its product after losing an *alpha particle are isodiapheres.

isodynamic line A line on a map or chart joining points or places at which the total strengths of the earth's magnetic field are equal (*see* GEOMAGNETISM).

isoelectronic Describing compounds that have the same numbers of valence electrons. For example, nitrogen (N_2) and carbon monoxide (CO) are isoelectronic molecules.

isogonal line A line on a map or chart joining points or places of equal magnetic declination (*see* GEOMAGNETISM).

isomerism The existence of atomic nuclei that have the same atomic number and the same mass number but different energy states.

isomers *See* ISOMERISM.

isometric 1. (in technical drawing) Denoting a projection in which the three axes are equally inclined to the surface of the drawing and lines are drawn to scale. **2.** (in crystallography) Denoting a system in which the axes are perpendicular to each other, as in cubic crystals. **3.** (in physics) Denoting a line on a graph illustrating the way in which temperature and pressure are interrelated at constant volume.

isomorphism The existence of two or more substances (**isomorphs**) that have the same crystal structure, so that they are able to form *solid solutions.

isospin *See* ISOTOPIC SPIN.

isotherm **1.** A line on a map or chart joining points or places of equal temperature. **2.** A curve on a graph representing readings taken at constant temperature (e.g. the relationship between the pressure and volume of a gas at constant temperature).

isothermal process Any process that takes place at constant temperature. In such a process heat is, if necessary, supplied or removed from the system at just the right rate to maintain constant temperature. *Compare* ADIABATIC PROCESS.

isotone One of two or more nuclides that contain the same number of neutrons but different numbers of protons. The naturally occurring isotones, for example, strontium–88 and yttrium–89 (both with 50 neutrons), give an indication of the stability of certain nuclear configurations.

isotonic Describing solutions that have the same osmotic pressure.

isotope One of two or more atoms of the same element that have the same number of protons in their nucleus but different numbers of neutrons. Hydrogen (1 proton, no neutrons), deuterium (1 proton, 1 neutron), and tritium (1 proton, 2 neutrons) are isotopes of hydrogen. Most elements in nature consist of a mixture of isotopes. *See* ISOTOPE SEPARATION.

isotope separation The separation of the *isotopes of an element from each other on the basis of slight differences in their physical properties. For laboratory quantities the most suitable device is often the mass spectrometer. On a larger scale the methods used include gaseous diffusion (widely used for separating isotopes of uranium in the form of the gas uranium hexafluoride), distillation (formerly used to produce heavy water), electrolysis (requiring cheap electrical power), thermal diffusion (formerly used to separate uranium isotopes, but now considered uneconomic), centrifuging, and laser methods (involving the excitation of one isotope and its subsequent separation by electromagnetic means).

isotopic number (**neutron excess**) The difference between the number of neutrons in an isotope and the number of protons.

isotopic spin (isospin; isobaric spin) A quantum number applied to hadrons (*see* ELEMENTARY PARTICLES) to distinguish between members of a set of particles that differ in their electromagnetic properties but are otherwise apparently identical. For example if electromagnetic interactions and weak interactions are ignored, the proton cannot be distinguished from the neutron in their strong interactions: isotopic spin was introduced to make a distinction between them. The use of the word 'spin' implies only an analogy to angular momentum, to which isotopic spin has a formal resemblance.

isotropic Denoting a medium whose physical properties are independent of direction. *Compare* ANISOTROPIC.

IT (information technology) The use of computers and telecommunications equipment (with their associated microelectronics) to send, receive, store and manipulate data. The data may be textual, numerical, audio or video, or any combination of these. *See also* INTERNET; WORLD WIDE WEB.

iteration The process of successive approximations used as a technique for solving a mathematical problem. The technique can be used manually but is widely used by computers.

Jahn–Teller effect A distortion of a molecule that forces the molecule to adopt a shape in which valence molecular orbitals are not degenerate. This effect is sometimes regarded as an example of spontaneously broken symmetry. Examples of the Jahn–Teller effect include certain transition metal complexes in which metal ions surrounded by six ligands are distorted octahedra rather than regular octahedra. H. A. Jahn and Edward Teller predicted this effect in 1937 using group theory.

Jeans instability Instability in a cloud of gas in space due to fluctuations in the density of the gas, causing the matter in the cloud to clump together and lead to gravitational collapse. The conditions under which this occurs were worked out by Sir James Hopwood Jeans (1877–1946) in terms of Newtonian gravity. The analogous analysis of this problem using general relativity theory is the basis of the theory of *structure formation.

jet A set of particles produced from the *vacuum state by the movement of *quarks and *gluons with high momentum found in electron–positron annihilation. The energy associated with electron–positron annihilation produces quark–antiquark pairs. The quark and antiquark move away from each other very rapidly if the electron and positron had high energies. The potential energy between the quark and antiquark grows as they separate, thus causing further quarks and antiquarks to emerge from the vacuum. These are swept along with the original quark and antiquark. The overall effect is that two jets of hadrons emerge, one in the direction of the first quark, the other in the direction of the first antiquark. *Quantum chromodynamics predicts that about 10% of the events have three jets, the third jet being due to gluons. This prediction has been confirmed experimentally.

JET (Joint European Torus) A large Tokomak (*see* THERMONUCLEAR REACTOR) experiment at Culham, England, carried out in collaboration by various European nations in the search for thermonuclear energy by *nuclear fusion.

jet propulsion (reaction propulsion) The propulsion of a body by means of a force produced by discharging a fluid in the form of a jet. The backward-moving jet of fluid reacts on the body in which it was produced, in accordance with Newton's third law of motion, to create a reactive force that drives the body forward. Jet propulsion occurs in nature, the squid using a form of it to propel itself through water. Although jet-propelled boats and cars have been developed, the main use of jet propulsion is in aircraft and spacecraft. Jet propulsion is the only known method of propulsion in space. In the atmosphere, jet propulsion becomes more efficient at higher altitudes, as efficiency is inversely proportional to the density of the medium through which a body is flying. The three principal means of providing jet propulsion are the turbojet, the ramjet, and the rocket. The **turbojet** is an air-breathing *heat engine based on the *gas turbine, used to power jet aircraft. The **ramjet** is also an air-breathing engine, but compression of the oxidant is achieved by the forward motion of the device through the atmosphere. This enables the compressor and turbine of the gas turbine to be dispensed with and the remaining system consists simply of an inlet diffuser, a combustion chamber in which fuel is burnt, and a jet nozzle through which the products of combustion are discharged. Used in guided missiles, the ramjet must be accelerated to its operating velocity before it can fly (*see also* PULSE JET). These two forms of jet propulsion, being air-breathing engines, can only be used in the earth's atmosphere. The *rocket, however, carries its own oxidant and can therefore be used in space. *See also* ION ENGINE.

j-j coupling A type of *coupling in many-fermion systems, such as electrons in atoms and nucleons in nuclei, in which the energies associated with the spin–orbit interactions are much higher than the energies associated with electrostatic repulsion. *Multiplets of many-electron atoms having a large atomic number are characterized by j-j coupling. Multiplets in the *shell model of nuclei characterized by j-j coupling are invoked to explain the *magic numbers of nuclei. The multiplets of many atoms are intermediate between j-j coupling and *Russell–Saunders coupling, a state of affairs known as **intermediate coupling**.

Joliot-Curie, Irène (1897–1956) French physicist, daughter of Marie and Pierre *Curie, who was educated by her mother and her scientist associates. In 1921 she began work at the Radium Institute, becoming director in 1946. In 1926 she married Frédéric Joliot (1900–58). They shared the 1935 Nobel Prize for chemistry for their discovery of artificial radioactivity the previous year.

Joly's steam calorimeter An apparatus invented by John Joly (1857–1933) to measure the specific heat capacity of a gas at constant volume. Two equal spherical containers are suspended from the opposite ends of a balance arm. One sphere is evacuated and the other contains the sample gas. The whole apparatus is enclosed in a steam bath, the specific heat capacity of the sample gas being calculated from the difference between the masses of the water that condenses on each sphere.

Josephson effects Electrical effects observed when two superconducting materials (at low temperature) are separated by a thin layer of insulating material (typically a layer of oxide less than 10^{-8} m thick). If normal metallic conductors are separated by such a barrier it is possible for a small current to flow between the conductors by the *tunnel effect. If the materials are superconductors (*see* SUPERCONDUCTIVITY), several unusual phenomena occur:

(1) A supercurrent can flow through the barrier; i.e. it has zero resistance.

(2) If this current exceeds a critical value, this conductivity is lost; the barrier then only passes the 'normal' low tunnelling current and a voltage develops across the junction.

(3) If a magnetic field is applied below the critical current value, the current density changes regularly with distance across the junction. The net current through the barrier depends on the magnetic field applied. As the field is increased the net current increases from zero to a maximum, decreases to zero, increases again to a (lower) maximum, decreases, and so on. If the field exceeds a critical value the superconductivity in the barrier vanishes and a potential difference develops across the junction. (4) If a potential difference is applied across the junction, a high-frequency alternating current flows through the junction. The frequency of this current depends on the size of the potential difference.

A junction of this type is called a **Josephson junction**; two or more junctions joined by superconducting paths form a **Josephson interferometer**. Such junctions can be used in measuring fundamental constants, in defining a voltage standard, and in the highly accurate measurement of magnetic fields. An important potential use is in logic components in high-speed computers. Josephson junctions can switch states very quickly (as low as 6 picoseconds). Moreover they have very low power consumption and can be packed closely without generating too much heat. The effects are named after B. D. Josephson (1940–), who predicted them theoretically in 1962.

joule Symbol J. The *SI unit of work and energy equal to the work done when the point of application of a force of one newton moves, in the direction of the force, a distance of one metre. 1 joule = 10^7 ergs = 0.2388 calorie. It is named after James Prescott Joule.

Joule, James Prescott (1818–89) British physicist. In 1840 he discovered the relationship between electric current passing through a wire, its resistance, and the amount of heat produced. In 1849 he gave an account of the *kinetic theory of gases, and a year later announced his best-known finding, the *mechanical equivalent of heat. Later, with William Thomson (Lord *Kelvin), he discovered the *Joule–Thomson effect.

Joule heating The production of heat in a conductor as a result of the passage of an electric current through the conductor. The quantity of heat produced is given by *Joule's law.

Joule's laws 1. The heat (Q) produced when an electric current (I) flows through a resistance (R) for a time (t) is given by $Q = I^2Rt$. **2.** The *internal energy of a given mass of gas is independent of its volume and pressure, being a function of temperature alone. This law applies only to *ideal gases (for which it provides a definition of thermodynamic temperature) as in a real gas intermolecular forces would cause changes in the internal energy should a change of volume occur. *See also* JOULE–THOMSON EFFECT.

Joule–Thomson effect (Joule–Kelvin effect) The change in temperature that occurs when a gas expands through a porous plug into a region of lower pressure. For most real gases the temperature falls under these circumstances as the gas has to do internal work in overcoming the intermolecular forces to enable the expansion to take place. This is a deviation from *Joule's law. There is usually also a deviation from *Boyle's law, which can cause either a rise or a fall in temperature since any increase in the product of pressure and volume is a measure of external work done. At a given pressure, there is a particular temperature, called the **inversion temperature** of the gas, at which the rise in temperature from the Boyle's law deviation is balanced by the fall from the Joule's law deviation. There is then no temperature change. Above the inversion temperature the gas is heated by expansion, below it, it is cooled. The effect was discovered by James Joule working in collaboration with William Thomson (later Lord *Kelvin).

Jovian Relating to the planet Jupiter.

JUGFET *See* TRANSISTOR.

junction detector (solid-state detector) A sensitive detector of *ionizing radiation in which the output is a current pulse proportional to the energy falling in or near the depletion region of a reverse-biased *semiconductor junction. The first types were made by evaporating a thin layer of gold on to a polished wafer of n-type germanium; however, gold–silicon devices can be operated at room temperature and these have superseded the germanium type, which have to be operated at the temperature of liquid nitrogen to reduce noise. When the gold–silicon junction is reverse-biased a depletion region, devoid of charge carriers (electrons and holes), forms in the silicon. Incoming ionizing radiation falling in this depletion region creates pairs of electrons and holes, which both have to be collected in order to give an output pulse proportional to the energy of the detected particle.

Junction detectors are used in medicine and biology as well as in space systems.

junction transistor *See* TRANSISTOR.

Kaluza–Klein theory A type of *unified-field theory that postulates a generalization of the general theory of relativity to higher than four space–time dimensions. In five space–time dimensions this gives general relativity and electromagnetic interactions. In higher space–time dimensions Kaluza–Klein theories give general relativity and more general *gauge theories. A combination of Kaluza–Klein theory and *supersymmetry gives rise to *supergravity, which needs eleven space–time dimensions. In these theories it is proposed that the higher dimensions are 'rolled up' to become microscopically small (a process known as **spontaneous compactification**) with four macroscopic space–time dimensions remaining. It is named after Theodor Kaluza (1885–1954) and Oskar Klein (1894–1977).

KAM theorem (Kolmogorov–Arnold–Moser theorem) A theorem in *dynamical systems of importance in celestial mechanics and *statistical mechanics. The KAM theorem (named after the Russian mathematicians A. N. Kolmogorov and V. I Arnold and the German mathematician J. Moser) states that a non-negligible fraction of orbits in the *phase space of a dynamical system remain indefinitely in a specific region of phase space, even in the presence of small *perturbations in the system. This result is a step towards resolving the unsolved problem of the stability of motion of planets. The KAM theorem is also of interest to *ergodicity in statistical mechanics, since the majority of orbits lead to ergodic motion going through all available phase space. As the number of *degrees of freedom of the system increases, the dominance of ergodicity over stable orbits becomes more pronounced.

kaon A K-meson. *See* MESON.

Kater's pendulum A complex *pendulum designed by Henry Kater (1777–1835) to measure the acceleration of free fall. It consists of a metal bar with knife edges attached near the ends and two weights that can slide between the knife edges. The bar is pivoted from each knife edge in turn and the positions of the weights are adjusted so that the period of the pendulum is the same with both pivots. The period is then given by the formula for a simple pendulum, which enables g to be calculated.

katharometer An instrument for comparing the thermal conductivities of two gases by comparing the rate of loss of heat from two heating coils surrounded by the gases. The instrument can be used to detect the presence of a small amount of an impurity in air and is also used as a detector in gas chromatography.

keeper A piece of soft iron used to bridge the poles of a permanent magnet when it is not in use. It reduces the leakage field and thus preserves the magnetization.

kelvin Symbol K. The *SI unit of thermodynamic *temperature equal to the fraction 1/273.16 of the thermodynamic temperature of the *triple point of water. The magnitude of the kelvin is equal to that of the degree celsius (centigrade), but a temperature expressed in degrees celsius is numerically equal to the temperature in kelvins less 273.15 (i.e. °C = K – 273.15). The *absolute zero of temperature has a temperature of 0 K (–273.15°C). The former name **degree kelvin** (symbol °K) became obsolete by international agreement in 1967. The unit is named after Lord *Kelvin.

Kelvin, Lord (William Thomson; 1824–1907) British physicist, born in Belfast, who became professor of natural philosophy at Glasgow University in 1846. He carried out important experimental and theoretical work on electromagnetism, inventing the mirror *galvanometer, contributing to the development of telegraphy, and pioneering theoretical work by expressing electricity and magnetism in terms of fields. He also worked with James *Joule on the *Joule–Thomson effect. His main theoretical work was in *thermodynamics, in which he stressed the importance of the conservation of energy (*see* CONSERVATION LAW). He also introduced the concept of *absolute zero and the Kelvin temperature scale based on it. In 1896 he was created Baron Kelvin of Largs.

Kelvin effect *See* THOMSON EFFECT.

Kepler, Johannes (1571–1630) German astronomer, who in 1594 became a mathematics teacher in Graz, where he learned of the work of *Copernicus. In 1600 he went to work for Tycho Brahe in Prague. It was there that he worked out *Kepler's laws of planetary motion, for which he is best remembered.

Kepler's laws Three laws of planetary motion formulated by Johannes Kepler in about 1610 on the basis of observations made by Tycho Brahe (1546–1601). They state that: (1) the orbits of the planets are elliptical with the sun at one *focus of the ellipse; (2) each planet revolves around the sun so that an imaginary line (the **radius vector**) connecting the planet to the sun sweeps out equal areas in equal times; (3) the ratio of the square of each planet's *sidereal period to the cube of its distance from the sun is a constant for all the planets.

Kerr effect The ability of certain substances to refract differently light waves whose vibrations are in two directions (*see* DOUBLE REFRACTION) when the substance is placed in an electric field. The effect, discovered in 1875 by John Kerr (1824–1907), is caused by the fact that certain molecules have electric *dipoles, which tend to be orientated by the applied field; the normal random motions of the molecules tends to destroy this orientation and the balance is struck by the relative magnitudes of the field strength, the temperature, and the magnitudes of the dipole moments.

The Kerr effect is observed in a **Kerr cell**, which consists of a glass cell containing the liquid or gaseous substance; two capacitor plates are inserted into the cell and light is passed through it at right angles to the electric field. There are two principal indexes of refraction: n_o (the ordinary index) and n_e (the extraordinary index). The difference in the velocity of propagation in the cell causes a phase difference, δ, between the two waves formed from a beam of monochromatic light, wavelength λ, such that

$$\delta = (n_o - n_e)x/\lambda,$$

where x is the length of the light path in the cell. Kerr also showed empirically that the ratio

$$(n_o - n_e)\lambda = BE^2,$$

where E is the field strength and B is a constant, called the **Kerr constant**, which is characteristic of the substance and approximately inversely proportional to the thermodynamic temperature.

The **Kerr shutter** consists of a Kerr cell filled with a liquid, such as nitrobenzene, placed between two crossed polarizers; the electric field is arranged to be perpendicular to the axis of the light beam and at 45° to the axis of the polarizers. In the absence of a field there is no optical path through the device. When the field is switched on the nitrobenzene becomes doubly refracting and a path opens between the crossed polarizers.

kibi- *See* BINARY PREFIXES.

kilo- Symbol k. A prefix used in the metric system to denote 1000 times. For example, 1000 volts = 1 kilovolt (kV).

kilogram Symbol kg. The *SI unit of mass defined as a mass equal to that of the international platinum–iridium prototype kept by the International Bureau of Weights and Measures at Sèvres, near Paris.

kiloton weapon A nuclear weapon with an explosive power equivalent to one thousand tons of TNT. *Compare* MEGATON WEAPON.

kilowatt-hour Symbol kWh. The commercial unit of electrical energy. It is equivalent to a power consumption of 1000 watts for 1 hour.

kinematic equation *See* EQUATION OF MOTION.

kinematics The branch of mechanics concerned with the motions of objects without being concerned with the forces that cause the motion. In this latter respect it differs from *dynamics, which is concerned with the forces that affect motion. *See also* EQUATION OF MOTION.

kinematic viscosity Symbol v. The ratio of the *viscosity of a liquid to its density. The SI unit is $m^2\ s^{-1}$.

kinetic effect A chemical effect that depends on reaction rate rather than on thermodynamics. For example, diamond is thermodynamically less stable than graphite; its apparent stability depends on the vanishingly slow rate at which it is converted. Overvoltage in electrolytic cells is another example of a kinetic effect. **Kinetic isotope effects** are changes in reaction rates produced by isotope substitution. For example, if the slow step in a chemical reaction is the breaking of a C–H bond, the rate for the deuterated compound would be slightly lower because of the lower vibrational frequency of the C–D bond.

kinetic energy *See* ENERGY.

kinetic equations Equations used in *kinetic theory. The *Boltzmann equation is an example of a kinetic equation. An important application of kinetic equations is to calculate *transport coefficients (and inverse transport coefficients), such as *conductivity and *viscosity in *nonequilibrium statistical mechanics. In general, kinetic equations do not have exact solutions for interacting systems. If the system is near to equilibrium an approximation technique can be used by regarding the deviation from equilibrium as a *perturbation.

kinetics The branch of physical chemistry concerned with measuring and studying the rates of chemical reactions. The main aim of chemical kinetics is to determine the mechanism of reactions by studying the rate under different conditions (temperature, pressure, etc.).

kinetic theory A theory, largely the work of Count *Rumford, James *Joule, and James Clerk *Maxwell, that explains the physical properties of matter in terms of the motions of its constituent particles. In a gas, for example, the pressure is due to the incessant impacts of the gas molecules on the walls of the container. If it is assumed that the molecules occupy negligible space, exert negligible forces on each other except during collisions, are perfectly elastic, and make only brief collisions with each other, it can be shown that the pressure p exerted by one mole of gas containing n molecules each of mass m in a container of volume V, will be given by:

$$p = nm\bar{c}^2/3V,$$

where $\bar{c}^2$ is the mean square speed of the molecules. As according to the *gas laws for one mole of gas: $pV = RT$, where T is the thermodynamic temperature, and R is the molar *gas constant, it follows that:

$$RT = nm\bar{c}^2/3$$

Thus, the thermodynamic temperature of a gas is proportional to the mean square speed of its molecules. As the average kinetic *energy of translation of the molecules is $m\bar{c}^2/2$, the temperature is given by:

$$T = (m\bar{c}^2/2)(2n/3R)$$

The number of molecules in one mole of any gas is the *Avogadro constant, N_A; therefore in this equation $n = N_A$. The ratio R/N_A is a constant called the *Boltzmann constant (k). The average kinetic energy of translation of the molecules of one mole of any gas is therefore $3kT/2$. For monatomic gases this is proportional to the *internal energy (U) of the gas, i.e.

$U = N_A 3kT/2$ and as $k = R/N_A$

$U = 3RT/2$

For diatomic and polyatomic gases the rotational and vibrational energies also have to be taken into account (see DEGREES OF FREEDOM).

In liquids, according to the kinetic theory, the atoms and molecules still move around at random, the temperature being proportional to their average kinetic energy. However, they are sufficiently close to each other for the attractive forces between molecules to be important. A molecule that approaches the surface will experience a resultant force tending to keep it within the liquid. It is, therefore, only some of the fastest moving molecules that escape; as a result the average kinetic energy of those that fail to escape is reduced. In this way evaporation from the surface of a liquid causes its temperature to fall.

In a crystalline solid the atoms, ions, and molecules are able only to vibrate about the fixed positions of a *crystal lattice; the attractive forces are so strong at this range that no free movement is possible.

Kirchhoff, Gustav Robert (1824–87) German physicist, who in 1850 became a professor at Breslau and four years later joined Robert *Bunsen at Heidelberg. In 1845, while still a student, he formulated *Kirchhoff's laws concerning electric circuits. With Bunsen he worked on spectroscopy, a technique that led them to discover the elements caesium (1860) and rubidium (1861).

Kirchhoff's law of radiation A law stating that the emissivity of a body is equal to its absorptance at the same temperature. The law was formulated by Gustav Kirchhoff.

Kirchhoff's laws Two laws relating to electric circuits, first formulated by Gustav Kirchhoff. (a) The current law states that the algebraic sum of the currents flowing through all the wires in a network that meet at a point is zero. (b) The voltage law states that the algebraic sum of the e.m.f.s within any closed circuit is equal to the sum of the products of the currents and the resistances in the various portions of the circuit.

Klein–Gordon equation An equation in relativistic quantum mechanics for spin-zero particles. The Klein–Gordon equation has the form

$$[\Box - (mc/\hbar)^2]\psi = 0,$$

where $\Box = \nabla^2 - (1/c^2)(\partial/\partial t^2)$, and ∇^2 is the Laplace operator (see LAPLACE EQUATION), m is the mass of the particle, c is the *speed of light, $\hbar$ is the Dirac constant (see PLANCK CONSTANT), and ψ is the *wave function of the particle. This equation was discovered by several authors independently, including Oskar Klein (1894–1977) and Walter Gordon (1893–1940) in 1926. The Klein–Gordon equation is not viable as an equation in single-particle relativistic quantum mechanics but does describe spin-zero particles, such as *mesons; it includes the prediction of the existence of *antiparticles for such particles. *See also* DIRAC EQUATION.

klystron An electron tube that generates or amplifies microwaves by **velocity modulation**. Several types are used; in the simple two-cavity klystron a beam of high-energy electrons from an electron gun is passed through a *resonant cavity, where it interacts with high-frequency radio waves. This microwave energy modulates the velocities of the electrons in the beam, which then enters a drift space where the faster electrons overtake the slower ones to form bunches. The bunched beam now has an alternating component, which is transferred to an output cavity and thence to an output waveguide.

knocking The metallic sound produced by a spark-ignition petrol engine under certain conditions. It is caused by rapid combustion of the unburnt explosive mixture in the combustion chambers ahead of the flame front. As the flame travels from the sparking plug towards the piston it compresses and heats the unburnt gases ahead of it. If the flame front moves fast enough, normal combustion occurs and the explosive mixture is ignited progressively by the flame. If it moves too slowly, ignition of the last part of the unburnt gas can occur very rapidly before the flame reaches it, producing a shock wave that travels back and forth across the combustion chamber. The result is overheating, possible damage to the plugs, an undesirable noise, and loss of power (probably due to preignition caused by overheated plugs). Knocking can be avoided by an engine design that increases turbulence in the combustion chamber and thereby increases flame speed. It also can be avoided by reducing the compression ratio, but this involves loss of efficiency. The most effective method is to use high-octane fuel (*see* OCTANE NUMBER), which has a longer self-ignition delay than low-octane fuels. This can be achieved by the addition of an **antiknock agent**, such as lead(IV) tetraethyl, to the fuel, which retards the combustion chain reactions. However, lead-free petrol is now preferred to petrol containing lead tetraethyl owing to environmental dangers arising from lead in the atmosphere. In the USA the addition of lead compounds is now forbidden. New formulae for petrol are designed to raise the octane number without polluting the atmosphere. These new formulae include increasing the content of aromatics and oxygenates (oxygen-containing compounds, such as alcohols). However, it is claimed that the presence in the atmosphere of incompletely burnt aromatics constitutes a cancer risk.

knot theory A branch of mathematics used to classify knots and entanglements. Knot theory has applications to the study of the properties of polymers (including polymers of biological importance, such as DNA) and the *statistical mechanics of certain models of *phase transitions. Knot theory was developed extensively in the late 19th century in an attempt to explain the structure of atoms. Some quantities in knot theory can be expressed in terms of *quantum field theories, enabling them to be used to formulate *gauge theories and *quantum gravity.

Knudsen flow *See* MOLECULAR FLOW.

Kohlrausch's law If a salt is dissolved in water, the conductivity of the (dilute) solution is the sum of two values – one depending on the positive ions and the other on the negative ions. The law, which depends on the independent migration of ions, was deduced experimentally by the German chemist Friedrich Kohlrausch (1840–1910).

Kovar A tradename for an alloy of iron, cobalt, and nickel with an *expansivity similar to that of glass. It is therefore used in making glass-to-metal seals, especially in circumstances in which a temperature variation can be expected.

Kramers theorem The energy levels of a system, such as an atom that contains an odd number of spin-1/2 particles (e.g. electrons), are at least double degenerate in the absence of an external *magnetic field. This degeneracy, known as **Kramers degeneracy**, is a consequence of *time reversal invariance. Kramers theorem was stated by the Dutch physicist Hendrick Anton Kramers (1894–1952) in 1930. Kramers degeneracy is removed by placing the system in an external magnetic field. Kramers theorem holds even in the presence of crystal fields and *spin–orbit coupling.

Kuiper belt A band of a large number of objects (**Kuiper belt objects**, or **KBOs**) orbiting the sun beyond the orbits of Neptune and Pluto. The Kuiper belt is thought to be the source of short-period *comets. Its existence was first suggested by the US astronomer Gerard Kuiper (1905–73) in 1951. It was discovered in 1992.

Kundt's tube An apparatus designed by August Kundt (1839–94) in 1866 to measure the speed of sound in various fluids. It consists of a closed glass tube into which a dry powder (such as lycopodium) has been sprinkled. The source of sound in the original device was a metal rod clamped at its centre with a piston at one end, which is inserted into the tube. When the rod is stroked, sound waves generated by the piston enter the tube. If the position of the piston in the tube is adjusted so that the gas column is a whole number of half wavelengths long, the dust will be disturbed by the resulting *stationary waves forming a series of striations, enabling distances between *nodes to be measured. The vibrating rod can be replaced by a small loudspeaker fed by an oscillator.

labelling The process of replacing a stable atom in a compound with a radioisotope of the same element to enable its path through a biological or mechanical system to be traced by the radiation it emits. In some cases a different stable isotope is used and the path is detected by means of a mass spectrometer. A compound containing either a radioactive or stable isotope is called a **labelled compound** and the atom used is a **label**. If a hydrogen atom in each molecule of the compound has been replaced by a tritium atom, the compound is called a **tritiated compound**. A radioactive labelled compound will behave chemically and physically in the same way as an otherwise identical stable compound, and its presence can easily be detected using a *Geiger counter. This process of **radioactive tracing** is widely used in chemistry, biology, medicine, and engineering. For example, it can be used to follow the course of the reaction of a carboxylic acid with an alcohol to give an ester, e.g.

$$CH_3COOH + C_2H_5OH \rightarrow C_2H_5COOCH_3 + H_2O$$

To determine whether the noncarbonyl oxygen in the ester comes from the acid or the alcohol, the reaction is performed with the labelled compound $CH_3CO^{18}OH$, in which the oxygen in the hydroxyl group of the acid has been 'labelled' by using the ^{18}O isotope. It is then found that the water product is $H_2^{18}O$; i.e. the oxygen in the ester comes from the alcohol, not the acid.

laevorotatory Designating a chemical compound that rotates the plane of plane-polarized light to the left (anticlockwise for someone facing the oncoming radiation). *See* OPTICAL ACTIVITY.

lag *See* PHASE ANGLE.

Lagrange multipler (**undetermined multipler**) A method used in solving problems in the calculus of variations. One wants to find the maximum or minimum values of a function $f(x,y)$, where x and y are related by some equation $\phi(x,y) = c$, where c is a constant. The function $F(x,y) = f(x,y) + L\phi(x,y)$ is formed, with L being a Lagrange multiplier. The partial derivatives of F with respect to x and y are set to zero. Together with the equation $\phi(x,y) = c$, this enables the maximum or minimum value of f to be determined. The method can be extended to more than two variables.

Lagrangian Symbol L. A function used to define a dynamical system in terms of functions of coordinates, velocities, and times given by:

$$L = T - V$$

where T is the kinetic energy of the system and V is the potential energy of the system. The Lagrangian formulation of dynamics has the advantage that it does not deal with many vector quantities, such as forces and accelerations, but only with two scalar functions, T and V. This leads to great simplifications in calculations. **Lagrangian dynamics** was formulated by J. L. Lagrange (1736–1813).

lambda particle A spin-½ electrically neutral *baryon made up of one up quark, one down quark, and one strange quark. The mass of the lambda particle is 1115.60 MeV and its average lifetime is 2.6×10^{-10} s.

lambda point Symbol λ. The temperature of 2.186 K below which helium–4 becomes a superfluid. The name derives from the shape of the curve of specific heat capacity against temperature, which is shaped like a Greek letter lambda (λ) at this point. *See* SUPERFLUIDITY.

lambert A former unit of *luminance equal to the luminance of a uniformly diffusing surface that emits or reflects one lumen per square centimetre. It is approximately equal to 3.18×10^3 Cd m^{-2}. It is named after Johann H. Lambert (1728–77).

Lambert's laws (1) The *illuminance of a surface illuminated by light falling on it perpendicularly from a point source is inversely proportional to the square of the distance between the surface and the source.

(2) If the rays make an angle θ with the normal to the surface, the illuminance is proportional to $\cos\theta$.

(3) (Also called **Bouquer's law**) The *luminous intensity (I) of light (or other electromagnetic radiation) decreases exponentially with the distance d that it enters an absorbing medium, i.e.

$$I = I_0 \exp(-\alpha d)$$

where I_0 is the intensity of the radiation that enters the medium and α is its **linear absorption coefficient**. These laws were first stated (for light) by Johann H. Lambert.

Lamb shift A small energy difference between two levels ($^2S_{1/2}$ and $^2P_{1/2}$) in the *hydrogen spectrum. The shift results from the quantum interaction between the atomic electron and the electromagnetic radiation. It was first discovered by Willis Eugene Lamb (1913–). It was calculated and explained using *renormalization in *quantum electrodynamics by the US physicists Richard *Feynman and Julian Seymour Schwinger (1918–94) and the Japanese physicist Sinitiro Tomonaga (1906–79).

laminar flow *Streamline flow of a fluid in which the fluid moves in layers without fluctuations or turbulence so that successive particles passing the same point have the same velocity. It occurs at low *Reynolds numbers, i.e. low velocities, high viscosities, low densities or small dimensions. The flow of lubricating oil in bearings is normally laminar because of the thinness of the lubricant layer.

laminated core A core for a transformer or other electrical machine in which the ferromagnetic alloy is made into thin sheets (laminations), which are oxidized or varnished to provide a relatively high resistance between them. This has the effect of reducing *eddy currents, which occur when alternating currents are used.

Landau damping A damping mechanism for collective oscillations in *plasmas. If the velocity of the wave associated with the *collective oscillation is comparable to the velocities of individual electrons in plasmas, with the velocities of the electrons being determined by their thermal motion, there is a transfer of energy from the wave motion to the electrons. This causes damping (attenuation) and eventually destruction of the wave motion. The process is named after the Soviet physicist Lev Davidovich Landau (1908–68), who postulated this type of damping in 1946. A similar mechanism occurs for the damping of *collective excitations in the quantum theory of many-body systems by energy transfer to individual *quasiparticles.

Landauer's principle The principle put forward by Rolf Landauer in the 1960s that energy has to be expended to erase information. This principle links thermodynamics and information theory.

Landau ghost A possible inconsistency in the *renormalization procedure that appears at very high energies in *quantum electrodynamics (QED) and other *quantum field theories in which there is not *asymptotic freedom. It was shown in the mid-1950s by Lev Davidovich Landau and others, using *approximation techniques, that in the limit of extremely high energies the relation between the unrenormalized and actual electric charges is such that the observed charge should fall to zero, thus making the procedure of renormalization impossible. This difficulty is called the **Landau ghost** (or sometimes the **Moscow zero**). It has not been established whether or not it is true in general, for the consistency of QED. This difficulty is theoretical rather than practical for QED as, at the energies at which the problem arises, electromagnetic interactions are thought to unify with weak interactions and strong interactions. The opposite situation to the Landau ghost is *asymptotic freedom, which occurs in *quantum chromodynamics.

Landé interval rule A rule in atomic spectra stating that if the *spin–orbit coupling is weak in a given multiplet, the energy differences between two successive J levels (where J is the total resultant angular momentum of the coupled electrons) are proportional to the larger of the two values of J. The rule was stated by the German-born US physicist Alfred Landé (1888–1975) in 1923. It can be deduced from the quantum theory of angular momentum. In addition to assuming *Russell–Saunders coupling, the Landé interval rule assumes that the interactions between spin magnetic moments can be ignored, an assumption that is not correct for very light atoms, such as helium. Thus the Landé interval rule is best obeyed by atoms with medium atomic numbers.

Langevin equation A type of random equation of motion (*see* STOCHASTIC PROCESS) used to study *Brownian movement. The Langevin equation can be written in the form $\dot{v} = \xi v + A(t)$, where v is the velocity of a particle of mass m immersed in a fluid and $\dot{v}$ is the acceleration of the particle; ξv is a frictional force resulting from the *viscosity of the fluid, with ξ being a constant friction coefficient, and $A(t)$ is a random force describing the average effect of the Brownian motion. The Langevin equation is named after the French physicist Paul Langevin (1872–1946). It is necessary to use statistical methods and the theory of probability to solve the Langevin equation.

Langmuir frequency (plasma frequency) Symbol ω_p. The frequency of *plasma oscillations in an equilibrium charge distribution. The Langmuir frequency is given by $\omega_p = (4\pi n_e q^2/m)^{1/2} = 5.7 \times 10^5 \, n_e^{1/2}$ (radians/sec), where n_e is the density of electrons and m and q are the mass and charge, respectively, of the electron. The Langmuir frequency is named after the US chemist Irving Langmuir (1881–1957), who derived the expression in 1928 in his analysis of plasma oscillations by combining Poisson's equation with *Newton's law of motion. The Langmuir frequency can also be derived using *quantum mechanics for the electron gas.

Laplace equation The partial differential equation:

$\partial^2 u/\partial x^2 + \partial^2 u/\partial y^2 + \partial^2 u/\partial z^2 = 0$

It may also be written in the form $\nabla^2 u = 0$, where ∇^2 is called the **Laplace operator**. It was formulated by the French mathematician P. S. Laplace (1749–1827).

Laplace's correction A correction to the calculation of the speed of sound in a gas. Newton assumed that the pressure–volume changes that occur when a sound wave travels through the gas are isothermal. Laplace was subsequently able to obtain agreement between theory and experiment by assuming that pressure–volume changes are adiabatic.

large-scale structure The structure of the distribution of visible matter in the universe at very large scales. This structure includes galaxies, clusters of galaxies, superclusters, and voids. *See also* STRUCTURE FORMATION.

Larmor precession A precession of the motion of charged particles in a magnetic field. It was first deduced in 1897 by Sir Joseph Larmor (1857–1942). Applied to the orbital motion of an electron around the nucleus of an atom in a magnetic field of flux density B, the frequency of precession is given by $eB/4\pi m v \mu$, where e and m are the electronic charge and mass respectively, μ is the permeability, and v is the velocity of the electron. This is known as the **Larmor frequency**.

laser (light *a*mplification by *s*timulated *e*mission of *r*adiation) A light amplifier usually used to produce monochromatic coherent radiation in

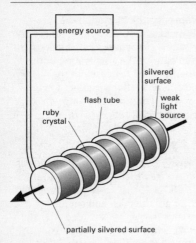

A simple ruby laser

the infrared, visible, and ultraviolet regions of the *electromagnetic spectrum.

Nonlaser light sources emit radiation in all directions as a result of the spontaneous emission of photons by thermally excited solids (filament lamps) or electronically excited atoms, ions, or molecules (fluorescent lamps, etc.). The emission accompanies the spontaneous return of the excited species to the *ground state and occurs randomly, i.e. the radiation is not coherent. In a laser, the atoms, ions, or molecules are first 'pumped' to an excited state and then stimulated to emit photons by collision of a photon of the same energy. This is called **stimulated emission**. In order to use it, it is first necessary to create a condition in the amplifying medium, called **population inversion**, in which the majority of the relevant entities are excited. Random emission from one entity can then trigger coherent emission from the others that it passes. In this way amplification is achieved.

The laser amplifier is converted to an oscillator by enclosing the amplifying medium within a resonator. Radiation then introduced along the axis of the resonator is reflected back and forth along its path by a mirror at one end and by a partially transmitting mirror at the other end. Between the mirrors the waves are amplified by stimulated emission. The radiation emerges through the semitransparent mirror at one end as a powerful coherent monochromatic parallel beam of light. The emitted beam is uniquely parallel because waves that do not bounce back and forth between the mirrors quickly escape through the sides of the oscillating medium without amplification.

Some lasers are solid, others are liquid or gas devices. Population inversion can be achieved by **optical pumping** with flashlights or with other lasers. It can also be achieved by such methods as chemical reactions, discharges in gases, and recombination emission in semiconducting materials (*see* RECOMBINATION PROCESS).

Lasers have found many uses since their invention in 1960, including laser welding, surgery, *holography, printing, optical communications, and the reading of digital information.

laser cooling A technique for producing extremely low temperatures using lasers to slow down and trap atoms. The basic method is to direct a set of crossed laser beams at a sample of gas, with the wavelength set so that photons are absorbed by the atoms. One atom moving towards the photon beam will lose momentum on absorbing a photon and be cooled. An atom moving away from the incident photons will gain energy on absorption. Atoms moving towards the incident photons 'see' the incident photons as having a slightly different frequency than those moving away because of the *Doppler effect, and it is possible to adjust the incident laser frequency by a small amount so that atoms are more likely to absorb when they are moving towards the oncoming photons. This results in a net cooling effect, a technique known as **Doppler cooling**. It produces a region of slow-moving atoms at the intersection of the laser beams – a state of matter sometimes called **optical molasses**. Further cooling, to temperatures below the theoretical limit for Doppler cooling, can be obtained by a mechanism known as **Sisyphus cooling**. Here the atom moves through a standing wave created by the laser. As it moves to the top of each 'hill' it loses energy and at the top it is optically pumped to a state at the bottom of the 'valley'. Consequently, the effect is of an atom always moving up a potential gradient and losing energy. The name comes from the character Sisyphus in Greek mythology, who was condemned by the Gods continuously to push a boulder to the top of a hill, only for it to roll back down again when he reached the summit.

Work on laser cooling has also involved methods of trapping atoms. The **magneto-optical trap** (**MOT**) uses six crossed laser beams together with an applied magnetic field to keep cooled atoms together. This allows a further method of cooling in which the height of the trap is lowered so as to let the more energetic atoms escape (a method known as **evaporative cooling**). Techniques of this type have led to temperatures less than 10^{-6} K and to the discovery of the Bose–Einstein condensate (*see* BOSE–EINSTEIN CONDENSATION).

laser printer *See* PRINTER.

latent heat Symbol L. The quantity of heat absorbed or released when a substance changes its physical phase at constant temperature (e.g. from solid to liquid at the melting point or from liquid to gas at the boiling point). For example, the latent heat of vaporization is the energy a substance absorbs from its surroundings in order to overcome the attractive forces between its molecules as it changes from a liquid to a gas and in order to do work against the external atmosphere as it expands. In thermodynamic terms the latent heat is the *enthalpy of evaporation

(ΔH), i.e. $L = \Delta H = \Delta U + p\Delta V$, where ΔU is the change in the internal energy, p is the pressure, and ΔV is the change in volume.

The **specific latent heat** (symbol l) is the heat absorbed or released per unit mass of a substance in the course of its isothermal change of phase. The **molar latent heat** is the heat absorbed or released per unit amount of substance during an isothermal change of state.

lateral inversion (perversion) The type of reversal that occurs with an image formed by a plane mirror. A person with a mole on his left cheek sees an image in a plane mirror of a person with a mole on his right cheek. Since, however, that is (correctly) to the observer's left, the real reversal is of front and back; the image is 'turned through' itself to face the object – hence the alternative name.

lateral velocity The component of a celestial body's velocity that is at 90° to its *line-of-sight velocity.

latitude and longitude **1.** (in geography) Imaginary lines on the earth's surface, enabling any point to be defined in terms of two angles subtended at its centre (see illustration). **Parallels of latitude** are circles drawn round the earth parallel to the equator; their diameters diminish as they approach the poles. These parallels are specified by the angle subtended at the centre of the earth by the arc formed between a point on the parallel and the equator. All points on the equator therefore have a latitude of 0°, while the north pole has a latitude of 90°N and the south pole of 90°S. Parallels of latitude 1° apart are separated on the earth's surface by about 100 km.

Meridians of longitude are half *great circles passing through both poles; they cross parallels of latitude at right angles. In 1884 the meridian through Greenwich, near London, was selected as the prime meridian and designated as 0°. Other meridians are defined by the angle between the

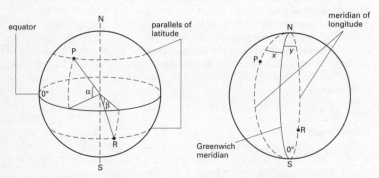

The latitude of P is given by the angle α. In this case it would be α° N. The latitude of R is β° S.

The longitude of P is given by the angle x. In this case it would be x° W. R has a longitude of y° E.

Latitude and longitude

plane of the meridian and the plane of the prime meridian specifying whether it is E or W of the prime meridian. At the equator meridians 1° apart are separated by about 112 km. **2.** (in astronomy) The **celestial latitude** of a star, or other celestial body, is its angular distance north (taken as positive) or south (taken as negative) of the ecliptic measured along the great circle through the body and the poles of the ecliptic. The **celestial longitude** is the angular distance from the vernal equinox measured eastwards along the ecliptic to the intersection of the body's circle of longitude; it is measured in the same direction as the sun's apparent annual motion.

lattice The regular arrangement of atoms, ions, or molecules in a crystalline solid. *See* CRYSTAL LATTICE.

lattice energy A measure of the stability of a *crystal lattice, given by the energy that would be released per mole if atoms, ions, or molecules of the crystal were brought together from infinite distances apart to form the lattice. *See* BORN–HABER CYCLE.

lattice gauge theory A formulation of *gauge theories in which space and time are taken to be discrete, rather than continuous. At the end of calculations in lattice gauge theories it is necessary to take the continuum limit. Lattice gauge theory is used to make calculations for gauge theories with strong coupling, such as *quantum chromodynamics, in which many of the important features of the theory cannot be obtained by *perturbation theory. Lattice gauge theory is particularly suitable for numerical and computational calculations. Techniques from *statistical mechanics can be used in lattice gauge theory. Difficulties arise with putting fermions on a lattice, although various remedies have been used to overcome these difficulties.

lattice vibrations The periodic vibrations of the atoms, ions, or molecules in a *crystal lattice about their mean positions. On heating, the amplitude of the vibrations increases until they are so energetic that the lattice breaks down. The temperature at which this happens is the melting point of the solid and the substance becomes a liquid. On cooling, the amplitude of the vibrations diminishes. At *absolute zero a residual vibration persists, associated with the *zero-point energy of the substance. The increase in the electrical resistance of a conductor is due to increased scattering of the free conduction electrons by the vibrating lattice particles.

latus rectum *See* ELLIPSE; HYPERBOLA; PARABOLA.

Laue, Max Theodor Felix von (1879–1960) German physicist, who became a professor at Berlin in 1919, moving in 1943 to the Max Planck Institute at Göttingen. He is best known for his discovery in 1912 of *X-ray diffraction, for which he was awarded the 1914 Nobel Prize for physics.

launch vehicle A rocket used to launch a satellite, spaceprobe, space station, etc. Multistage rockets are usually used, the empty tanks and engine of the first two stages being jettisoned before the desired orbit is reached. The **launch window** is the time interval during which the vehicle must be launched to achieve the orbit.

Lavoisier, Antoine Laurent (1743–94) French chemist, who collected taxes for the government in Paris. In the 1770s he discovered oxygen and nitrogen in air and demolished the *phlogiston theory of combustion by demonstrating the role of oxygen in the process. In 1783 he made water by burning hydrogen in oxygen (*see* CAVENDISH, HENRY). He also devised a rational nomenclature for chemical compounds. In 1794 he was tried by the Jacobins as an opponent of the Revolution (because of his tax-gathering), found guilty, and guillotined.

lawrencium Symbol Lr. A radioactive metallic transuranic element belonging to the actinoids; a.n. 103; mass number of only known isotope 257 (half-life 8 seconds). The element was identified by A. Ghiorso and associates in 1961. It is named after the US physicist E. O. Lawrence (1901–58). The alternative name **unniltrium** has been proposed.

Lawson criterion A condition for the release of energy from a *thermonuclear reactor first laid down by J. D. Lawson in 1957. It is usually stated as the minimum value for the product of the density (n_G) of the fusion-fuel particles and the *containment time (τ) for energy breakeven, i.e. it is a measure of the density of the reacting particles required and the time for which they need to react in order to produce more energy than was used in raising the temperature of the reacting particles to the *ignition temperature. For a 50:50 mixture of deuterium and tritium at the ignition temperature, the value of $n_G\tau$ is between 10^{14} and 10^{15} cm^{-3} s.

laws, theories, and hypotheses In science, a law is a descriptive principle of nature that holds in all circumstances covered by the wording of the law. There are no loopholes in the laws of nature and any exceptional event that did not comply with the law would require the existing law to be discarded or would have to be described as a miracle. Eponymous laws are named after their discoverers (e.g. *Boyle's law); some laws, however, are known by their subject matter (e.g. the law of conservation of mass), while other laws use both the name of the discoverer and the subject matter to describe them (e.g. *Newton's law of gravitation).

A description of nature that encompasses more than one law but has not achieved the uncontrovertible status of a law is sometimes called a **theory**. Theories are often both eponymous and descriptive of the subject matter (e.g. Einstein's theory of relativity and Darwin's theory of evolution).

A **hypothesis** is a theory or law that retains the suggestion that it may not be universally true. However, some hypotheses about which no doubt still lingers have remained hypotheses (e.g. Avogadro's hypothesis), for no clear reason. Clearly there is a degree of overlap between the three concepts.

L–D process *See* BASIC-OXYGEN PROCESS.

LDR *See* LIGHT-DEPENDENT RESISTOR.

lead (in physics) *See* PHASE ANGLE.

lead–acid accumulator An accumulator in which the electrodes are made of lead and the electrolyte consists of dilute sulphuric acid. The electrodes are usually cast from a lead alloy containing 7–12% of antimony (to give increased hardness and corrosion resistance) and a small amount of tin (for better casting properties). The electrodes are coated with a paste of lead(II) oxide (PbO) and finely divided lead; after insertion into the electrolyte a 'forming' current is passed through the cell to convert the PbO on the negative plate into a sponge of finely divided lead. On the positive plate the PbO is converted to lead(IV) oxide (PbO_2). The equation for the overall reaction during discharge is:

$$PbO_2 + 2H_2SO_4 + Pb \rightarrow 2PbSO_4 + 2H_2O$$

The reaction is reversed during charging. Each cell gives an e.m.f. of about 2 volts and in motor vehicles a 12-volt battery of six cells is usually used. The lead–acid battery produces 80–120 kJ per kilogram. *Compare* NICKEL–IRON ACCUMULATOR.

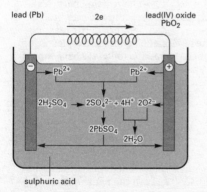

Lead–acid accumulator

lead equivalent A measure of the absorbing power of a radiation screen, expressed as the thickness of a lead screen in millimetres that would afford the same protection as the material being considered.

least action *See* PRINCIPLE OF LEAST ACTION.

least-squares method *See* REGRESSION ANALYSIS.

Le Chatelier's principle If a system is in equilibrium, any change imposed on the system tends to shift the equilibrium to nullify the effect of the applied change. The principle, which is a consequence of the law of conservation of energy, was first stated in 1888 by Henri Le Chatelier (1850–1936). It is applied to chemical equilibria. For example, in the gas reaction

$$2SO_2 + O_2 \rightleftharpoons 2SO_3$$

an increase in pressure on the reaction mixture displaces the equilibrium to the right, since this reduces the total number of molecules present and thus decreases the pressure. The standard enthalpy change for the forward reaction is negative (i.e. the reaction is exothermic). Thus, an increase in temperature displaces the equilibrium to the left since this tends to reduce the temperature. The equilibrium constant thus falls with increasing temperature.

Leclanché cell A primary *voltaic cell consisting of a carbon rod (the anode) and a zinc rod (the cathode) dipping into an electrolyte of a 10–20% solution of ammonium chloride. *Polarization is prevented by using a mixture of manganese dioxide mixed with crushed carbon, held in contact with the anode by means of a porous bag or pot; this reacts with the hydrogen produced. This wet form of the cell, devised in 1867 by Georges Leclanché (1839–82), has an e.m.f. of about 1.5 volts. The *dry cell based on it is widely used in torches, radios, and calculators.

LED *See* LIGHT-EMITTING DIODE.

LEED Low-energy electron diffraction. *See* ELECTRON DIFFRACTION.

lens A curved, ground, and polished piece of glass, moulded plastic, or other transparent material used for the refraction of light. A **converging lens** is one that brings the rays of a parallel beam of light to a real *principal focus. They include biconvex, planoconvex, and converging meniscus lenses. **Diverging lenses** cause the rays of a parallel beam to

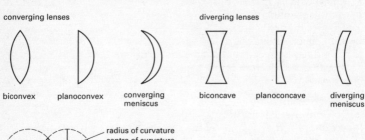

converging lenses

biconvex   planoconvex   converging meniscus

diverging lenses

biconcave   planoconcave   diverging meniscus

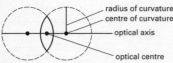

radius of curvature
centre of curvature
optical axis
optical centre

Lenses

diverge as if from a virtual principal focus; these include the biconcave, planoconcave, and diverging meniscus lenses. See illustrations.

The **centre of curvature** of a lens face is the centre of the sphere of which the surface of the lens is a part. The **optical axis** is the line joining the two centres of curvature of a lens or, in the case of a lens with one plane surface, the line through one centre of curvature that is normal to the plane surface. The **optical centre** of a lens is the point within a lens on the optical axis through which any rays entering the lens pass without deviation. The distance between the optical centre and the principal focus of a lens is called the **focal length** (f). The distance (v) between the lens and the image it forms is related to the distance (u) between the lens and the object by the **lens equation**:

$1/v + 1/u = 1/f$,

provided that the *real-is-positive convention is used. This takes distances to real objects, images, and foci as positive; those to virtual objects, images, and foci as negative. The equation does not always apply if the alternative New Cartesian convention (*see* SIGN CONVENTION) is used.

Lense–Thirring effect An effect predicted to occur in general relativity theory by J. Lense and Hans Thirring in 1918 in which a compact rotating body causes the space near it to rotate in the same direction. The phenomenon is also known as **frame dragging**. It has been reported in observations of neutron stars and black holes. Measurements have also been made using shifts in the orbits of satellites around the earth.

Lenz's law An induced electric current always flows in such a direction that it opposes the change producing it. This law, first stated by Heinrich Lenz (1804–65) in 1835, is a particular example of the law of conservation of energy.

lepton Any of a class of *elementary particles that consists of the *electron, muon, tau particle, and three types of *neutrino (one associated with each of the other types of lepton). For each lepton there is an equivalent antiparticle. The antileptons have a charge opposite that of the leptons; the antineutrinos, like the neutrinos, have no charge. The electron, muon, and tau particle all have a charge of –1. These three particles differ from each other only in mass: the muon is 200 times more massive than the electron and the tau particle is 3500 times more massive than the electron. Leptons interact by the electromagnetic interaction and the weak interaction (*see* FUNDAMENTAL INTERACTIONS).

lepton number *See* ELEMENTARY PARTICLES.

Leslie's cube A metal box in the shape of a cube in which each of the four vertical sides have different surface finishes. When hot water is placed in the cube, the emissivity of the finishes can be compared. The device was first used by Sir John Leslie (1766–1832).

level An instrument used in surveying to determine heights. It usually consists of a telescope and attached spirit level mounted on a tripod. The level is set up between a point of known height and a point for which the height is required. Before use it is adjusted until the line of sight is exactly horizontal. Sightings are then made onto a graduated levelling staff at the two points. The difference in elevation between the two points can then be calculated from the readings taken at these points.

lever A simple machine consisting of a rigid bar pivoted about a fulcrum. The mechanical advantage or *force ratio of a lever (the ratio of load to effort) is equal to the ratio of the perpendicular distance of the line of action of the effort from the fulcrum to the perpendicular distance of the line of action of the load from the fulcrum. In a first-order lever the fulcrum comes between load and effort. In a second-order lever the load comes between the fulcrum and the effort. In a third-order lever the effort comes between the fulcrum and the load. See illustrations.

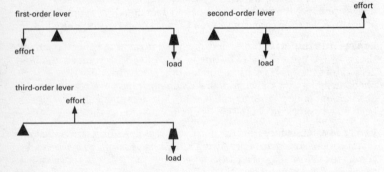

Levers

levitation The raising up of an object against gravity so that it is not in contact with a surface beneath it. This is frequently done using electric or magnetic fields. There are many technological applications of levitation, including **levitation melting**, in which a metal or other substance being melted is suspended in an electric or magnetic field and does not come into contact with the material of any container.

Leyden jar An early form of *capacitor consisting of a glass jar with a layer of metal foil on the outside and a similar layer on the inside. Contact to the inner foil is by means of a loose chain hanging inside the jar. It was invented in the Dutch town of Leyden in about 1745.

LF *See* LOW FREQUENCY.

***l*-form** *See* OPTICAL ACTIVITY.

libration The phenomenon that enables 59% of the moon's surface to be observed from earth over a 30-year period, in spite of its *synchronous rotation. **Physical libration** arises from slight variations in the rotation of the moon on its axis, caused by minor distortions in its physical shape. **Geometric librations** are apparent oscillations arising from the fact that the moon is observed from slightly different directions at different times. The geometric **libration in longitude** results from the nonuniform orbital motion of the moon. The geometric **libration in latitude** arises because the moon's axis of rotation is not perpendicular to its orbital plane; it enables more of the lunar polar regions to be observed.

lifetime The average time spent by an elementary particle or radioactive nucleus in a certain state before *decay occurs or before there is a transition from an atom in an excited state to a lower energy state.

Lifshitz–Slyozov law The relation $L \sim t^{1/3}$, between a length-scale L, characteristic of a system undergoing *spinodal decomposition, and the time t, for a long time after the system has been quenched. The Lifshitz–Slyozov law applies below the *spinodal curve to a system in which the *order parameter is conserved. This law is in accord with light-scattering experiments on binary liquid mixtures. The Lifshitz–Slyozov law was put forward by the Soviet physicists I. M. Lifshitz and V. V. Slyozov in 1961.

ligand-field theory An extension of crystal-field theory describing the properties of compounds of transition-metal ions or rare-earth ions in which covalent bonding between the surrounding molecules and the transition-metal ions is taken into account. This may involve using valence-bond theory or molecular-orbital theory. Ligand-field theory was developed extensively in the 1930s. As with crystal-field theory, ligand-field theory indicates that energy levels of the transition-metal ions are split by the surrounding ligands, as determined by *group theory. The theory has been very successful in explaining the optical, spectroscopic, and magnetic properties of the compounds of transition-metal and rare-earth ions.

light The form of *electromagnetic radiation to which the human eye is sensitive and on which our visual awareness of the universe and its contents relies (*see* COLOUR).

The finite velocity of light was suspected by many early experimenters in optics, but it was not established until 1676 when Ole Røemer (1644–1710) measured it. Sir Isaac *Newton investigated the optical *spectrum and used existing knowledge to establish a primarily **corpuscular theory** of light, in which it was regarded as a stream of particles that set up disturbances in the 'aether' of space. His successors adopted the corpuscles but ignored the wavelike disturbances until Thomas *Young rediscovered the *interference of light in 1801 and showed that a **wave theory** was essential to interpret this type of phenomenon. This view was accepted for most of the 19th century and it enabled James Clerk *Maxwell to show that light forms part of the *electromagnetic spectrum. He believed that waves of electromagnetic radiation required a special medium to travel through, and revived the name 'luminiferous ether' for such a medium. The *Michelson–Morley experiment in 1887 showed that, if the medium existed, it could not be detected; it is now generally accepted that the ether is an unnecessary hypothesis. In 1905 Albert *Einstein showed that the *photoelectric effect could only be explained on the assumption that light consists of a stream of discrete *photons of electromagnetic energy. This renewed conflict between the corpuscular and wave theories has gradually been resolved by the evolution of the *quantum theory and *wave mechanics. While it is not easy to construct a model that has both wave and particle characteristics, it is accepted, according to the theory of *complementarity proposed by Neils *Bohr, that in some experiments light will appear wavelike, while in others it will appear to be corpuscular. During the course of the evolution of wave mechanics it has also become evident that electrons and other elementary particles have dual wave and particle properties.

light amplifier *See* IMAGE INTENSIFIER.

light bulb *See* ELECTRIC LIGHTING.

light-dependent resistor (LDR) A resistor in which the value of the resistance depends on whether light is present or absent. Light-dependent resistors depend for their action on the photoconductive effect (*see* PHOTOELECTRIC EFFECT).

light-emitting diode (LED) A *semiconductor device that converts electrical energy into light or infrared radiation in the range 550 nm (green light) to 1300 nm (infrared radiation). The most commonly used LED (see illustration) emits red light and consists of gallium arsenide–phosphide on a gallium arsenide substrate, light being emitted at a p–n junction, when electrons and holes recombine (*see* RECOMBINATION PROCESS). LEDs are extensively used for displaying letters and numbers in digital instruments in which a self-luminous display is required.

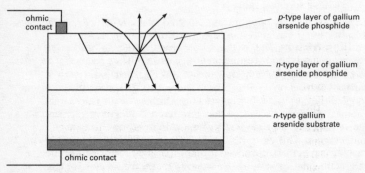

Light-emitting diode

lightning A high-energy luminous electrical discharge that passes between a charged cloud and a point on the surface of the earth, between two charged clouds, or between oppositely charged layers of the same cloud. In general, the upper parts of clouds are positively charged and the lower parts are negatively charged; the reasons for this separation of charge are complex.

Lightning usually occurs in the form of a downward step leader followed by an intensely luminous return stroke, which can produce instantaneous temperatures as high as 30 000°C. In the typical step leader a surge of electrons descends in approximately 50-metre steps with about 50-microsecond pauses between steps. When this leader reaches the earth a surge of charge returns up the preionized path taken by the leader. Cloud-to-cloud strokes also involve a leader and return stroke. The average current in a lightning stroke is about 10 000 amperes, but maximum currents in the return stroke can reach 20 000 A. *See also* BALL LIGHTNING.

light scattering by light A process predicted to occur in *quantum electrodynamics (QED) that does not occur in classical electrodynamics. Light scattering by light does not occur in classical physics because *Maxwell's equations are linear, whereas electrodynamics becomes nonlinear when quantum effects are taken into account. The process has never been conclusively observed but powerful *lasers may enable it to be observed in the future. Light scattering by light was first predicted in the 1930s, when calculations of the process were performed by the German physicist Werner *Heisenberg and others. Complete calculations of light scattering by light were not performed until the *renormalization of QED was developed in the late 1940s.

light year A unit of distance used in astronomy; the distance travelled by light in a vacuum during one year. It is equal to 9.4650×10^{15} metres or 5.8785×10^{12} miles.

limit The value that a function approaches as the independent variable approaches a specified value.

limit cycle *See* ATTRACTOR.

limiting friction The friction force that just balances a moving force applied to a solid body resting on a solid surface when the body fails to move. If the moving force exceeds the limiting friction, the body will begin to move.

linac *See* LINEAR ACCELERATOR.

Linde process A process for the *liquefaction of gases by the Joule–Thomson effect. In this process, devised by Carl von Linde (1842–1934) in the late 19th century for liquefying air, the air is freed of carbon dioxide and water and compressed to 150 atmospheres. The compressed gas is passed through a copper coil to an expansion nozzle within a Dewar flask. The emerging air is cooled by the Joule–Thomson effect as it expands and then passes back within a second copper coil that surrounds the first coil. Thus the expanded gas cools the incoming gas in a process that is said to be **regenerative**. Eventually the air is reduced to its *critical temperature and, at the pressure of 150 atmospheres (well above its critical pressure), liquefies. The process is also used for other gases, especially hydrogen and helium. Hydrogen has first to be cooled below its inversion temperature (*see* JOULE–THOMSON EFFECT) using liquid air; helium has first to be cooled below its inversion temperature using liquid hydrogen.

linear absorption coefficient *See* LAMBERT'S LAWS.

linear accelerator (linac) A type of particle *accelerator in which charged particles are accelerated in a straight line, either by a steady electric field or by means of radio-frequency electric fields. There are three main types:

Van de Graaff accelerator. This device accelerates charged particles by applying a high electrical potential difference generated by a *Van de Graaff generator. The potential difference can be kept steady to within one part in a thousand, enabling a very uniformly energetic beam of accelerated particles to be created. The maximum electrical potential attainable is about 10 MV and depends on the insulating properties of the gas around the Van de Graaff sphere. It is increased by enclosing the whole generator in a pressure vessel containing an inert gas at a pressure of about 20 atmospheres. A source, at the same potential as the sphere, produces charged particles that enter a column of cylindrical electrodes, each of which is at a lower potential than the one above it. The ions are accelerated as they pass through the gaps between the cylinders. The nonuniform electric fields between the gaps have the effect of focusing the beam of charged particles.

Drift-tube accelerator. In this device charged particles are accelerated along a line of hollow metal cylinders called **drift tubes**. The cylinders are connected alternately to opposite terminals of an alternating potential difference produced by either a *magnetron or a *klystron. The arrangement ensures that adjacent cylinders are always at opposite electrical potentials. For example, a proton beam may be injected into the first of the line of drift tubes from a Van de Graaff accelerator. Protons reaching the gap between the first two tubes will be accelerated into the second tube, when the alternating potential makes the first tube positive and the second tube negative. This enables the protons emerging into the gap between two cylinders to be accelerated into the next cylinder. All parts of a particular tube are at the same potential, since the metal acts as an equipotential surface. Therefore within a cylinder the particles travel at a constant speed (hence drift tube). It follows that the energy of the beam is increased every time the protons cross between drift tubes, and therefore a device with a large number of gaps can produce extremely high-energy beams using only moderate supply voltages. The Berkeley proton accelerator has a drift-tube arrangement of 47 cylinders, 19 km long. It accelerates protons up to 31.5 GeV.

Travelling-wave accelerator. This apparatus uses radio-frequency electromagnetic waves to accelerate charged particles. Charged particles are fed into the travelling-wave accelerator at close to the speed of light and are carried through a *wave guide by the electric field component of a radio wave. The very high initial speeds for charged particles are needed to match the phase velocity of radio signals propagating along the wave guide. However, this means that travelling-wave accelerators are suitable only for accelerating lighter particles, such as electrons. The electrons can be accelerated to initial speeds of 98% of the speed of light by a Van de Graaff accelerator. At such high initial speeds, there is little scope for further acceleration and any increase in electron energy provided by the accelerator results from the relativistic increase in mass. The Stanford linear accelerator (SLAC) uses the travelling-wave principle. SLAC is capable of accelerating electrons and positrons to 50 GeV in a tube 3 km long.

linear energy transfer (LET) The energy transferred per unit path length by a moving high-energy charged particle (such as an electron or a proton) to the atoms and molecules along its path. It is of particular importance when the particles pass through living tissue as the LET modifies the effect of a specific dose of radiation. LET is proportional to the square of the charge on the particle and increases as the velocity of the particle decreases.

linear equation An equation between two variables that gives a straight line when plotted on a graph. It has the general form $y = mx + c$, where m is the gradient of the line and c is the intercept of the line on the y-axis (in Cartesian coordinates).

linear expansivity *See* EXPANSIVITY.

linear momentum *See* MOMENTUM.

linear motor A form of induction motor in which the stator and armature are linear and parallel, rather than cylindrical and coaxial. In some experimental trains the magnetic force between the primary winding in the vehicle and the secondary winding on the ground support the vehicle on an air cushion thus eliminating track friction. However, because of the high cost of the installation and the low efficiency the device has not yet found commercial application.

line defect *See* DEFECT.

line integral If a vector function $V(x,y,z)$ is defined between two points A and B on a curve, the curve is approximately given by a series of equal directed chords $\Delta_1 l, \Delta_2 l, \dots \Delta_n l$. In each segment i of the curve it is possible to define the *scalar product $V_i \cdot \Delta_i l$. If the sum of the scalar products

$$\sum_{i}^{n} = V_i \cdot \Delta_i l$$

is considered, the line integral is defined by

$$\int_A^B \boldsymbol{V} \cdot d\boldsymbol{l} = \lim_{n \to \infty} \sum_{i=1}^n \boldsymbol{V} \cdot \Delta_i \boldsymbol{l}.$$

An example of a line integral is given by the case of a force $\boldsymbol{F}$ acting on a particle in the field of the force. The line integral $\int_A^B \boldsymbol{F} \cdot d\boldsymbol{l}$ is the *work done on the particle as it moves from A to B because of the force.

If the line integral is taken round a closed path (loop), the line integral is denoted by $\int_C \boldsymbol{V} \cdot d\boldsymbol{l}$ or $\oint \boldsymbol{V} \cdot d\boldsymbol{l}$. A line integral for a scalar function $\phi(x,y,z)$ is defined in a similar way and is denoted $\int_A^B \phi d\boldsymbol{l}$. It is also possible to define another type of line integral for a vector function $\boldsymbol{V}$ by $\int_A^B \boldsymbol{V} \times d\boldsymbol{l}$.

line-of-sight velocity (radial velocity) The component of a celestial body's velocity along the line of sight of the observer. It is usually given in relation to the sun to avoid complications arising from the earth's orbital motion. Line-of-sight velocity is normally calculated from the *Doppler effect on the body's spectrum, a *redshift indicating a receding body (taken as a positive velocity) and a blueshift indicating an approaching body (taken as negative).

line printer *See* PRINTER.

lines of force Imaginary lines in a *field of force that enable the direction and strength of the field to be visualized. They are used primarily in electric and magnetic fields; in electric fields they are sometimes called **tubes of force**, to express their characteristic of being perpendicular to a conducting surface. The tangent to a line of force at any point gives the direction of the field at that point and the number of lines per unit area perpendicular to the force represents the *intensity of the field.

line spectrum *See* SPECTRUM.

Linz–Donawitz process *See* BASIC-OXYGEN PROCESS.

liquation The separation of mixtures of solids by heating to a temperature at which lower-melting components liquefy.

liquefaction of gases The conversion of a gaseous substance into a liquid. This is usually achieved by one of four methods or by a combination of two of them:
(1) by vapour compression, provided that the substance is below its
   *critical temperature;
(2) by refrigeration at constant pressure, typically by cooling it with a
   colder fluid in a countercurrent heat exchanger;
(3) by making it perform work adiabatically against the atmosphere in a
   reversible cycle;
(4) by the *Joule–Thomson effect (*see also* LINDE PROCESS).
   Large quantities of liquefied gases are now used commercially, especially *liquefied petroleum gas and liquefied natural gas.

liquefied petroleum gas (LPG) Various petroleum gases, principally propane and butane, stored as a liquid under pressure. It is used as an engine fuel and has the advantage of causing very little cylinder-head deposits.

Liquefied natural gas (LNG) is a similar product and consists mainly of methane. However, it cannot be liquefied simply by pressure as it has a low critical temperature of 190 K and must therefore be cooled to below this temperature before it will liquefy. Once liquefied it has to be stored in well-insulated containers. It provides a convenient form in which to ship natural gas in bulk from oil wells or gas-only wells to users. It is also used as an engine fuel.

liquid A phase of matter between that of a crystalline solid and a *gas. In a liquid, the large-scale three-dimensional atomic (or ionic or molecular) regularity of the solid is absent but, on the other hand, so is the total disorganization of the gas. Although liquids have been studied for many years there is still no simple comprehensive theory of the liquid state. It is clear, however, from diffraction studies that there is a short-range structural regularity extending over several molecular diameters. These bundles of ordered atoms, molecules, or ions move about in relation to each other, enabling liquids to have almost fixed volumes, which adopt the shape of their containers.

liquid crystal A substance that flows like a liquid but has some order in its arrangement of molecules. **Nematic crystals** have long molecules all aligned in the same direction, but otherwise randomly arranged. **Cholesteric** and **smectic** liquid crystals also have aligned molecules, which are arranged in distinct layers. In cholesteric crystals, the axes of the molecules are parallel to the plane of the layers; in smectic crystals they are perpendicular.

liquid-crystal display A display unit used in digital watches, calculators, miniature TV screens, etc. It provides a source of clearly displayed digits for a very low power consumption. In the display unit a thin film of *liquid crystal is sandwiched between two transparent electrodes (glass with a thin metal or oxide coating). In the commonly used field-effect display, twisted nematic crystals are used. The nematic liquid crystal cell is placed between two crossed polarizers. Polarized light entering the cell follows the twist of the nematic liquid crystal, is rotated through 90°, and can therefore pass through the second polarizer. When an electric field is applied the molecular alignment in the liquid crystal is altered, the polarization of the entering light is unchanged, and no light is therefore transmitted. In these circumstances, a mirror placed behind the second polarizer will cause the display to appear black. One of the electrodes, shaped in the form of a digit, will then provide a black digit when the voltage is applied.

liquid-drop model A model of the atomic nucleus that has been successful in accounting for *nuclear fission and the variation of *nuclear stability with mass number. The density of a nucleus is independent of its size, which suggests that nuclear matter can be modelled on a drop of an incompressible liquid, such as water. Different excitation states of a drop-like nucleus can then be described in terms of *spherical harmonics (*see also* NORMAL MODES). The success of the model has been associated with the fact that the binding forces in both the nucleus and the liquid drop are essentially short-ranged (*see* FUNDAMENTAL INTERACTIONS). The liquid-drop model provides a subtle explanation for the variation of binding energy in different nuclei. Energy must be supplied to a drop of liquid in order to break it up into its component parts. For a drop of liquid, this energy is its latent heat of vaporization, which is proportional to the mass of liquid present. One would therefore expect a nucleus to have a binding energy proportional to its mass number. However, one needs also to offset this expectation by considering an effect analogous to the surface tension in the liquid drop. Nucleons (protons or neutrons) in the interior of a nucleus are attracted from all sides, whereas those at the surface are attracted only from within. The surface nucleons are therefore easier to remove and consequently have lower binding energies. In smaller nuclei, the surface nucleons form a greater proportion of the total volume than those in larger nuclei. It follows, therefore, that the binding energy is reduced by an amount proportional to the surface area of the nucleus. The binding energy may be further modified by the contribution of electrostatic repulsion between protons, which will be significant for the larger nuclei (in which there are more protons). The nuclear model becomes a drop of positively charged liquid whose stability is expected to be further modified by the internal motions of the protons and neutrons within it. Quantum mechanics can provide the form of this final contribution, which is often called a 'symmetry term' on account of its dependence on the difference of neutron and proton numbers (hence the tendency for the neutron number to equal the proton number in nuclei with low nucleon numbers).

Lissajous figures A curve in one plane traced by a point moving under the influence of two independent harmonic motions. In the common case the harmonic motions are simple, perpendicular to each other, and have a simple frequency ratio. They can be displayed by applying sinusoidal alternating potentials to the X- and Y-inputs of a *cathode-ray oscilloscope. they are named after Jules Lissajous (1822–80).

litre Symbol l or L. A unit of volume in the metric system regarded as a special name for the cubic decimetre. It was formerly defined as the volume of 1 kilogram of pure water at 4°C at standard pressure, which is equivalent to 1.000 028 dm^3.

Lloyd's mirror An optical arrangement for producing interference fringes. A slit is illuminated by monochromatic light and placed close to a plane mirror. Interference occurs between direct light from the slit and light reflected from the mirror. It was first used by Humphrey Lloyd (1800–81) in 1834.

loaded concrete Concrete containing elements (such as iron or lead) with a high mass number; it is used in making the radiation shield around nuclear reactors.

Local Group The group of *galaxies of which our own Galaxy is a member. It consists of some 30–40 known members, the most massive of which are the Galaxy and the Andromeda galaxy.

localization 1. The confinement of electrons to a particular atom in a molecule or to a particular chemical bond. **2.** In the theory of *disordered solids, the concept that an electron is concentrated around a specific site and cannot contribute to the solid's electrical conductivity (at *absolute zero) by moving through the system. In one dimension any amount of disorder makes all electron states localized. In three dimensions a small amount of disorder makes electron states near the top and the bottom of the *energy bands localized; states in the centre of the bands are called **extended** states because they can propagate through the system and hence contribute to electrical conductivity. The dividing energies between localized and extended states are called **mobility edges**. Given sufficient disorder all states become localized. In two dimensions all electron states in disordered solids are thought to be localized, with some states being strongly localized around specific sites while other states are weakly localized around specific sites. Localization also occurs in disordered solids for other *excitations, such as *phonons and *spin waves.

local oscillator An *oscillator in a *heterodyne or *superheterodyne receiver. It supplies the radio-frequency signal that beats with the incoming signal to produce the intermediate frequency.

locus A set of points whose location is specified by an equation. For example, if a point moves so that the sum of its distances from two fixed points is constant, the locus of the point is an *ellipse.

Lodge, Sir Oliver Joseph (1851–1940) British physicist, who became principal of the then-new Birmingham University in 1900. His best-known work was in *radio, particularly his invention in 1894 of the 'coherer', used as a detector in early radio receivers (see DEMODULATION). After 1910 he became increasingly interested in spiritualism and reconciling science and religion.

logarithm The power to which a number, called the **base**, has to be raised to give another number. Any number y can be written in the form $y = x^n$. n is then the logarithm to the base x of y, i.e. $n = \log_x y$. If the base is 10, the logarithms are called **common logarithms**. **Natural** (or **Napierian**) **logarithms** (named after John Napier; 1550–1617) are to the base $e = 2.718\ 28...$, written $\log_e y$ or $\ln y$. Logarithms were formerly used to facilitate calculations, before the advent of electronic calculators.

A logarithm contains two parts, an integer and a decimal. The integer is called the **characteristic**, and the decimal is called the **mantissa**. For example, the logarithm to the base 10 of 210 is 2.3222, where 2 is the characteristic and 0.3222 is the mantissa.

logarithmic scale 1. A scale of measurement in which an increase or decrease of one unit represents a tenfold increase or decrease in the quantity measured. Decibels and pH measurements are common examples of logarithmic scales of measurement. **2.** A scale on the axis of a graph in which an increase of one unit represents a tenfold increase in the variable quantity. If a curve $y = x^n$ is plotted on graph paper with logarithmic scales on both axes, the result is a straight line of slope n, i.e. $\log y = n \log x$, which enables n to be determined.

logarithmic series The expansion of a logarithmic function, such as $\log_e(1 + x)$, i.e. $x - x^2/2 + x^3/3 - \dots + (-1)^n x^n/n$, or $\log_e(1 - x)$, i.e. $-x - x^2/2 - x^3/3 \dots - x^n/n$.

logic circuits The basic switching circuits or *gates used in digital computers and other digital electronic devices. The output signal, using a *binary notation, is controlled by the logic circuit in accordance with the input system. The three basic logic circuits are the **AND**, **OR**, and **NOT** **circuits**. The AND circuit gives a binary 1 output if a binary 1 is present on each input circuit; otherwise the output is a binary 0. The OR circuit gives a binary 1 output if a binary 1 is present on at least one input circuit; otherwise the output is binary 0. The NOT circuit inverts the input signal, giving a binary 1 output for a binary 0 input or a 0 output for a 1 input.

Often these basic logic circuits are used in combination, e.g. a **NAND circuit** consists of NOT + AND circuits. In terms of electronic equipment, logic circuits are now almost exclusively embodied into *integrated circuits.

longitude *See* LATITUDE AND LONGITUDE.

longitudinal wave *See* WAVE.

long-sightedness *See* HYPERMETROPIA.

Lorentz, Hendrick Antoon (1853–1928) Dutch physicist who was one of the founders of the electron theory of matter. This gave a partial explanation of the *Zeeman effect; for this work Lorentz shared the 1902 Nobel Prize with Pieter Zeeman (1865–1943). Lorentz also solved one of the problems raised by the *Michelson–Morley experiment (*see* LORENTZ–FITZGERALD CONTRACTION) and suggested the *Lorentz transformations, both of which were used by Einstein in his special theory of relativity.

Lorentz–Dirac equation An equation for describing the motion of a point electric charge, derived by classical electrodynamics. This equation is in accord with special relativity theory and avoids the *self-energy problem. However, there are a number of difficulties, which are usually attributed to the fact that classical electrodynamics is not applicable to the point electron. The equation was derived by Paul Dirac in 1938, generalizing work in the late 19th century by H. A. Lorentz.

Lorentz–Fitzgerald contraction (Fitzgerald contraction) The contraction of a moving body in the direction of its motion. It was proposed independently by H. A. Lorentz and G. F. Fitzgerald (1851–1901) in 1892 to account for the null result of the *Michelson–Morley experiment. The contraction was given a theoretical background in Einstein's special theory of *relativity. In this theory, an object of length l_0 at rest in one *frame of reference will appear, to an observer in another frame moving with relative velocity v with respect to the first, to have length $l_0\sqrt{(1 - v^2/c^2)}$, where c is the speed of light. The original hypothesis regarded this contraction as a real one accompanying the absolute motion of the body. The contraction is in any case negligible unless v is of the same order as c.

Lorentz force The force exerted on a moving electric charge by a magnetic field. The force F exerted on a charge Q that is moving with a velocity v by a magnetic field with magnetic flux density B is given by the vector product: $F = Q(v \times B)$.

Lorentz group The set of **proper Lorentz transformations**, i.e. rotations in the four-dimensional space x, y, z, τ, where x, y, and z are space coordinates and $\tau = ict$, where t is the time and c is the speed of light in a vacuum. The Lorentz group is named after H. A. Lorentz. If the Lorentz group is combined with *translations in space and time the **proper Poincaré group** is formed; this is named after the French mathematician and scientist Henri Poincaré (1854–1912). If inversions in space and time are combined with the proper Poincaré group the **unrestricted Poincaré group** results. Analysis of the Lorentz group in relativistic quantum mechanics is used to classify *elementary particles.

Lorentz–Lorenz equation A relation between the *polarizability α of a molecule and the *refractive index n of a substance made up of molecules with this polarizability. The Lorentz–Lorenz equation can be written in the form $\alpha = (3/4\pi N)\,[(n^2 - 1/(n^2 + 2)]$, where N is the number of molecules per unit volume. The equation provides a link between a microscopic quantity (the polarizability) and a macroscopic quantity (the refractive index). It was derived using macroscopic electrostatics in 1880 by H. A. Lorentz and independently by the Danish physicist Ludwig Valentin Lorenz also in 1880. *Compare* CLAUSIUS–MOSSOTTI EQUATION.

Lorentz transformations A set of equations proposed by H. A. Lorentz for transforming the position and motion parameters from a frame of reference with origin at O and coordinates (x,y,z) to a frame moving relative to it with origin at O′ and coordinates $(x′,y′,z′)$. They replace the *Galilean transformations used in *Newtonian mechanics and are used in relativistic mechanics. They are:

$$x' = \beta(x - vt)$$

$$y' = y$$

$$z' = z$$

$$t' = \beta(t - vx/c^2),$$

where v is the relative velocity of separation of O and O', c is the speed of light, and $\beta = 1/\sqrt{(1 - v^2/c^2)}$. The above equations apply for constant v in the xx' direction with O and O' coinciding at $t = t' = 0$.

Loschmidt's constant (Loschmidt number) Symbol N_L. The number of particles per unit volume of an *ideal gas at STP. It has the value $2.686\ 763(23) \times 10^{25}$ m^{-3} and was first worked out by Joseph Loschmidt (1821–95).

loudness The physiological perception of sound intensity. As the ear responds differently to different frequencies, for a given intensity loudness is dependent on frequency. Sounds with frequencies between 1000 hertz and 5000 Hz are louder than sounds of the same intensity at higher or lower frequencies. Duration is also a factor in loudness, long bursts of sound being louder than short bursts. Loudness increases up to a duration of about 0.2 second; above this limit loudness does not increase with duration.

Relative loudness is usually measured on the assumption of proportionality to the logarithm of the intensity (for a given frequency), i.e. proportionality to the relative intensity on the *decibel scale. A subjective judgment is made of the relative intensity above threshold that a note of 1000 Hz must have to match the specimen sound; the loudness of this, in *phons, is then equal to that relative intensity in decibels.

loudspeaker A transducer for converting an electrical signal into an acoustic signal. Usually it is important to preserve as many characteristics of the electrical waveform as possible. The device must be capable of reproducing frequencies in the range 150–8000 hertz for speech and 20–20 000 Hz for music.

The most common loudspeaker consists of a moving-coil device. In this a cone-shaped diaphragm is attached to a coil of wire and made to vibrate in accordance with the electrical signal by the interaction between the current passing through the coil and a steady magnetic field from a permanent magnet surrounding it.

low-dimensional system A solid-state system in which the spatial dimension is less than three. In practice, a **two-dimensional system** is a thin film or layer and a **one-dimensional system** is a thin wire. Two-dimensional systems have applications to *semiconductor technology, in such devices as MOSFETs (*see* TRANSISTORS). The behaviour of low-dimensional systems is of interest because the problems for low-dimensional systems (particularly one-dimensional systems) are much easier to solve than the corresponding problems in three dimensions.

Clusters of atoms and very small crystals can be considered as **zero-dimensional systems**.

lowering of vapour pressure A reduction in the saturated vapour pressure of a pure liquid when a solute is introduced. If the solute is a solid of low vapour pressure, the decrease in vapour pressure of the liquid is proportional to the concentration of particles of solute; i.e. to the number of dissolved molecules or ions per unit volume. To a first approximation, it does not depend on the nature of the particles. *See* COLLIGATIVE PROPERTIES; RAOULT'S LAW.

low frequency (LF) A radio frequency in the range 30–300 kilohertz; i.e. having a wavelength in the range 1–10 kilometre.

lubrication The use of a substance to prevent contact between solid surfaces in relative motion in order to reduce friction, wear, overheating, and rusting. Liquid hydrocarbons (oils), either derived from petroleum or made synthetically, are the most widely used lubricants as they are relatively inexpensive, are good coolants, provide the appropriate range of viscosities, and are thermally stable. Additives include polymeric substances that maintain the desired viscosity as the temperature increases, antioxidants that prevent the formation of a sludge, and alkaline-earth phenates that neutralize acids and reduce wear.

At high temperatures, solid lubricants, such as graphite or molybdenum disulphide, are often used. Semifluid lubricants (greases) are used to provide a seal against moisture and dirt and to remain attached to vertical surfaces. They are made by adding gelling agents, such as metallic soaps, to liquid lubricants.

Recent technology has made increasing use of gases as lubricants, usually in air bearings. Their very low viscosities minimize energy losses at the bearings but necessitate some system for pumping the gas continuously to the bearings. The principle is that of the hovercraft.

lumen Symbol lm. The SI unit of *luminous flux equal to the flux emitted by a uniform point source of 1 candela in a solid angle of 1 steradian.

luminance (photometric brightness) Symbol L. The *luminous intensity of any surface in a given direction per unit projected area of the surface, viewed from that direction. It is given by the equation $L = dI/(dA\cos\theta)$, where I is the luminous intensity and θ is the angle between the line of sight and the normal to the surface area A being considered. It is measured in candela per square metre.

luminescence The emission of light by a substance for any reason other than a rise in its temperature. In general, atoms of substances emit *photons of electromagnetic energy when they return to the *ground state after having been in an excited state (*see* EXCITATION). The causes of the excitation are various. If the exciting cause is a photon, the process is called **photoluminescence**; if it is an electron it is called **electroluminescence**. **Chemiluminescence** is luminescence resulting from a chemical reaction (such as the slow oxidation of phosphorus); *bioluminescence is the luminescence produced by a living organism (such as a firefly). If the luminescence persists significantly after the exciting cause is removed it is called **phosphorescence**; if it does not it is called **fluorescence**. There must always be some delay; in some definitions a persistence of more than 10 nanoseconds (10^{-8} s) is treated as phosphorescence.

luminosity 1. *Luminous intensity in a particular direction; the apparent brightness of an image. **2.** The brightness of a star defined as the total energy radiated in unit time. It is related to the surface area (A) and the **effective temperature** (T_e; the temperature of a black body having the same radius as the star and radiating the same amount of energy per unit area in one second) by a form of *Stefan's law, i.e.

$$L = A\sigma T_e^4$$

where σ is the Stefan constant and L is the luminosity.

luminous exitance *See* EXITANCE.

luminous flux Symbol Φ_v. A measure of the rate of flow of light, i.e. the radiant flux in the wavelength range 380–760 nanometres, corrected for the dependence on wavelength of the sensitivity of the human eye. It is measured by reference to emission from a standard source, usually in lumens.

luminous intensity Symbol I_v. A measure of the light-emitting ability of a light source, either generally or in a particular direction. It is measured in candelas.

lunar eclipse *See* ECLIPSE.

lunation *See* SYNODIC MONTH.

lux Symbol lx. The SI unit of *illuminance equal to the illumination produced by a *luminous flux of 1 lumen distributed uniformly over an area of 1 square metre.

Lyman series *See* HYDROGEN SPECTRUM.

machine A device capable of making the performance of mechanical work easier, usually by overcoming a force of resistance (the load) at one point by the application of a more convenient force (the effort) at some other point. In physics, the six so-called **simple machines** are the lever, wedge, inclined plane, screw, pulley, and wheel and axle.

Mach number The ratio of the relative speeds of a fluid and a rigid body to the speed of sound in that fluid under the same conditions of temperature and pressure. If the Mach number exceeds 1 the fluid or body is moving at a **supersonic speed**. If the Mach number exceeds 5 it is said to be **hypersonic**. The number is named after Ernst Mach (1838–1916).

Mach's principle The *inertia of any particular piece of matter is attributable to the interaction between that piece of matter and the rest of the universe. A body in isolation would have zero inertia. This principle was stated by Ernst Mach in the 1870s and was made use of by Einstein in his general theory of *relativity. The significance of Mach's principle in general relativity theory and the foundations of physics continues to be a contentious issue.

Maclaurin's series *See* TAYLOR SERIES.

macroscopic Designating a size scale very much larger than that of atoms and molecules. Macroscopic objects and systems are described by *classical physics although *quantum mechanics can have macroscopic consequences. *Compare* MESOSCOPIC; MICROSCOPIC.

Madelung's rule An approximate rule about atomic orbitals stating that the order of their being filled is the order of increasing $n + l$, and that when two or more orbitals have the same value of $n + l$ then the order of filling is increasing n. This rule was first put forward by Erwin Madelung (1881–1972) in 1936. There are some exceptions to it but it does have some theoretical justification.

Magellanic clouds Two small galaxies situated close to the Milky Way that are only visible from the southern hemisphere. They were first recorded by Ferdinand Magellan (1480–1521) in 1519.

magic numbers Numbers of neutrons or protons that occur in atomic nuclei to produce very stable structures. The magic numbers for both protons and neutrons are 2, 8, 20, 28, 50, and 82. For neutrons 126 and 184 are also magic numbers and for protons 114 is a magic number. The relationship between stability and magic numbers led to a nuclear *shell model in analogy to the electron shell model of the atom.

Magnadur A tradename for a ceramic material used to make permanent magnets. It consists of sintered iron oxide and barium oxide.

Magnalium A tradename for an aluminium-based alloy of high reflectivity for light and ultraviolet radiation that contains 1–2% of copper and between 5% and 30% of magnesium. Strong and light, these alloys also sometimes contain other elements, such as tin, lead, and nickel.

magnet A piece of magnetic material (*see* MAGNETISM) that has been magnetized and is therefore surrounded by a *magnetic field. A magnet, often in the shape of a bar or horseshoe, that retains appreciable magnetization indefinitely (provided it is not heated, beaten, or exposed to extraneous magnetic fields) is called a **permanent magnet**. *See also* ELECTROMAGNET.

magnetic bottle A nonuniform *magnetic field used to contain the *plasma in a thermonuclear experimental device. At the temperature of a thermonuclear reaction (10^8 K) any known substance would vaporize and the plasma has therefore to be contained in such a way that it does not come into contact with a material surface. The magnetic bottle provides a means of achieving this, by deflecting away from its boundaries the moving charged particles that make up the plasma.

magnetic bubble memory A form of computer memory in which a small magnetized region of a substance is used to store information. Bubble memories consist of materials, such as magnetic garnets, that are easily magnetized in one direction but hard to magnetize in the perpendicular direction. A thin film of these materials deposited on a nonmagnetic substrate constitutes a bubble-memory chip. When a magnetic field is applied to such a chip, by placing it between two permanent magnets, cylindrical domains (called magnetic bubbles) are formed. These bubbles constitute a magnetic region of one polarity surrounded by a magnetic region of the opposite polarity. Information is represented as the presence or absence of a bubble at a specified storage location and is retrieved by means of a rotating magnetic field. Typically a chip measures 15 mm^2, or 25 mm^2 enclosed in two permanent magnets and two rotating field coils; each chip can store up to one million bits.

magnetic circuit A closed path containing a *magnetic flux. The path is clearly delimited only if it consists mainly or wholly of ferromagnetic or other good magnetic materials; examples include transformer cores and iron parts in electrical machines. The design of these parts can often be assisted by analogy with electrical circuits, treating the *magnetomotive force as the analogue of e.m.f., the magnetic flux as current, and the *reluctance as resistance. There is, however, no actual flow around a magnetic circuit.

magnetic compass *See* COMPASS.

magnetic constant *See* PERMEABILITY.

magnetic containment *See* THERMONUCLEAR REACTOR.

magnetic declination *See* GEOMAGNETISM.

magnetic dip *See* GEOMAGNETISM.

magnetic disk A smooth aluminium disk, usually 35.6 cm in diameter, both surfaces of which are coated with magnetic iron oxide. The disks are used as a recording medium in computers, up to ten such disks being mounted in a disk pack. Data is recorded in concentric tracks on both surfaces with up to 236 tracks per centimetre. The disks rotate at 3600 revolutions per minute, information being put onto the disk and removed from it by means of a record-playback head. *See also* FLOPPY DISK.

magnetic domain *See* MAGNETISM.

magnetic elements *See* GEOMAGNETISM.

magnetic equator *See* EQUATOR; GEOMAGNETISM.

magnetic field A *field of force that exists around a magnetic body (*see* MAGNETISM) or a current-carrying conductor. Within a magnetic field a magnetic dipole may experience a torque and a moving charge may experience a force. The strength and direction of the field can be given in terms of the **magnetic flux density** (or **magnetic induction**), symbol B; it can also be given in terms of the **magnetic field strength** (**magnetizing force** or **magnetic intensity**), symbol H.

The magnetic flux density is a vector quantity and is the *magnetic flux per unit area of a magnetic field at right angles to the magnetic force. It can be defined in terms of the effects the field has, for example by $B = F/qv\sin\theta$, where F is the force a moving charge q would experience if it was travelling at a velocity v in a direction making an angle θ with that of the field. The *SI unit is the tesla.

The magnetic field strength is also a vector quantity and is related to B by:

$$H = B/\mu,$$

where μ is the *permeability of the medium. The SI unit of field strength is the ampere per metre (A m^{-1}).

magnetic field strength *See* MAGNETIC FIELD.

magnetic flux Symbol Φ. A measure of quantity of magnetism, taking account of the strength and the extent of a *magnetic field. The flux $d\Phi$ through an element of area dA perpendicular to B is given by $d\Phi = BdA$. The *SI unit of magnetic flux is the weber.

magnetic flux density *See* MAGNETIC FIELD.

magnetic force The attractive or repulsive force exerted on a *magnetic pole or a moving electric charge in a *magnetic field.

magnetic induction *See* MAGNETIC FIELD.

magnetic intensity *See* MAGNETIC FIELD.

magnetic meridian *See* MERIDIAN.

magnetic mirror A device used to contain *plasma in thermonuclear experimental devices. It consists of a region of high magnetic field strength at the end of a containment tube. Ions entering the region reverse their motion and return to the plasma from which they have emerged. *See also* MAGNETIC BOTTLE.

magnetic moment The ratio between the maximum torque (T_{max}) exerted on a magnet, current-carrying coil, or moving charge situated in a *magnetic field and the strength of that field. It is thus a measure of the strength of a magnet or current-carrying coil. In the Sommerfeld approach this quantity (also called **electromagnetic moment** or **magnetic area moment**) is T_{max}/B. In the Kennelly approach the quantity (also called **magnetic dipole moment**) is T_{max}/H.

In the case of a magnet placed in a magnetic field of field strength H, the maximum torque T_{max} occurs when the axis of the magnet is perpendicular to the field. In the case of a coil of N turns and area A carrying a current I, the magnetic moment can be shown to be $m = T/B = NIA$ or $m = T/H = \mu NIA$. Magnetic moments are measured in *SI units in $A\,m^2$.

An orbital electron has an orbital magnetic moment IA, where I is the equivalent current as the electron moves round its orbit. It is given by $I = q\omega/2\pi$, where q is the electronic charge and ω is its angular velocity. The orbital magnetic moment is therefore $IA = q\omega A/2\pi$, where A is the orbital area. If the electron is spinning there is also a spin magnetic moment (*see* SPIN); atomic nuclei also have magnetic moments (*see* NUCLEAR MOMENT).

magnetic monopole A hypothetical magnetic entity consisting of an isolated elementary north or south pole. It has been postulated as a source of a *magnetic field by analogy with the way in which an electrically charged particle produces an electric field. Numerous ingenious experiments have been designed to detect the monopole but so far none has produced an unequivocal result. Magnetic monopoles are predicted to exist in certain *gauge theories with *Higgs bosons. In particular, some *grand unified theories predict very heavy monopoles (with mass of order 10^{16} GeV). Magnetic monopoles are also predicted to exist in *Kaluza–Klein theories and *superstring theory.

magnetic permeability *See* PERMEABILITY.

magnetic poles **1.** *See* GEOMAGNETISM. **2.** The regions of a *magnet from which the magnetic forces appear to originate. A magnetized bar has a pole at each end; if it is freely suspended in the earth's magnetic field (*see* GEOMAGNETISM) it will rotate so that one end points approximately towards the earth's geographical north pole. This end is called the north-seeking end or the north pole of the magnet. The other end is accordingly called the south-seeking end or south pole. In the obsolete theory associated with the c.g.s. electromagnetic system of units, a **unit magnetic pole** was treated as one of a pair, which repelled each other with a force of 1 dyne when separated by 1 cm in space.

magnetic potential *See* MAGNETOMOTIVE FORCE.

magnetic quantum number *See* ATOM.

magnetic resonance imaging *See* NUCLEAR MAGNETIC RESONANCE (Feature).

magnetic susceptibility *See* SUSCEPTIBILITY.

magnetic tape A plastic tape coated with a ferromagnetic material – iron oxide powder, chromium dioxide, or, for the best results, particles of pure iron. The tape is used for recording data in tape recorders and computers. To record, the tape is passed over a recording head containing a gap in a magnetic circuit whose magnetization is modulated by the information to be recorded; the information is imprinted on the tape in the form of the direction of magnetization of the individual particles of iron oxide. The particles themselves are not rotated by the magnetizing field: it is their directions of magnetization that are orientated in accordance with the information. In audio-frequency recorders a high-frequency bias (in the range 75–100 kHz) is used to reduce distortion by facilitating the re-orientation. The playback procedure is the reverse of recording; the tape containing its orientation of tiny magnets is fed over the gap of the same (now the playback) head, in whose coil corresponding e.m.f.s are generated by induction.

magnetic variation (secular magnetic variation) *See* GEOMAGNETISM.

magnetism A group of phenomena associated with *magnetic fields. Whenever an electric current flows a magnetic field is produced; as the orbital motion and the *spin of atomic electrons are equivalent to tiny current loops, individual atoms create magnetic fields around them, when their orbital electrons have a net *magnetic moment as a result of their angular momentum. The magnetic moment of an atom is the vector sum of the magnetic moments of the orbital motions and the spins of all the electrons in the atom. The macroscopic magnetic properties of a substance arise from the magnetic moments of its component atoms and molecules. Different materials have different characteristics in an applied magnetic field; there are four main types of magnetic behaviour:

(a) In **diamagnetism** the magnetization is in the opposite direction to that of the applied field, i.e. the *susceptibility is negative. Although all substances are diamagnetic, it is a weak form of magnetism and may be masked by other, stronger, forms. It results from changes induced in the orbits of electrons in the atoms of a substance by the applied field, the direction of the change (in accordance with *Lenz's law) opposing the applied flux. There is thus a weak negative susceptibility (of the order of -10^{-8} m^3 mol^{-1}) and a relative permeability of slightly less than one.

(b) In **paramagnetism** the atoms or molecules of the substance have net orbital or spin magnetic moments that are capable of being aligned in the direction of the applied field. They therefore have a positive (but small) susceptibility and a relative permeability slightly in excess of one. Paramagnetism occurs in all atoms and molecules with unpaired electrons; e.g. free atoms, free radicals, and compounds of transition metals containing ions with unfilled electron shells. It also occurs in metals as a result of the magnetic moments associated with the spins of the conducting electrons.

(c) In **ferromagnetic** substances, within a certain temperature range, there are net atomic magnetic moments, which line up in such a way that magnetization persists after the removal of the applied field. Below a certain temperature, called the *Curie point (or Curie temperature) an increasing magnetic field applied to a ferromagnetic substance will cause increasing magnetization to a high value, called the **saturation magnetization**. This is because a ferromagnetic substance consists of small (1–0.1 mm across) magnetized regions called **domains**. The total magnetic moment of a sample of the substance is the vector sum of the magnetic moments of the component domains. Within each domain the individual atomic magnetic moments are spontaneously aligned by **exchange forces**, related to whether or not the atomic electron spins are parallel or antiparallel. However, in an unmagnetized piece of ferromagnetic material the magnetic moments of the domains themselves are not aligned; when an external field is applied those domains that are aligned with the field increase in size at the expense of the others. In a very strong field all the domains are lined up in the direction of the field and provide the high observed magnetization. Iron, nickel, cobalt, and their alloys are ferromagnetic. Above the Curie point, ferromagnetic materials become paramagnetic.

(d) Some metals, alloys, and transition-element salts exhibit another form of magnetism called **antiferromagnetism**. This occurs below a certain temperature, called the *Néel temperature, when an ordered array of atomic magnetic moments spontaneously forms in which alternate moments have opposite directions. There is therefore no net resultant magnetic moment in the absence of an applied field. In manganese fluoride, for example, this antiparallel arrangement occurs below a Néel temperature of 72 K. Below this temperature the spontaneous ordering opposes the normal tendency of the magnetic moments to align with the applied field. Above the Néel temperature the substance is paramagnetic.

A special form of antiferromagnetism is **ferrimagnetism**, a type of magnetism exhibited by the *ferrites. In these materials the magnetic moments of adjacent ions are antiparallel and of unequal strength, or the number of magnetic moments in one direction is greater than those in the opposite direction. By suitable choice of rare-earth ions in the ferrite lattices it is possible to design ferrimagnetic substances with specific magnetizations for use in electronic components. *See also* GEOMAGNETISM.

magneto An alternating-current generator used as a high-tension source in the ignition systems of petrol engines in which there are no batteries, e.g. in some tractor, marine, and aviation engines. Most modern magnetos consist of a permanent-magnet rotor revolving within a primary (low-voltage) winding around which a secondary winding is placed in which to induce the high voltage needed to produce the spark across the points of the plugs. Magnetos are geared to the engine shaft, the speed depending on the number of poles of the magneto and the number of engine cylinders. A make-and-break device is incorporated in the primary winding; when the primary current stops the change of flux within the secondary induces in it a large e.m.f.

magnetobremsstrahlung *See* SYNCHROTRON RADIATION.

magnetocaloric effect A reversible change of temperature resulting from a change in the magnetization of a ferromagnetic or paramagnetic substance (*see* MAGNETISM). The change in temperature ΔT, accompanying an adiabatic change of magnetic field ΔH, is:

$\Delta T / \Delta H = -T / C_H (\partial M / \partial T)_H$

C_H is the specific heat capacity per unit volume at constant H and M is the magnetization.

magnetochemistry The branch of physical chemistry concerned with measuring and investigating the magnetic properties of compounds. It is used particularly for studying transition-metal complexes, many of which are paramagnetic because they have unpaired electrons. Measurement of the magnetic susceptibility allows the magnetic moment of the metal atom to be calculated, and this gives information about the bonding in the complex.

magnetohydrodynamics (MHD) The study of the interactions between a conducting fluid and a *magnetic field. MHD is important in the study of controlled thermonuclear reactions in which the conducting fluid is a *plasma confined by a magnetic field. Other important applications include the **magnetohydrodynamic power generator** (see illustration). In the open-cycle MHD generator a fossil fuel, burnt in oxygen or preheated compressed air, is seeded with an element of low *ionization potential (such as potassium or caesium). This element is thermally ionized at the combustion temperature (usually over 2500 K) producing sufficient free electrons (e.g. $K \rightarrow K^+ + e$) to provide adequate electrical conductivity. The interaction between the moving conducting fluid and the strong applied magnetic field across it generates an e.m.f. on the Faraday principle, except that the solid conductor of the conventional generator is replaced by a fluid conductor. The power output per unit fluid volume (W) is given by $W = k\sigma v^2 B^2$, where σ is the conductivity of the fluid, v is its velocity, B is the magnetic flux density, and K is a constant.

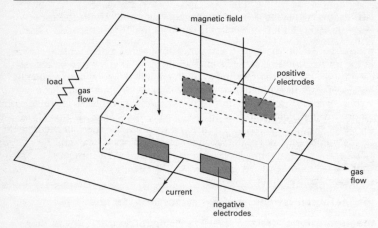

magnetic field

load

gas flow

positive electrodes

gas flow

current

negative electrodes

Magnetohydrodynamic generator

Devices of this kind are in use in some power stations, where they are suitable for helping to meet high short-term demands and have the ability of increasing the thermal efficiency of a steam-turbine generator from about 40% to 50%. In experimental closed-cycle systems the fluid is continuously recirculated through a compressor; the fluid consists of a heated and seeded noble gas or a liquid metal.

magnetomechanical ratio *See* GYROMAGNETIC RATIO.

magnetometer An instrument for measuring the magnitude, and sometimes the direction, of a magnetic field. **Absolute magnetometers** measure the field without reference to a standard magnetic instrument. The most widely used are the **vibration magnetometer**, the **deflection galvanometer**, and the more modern **nuclear magnetometer**. The vibration instrument was devised by Gauss in 1832 and depends on the rate of oscillation of a small bar magnet suspended in a horizontal plane. The same magnet is then used as a fixed deflector to deflect a second similarly suspended magnet. The deflection galvanometer uses a Helmholtz coil system of known dimensions with a small magnet suspended at its centre. The deflected magnet comes to rest at a position controlled by the earth's magnetic field, the coil's magnetic field, and the angle through which the coil must be turned to keep the magnet and the coil in alignment. The sensitive nuclear magnetometers are based on measuring the audiofrequency voltage induced in a coil by the precessing protons in a sample of water. Various **relative magnetometers** are also in use, especially for measuring the earth's magnetic field and in calibrating other equipment.

magnetomotive force (m.m.f.) The analogue of *electromotive force in a *magnetic circuit. Mathematically, it is the circular integral of

$H\cos\theta$ ds, where $H\cos\theta$ is the component of the magnetic field strength in the direction of a path of length ds. The m.m.f. is measured in *SI units in ampere-turns. It was formerly called the **magnetic potential**.

magneton A unit for measuring *magnetic moments of nuclear, atomic, or molecular magnets. The **Bohr magneton** μ_B has the value of the classical magnetic moment of an electron, given by

$\mu_B = eh/4\pi m_e = 9.274 \times 10^{-24}$ A m^2,

where e and m_e are the charge and mass of the electron and h is the Planck constant. The **nuclear magneton** μ_N is obtained by replacing the mass of the electron by the mass of the proton and is therefore given by

$\mu_N = \mu_B m_e/m_p = 5.05 \times 10^{-27}$ A m^2.

magneto-optical effects Effects resulting from the influence of a *magnetic field upon matter that is in the process of emitting or absorbing light. Examples are the *Faraday effect and the *Zeeman effect.

magneto-optical trap *See* LASER COOLING.

magnetoresistance An increase in the resistance of a metal due to the presence of a magnetic field, which alters the paths of the electrons. At normal temperatures the change in resistance resulting from the magnetic field is small but at very low temperatures the increase is considerable. The theory of magnetoresistance is too complicated to be explained quantitatively by the simple model of electrical *conductivity in metals, which assumes that it results from the movement of free electrons. To obtain a quantitative explanation, it is necessary to take into account the *energy-band structure of metals.

magnetosphere A comet-shaped region surrounding the earth and other magnetic planets in which the charged particles of the *solar wind

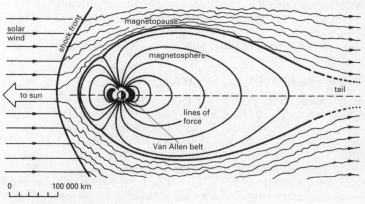

Magnetosphere and magnetopause

are controlled by the planet's magnetic field rather than the sun's magnetic field. It extends for some 60 000 km on the side facing the sun but on the opposite side it extends to a much greater extent. The boundary of the magnetosphere is known as the **magnetopause** (see illustration). The magnetosphere of the earth includes the *Van Allen belts. The detailed arrangements of magnetospheres are different in each magnetic planet.

magnetostriction The change in length of a ferromagnetic material (*see* MAGNETISM) when it is magnetized. It results from changes in the boundaries of the domains. A ferromagnetic rod exposed to an alternating field will vibrate along its length. This appears to be a major source of transformer hum, which can be removed by using a magnetic steel containing 6.5% silicon. Magnetostriction of a nickel transducer is used to generate and receive ultrasonic waves.

magnetron A microwave generator in which electrons, generated by a heated cathode, move under the combined force of an electric field and a magnetic field. The cathode consists of a central hollow cylinder, the outer surface of which carries the barium and strontium oxide electron emitters. The anode is also a cylinder, arranged concentrically around the cathode, and it contains a series of quarter-wavelength *resonant cavities arranged around its inner surface. The electric field is applied radially between anode and cathode, the magnetic field is coaxial with the cathode. The whole device is maintained in a vacuum enclosure. The magnetron is extensively used as a generator for radar installations and can produce microsecond pulses of up to 10 MW.

magnification A measure of the extent to which an optical system enlarges or reduces an image. The **linear magnification**, m, is the ratio of the image height to the object height. If this ratio is greater than one the system is enlarging, if it is less than one, it is reducing. The **angular magnification**, M or γ, is the ratio of the angles formed by the final image and the object (when viewed directly, in the most favourable position available) at the eye. This is also sometimes called the **magnifying power** of an optical system.

magnifying power *See* MAGNIFICATION.

magnitude A measure of the relative brightness of a star or other celestial object. The **apparent magnitude** depends on the star's *luminosity, its distance, and the absorption of light between the object and the earth. In 1856 the astronomer N. R. Pogson devised a scale in which a difference of five magnitudes corresponds to a brightness ratio of 100 to 1. Two stars that differ by one magnitude therefore have a brightness ratio of $(100)^{0.2}:1 = 2.512$, known as the **Pogson ratio**. This scale is now universally adopted. Apparent magnitudes are not a measure of luminosity, which is defined in terms of the **absolute magnitude**. This is the apparent magnitude of a body if it was situated at a standard distance of 10 parsecs.

magnon *See* SPIN WAVE.

Magnox A group of magnesium alloys used to enclose uranium fuel elements in *nuclear reactors. They usually contain some aluminium as well as other elements, such as beryllium.

mainframe computer *See* COMPUTER.

main-sequence stars *See* HERTZSPRUNG–RUSSELL DIAGRAM.

Main-Smith–Stoner rule A rule that if a subshell in an atom has an orbital quantum number l then that subshell is full if it contains $2(2l + 1)$ electrons. This rule, which was formulated in 1924 by J. D. Main-Smith and independently by Edmund Stoner, can be derived from the *Pauli exclusion principle.

majority carrier *See* SEMICONDUCTOR.

Maksutov telescope *See* TELESCOPE.

malleability The ability of metals to be beaten into sheets and subsequently rolled and shaped. The malleability of metals is one of their characteristic properties.

Malus's law A law concerning the polarization of light stating that the intensity of light passing through an analyser and polarizer is proportional to $\cos^2\theta$, where θ is the angle through which the analyser has been rotated with respect to the polarizer. The law was discovered by the French scientist Étienne Louis Malus (1775–1812) in 1809.

Mandelbrot set A *fractal that produces complex self-similar patterns. In mathematical terms, it is the set of values of c that make the series $z_n + 1 = (z_n)^2 + c$ converge, where c and z are complex numbers and z begins at the origin (0,0). It was discovered by and named after the Polish-born French mathematician Benoit Mandelbrot (1924–).

manganin A copper alloy containing 13–18% of manganese and 1–4% of nickel. It has a high electrical resistance, which is relatively insensitive to temperature changes. It is therefore suitable for use as a resistance wire.

manometer A device for measuring pressure differences, usually by the difference in height of two liquid columns. The simplest type is the U-tube manometer, which consists of a glass tube bent into the shape of a U. If a pressure to be measured is fed to one side of the U-tube and the other is open to the atmosphere, the difference in level of the liquid in the two limbs gives a measure of the unknown pressure.

mantissa *See* LOGARITHM.

many-body problem The problem that it is very difficult to obtain exact solutions to systems involving interactions between more than two bodies – using either classical mechanics or quantum mechanics. To understand the physics of many-body systems it is necessary to make use of *approximation techniques or *model systems that capture the essential physics of the problem. For some problems, such as the **three-body problem** in classical mechanics, it is possible to obtain qualitative information about the system. Useful concepts in the quantum theory of many-body systems are *quasiparticles and *collective excitations. If there are a great many bodies interacting, such as the molecules in a gas, the problem can be analysed using the techniques of *statistical mechanics.

many-worlds interpretation An interpretation of quantum mechanics in which it is supposed that any 'measurement' leads to all the possible outcomes of the measurement coexisting in different versions of the universe. One common way of thinking about the many-worlds view is to assume that every quantum event results in a split. For example, the spin of a particle may have two states, say 'up' and 'down'. In the standard *Copenhagen interpretation of quantum mechanics the state is indeterminate until a measurement is made, at which point the particle becomes either 'up' or 'down'. In the many-worlds interpretation the measurement creates two universes – one in which the spin is 'up' and the other in which the spin is 'down'. The idea is that the universe is constantly multiplying into an enormous number of **parallel worlds**, which all exist and together form a **multiverse**. The theory was first put forward in 1957 by the US physicist Hugh Everett III (1930–82).

Marconi, Guglielmo (1874–1937) Italian electrical engineer, who in 1894 began experimenting with Hertzian waves (*see* HERTZ, HEINRICH), making the first *radio transmissions. Moving to London in 1896, for the next few years he worked on improving the range and reliability of his equipment. This enabled him in late 1901 to transmit Morse signals across the Atlantic Ocean, establishing radio telegraphy and more importantly the use of radio waves as a communications medium. In 1909 he and Karl *Braun were awarded the Nobel Prize for physics.

Markoffian process (Markov process) A random process (*see* STOCHASTIC PROCESS) in which the rate of change of a time-dependent quantity $\partial a(t)/\partial t$ depends on the instantaneous value of the quantity $a(t)$, where t is the time, but not on its previous history. If a random process can be assumed to be a Markov process, an analysis of the process is greatly simplified enabling many problems in *nonequilibrium statistical mechanics and *disordered solids to be solved. Problems involving Markov processes are solved using statistical methods and the theory of probability. Markov processes are named after the Russian mathematician Andrei Andreevich Markov (1856–1922).

martensite A solid solution of carbon in alpha-iron formed when *steel is cooled too rapidly for pearlite to form from austenite. It is responsible for the hardness of quenched steel.

mascon A gravitational anomaly on the surface of the moon resulting from a concentration of mass below the lunar surface. They occur in circular lunar maria and were caused either by the mare basalt as it flooded the basins or by uplift of high-density mantle material when the basins were formed.

maser (*m*icrowave *a*mplification by *s*timulated *e*mission of *r*adiation) A device for amplifying or generating *microwaves by means of stimulated emission (*see* LASER). As oscillators, masers are used in *atomic clocks, while they are used as amplifiers in *radio astronomy, being especially suitable for amplifying feeble signals from space.

In the **ammonia gas maser** (devised in 1954) a molecular beam of ammonia passes through a small orifice into a vacuum chamber, where it is subjected to a nonuniform electric field. This field deflects ground-state ammonia molecules, shaped like a pyramid with the three hydrogen atoms forming the plane of the base and the single nitrogen atom forming the apex. The ground-state molecule has a dipole moment on account of its lack of symmetry and it is for this reason that it suffers deflection. Excited molecules, in which the nitrogen atom vibrates back and forth through the plane of the hydrogen atoms, have no resultant dipole moment and are not deflected. The beam, now consisting predominately of excited molecules, is passed to a resonant cavity fed with the microwave radiation corresponding to the energy difference between the excited and the ground states. This causes stimulated emission as the excited molecules fall to the ground state and the input microwave radiation is amplified coherently. This arrangement can also be made to oscillate and in this form is the basis of the *ammonia clock.

In the more versatile **solid-state maser** a magnetic field is applied to the electrons of paramagnetic (*see* MAGNETISM) atoms or molecules. The energy of these electrons is quantized into two levels, depending on whether or not their spins are parallel to the magnetic field. The situation in which there are more parallel magnetic moments than antiparallel can be reversed by sudden changes in the magnetic field. This electron-spin resonance in paramagnetic materials allows amplification over broader bandwidths than gas masers.

mass A measure of a body's *inertia, i.e. its resistance to acceleration. According to Newton's laws of motion, if two unequal masses, m_1 and m_2, are allowed to collide, in the absence of any other forces both will experience the same force of collision. If the two bodies acquire accelerations a_1 and a_2 as a result of the collision, then $m_1a_1 = m_2a_2$. This equation enables two masses to be compared. If one of the masses is regarded as a standard of mass, the mass of all other masses can be measured in terms of this standard. The body used for this purpose is a 1-kg cylinder of platinum–iridium alloy, called the international standard of mass. Mass defined in this way is called the **inertial mass** of the body.

Mass can also be defined in terms of the gravitational force it produces. Thus, according to Newton's law of gravitation, $m_g = Fd^2/MG$, where M is the mass of a standard body situated a distance d from the body of mass m_g; F is the gravitational force between them and G is the *gravitational constant. The mass defined in this way is the **gravitational mass**. In the 19th century Roland Eötvös (1848–1919) showed experimentally that gravitational and inertial mass are indistinguishable, i.e. $m_i = m_g$. Experiments performed in the 20th century have confirmed this conclusion to even greater accuracy.

Although mass is formally defined in terms of its inertia, it is usually measured by gravitation. The **weight** (W) of a body is the force by which a body is gravitationally attracted to the earth corrected for the effect of rotation and equals the product of the mass of the body and the *acceleration of free fall (g), i.e. $W = mg$. In the general language, weight and mass are often used synonymously; however, for scientific purposes they are different. Mass is measured in kilograms; weight, being a force, is measured in newtons. Weight, moreover, depends on where it is measured, because the value of g varies at different localities on the earth's surface. Mass, on the other hand, is constant wherever it is measured, subject to the special theory of *relativity. According to this theory, announced by Albert Einstein in 1905, the mass of a body is a measure of its total energy content. Thus, if the energy of a body increases, for example by an increase in kinetic energy or temperature, then its mass will increase. According to this law an increase in energy ΔE is accompanied by an increase in mass Δm, according to the **mass–energy equation** $\Delta m = \Delta E/c^2$, where c is the speed of light. Thus, if 1 kg of water is raised in temperature by 100 K, its internal energy will increase by 4×10^{-12} kg. This is, of course, a negligible increase and the mass–energy equation is only significant for extremely high energies. For example, the mass of an electron is increased sevenfold if it moves relative to the observer at 99% of the speed of light.

mass action The law of mass action states that the rate at which a chemical reaction takes place at a given temperature is proportional to the product of the **active masses** of the reactants. The active mass of a reactant is taken to be its molar concentration. For example, for a reaction

$$xA + yB \rightarrow \text{products}$$

the rate is given by

$$R = k[A]^x[B]^y$$

where k is the rate constant. The principle was introduced by C. M. Guldberg and P. Waage in 1863. It is strictly correct only for ideal gases. In real cases activities can be used.

mass decrement *See* MASS DEFECT.

mass defect 1. The difference between the rest mass of an atomic nucleus and the sum of the rest masses of its individual nucleons in the unbound state. It is thus the mass equivalent of the *binding energy on the basis of the mass–energy equation (*see* MASS; RELATIVITY). **2. (mass decrement)** The difference between the rest mass of a radioactive nucleus before decay and the total rest mass of the decay products.

mass–energy equation *See* MASS; RELATIVITY.

mass number *See* NUCLEON NUMBER.

mass spectrum *See* SPECTRUM.

matrix (*pl.* **matrices**) **1.** (in mathematics) A set of quantities in a rectangular array, used in certain mathematical operations. The array is usually enclosed in large parentheses or in square brackets. **2.** (in geology) The fine-grained material of rock in which the coarser-grained material is embedded.

matrix mechanics A formulation of *quantum mechanics using matrices (*see* MATRIX) to represent states and operators. Matrix mechanics was the first formulation of quantum mechanics to be stated (by Werner *Heisenberg in 1925) and was developed by Heisenberg and Max *Born and the German physicist Pascual Jordan (1902–80). It was shown by Erwin *Schrödinger in 1926 to be equivalent to the *wave mechanics formulation of quantum mechanics.

maximum and minimum thermometer A thermometer designed to record both the maximum and minimum temperatures that have occurred over a given time period. It usually consists of a graduated capillary tube at the base of which is a bulb containing ethanol. The capillary contains a thin thread of mercury with a steel index at each end. As the temperature rises the index is pushed up the tube, where it remains in position to show the maximum temperature reached; as the temperature falls the lower index is pushed down the tube and similarly remains in position at the lowest temperature. The indexes are reset by means of a permanent magnet.

maximum permissible dose *See* DOSE.

maxwell A c.g.s. unit of magnetic flux, equal to the flux through 1 square centimetre perpendicular to a magnetic field of 1 gauss. 1 maxwell is equal to 10^{-8} weber. It is named after James Clerk Maxwell.

Maxwell, James Clerk (1831–79) British physicist, born in Edinburgh, who held academic posts at Aberdeen, London, and Cambridge. In the 1860s he was one of the founders of the *kinetic theory of gases, but his best-known work was a mathematical analysis of electricity, magnetism, and *electromagnetic radiation, published in 1865.

Maxwell–Boltzmann distribution A law describing the distribution of speeds among the molecules of a gas. In a system consisting of N molecules that are independent of each other except that they exchange energy on collision, it is clearly impossible to say what velocity any particular molecule will have. However, statistical statements regarding certain functions of the molecules were worked out by James Clerk Maxwell and Ludwig *Boltzmann. One form of their law states that $n = N\exp(-E/RT)$, where n is the number of molecules with energy in excess of E, T is the thermodynamic temperature, and R is the *gas constant.

Maxwell relations A set of relations in thermodynamics:

$(\delta T/\delta V)_S = (\delta P/\delta S)_V$,

$(\delta T/\delta P)_S = (\delta V/\delta S)_P$,

$(\delta P/\delta T)_V = (\delta S/\delta V)_T$,

$(\delta V/\delta T)_P = -(\delta S/\delta P)_T$,

where T is the absolute temperature, V is the volume, P is the pressure, and S is the entropy. The Maxwell relations are useful for converting quantities that cannot readily be measured into quantities that can.

Maxwell's demon An imaginary creature that is able to open and shut a partition dividing two volumes of a gas in a container, when the two volumes are initially at the same temperature. The partition operated by the demon is only opened to allow fast molecules through. Such a process would make the volume of gas containing the fast molecules hotter than it was at the start; the volume of gas remaining would accordingly become cooler. This process would be a violation of the second law of *thermodynamics and therefore cannot occur. Maxwell's demon was invented by James Clerk Maxwell in a letter written in 1867 to show that the second law of thermodynamics has its origins in *statistical mechanics, although the name was suggested by Sir William Thomson, subsequently Lord *Kelvin. The problem of Maxwell's demon was solved in the 1970s in terms of the relation between information theory and thermodynamics.

Maxwell's equations A set of differential equations describing the space and time dependence of the electromagnetic field and forming the basis of classical electrodynamics. They were proposed in 1864 by James Clerk Maxwell. In *SI units the equations are:

(1) $\text{div}\mathbf{D} = \rho$

(2) $\text{curl}\mathbf{E} = -\partial\mathbf{B}/\partial t$

(3) $\text{div}\mathbf{B} = 0$

(4) $\text{curl}\mathbf{H} = \partial\mathbf{D}/\partial t + \mathbf{J}$

where $\mathbf{D}$ is the electric displacement, $\mathbf{E}$ is the electric field strength, $\mathbf{B}$ is the magnetic flux density, $\mathbf{H}$ is the magnetic field strength, ρ is the volume charge density, and $\mathbf{J}$ is the electric current density. Note that in relativity and particle physics it is common to use *Gaussian units or *Heaviside–Lorentz units, in which case Maxwell's equations include 4π and the speed of light c. Maxwell's equations have the following interpretation. Equation (1) represents *Coulomb's law; equation (2) represents *Faraday's laws of electromagnetic induction; equation (3) represents the absence of *magnetic monopoles; equation (4) represents a generalization of *Ampère's law.

McLeod gauge A vacuum pressure gauge in which a relatively large volume of a low-pressure gas is compressed to a small volume in a glass apparatus (see illustration). The volume is reduced to an extent that causes the pressure to rise sufficiently to support a column of fluid high enough to read. This simple device, which relies on *Boyle's law, is suitable for measuring pressures in the range 10^3 to 10^{-3} pascal. It is named after H. McLeod (1841–1923).

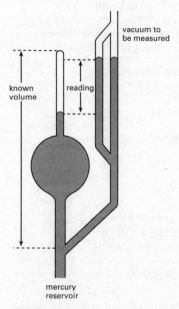

vacuum to be measured

known volume

reading

mercury reservoir

McLeod gauge

mean *See* AVERAGE.

mean free path The average distance travelled between collisions by the molecules in a gas, the electrons in a metallic crystal, the neutrons in a moderator, etc. According to the *kinetic theory the mean free path between elastic collisions of gas molecules of diameter d (assuming they are rigid spheres) is $1/\sqrt{2}n\pi d^2$, where n is the number of molecules per unit volume in the gas. As n is proportional to the pressure of the gas, the mean free path is inversely proportional to the pressure.

mean free time The average time that elapses between the collisions of the molecules in a gas, the electrons in a crystal, the neutrons in a moderator, etc. *See* MEAN FREE PATH.

mean life *See* DECAY.

mean solar day *See* DAY.

measurement error The difference between the true value of a quantity and the value obtained by measurement. The two main types of error are **random errors** and **systematic errors**.

In any measurement there is a random error, which arises when the experimenter has to make an estimate in the last figure of the measurement, i.e. when the instrument is at the limit of its accuracy. Random errors of this type can be analysed, using statistics; for this reason they are sometimes called **statistical errors**. Thus, for this type of error repeating the measurement can improve the accuracy of the quantity being measured.

A systematic error cannot be analysed using statistics, however, as it results from an incorrect calibration of the measuring instrument or an incorrect position of the zero point. Repeating the experiment in these cirumstances, will not improve the accuracy of the measurement.

In *quantum mechanics, the product of the uncertainties of certain quantities, such as the position and momentum of a particle, cannot be eliminated or reduced beyond the limit given by the *Heisenberg uncertainty principle.

mebi- *See* BINARY PREFIXES.

mechanical advantage *See* FORCE RATIO.

mechanical equivalent of heat Symbol J. The ratio of a unit of mechanical energy to the equivalent unit of thermal energy, when a system of units is used in which they differ. J has the value 4.1868×10^7 ergs per calorie. The concept loses its usefulness in *SI units in which all forms of energy are expressed in joules and J therefore has a value of 1.

mechanics The study of the interactions between matter and the forces acting on it. *Statics is broadly concerned with the action of forces when no change of momentum is concerned, while *dynamics deals with cases in which there is a change of momentum. *Kinematics is the study of the motion of bodies without reference to the forces affecting the motion. These classical sciences (*see* NEWTONIAN MECHANICS; RELATIVISTIC MECHANICS) are concerned with macroscopic bodies in the solid state, while *fluid mechanics is the science of the interaction between forces and fluids.

median 1. The middle number or value in a series of numbers or values. *See also* PERCENTILE. **2.** A straight line in a triangle that joins the vertex to the mid-point of the base.

medium frequency (MF) A radio frequency in the range 0.3–3 megahertz; i.e. having a wavelength in the range 100–1000 metres.

mega- Symbol M. A prefix used in the metric system to denote one million times. For example, 10^6 volts = 1 megavolt (MV).

megaton weapon A nuclear weapon with an explosive power equivalent to one million tons of TNT. *Compare* KILOTON WEAPON.

Meissner effect The falling off of the magnetic flux within a superconducting metal when it is cooled to a temperature below the critical temperature in a magnetic field. It was discovered by Walther Meissner (1882–1974) in 1933 when he observed that the earth's magnetic field was expelled from the interior of tin crystals below 3.72 K, indicating that as *superconductivity appeared the material became perfectly diamagnetic. *See* MAGNETISM.

meitnerium Symbol Mt. A radioactive *transactinide element; a.n. 109. It was first made in 1982 by Peter Armbruster and a team in Darmstadt, Germany, by bombarding bismuth-209 nuclei with iron-58 nuclei. Only a few atoms have ever been detected. It is named after Lise Meitner.

Meitner, Lise (1878–1968) Austrian-born Swedish physicist, who went to Berlin to study *radioactivity with Otto *Hahn. In 1917 they discovered *protactinium. After World War I Meitner and Hahn returned to Berlin, where in 1935 they bombarded uranium with neutrons. In 1938 she left Germany, with other Jewish scientists, and went to the Nobel Institute in Stockholm. In 1939 she and Otto Frisch (1904–79) explained Hahn's results in terms of *nuclear fission.

melting point (m.p.) The temperature at which a solid changes into a liquid. A pure substance under standard conditions of pressure (usually 1 atmosphere) has a single reproducible melting point. If heat is gradually and uniformly supplied to a solid the consequent rise in temperature stops at the melting point until the fusion process is complete.

membrane **1.** A two-dimensional structure at the boundary between two distinct three-dimensional phases that is physically or chemically distinct from the three-dimensional phases. A membrane resists the movement of molecules between the two phases to a greater or lesser extent. **2.** A two-dimensional structure postulated as an alternative to *unified field theory and *string theory, as a *theory of everything. Membrane theory suffers from the difficulty that the procedure of *renormalization does not appear to work for it. Membranes, and other extended objects in higher dimensions, occur as solutions of string theory. *See also* SUPERSYMMETRY.

memory (in computing) The part of a computer in which data is stored while it is being worked on. A typical microcomputer, for example, has a comparatively small amount of read-only memory (*see* ROM) and a large amount of random-access memory (*see* RAM). Only data in ROM is preserved when the machine is switched off; any data in RAM must be saved to disk if it is wanted again.

Mendeleev, Dmitri Ivanovich (1834–1907) Russian chemist, who became professor of chemistry at St Petersburg in 1866. His most famous work, published in 1869, was the compilation of the periodic table of the elements, based on the *periodic law.

mendelevium Symbol Md. A radioactive metallic transuranic element belonging to the actinoids; a.n. 101; mass number of the only known nuclide 256 (half-life 1.3 hours). It was first identified by A. Ghiorso, G. T. Seaborg (1912–99), and associates in 1955. The alternative name **unnilunium** has been proposed. It is named after D. I. Mendeleev.

meniscus 1. A concave or convex upper surface that forms on a liquid in a tube as a result of *surface tension. **2.** *See* CONCAVE.

mercury cell A primary *voltaic cell consisting of a zinc anode and a cathode of mercury(II) oxide (HgO) mixed with graphite. The electrolyte is potassium hydroxide (KOH) saturated with zinc oxide, the overall reaction being:

$$Zn + HgO \rightarrow ZnO + Hg$$

The e.m.f. is 1.35 volts and the cell will deliver about 0.3 ampere-hour per cm^3.

mercury-vapour lamp A type of discharge tube in which a glow discharge takes place in mercury vapour. The discharge takes place in a transparent tube of fused silica or quartz into the ends of which molybdenum and tungsten electrodes are sealed; this tube contains argon and a small amount of pure mercury. A small arc is struck between a starter electrode and one of the main electrodes causing local ionization of some argon atoms. The ionized atoms diffuse through the tube causing the main discharge to strike; the heat from this vaporizes the mercury droplets, which become ionized current carriers. Radiation is confined to four visible wavelengths in the visible spectrum and several strong ultraviolet lines. The light is bluish but can be changed by the use of phosphors on an outer tube. The outer tube is also usually used to filter out excessive ultraviolet radiation. The lamp is widely used for street lighting on account of its low cost and great reliability and as a source of ultraviolet radiation.

meridian 1. *See* LATITUDE AND LONGITUDE. **2. (magnetic meridian)** An imaginary great circle on the earth's surface that passes through the north and south magnetic poles. A compass needle on the earth's surface influenced only by the earth's magnetic field (*see* GEOMAGNETISM) comes to rest along a magnetic meridian. **3. (celestial meridian)** A great circle of the *celestial sphere that passes through the zenith and the celestial poles. It meets the horizon at the north and south points.

meso form *See* OPTICAL ACTIVITY.

meson Any of a class of *elementary particles that are a subclass of the *hadrons. According to current quark theory mesons consist of quark–antiquark pairs. They exist with positive, negative, and zero charges, but when charged the charge has the same magnitude as that of the electron. They include the **kaon**, **pion**, and **psi** particles. Mesons are believed to participate in the forces that hold nucleons together in the nucleus, as described by *quantum chromodynamics. The muon, originally called a mu-meson, was thought to be a meson but is now recognized as a *lepton.

meson-catalysed fusion *See* NUCLEAR FUSION.

mesoscopic Designating a size scale intermediate between those of the *microscopic and the *macroscopic states. Mesoscopic objects and systems require *quantum mechanics to describe them. Many devices in *electronics are mesoscopic.

metal fatigue A cumulative effect causing a metal to fail after repeated applications of *stress, none of which exceeds the ultimate *tensile strength. The **fatigue strength** (or **fatigue limit**) is the stress that will cause failure after a specified number (usually 10^7) of cycles. The number of cycles required to produce failure decreases as the level of stress or strain increases. Other factors, such as corrosion, also reduce the fatigue life.

metal–insulator transition A transition between a metal and an insulator. There are several ways in which a metal–insulator transition (or the reverse transition) can occur, including changing the distance between the atoms or the amount of disorder in the system. *See also* LOCALIZATION; MOTT TRANSITION.

metallography The microscopic study of the structure of metals and their alloys. Both optical *microscopes and *electron microscopes are used in this work.

metallurgy The branch of engineering concerned with the production of metals from their ores, the purification of metals, the manufacture of alloys, and the use and performance of metals in engineering practice. **Process metallurgy** is concerned with the extraction and production of metals, while **physical metallurgy** concerns the mechanical behaviour of metals.

metamagnet A material that is an antiferromagnet in the absence of an external magnetic field but undergoes a first-order transition to a phase in which there is a nonzero ferromagnetic moment when the external magnetic field becomes sufficiently large. Iron(II) chloride is an example of a material that can act as a metamagnet.

metamict state The amorphous state of a substance that has lost its crystalline structure as a result of the radioactivity of uranium or thorium. **Metamict minerals** are minerals whose structure has been disrupted by this process. The metamictization is caused by alpha-particles and the recoil nuclei from radioactive disintegration.

metastable state A condition of a system in which it has a precarious stability that can easily be disturbed. It is unlike a state of stable equilibrium in that a minor disturbance will cause a system in a metastable state to fall to a lower energy level. A book lying on a table is in a state of stable equilibrium; a thin book standing on edge is in metastable equilibrium. Supercooled water is also in a metastable state. It is liquid below 0°C; a grain of dust or ice introduced into it will cause it to freeze. An excited state of an atom or nucleus that has an appreciable lifetime is also metastable.

meteor A streak of light observable in the sky when a particle of matter enters the earth's atmosphere and becomes incandescent as a result of friction with atmospheric atoms and molecules. These particles of matter are known collectively as **meteoroids**. Meteoroids that survive their passage through the atmosphere and strike the earth's surface are known as **meteorites**. Only some 2500 meteorites are known, excluding the **micrometeorites** (bodies less than 1 mm in diameter). Meteorites consist mainly of silicate materials (stony meteorites) or iron (iron meteorites). It is estimated that the earth collects over 10^8 kg of meteoritic material every year, mostly in the form of micrometeorites. Micrometeorites survive atmospheric friction because their small size enables them to radiate away the heat generated by friction before they vaporize.

meteorite *See* METEOR.

method of mixtures A method of determining the specific heat capacities of liquids or a liquid and a solid by mixing known masses of the substances at different temperatures and measuring the final temperature of the mixture.

metre Symbol m. The SI unit of length, being the length of the path travelled by light in vacuum during a time interval of $1/(2.99\ 792\ 458 \times 10^8)$ second. This definition, adopted by the General Conference on Weights and Measures in October, 1983, replaced the 1967 definition based on the krypton lamp, i.e. 1 650 763.73 wavelengths in a vacuum of the radiation corresponding to the transition between the levels $2p^{10}$ and $5d^5$ of the nuclide krypton–86. This definition (in 1958) replaced the older definition of a metre based on a platinum–iridium bar of standard length. When the *metric system was introduced in 1791 in France, the metre was intended to be one ten-millionth of the earth's meridian quadrant passing through Paris. However, the original geodetic surveys proved the impractibility of such a standard and the original platinum metre bar, the *mètre des archives*, was constructed in 1793.

metre bridge *See* WHEATSTONE BRIDGE.

metric system A decimal system of units originally devised by a committee of the French Academy, which included J. L. Lagrange and P. S. Laplace, in 1791. It was based on the *metre, the gram defined in terms of the mass of a cubic centimetre of water, and the second. This centimetre-gram-second system (*see* C.G.S. UNITS) later gave way for scientific work to the metre-kilogram-second system (*see* M.K.S. UNITS) on which *SI units are based.

metric ton (tonne) A unit of mass equal to 1000 kg or 2204.61 lb. 1 tonne = 0.9842 ton.

metrology The scientific study of measurement, especially the definition and standardization of the units of measurement used in science.

MHD *See* MAGNETOHYDRODYNAMICS.

mho A reciprocal ohm, the former name of the unit of electrical *conductance now known as the siemens.

Michelson–Morley experiment An experiment, conducted in 1887 by Albert Michelson (1852–1931) and Edward Morley (1838–1923), that attempted to measure the velocity of the earth through the *ether. Using a modified **Michelson interferometer** (see illustration) they expected to observe a shift in the interference fringes formed when the instrument was rotated through 90°, showing that the speed of light measured in the direction of the earth's rotation, or orbital motion, is not identical to its speed at right angles to this direction. No shift was observed. An explanation was subsequently provided by the *Lorentz–Fitzgerald contraction, which provided an important step in the formulation of Einstein's special theory of *relativity and the abandonment of the ether concept.

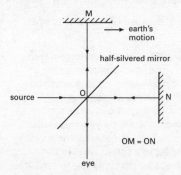

Michelson interferometer

micro- Symbol μ. A prefix used in the metric system to denote one millionth. For example, 10^{-6} metre = 1 micrometre (μm).

microbalance A sensitive *balance capable of weighing masses of the order 10^{-6} to 10^{-9} kg.

microcomputer A *computer in which the central processing unit is implemented by means of a semiconductor chip or chip set, known as a **microprocessor**. The power of a microcomputer is determined not only by the speed and power of the processor but also by features of the other components in the computer, such as the storage capacity of the main (*RAM) memory and the disks used as *backing store, as well as the operating system and other software used. Microcomputers are used in a wide variety of forms, including *personal computers, electronic point-of-sales terminals, and cash-dispensing automated teller machines.

microelectronics The techniques of designing and making electronic circuits of very small size. As a result of these techniques a single *silicon chip measuring less than a centimetre in either direction can contain many thousands of transistors and may constitute the central processing unit of a microcomputer. In addition to an enormous drop in size, compared to an equivalent valve-operated device, these microelectronic circuits are some 100 000 times more reliable than their thermionic predecessors.

microgravity The very weak gravitational force that occurs on board a spacecraft in orbit round the earth.

micrometeorite *See* METEOR.

micrometer A gauge for measuring small diameters, thicknesses, etc., accurately. It consists of a G-shaped device in which the gap between the measuring faces is adjusted by means of an accurately calibrated screw, the end of which forms one of the measuring faces.

micron The former name for the *SI unit now called the micrometre, i.e. 10^{-6} m.

microphone A *transducer in which sound waves are converted into corresponding variations in an electrical signal for amplification, transmission to a distant point, or recording. Various types of device are used. In the **dynamic microphone** the sound waves impinge on a conductor of low mass supported in a magnetic field and cause it to oscillate at the frequency of the sound waves. These movements induce an e.m.f. in the conductor that is proportional to its velocity. The moving conductor consists of a metal ribbon, a wire, or a coil of wire. In the **moving-iron microphone**, sound waves cause a light armature to oscillate so that it varies the reluctance of a magnetic circuit. In a coil surrounding this path the varying reluctance is experienced as a variation in the magnetic flux within it, which induces a corresponding e.m.f. In the **carbon microphone**, widely used in telephones, a diaphragm constitutes a movable electrode in contact with carbon granules, which are also in contact with a fixed electrode. The movement of the diaphragm, in response to the sound waves, varies the resistance of the path through the granules to the fixed electrode. *See also* CAPACITOR MICROPHONE; CRYSTAL MICROPHONE.

microprocessor *See* COMPUTER.

microscope A device for forming a magnified image of a small object. The **simple microscope** consists of a biconvex magnifying glass or an equivalent system of lenses, either hand-held or in a simple frame. The **compound microscope** (see illustration) uses two lenses or systems of lenses, the second magnifying the real image formed by the first. The lenses are usually mounted at the opposite ends of a tube that has mechanical controls to move it in relation to the object. An optical condenser and mirror, often with a separate light source, provide illumination of the object. The widely used **binocular microscope** consists of two separate instruments fastened together so that one eye looks through one while the other eye looks through the other. This gives stereoscopic vision and reduces eye strain. *See also* ATOMIC FORCE MICROSCOPE; ELECTRON MICROSCOPE; FIELD-EMISSION MICROSCOPE; FIELD-IONIZATION MICROSCOPE; PHASE-CONTRAST MICROSCOPE; SCANNING TUNNELLING MICROSCOPE; ULTRAVIOLET MICROSCOPE. See also Chronology: Microscopy.

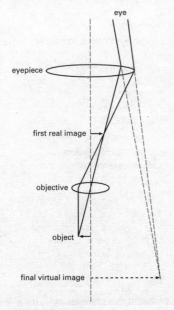

Compound microscope

microscopic Designating a size scale comparable to the subatomic

---

MICROSCOPY

c.1590   Dutch spectacle-makers Hans and Zacharias Janssen invent the compound microscope.

1610   Johannes Kepler invents the modern compound microscope.

1675   Anton van Leeuwenhoek invents the simple microscope.

1826   British biologist Dames Smith (d. 1870) constructs a microscope with much reduced chromatic and spherical aberrations.

1827   Italian scientist Giovanni Amici (1786–1863) invents the reflecting achromatic microscope.

1861   British chemist Joseph Reade (1801–70) invents the kettledrum microscope condenser.

1912   British microscopist Joseph Barnard (1870–1949) invents the ultramicroscope.

1932   Dutch physicist Frits Zernike (1888–1966) invents the phase-contrast microscope.

1936   German-born US physicist Erwin Mueller (1911–77) invents the field-emission microscope.

1938   German engineer Ernst Ruska (1906–88) develops the electron microscope.

1940   Canadian scientist James Hillier (1915–) makes a practical electron microscope.

1951   Erwin Mueller invents the field-ionization microscope.

1978   US scientists of the Hughes Research Laboratory invent the scanning ion microscope.

1981   Swiss physicists Gerd Binning (1947–) and Heinrich Rohrer (1933–) invent the scanning tunnelling microscope.

1985   Gerd Binning invents the atomic force microscope.

1987   James van House and Arthur Rich invent the positron microscope.

---

particles, atoms, and molecules. Microscopic objects and systems are described by *quantum mechanics. *Compare* MACROSCOPIC; MESOSCOPIC.

microwave background radiation A cosmic background of radiation in the frequency range 3×10^{11} hertz to 3×10^8 hertz discovered in 1965. Believed to have emanated from the primordial fireball of the big bang with which the universe is thought to have originated (*see* BIG-BANG THEORY), the radiation has an energy density in intergalactic space of some 4×10^{-14} J m^{-3}. Detailed study of this radiation has been very important in modern cosmology. *See* COBE; WMAP.

microwave optics The study of the behaviour of microwaves by analogy with the behaviour of light waves. On the large scale microwaves are propagated in straight lines and, like light waves, they undergo reflection, refraction, diffraction, and polarization.

microwaves Electromagnetic waves with wavelengths in the range 10^{-3} to 0.03 m.

Milky Way *See* GALAXY.

milli- Symbol m. A prefix used in the metric system to denote one thousandth. For example, 0.001 volt = 1 millivolt (mV).

millibar *See* BAR.

Millikan, Robert Andrews (1868–1953) US physicist, who after more than 20 years at the University of Chicago went to the California Institute of Technology in 1921. His best-known work, begun in 1909, was to determine the charge on the *electron in his oil-drop experiment, which led to the award of the 1923 Nobel Prize for physics. He then went on to determine the value of the *Planck constant and to do important work on *cosmic radiation.

Millikan's oil-drop experiment An experiment begun in 1909 by Robert Millikan to determine the charge on an electron. Between two horizontal oppositely charged plates Millikan introduced a fine spray of oil droplets. He first measured the mass of the oil drops by measuring their rate of fall under the influence of gravity and against the air resistance. Then using X-rays to ionize the air he noted the changed rate of fall of droplets that were attracted to the lower plate after these droplets had acquired a charge from the ionized air. These figures enabled Millikan to conclude that the acquired charge was always a simple multiple of a basic unit. He found this basic unit to be 4.774×10^{-10} esu, which was the value for the charge on an electron used until it was improved on in 1928.

Minkowski space *See* SPACE–TIME.

minority carrier *See* SEMICONDUCTOR.

minor planets *See* ASTEROIDS.

minute 1. One sixtieth of an hour. **2.** One sixtieth of a degree (angle).

mirage An optical phenomenon that occurs as a result of the bending of light rays through layers of air having very large temperature gradients. An **inferior mirage** occurs when the ground surface is strongly heated and the air near the ground is much warmer that the air above. Light rays from the sky are strongly refracted upwards near the surface giving the appearance of a pool of water. A **superior mirage** occurs if the air close to the ground surface is much colder than the air above. Light is bent downwards from the object towards the viewer so that it appears to be elevated or floating in the air.

mirror A surface that reflects most of the light falling on it. A **plane mirror** is a flat surface that produces an erect virtual *image of a real

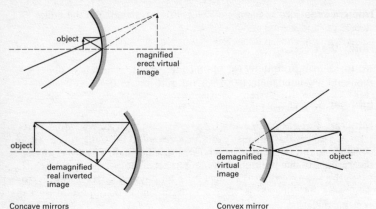

Concave mirrors   Convex mirror

Mirrors

object, in which front and back are reversed. **Spherical mirrors** are formed from the surfaces of spheres and form images of real objects in much the same way as lenses. A convex mirror forms erect virtual images. They are commonly used as rear-view mirrors in road vehicles, and give a diminished wide-angle image. A concave mirror can form either inverted real images or erect virtual images. (See illustrations.) Spherical mirrors obey the *lens equation (using the real-positive sign convention) and are subject to some *aberrations similar to those of lenses.

misch metal An alloy of cerium (50%), lanthanum (25%), neodymium (18%), praseodymium (5%), and other rare earths. It is used alloyed with iron (up to 30%) in lighter flints, and in small quantities to improve the malleability of iron. It is also added to copper alloys to make them harder, to aluminium alloys to make them stronger, to magnesium alloys to reduce creep, and to nickel alloys to reduce oxidation.

missing mass The mass of matter in the universe that cannot be observed by direct observations of its emitted or absorbed electromagnetic radiation. There are a number of astrophysical observations that suggest that the actual mass of the universe is much greater than that estimated by observations using optical telescopes, radiotelescopes, etc. It is thought that there is a considerable amount of **dark matter** (or **hidden matter**) causing this discrepancy. Various explanations have been put forward for the missing mass, including black holes, brown dwarfs, cosmic strings, axions, neutrinos, monopoles, and various exotic particles, such as **weakly interacting massive particles** (**WIMPS**). The universe contains far more missing matter than directly observable matter.

m.k.s. units A *metric system of units devised by A. Giorgi (and sometimes known as **Giorgi units**) in 1901. It is based on the metre, kilogram, and second and grew from the earlier *c.g.s. units. The electrical unit chosen to augment these three basic units was the ampere and the *permeability of space (magnetic constant) was taken as 10^{-7} H m^{-1}. To simplify electromagnetic calculations the magnetic constant was later changed to $4\pi \times 10^{-7}$ H m^{-1} to give the **rationalized MKSA system**. This system, with some modifications, formed the basis of *SI units, now used in most scientific work.

m.m.f. *See* MAGNETOMOTIVE FORCE.

mmHg A unit of pressure equal to that exerted under standard gravity by a height of one millimetre of mercury, or 133.322 pascals.

mobility Symbol μ. A quantity that gives a measure of how readily a particle moves. The term is frequently used to characterize the motion of a charged entity, such as an ion, electron, or hole, in an electric field.

mobility edge The energies in an *energy band in a disordered solid, denoted E_c, at which there is a transition between localized states (*see* LOCALIZATION), which do not contribute to the electrical conductivity of the system at *absolute zero, and extended states, which can contribute to the electrical conductivity. For a given band there are two mobility edges, with the extended states towards the middle of the band and the localized states in the 'tails' of the band. In a three-dimensional disordered solid the mobility edges move towards the centre of the band as the disorder increases, until eventually the mobility edges meet and all states become localized. A mobility edge is analogous to a transition temperature in a *phase transition.

mode The pattern of motion in a vibrating body. If the body has several component particles, such as a molecule consisting of several atoms, the modes of vibration are the different types of molecular vibrations possible. In the case of a vibrating string, the different modes of vibration of the system are characterized by the number of *nodes in the string (*see* HARMONIC).

model A simplified description of a physical system intended to capture the essential aspects of the system in a sufficiently simple form to enable the mathematics to be solved. In practice some models require *approximation techniques to be used, rather than being exactly soluble. When exact solutions are available they can be used to examine the validity of the approximation techniques.

modem (derived from *mo*dulator/*dem*odulator) A device that can convert a digital signal (consisting of a stream of *bits) into an analogue (smoothly varying) signal, and vice versa. Modems are therefore required to link digital devices, such as computers, over an analogue telephone line.

moderator A substance that slows down free neutrons in a *nuclear reactor, making them more likely to cause fissions of atoms of uranium–235 and less likely to be absorbed by atoms of uranium–238. Moderators are light elements, such as deuterium (in heavy water), graphite, and beryllium, to which neutrons can impart some of their kinetic energy on collision without being captured. Neutrons that have had their energies reduced in this way (to about 0.025 eV, equivalent to a speed of 2200 m s^{-1}) are said to have been **thermalized** or to have become **thermal neutrons**.

modulation The process of changing an electrical signal. In radio transmission, it is the process of superimposing the characteristics of a periodic signal onto a *carrier wave so that the information contained in the signal can be transmitted by the wave. The simplest form of modulation is **amplitude modulation** (**AM**), in which the amplitude of the carrier is increased or diminished as the signal amplitude increases and diminishes. The modulated wave is composed of the carrier wave plus upper and lower sidebands. In **single-sideband modulation** (**SSB**) the carrier and one of the sidebands of an amplitude-modulated waveform are suppressed. This saves on bandwidth occupancy and signal power. In **frequency modulation** (**FM**), the frequency of the carrier is increased or diminished as the signal amplitude increases and diminishes but the carrier amplitude remains constant. In **phase modulation**, the relative phase of the carrier is varied in accordance with the signal amplitude. (See illustrations.) Both frequency modulation and phase modulation are forms of **angle modulation**.

In **pulse modulation** the information is transmitted by controlling the amplitude, duration, position, or presence of a series of pulses. Morse code is a simple form of a pulse modulation.

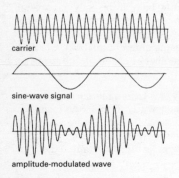

carrier

sine-wave signal

phase-modulated wave

amplitude-modulated wave

frequency-modulated wave

Modulation

modulus *See* ABSOLUTE VALUE.

modulus of elasticity *See* ELASTIC MODULUS.

Moho (Mohorovičić discontinuity) A discontinuity within the *earth that marks the junction between the crust and the underlying mantle. Below the discontinuity earthquake seismic waves undergo a sudden increase in velocity, a feature that was first observed in 1909 by the Yugoslavian geophysicist Andrija Mohorovičić (1857–1936), after whom the discontinuity was named. The Moho lies at a depth of about 10–12 km below the oceans and about 33–35 km below the continents.

Mohs' scale A hardness scale in which a series of ten minerals are arranged in order, each mineral listed being scratched by and therefore softer than those below it. The minerals are: (1) talc; (2) gypsum; (3) calcite; (4) fluorite; (5) apatite; (6) orthoclase; (7) quartz; (8) topaz; (9) corundum; (10) diamond. As a rough guide a mineral with a value up to 2.5 on this scale can be scratched by a fingernail, up to 4 can be scratched by a coin, and up to 6 by a knife. The scale was devised by Friedrich Mohs (1773–1839).

moiré pattern A pattern produced when two sets of line grids are slightly inclined to one another. This results in the appearance of dark bands. In optics the **moiré fringe** is the series of *interference fringes formed when two patterns of line grids on screens are superposed on each other and slightly inclined to one another. This type of interference pattern can occur in *television. Moiré fringes can be used to measure displacements of one grid against another either to measure the deformation of a surface or to check the accuracy of *diffraction gratings. The use of moiré fringes to measure displacements of one grid against another is called **moiré interferometry**.

molar Denoting that an extensive physical property is being expressed per *amount of substance, usually per mole. For example, the molar heat capacity of a compound is the heat capacity of that compound per unit amount of substance, i.e. it is usually expressed in $J\,K^{-1}\,mol^{-1}$.

molar conductivity Symbol Λ. The conductivity of that volume of an electrolyte that contains one mole of solution between electrodes placed one metre apart.

molar heat capacity *See* HEAT CAPACITY.

molar latent heat *See* LATENT HEAT.

molar volume (molecular volume) The volume occupied by a substance per unit amount of substance.

mole Symbol mol. The SI unit of *amount of substance. It is equal to the amount of substance that contains as many elementary units as there are atoms in 0.012 kg of carbon-12. The elementary units may be atoms, molecules, ions, radicals, electrons, etc., and must be specified. 1 mole of a compound has a mass equal to its *relative molecular mass expressed in grams.

molecular beam A beam of atoms, ions, or molecules at low pressure, in which all the particles are travelling in the same direction and there are few collisions between them. They are formed by allowing a gas or vapour to pass through an aperture into an enclosure, which acts as a collimator by containing several additional apertures and vacuum pumps to remove any particles that do not pass through the apertures. Molecular beams are used in studies of surfaces and chemical reactions and in spectroscopy.

molecular distillation Distillation in high vacuum (about 0.1 pascal) with the condensing surface so close to the surface of the evaporating liquid that the molecules of the liquid travel to the condensing surface without collisions. This technique enables very much lower temperatures to be used than are used with distillation at atmospheric pressure and therefore heat-sensitive substances can be distilled. Oxidation of the distillate is also eliminated as there is no oxygen present.

molecular flow (Knudsen flow) The flow of a gas through a pipe in which the mean free path of gas molecules is large compared to the dimensions of the pipe. This occurs at low pressures; because most collisions are with the walls of the pipe rather than other gas molecules, the flow characteristics depend on the relative molecular mass of the gas rather than its viscosity. The effect was studied by M. H. C. Knudsen (1871–1949).

molecular formula *See* FORMULA.

molecular orbital *See* ORBITAL.

molecular sieve Porous crystalline substances, especially aluminosilicates, that can be dehydrated with little change in crystal structure. As they form regularly spaced cavities, they provide a high surface area for the adsorption of smaller molecules.

The general formula of these substances is $M_nO.Al_2O_3.xSiO_2.yH_2O$, where M is a metal ion and n is twice the reciprocal of its valency. Molecular sieves are used as drying agents and in the separation and purification of fluids. They can also be loaded with chemical substances, which remain separated from any reaction that is taking place around them, until they are released by heating or by displacement with a more strongly adsorbed substance. They can thus be used as cation exchange mediums and as catalysts and catalyst supports. They are also used as the stationary phase in certain types of chromatography (**molecular-sieve chromatography**).

molecular volume *See* MOLAR VOLUME.

molecular weight *See* RELATIVE MOLECULAR MASS.

molecule One of the fundamental units forming a chemical compound; the smallest part of a chemical compound that can take part in a chemical reaction. In most covalent compounds, molecules consist of groups of atoms held together by covalent or coordinate bonds. Covalent substances that form macromolecular crystals have no discrete molecules (in a sense, the whole crystal is a molecule). Similarly, ionic compounds do not have single molecules, being collections of oppositely charged ions.

mole fraction Symbol X. A measure of the amount of a component in a mixture. The mole fraction of component A is given by $X_A = n_A/N$, where n_A is the amount of substance of A (for a given entity) and N is the total amount of substance of the mixture (for the same entity).

moment of a force A measure of the turning effect produced by a force about an axis. The magnitude of the moment is the product of the force and the perpendicular distance from the axis to the line of action of the force. An object will be in rotational equilibrium if the algebraic sum of all the moments of the forces on it about any axis is zero. See illustration.

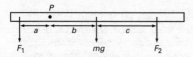

Moment of a force. For equilibrium $mgb + F_2 (b + c) = F_1 a$, where mg is the weight of the beam acting through its centre of mass.

moment of inertia Symbol I. The moment of inertia of a massive body about an axis is the sum of all the products formed by multiplying the magnitude of each element of mass (δm) by the square of its distance (r) from the line, i.e. $I_m = \Sigma r^2 \delta m$. It is the analogue in rotational dynamics of mass in linear dynamics. The basic equation is $T = I\alpha$, where T is the torque causing angular acceleration α about the specified axis.

momentum The **linear momentum** (p) of a body is the product of its mass (m) and its velocity (v), i.e. $p = mv$. See also ANGULAR MOMENTUM.

momentum space See ELECTRICAL CONDUCTIVITY IN METALS.

monochromatic radiation Electromagnetic radiation, especially visible radiation, of only one frequency or wavelength. Completely monochromatic radiation cannot be produced, but *lasers produce radiation within a very narrow frequency band. Compare POLYCHROMATIC RADIATION.

monochromator A device that provides monochromatic radiation from a polychromatic source. In the case of visible radiation, for example, a prism can be used together with slits to select a small range of wavelengths.

monoclinic See CRYSTAL SYSTEM.

monomode fibre An *optical fibre in which there is a narrow glass core of about 5 μm in diameter with a large amount of cladding of a material with a smaller refractive index than the core.

Monte Carlo simulation A type of calculation that involves random sampling for the mathematical simulation of physical systems. Monte Carlo calculations are applied to problems that can be formulated in terms of probability and are usually carried out by computer. Such calculations have been performed for nuclei, atoms, molecules, solids, liquids, and nuclear reactors. The technique is named after the gambling centre in Monaco, renowned for its casino.

moon The earth's only natural satellite, which orbits the earth at a mean distance of 384 400 km. It has a diameter of 3476 km. It has no atmosphere or surface water. Its surface temperature varies between 80 K (night minimum) and 400 K (noon at the equator). It is the only celestial body outside the earth to have been reached by man (1969).

Moscow zero *See* LANDAU GHOST.

Moseley's law The frequencies of the lines in the X-ray spectra of the elements are related to the atomic numbers of the elements. If the square roots of the frequencies of corresponding lines of a set of elements are plotted against the atomic numbers a straight line is obtained. The law was discovered by H. G. Moseley (1887–1915) in 1913–14 and provided an experimental basis for ideas about the structure of the atom.

Mössbauer effect The emission without recoil of a gamma-ray photon from a nucleus embedded in a solid. The emission of a gamma ray by a single atom in a gas causes the atom to recoil and reduces the energy of the gamma ray from its usual transition energy E_0 to $E_0 - R$, where R is the recoil energy. In 1957 R. L. Mössbauer (1929–) discovered that if the emitting nucleus is held by strong forces in the lattice of a solid, the recoil energy is shared by all the nuclei in the lattice. As there may typically be 10^{10}–10^{20} atoms in the lattice the recoil will be negligible and the gamma-ray photon has the energy E_0. The same principle applies to the absorption of gamma rays and is used in **Mössbauer-effect spectroscopy** to elucidate problems in nuclear physics, solid-state physics, and chemistry.

MOT *See* LASER COOLING.

motion A change in the position of a body or system with respect to time, as measured by a particular observer in a particular *frame of reference. Only relative motion can be measured; absolute motion is meaningless. *See also* EQUATION OF MOTION; NEWTON'S LAWS OF MOTION.

motor Any device for converting chemical energy or electrical energy into mechanical energy. *See* ELECTRIC MOTOR; INTERNAL-COMBUSTION ENGINE; LINEAR MOTOR.

motor generator An electric motor mechanically coupled to an electric generator. The motor is driven by a supply of specified voltage, frequency, or number of phases and the generator provides an output in which one or more of these parameters is different to suit a particular purpose.

Mott insulator A substance that is an insulator because of electron correlation and in which the highest occupied energy band is not necessarily full. Certain transition metal oxides in which there are narrow bands are Mott insulators. The concept was put forward and developed by the British physicist Sir Nevill Francis Mott (1905–96), starting in 1949.

Mott transition A type of *metal–insulator transition in which the key parameter is the ratio of electron repulsion to bandwidth. Above a certain critical value of this parameter the metal becomes a *Mott insulator, even though the uppermost band is not full. It was proposed by the physicist Sir Nevill Mott in 1949.

mouse A simple device that is connected to a personal computer by cable and can be moved by hand over a flat surface, its movements being sensed by the rotation of a ball in its base. These movements are communicated to the computer and cause corresponding movements of the cursor on the screen; the cursor indicates the active position on the screen. One or more buttons on the mouse can be pressed to initiate an action, for example to indicate a desired cursor position for typing or to select an item from a menu of options.

moving-coil instrument A measuring instrument in which current or voltage is determined by the couple on a small coil pivoted between the poles of a magnet with curved poles, giving a radial magnetic field. When a current flows through the coil it turns against a return spring. If the angle through which it turns is α, the current I is given by $I = k\alpha/BAN$, where B is the magnetic flux density, A is the area of the coil and N is its number of turns; k is a constant depending on the strength of the return spring. The instrument is suitable for measuring d.c. but can be converted for a.c. by means of a rectifier network. It is usually made as a *galvanometer and converted to an ammeter or voltmeter by means of a *shunt or a *multiplier.

moving-iron instrument A measuring instrument in which current or voltage is determined by the force of attraction on a bar of soft iron pivoted within the magnetic field of a fixed coil or by the repulsion between the poles induced in two soft iron rods within the coil. As the deflection caused by the passage of a current through the coil does not depend on the direction of the current, moving-iron instruments can be used with either d.c. or a.c. without a rectifier. They are, however, less sensitive than *moving-coil instruments.

moving-iron microphone *See* MICROPHONE.

M-theory *See* SUPERSTRING THEORY.

multiaccess system A system allowing several users of a computer, at different terminals, to make apparently simultaneous use of the computer without being aware of each other.

multimedia A combination of various media, such as text, sound, and moving and still images, now often held on *CD-ROM. The user can make use of the different media in an integrated way.

multimeter An electrical measuring instrument designed to measure potential differences or currents over a number of ranges. It also usually has an internal dry cell enabling resistances to be measured. Most multimeters are moving-coil instruments with a switch to enable series resistors or parallel resistors to be incorporated into the circuit.

multimode fibre An *optical fibre in which there is a relatively large glass core of about 50 μm in diameter surrounded by material with a smaller refractive index than the core. The value of the refractive index in the surrounding material can either suddenly drop or gradually drop to a lower value than that of the core.

multiplet 1. A spectral line formed by more than two (*see* DOUBLET) closely spaced lines. **2.** A group of *elementary particles that are identical in all respects except that of electric charge.

multiplication factor Symbol k. The ratio of the average number of neutrons produced in a *nuclear reactor per unit time to the number of neutrons lost by absorption or leakage in the same time. If $k = 1$, the reactor is said to be **critical**. If $k > 1$ it is **supercritical** and if $k < 1$ it is **subcritical**. *See also* CRITICAL REACTION.

multiplicity A quantity used in atomic *spectra to describe the energy levels of many-electron atoms characterized by *Russell–Saunders coupling given by $2S + 1$, where S is the total electron *spin quantum number. The multiplicity of an energy level is indicated by a left superscript to the value of L, where L is the resultant electron *orbital angular momentum of the individual electron orbital angular momenta l.

multiplier A fixed resistance used with a voltmeter, usually a *moving-coil instrument, to vary its range. Many voltmeters are provided with a series of multipliers from which the appropriate value can be selected. If the original instrument requires i amperes for full-scale deflection and the resistance of the moving coil is r ohms, the value R of the resistance of the multiplier required to give a full-scale deflection when a voltage V is applied across the terminals is given by $R = V/i - r$.

multiverse *See* MANY-WORLDS INTERPRETATION.

multivibrator An electronic *oscillator consisting of two active devices, usually transistors, interconnected in an electrical network. The purpose of the device is to generate a continuous square wave with which to store information in binary form in a logic circuit. This is achieved by applying a portion of the output voltage or current of each active device to the input of the other with the appropriate magnitude and polarity, so that the devices are conducting alternately for controllable periods.

mu-mesic atom *See* MUONIC ATOM.

Mumetal The original trade name for a ferromagnetic alloy, containing 78% nickel, 17% iron, and 5% copper, that had a high *permeability and a low *coercive force. More modern versions also contain chromium and molybdenum. These alloys are used in some transformer cores and for shielding various devices from external magnetic fields.

Muntz metal A form of *brass containing 60% copper, 39% zinc, and small amounts of lead and iron. Stronger than alpha-brass, it is used for hot forgings, brazing rods, and large nuts and bolts. It is named after G. F. Muntz (1794–1857).

muon *See* LEPTON; ELEMENTARY PARTICLES.

muonic atom (mu-mesic atom) An atom in which one of the electrons has been replaced by a muon. Since the mass of a muon is 207 times that of an electron, the average radius of the orbit of a muon is much smaller than that of a corresponding electron. Muonic atoms provide tests for quantum electrodynamics. They are also used in research into muon-catalysed fusion (*see* NUCLEAR FUSION).

musical scale A series of musical notes arranged in order of ascending pitch according to a scheme of intervals making them suitable for music. *See* TEMPERAMENT.

mutarotation Change of optical activity with time as a result of spontaneous chemical reaction.

mutual inductance *See* INDUCTANCE.

myopia Short-sightedness. It results from the lens of the eye refracting the parallel rays of light entering it to a focus in front of the retina generally because of an abnormally long eyeball. The condition is corrected by using diverging spectacle lenses to move the image back to the retina.

nabla *See* DEL.

nadir The point opposite the *zenith on the *celestial sphere.

NAND circuit *See* LOGIC CIRCUIT.

nano- Symbol n. A prefix used in the metric system to denote 10^{-9}. For example, 10^{-9} second = 1 nanosecond (ns).

nanotechnology The development and use of devices that have a size of only a few nanometres. Research has been carried out into very small components, which depend on electronic effects and may involve movement of a countable number of electrons in their action. Such devices would act much faster than larger components. Considerable interest has been shown in the production of structures on a molecular level by suitable sequences of chemical reactions. It is also possible to manipulate individual atoms on surfaces using a variant of the *atomic force microscope.

nanotube *See* BUCKMINSTERFULLERENE.

Napierian logarithm *See* LOGARITHM.

natural abundance *See* ABUNDANCE.

natural frequency 1. The frequency of the free oscillation of a system. **2.** The frequency at which resonance occurs in an electrical circuit.

natural logarithm *See* LOGARITHM.

natural units A system of units, used principally in particle physics, in which all quantities that have dimensions involving length, mass, and time are given dimensions of a power of energy (usually expressed in electronvolts). This is equivalent to setting the rationalized *Planck constant and the speed of light both equal to unity. *See also* GAUSSIAN UNITS; GEOMETRIZED UNITS; HEAVISIDE–LORENTZ UNITS; PLANCK UNITS.

nautical mile A measure of distance used at sea. In the UK it is defined as 6080 feet but the international definition is 1852 metres. 1 international nautical mile is therefore equivalent to 1.15078 land (statute) miles.

Navier–Stokes equation An equation describing the flow of a *Newtonian fluid. The Navier–Stokes equation can be written in the form

$$\partial \boldsymbol{v}/\partial t + (\boldsymbol{v}\ \mathrm{grad})\boldsymbol{v} = 1/\rho\ \mathrm{grad}p + \eta\nabla^2\boldsymbol{v},$$

where v is the velocity, ρ is the density, η is the viscosity, and p is the pressure, respectively, of the fluid; ∇^2 is the Laplace operator (*see* LAPLACE EQUATION). The Navier–Stokes equation describes the flow of fluids, such as air and water, but is not suitable for describing the flow of non-Newtonian fluids. The equation can be derived using *fluid mechanics or (in certain cases) from *kinetic theory. It requires *approximation techniques for a solution in all but the simplest problems. The Navier–Stokes equation was derived by the French engineer and scientist Claude-Louis-Marie-Henri Navier (1785–1836) and the British mathematician and physicist Sir George Gabriel Stokes (1819–1903) and also by the French mathematician Siméon-Denis Poisson (1781–1840) in the first half of the 19th century.

near point The nearest point at which the human eye can focus an object. As the lens becomes harder with age, the extent to which accommodation can bring a near object into focus decreases. Therefore with advancing age the near point recedes – a condition known as *presbyopia.

nebula Originally a fixed, extended, and somewhat fuzzy white haze observed in the sky with a telescope. Many of these objects can now be resolved into clouds of individual stars and have been identified as *galaxies. They are still sometimes referred to as **extragalactic nebulae**. The **gaseous nebulae**, however, cannot be resolved into individual stars and consist, for the most part, of interstellar dust and gas. In some of these gaseous nebulae the gas atoms have been ionized by ultraviolet radiation from nearby stars and light is emitted as these ions interact with the free electrons in the gas. These are called **emission nebulae**. In the **dark nebulae**, there are no nearby stars and these objects are consequently dark; they can only be detected by what they obscure.

Néel temperature The temperature above which an antiferromagnetic substance becomes paramagnetic (*see* MAGNETISM). The *susceptibility increases with temperature, reaching a maximum at the Néel temperature, after which it abruptly declines. The phenomenon was discovered around 1930 by L. E. F. Néel (1904–2000).

Ne'eman, Yuval *See* GELL-MANN, MURRAY.

negative charge *See* CHARGE.

negative feedback *See* FEEDBACK.

nematic crystal *See* LIQUID CRYSTAL.

neon lamp A small lamp consisting of a pair of electrodes, treated to emit electrons freely, sealed in a glass bulb containing neon gas at low pressure. When a minimum voltage of between 60 and 90 volts is applied across the electrodes, the kinetic energy of the electrons is sufficient to ionize the neon atoms around the cathode, causing the emission of a reddish light. With d.c. the glow is restricted to the cathode; with a.c. both electrodes act alternately as cathodes and a glow appears to emanate from both electrodes. The device consumes a very low power and is widely used as an indicator light showing that a circuit is live.

neper A unit used to express a ratio of powers, currents, etc., used especially in telecommunications to denote the attenuation of an amplitude A_1 to an amplitude A_2 as N nepers, where

$N = \ln(A_2/A_1)$.

1 neper = 8.686 decibels. The unit is named after John Napier (1550–1617), the inventor of natural logarithms.

neptunium Symbol Np. A radioactive metallic transuranic element belonging to the actinoids; a.n. 93; r.a.m. 237.0482. The most stable isotope, neptunium-237, has a half-life of 2.2×10^6 years and is produced in small quantities as a by-product by nuclear reactors. Other isotopes have mass numbers 229–236 and 238–241. The only other relatively long-lived isotope is neptunium-236 (half-life 5×10^3 years). The element was first produced by Edwin McMillan (1907–91) and Philip Abelson (1913–) in 1940.

neptunium series *See* RADIOACTIVE SERIES.

Nernst effect An effect in which a temperature gradient along an electric conductor or semiconductor placed in a perpendicular magnetic field, causes a potential difference to develop in the third perpendicular direction between opposite edges of the conductor. This effect, an analogue of the *Hall effect, was discovered in 1886 by Walter Nernst (1864–1941).

Nernst heat theorem A statement of the third law of *thermodynamics made by Walter Nernst in a restricted form: if a chemical change takes place between pure crystalline solids at *absolute zero there is no change of entropy.

Net *See* INTERNET.

Neumann's law The magnitude of an electromagnetically induced e.m.f. (E) is given by $E = -d\Phi/dt$, where Φ is the magnetic flux. This is a quantitative statement of *Faraday's second law of electromagnetic induction and is sometimes known as the Faraday–Neumann law. This law was stated quantitatively by the German physicist Franz Neumann (1798–1895) in the 1840s.

neural network A network of processors designed to mimic the transmission of impulses in the human brain. Neural networks are either electronic constructions or, often, computer-simulated structures. Each processor ('neurone') multiplies its input signal by a weighting factor and the final output signal depends on these factors, which can be adjusted. Such networks can be 'taught' to recognize patterns in large amounts of data. They are used in research into artificial intelligence and have also been applied in predicting financial market trends.

neutrino A *lepton (*see also* ELEMENTARY PARTICLES) that exists in three forms, one in association with the electron, one with the muon, and one with the tau particle. Each form has its own antiparticle. The neutrino, which was postulated in 1930 to account for the 'missing' energy in *beta decay, was identified tentatively in 1953 and, more definitely, in 1956. Neutrinos have no charge, very low masses, and travel very close to the speed of light. In some *grand unified theories they are predicted to have nonzero mass. There is now much indirect evidence that neutrinos have nonzero masses but the values have not been determined.

neutron A neutral hadron (*see* ELEMENTARY PARTICLES) that is stable in the atomic nucleus but decays into a proton, an electron, and an antineutrino with a mean life of 12 minutes outside the nucleus. Its rest mass (symbol m_n) is slightly greater than that of the proton, being $1.674\,9286(10) \times 10^{-27}$ kg. Neutrons occur in all atomic nuclei except normal hydrogen. The neutron was first reported in 1932 by James *Chadwick.

neutron bomb *See* NUCLEAR WEAPONS.

neutron diffraction The scattering of neutrons by atoms in solids, liquids, or gases. This process has given rise to a technique, analogous to *X-ray diffraction techniques, using a flux of thermal neutrons from a nuclear reactor to study solid-state structure and phenomena. Thermal neutrons have average kinetic energies of about 0.025 eV (4×10^{-21} J) giving them an equivalent wavelength of about 0.1 nanometre, which is suitable for the study of interatomic interference. There are two types of interaction in the scattering of neutrons by atoms: one is the interaction between the neutrons and the atomic nucleus, the other is the interaction between the *magnetic moments of the neutrons and the spin and orbital magnetic moments of the atoms. The latter interaction has provided valuable information on antiferromagnetic and ferrimagnetic materials (*see* MAGNETISM). Interaction with the atomic nucleus gives diffraction patterns that complement those from X-rays. X-rays, which interact with the extranuclear electrons, are not suitable for investigating light elements (e.g. hydrogen), whereas neutrons do give diffraction patterns from such atoms because they interact with nuclei.

neutron drip *See* NEUTRON STAR.

neutron excess *See* ISOTOPIC NUMBER.

neutron halo *See* HALO.

neutron number Symbol N. The number of neutrons in an atomic nucleus of a particular nuclide. It is equal to the difference between the *nucleon number and the *atomic number.

neutron optics The study of those aspects of neutron behaviour that have wavelike characteristics, e.g. *neutron diffraction.

neutron scattering *See* NEUTRON DIFFRACTION.

neutron star A compact stellar object that is supported against collapse under self-gravity by the *degeneracy pressure of the neutrons of which it is primarily composed. Neutron stars are believed to be formed as the end products of the evolution of stars of mass greater than a few (4–10) solar masses. The core of the evolved star collapses and (assuming that its mass is greater than the *Chandrasekhar limit for a *white dwarf), at the very high densities involved (about 10^{14} kg m^{-3}), electrons react with protons in atomic nuclei to produce neutrons. The neutron-rich nuclei thus formed release free neutrons in a process known as **neutron drip**. The density increases to about 10^{17} kg m^{-3}, at which most of the electrons and protons have been converted to a *degenerate gas of neutrons and the atomic nuclei have lost their separate identities. If the mass of the core exceeds the *Oppenheimer–Volkoff limit, then further collapse will occur, leading to the formation of a *black hole.

*Pulsars are believed to be rapidly rotating magnetized neutron stars and many X-ray sources are thought to be neutron stars in binary systems with another star, from which material is drawn into an accretion disc. This material, heated to a very high temperature, emits radiation in the X-ray region.

neutron temperature A concept used to express the energies of neutrons that are in thermal equilibrium with their surroundings, assuming that they behave like a monatomic gas. The neutron temperature T, on the Kelvin scale, is given by $T = 2E/3k$, where E is average neutron energy and k the *Boltzmann constant.

newton Symbol N. The *SI unit of force, being the force required to give a mass of one kilogram an acceleration of 1 m s^{-2}. It is named after Sir Isaac Newton.

Newton, Sir Isaac (1642–1727) English mathematician and physicist, one of the world's greatest scientists. He went to Cambridge University in 1661 and stayed for nearly 40 years except for 1665–67, when he returned to his home at Woolsthorpe in Lincolnshire (because of the Plague), where some of his best work was done. In 1699 he was made Master of the Royal Mint. He was reluctant to publish his work and his great mathematical masterpiece, the *Principia*, did not appear until 1687. In it he introduced *calculus and formulated *Newton's laws of motion. In 1665 he derived *Newton's law of gravitation, and in optics he produced *Newton's formula for a lens and, in 1672, his theories about *light and the spectrum (*see also* NEWTON'S RINGS); these were summarized in his *Opticks* of 1704. Also in the late 1660s he constructed a reflecting *telescope. The SI unit of force is named after him.

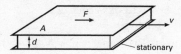

Newtonian fluid

Newtonian fluid A fluid in which the velocity gradient is directly proportional to the shear stress. If two flat plates of area A are separated by a layer of fluid of thickness d and move relative to each other at a velocity v, then the rate of shear is v/d and the shear stress is F/A, where F is the force applied to each (see illustration). For a Newtonian fluid $F/A = \mu v/d$, where μ is the constant of proportionality and is called the Newtonian *viscosity. Many liquids are Newtonian fluids over a wide range of temperatures and pressures. However, some are not; these are called **non-Newtonian fluids**. In such fluids there is a departure from the simple Newtonian relationships. For example, in some liquids the viscosity increases as the velocity gradient increases, i.e. the faster the liquid moves the more viscous it becomes. Such liquids are said to be **dilatant** and the phenomenon they exhibit is called **dilatancy**. It occurs in some pastes and suspensions. More common, however, is the opposite effect in which the viscosity depends not only on the velocity gradient but also on the time for which it has been applied. These liquids are said to exhibit **thixotropy**. The faster a **thixotropic liquid** moves the less viscous it becomes. This property is used in nondrip paints (which are more viscous on the brush than on the wall) and in lubricating oils (which become thinner when the parts they are lubricating start to move). Another example is the non-Newtonian flow of macromolecules in solution or in polymer melts. In this case the shearing force F is not parallel to the shear planes and the linear relationship does not apply. In general, the many types of non-Newtonian fluid are somewhat complicated and no theory has been developed to accommodate them fully.

Newtonian mechanics The system of *mechanics that relies on *Newton's laws of motion. Newtonian mechanics is applicable to bodies moving at speeds relative to the observer that are small compared to the speed of light. Bodies moving at speeds comparable to the speed of light require an approach based on *relativistic mechanics, in which the mass of a body changes with its speed.

Newtonian telescope *See* TELESCOPE.

Newton's formula For a lens, the distances p and q between two conjugate points and their respective foci is given by $pq = f^2$, where f is the focal length of the lens.

Newton's law of cooling The rate at which a body loses heat is proportional to the difference in temperature between the body and the surroundings. It is an empirical law that is only true for substantial temperature differences if the heat loss is by forced convection or conduction.

Newton's law of gravitation There is a force of attraction between any two massive particles in the universe. For any two point masses m_1 and m_2, separated by a distance d, the force of attraction F is given by $F = m_1 m_2 G/d^2$, where G is the *gravitational constant. Real bodies having spherical symmetry act as point masses positioned at their centres of mass.

Newton's laws of motion The three laws of motion on which *Newtonian mechanics is based. (1) A body continues in a state of rest or uniform motion in a straight line unless it is acted upon by external forces. (2) The rate of change of momentum of a moving body is proportional to and in the same direction as the force acting on it, i.e. $F = d(mv)/dt$, where F is the applied force, v is the velocity of the body, and m its mass. If the mass remains constant, $F = m \, dv/dt$ or $F = ma$, where a is the acceleration. (3) If one body exerts a force on another, there is an equal and opposite force, called a **reaction**, exerted on the first body by the second.

Newton's rings 1. (in optics) *Interference fringes formed by placing a slightly convex lens on a flat glass plate. If monochromatic light is reflected by the two close surfaces into the observer's eye at a suitable angle, the point of contact of the lens is seen as a dark spot surrounded by a series of bright and dark rings. The radius of the nth dark ring is given by $r_n = \sqrt{nR\lambda}$, where λ is the wavelength and R is the radius of curvature of the lens. The phenomenon is used in the quality testing of lens surfaces. With white light, coloured rings are formed. **2.** (in photography) The irregular patterns produced by thin film interference between a projected transparency and its cover glass.

Nichrome Trade name for a group of nickel–chromium alloys used for wire in heating elements as they possess good resistance to oxidation and have a high resistivity. Typical is Nichrome V containing 80% nickel and 19.5% chromium, the balance consisting of manganese, silicon, and carbon.

nickel–cadmium cell *See* NICKEL–IRON ACCUMULATOR.

nickel–iron accumulator (Edison cell; NIFE cell) A *secondary cell devised by Thomas Edison (1847-1931) having a positive plate of nickel oxide and a negative plate of iron both immersed in an electrolyte of potassium hydroxide. The reaction on discharge is

$$2NiOOH.H_2O + Fe \rightarrow 2Ni(OH)_2 + Fe(OH)_2,$$

the reverse occurring during charging. Each cell gives an e.m.f. of about 1.2 volts and produces about 100 kJ per kilogram during each discharge. The **nickel–cadmium cell** is a similar device with a negative cadmium electrode. It is often used as a *dry cell. *Compare* LEAD–ACID ACCUMULATOR.

nickel silver *See* GERMAN SILVER.

Nicol prism A device for producing plane-polarized light (*see* POLARIZER). It consists of two pieces of calcite cut with a 68° angle and stuck together with Canada balsam. The extraordinary ray (*see* DOUBLE REFRACTION) passes through the prism while the ordinary ray suffers total internal reflection at the interface between the two crystals, as the refractive index of the calcite is 1.66 for the ordinary ray and that of the Canada balsam is 1.53. Modifications of the prism using different shapes and cements are used for special purposes. It was devised in 1828 by William Nicol (1768–1851).

NIFE cell *See* NICKEL–IRON ACCUMULATOR.

nit A unit of *luminance equal to one *candela per square metre.

NMR *See* NUCLEAR MAGNETIC RESONANCE.

nobelium Symbol No. A radioactive metallic transuranic element belonging to the actinoids; a.n. 102; mass number of most stable element 254 (half-life 55 seconds). Seven isotopes are known. The element was first identified with certainty by A. Ghiorso and G. T. Seaborg in 1966. The alternative name **unnilbium** has been proposed.

no-cloning theorem A result stating that it is not possible to copy quantum information perfectly. This is the case because the Heisenberg uncertainty principle means that it is not possible to obtain complete information about a quantum state and because examining a quantum state alters that state. This theorem was proved by William Wooters and Wojciech Zurek in 1982.

nodal points Two points on the axis of a system of lenses; if the incident ray passes through one, the emergent ray will pass through the other.

node 1. (in physics) A point of minimum disturbance in a *stationary-wave system. **2.** (in astronomy) Either of two points at which the orbit of a celestial body intersects a reference plane, usually the plane of the *ecliptic or the celestial equator (*see* CELESTIAL SPHERE).

Noether's theorem Every continuous symmetry under which the *Lagrangian (or *Hamiltonian) is invariant in form is associated with a *conservation law. For example, invariance of a Langrangian under time displacement implies the conservation of energy. Not all conservation laws are associated with continuous symmetries, since some conservation laws are associated with *topology, particularly for *solitons. It does not automatically follow that if there is a conserved quantity in a *classical field theory associated with Noether's theorem, there is a conserved quantity in the corresponding *quantum field theory; this raises the possibility of an anomaly. Noether's theorem was stated in 1918 by the German mathematician Amalie Emmy Noether (1882–1935).

no-hair theorem *See* BLACK HOLE.

noise 1. Any undesired sound. It is measured on a *decibel scale ranging from the threshold of hearing (0 dB) to the threshold of pain (130 dB). Between these limits a whisper registers about 20 dB, heavy urban traffic about 90 dB, and a heavy hammer on steel plate about 110 dB. A high noise level (industrial or from overamplified music, for example) can cause permanent hearing impairment. **2.** Any unwanted disturbance within a useful frequency band in a communication channel.

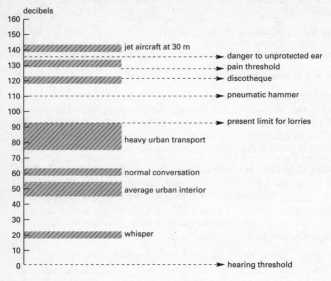

Noise: decibel scale

nomogram A graph consisting of three lines, each with its own scale, each line representing the values of a variable over a specified range. A ruler laid between two points on two of the lines enables the value of the third variable to be read off the third line.

noncommutative geometry A type of geometry in which the basic elements are noncommutative objects, such as matrices, rather than points. There are many applications of the subject in theoretical physics.

nonequilibrium statistical mechanics The statistical mechanics of systems not in thermal equilibrium. One of the main purposes of nonequilibrium statistical mechanics is to calculate *transport coefficients and inverse transport coefficients, such as *conductivity and *viscosity, from first principles and to provide a basis for *transport theory. The nonequilibrium systems easiest to understand are those near thermal

equilibrium. For systems far from equilibrium, such possibilities as
*chaos, *turbulence, and *self-organization can arise due to nonlinearity.

nonequilibrium thermodynamics The thermodynamics of systems
not in thermal *equilibrium. The nonequilibrium systems easiest to
understand are those near thermal equilibrium; these systems are
described by the Onsager relations. For systems far from equilibrium,
more complicated patterns, such as *chaos and *self-organization, can
arise due to nonlinearity. Which behaviour is observed depends on the
value of certain parameters in the system. The transition from one type of
behaviour to another as the parameters are altered occurs at bifurcations.

non-Euclidean geometry A type of geometry that does not comply
with the basic postulates of Euclidean geometry, particularly a form of
geometry that does not accept Euclid's postulate that only one straight
line can be drawn through a point in space parallel to a given straight
line. Several types of non-Euclidean geometry exist.

nonlinear optics A branch of optics concerned with the optical
properties of matter subjected to intense electromagnetic fields. For
nonlinearity to manifest itself, the external field should not be negligible
compared to the internal fields of the atoms and molecules of which the
matter consists. *Lasers are capable of generating external fields
sufficiently intense for nonlinearity to occur. Indeed, the subject of
nonlinear optics has been largely developed as a result of the invention of
the laser. In nonlinear optics the induced electric polarization (*see*
DIELECTRIC) of the medium is not a linear function of the strength of the
external *electromagnetic radiation. This leads to more complicated
phenomena than can occur in **linear optics**, in which the induced
polarization is proportional to the strength of the external
electromagnetic radiation.

nonmetal An element that is not a metal. Nonmetals can either be
*insulators or *semiconductors. At low temperatures nonmetals are poor
conductors of both electricity and heat as few free electrons move
through the material. If the conduction band is near to the valence band
(*see* ENERGY BANDS) it is possible for nonmetals to conduct electricity at
high temperatures but, in contrast to metals, the conductivity increases
with increasing temperature. Nonmetals are electronegative elements,
such as carbon, nitrogen, oxygen, phosphorus, sulphur, and the halogens.
They form compounds that contain negative ions or covalent bonds. Their
oxides are either neutral or acidic.

non-Newtonian fluid *See* NEWTONIAN FLUID.

nonpolar compound A compound that has covalent molecules with
no permanent dipole moment. Examples of nonpolar compounds are
methane and benzene.

nonrelativistic quantum theory *See* QUANTUM THEORY.

nonrenormalization theorems Theorems in *quantum field theory proving that the correct result of a calculation is given by the first term in *perturbation theory, with Feynman diagrams (*see* QUANTUM ELECTRODYNAMICS) used to perform calculations involving *renormalization of the theory, with no corrections due to higher-order terms in the perturbation series. Nonrenormalization theorems occur when perturbation theory is capable of giving a general nonperturbative result for the theory. Examples of nonrenormalization theorems occur in calculations involving chiral anomalies and in certain theories involving *supersymmetry.

nonstoichiometric compound (Berthollide compound) A chemical compound in which the elements do not combine in simple ratios. For example, rutile (titanium(IV) oxide) is often deficient in oxygen, typically having a formula $TiO_{1.8}$.

normal 1. (in mathematics) A line drawn at right angles to a surface. **2.** (in chemistry) Having a concentration of one gram equivalent per dm^3.

normalizing The process of heating steel to above an appropriate critical temperature followed by cooling in still air. The process promotes the formation of a uniform internal structure and the elimination of internal stress.

normal modes A convenient set of simple vibrational motions of an oscillatory system. The normal modes of an oscillating system are labelled by their characteristic frequencies. Modes of the same frequency are said to be **degenerate** in that frequency. More generally, when vibrational wave energy propagates through any physical system, the waves interact with the medium through which they are travelling. Any periodic response of the system may be deconstructed as a *Fourier series of normal modes by the *superposition principle. Oscillating systems possesses as many normal modes as they have degrees of freedom. For example, masses connected by electric strings and set to vibrate may be analysed in terms of their normal modes. The diagram below shows

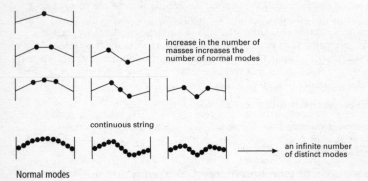

increase in the number of masses increases the number of normal modes

continuous string

an infinite number of distinct modes

Normal modes

systems containing 1, 2, 3, 4 and an infinite number (a continuous string) of masses. If these simple oscillations are confined to the vertical plane, the numbers of degrees of freedom for each of the five cases are equal to the numbers of masses present. The second diagram shows an application of the powerful superposition principle to the deconstruction of a square wave response by continuous sine waves. As with the normal modes of the continuous string, sine waves constitute an infinite set of normal modes, the addition of more and more members of this infinite set leads to better and better approximations of the periodic square wave (*see also* SPHERICAL HARMONICS).

NOT circuit *See* LOGIC CIRCUIT.

note 1. A musical sound of specified pitch. **2.** A representation of such a sound in a musical score. Such a representation has a specified duration as well as a specified pitch.

nova A star that, over a period of only a few days, becomes 10^3–10^4 times brighter than it was. Some 10–15 such events occur in the Milky Way each year. Novae are believed to be close *binary stars, one component of which is usually a *white dwarf and the other a *red giant. Matter is transferred from the red giant to the white dwarf, on whose surface it accumulates, eventually leading to a thermonuclear explosion. *See also* SUPERNOVA.

N.T.P. *See* S.T.P.

***n*-type conductivity** *See* SEMICONDUCTOR; TRANSISTOR.

nuclear battery A single cell, or battery of cells, in which the energy of particles emitted from the atomic nucleus is converted internally into electrical energy. In the high-voltage type, a beta-emitter, such as strontium–90, krypton–85, or tritium, is sealed into a shielded glass vessel, the electrons being collected on an electrode that is separated from the emitter by a vacuum or by a solid dielectric. A typical cell delivers some 160 picoamperes at a voltage proportional to the load resistance. It can be used to maintain the voltage of a charged capacitor. Of greater use, especially in space technology, are the various types of low-voltage nuclear batteries. Typical is the gas-ionization device in which a beta-emitter ionizes a gas in an electric field. Each beta-particle produces about 200 ions, thus multiplying the current. The electric field is obtained by the contact potential difference between two electrodes, such as lead dioxide and magnesium. Such a cell, containing argon and tritium, gives about 1.6 nanoamperes at 1.5 volts. Other types use light from a phosphor receiving the beta-particles to operate photocells or heat from the nuclear reaction to operate a thermopile.

nuclear energy Energy obtained as a result of *nuclear fission or *nuclear fusion. The nuclear fission of one uranium atom yields about 3.2×10^{-11} joule, whereas the combustion of one carbon atom yields about

6.4×10^{-19} joule. Mass for mass, uranium yields about 2 500 000 times more energy by fission than carbon does by combustion. The nuclear fusion of deuterium to form helium releases about 400 times as much energy as the fission of uranium (on a mass basis).

nuclear fission A nuclear reaction in which a heavy nucleus (such as uranium) splits into two parts (**fission products**), which subsequently emit either two or three neutrons, releasing a quantity of energy equivalent to the difference between the rest mass of the neutrons and the fission products and that of the original nucleus. Fission may occur spontaneously or as a result of irradiation by neutrons. For example, the fission of a uranium–235 nucleus by a *slow neutron may proceed thus:

$$^{235}U + n \rightarrow {}^{148}La + {}^{85}Br + 3n$$

The energy released is approximately 3×10^{-11} J per ^{235}U nucleus. For 1 kg of ^{235}U this is equivalent to 20 000 megawatt-hours – the amount of energy produced by the combustion of 3×10^{6} tonnes of coal. Nuclear fission is the process used in *nuclear reactors and atom bombs (*see* NUCLEAR WEAPONS).

nuclear force A strong attractive force between *nucleons in the atomic nucleus that holds the nucleus together. At close range (up to about 2×10^{-15} metre) these forces are some 100 times stronger than electromagnetic forces. *See* FUNDAMENTAL INTERACTIONS.

nuclear fuel A material that will sustain a fission chain reaction so that it can be used as a source of *nuclear energy. The *fissile materials are uranium–235, uranium–233, plutonium–241, and plutonium–239. The first occurs in nature as 1 part in 140 of natural uranium, the others have to be made artificially. ^{233}U is produced when thorium–232 captures a neutron and ^{239}Pu is produced by neutron capture in ^{238}U. ^{232}Th and ^{238}U are called *fertile isotopes (*see* FERTILE MATERIAL).

nuclear fusion 1. A type of *nuclear reaction in which atomic nuclei of low atomic number fuse to form a heavier nucleus with the release of large amounts of energy. In *nuclear fission reactions a neutron is used to break up a large nucleus, but in nuclear fusion the two reacting nuclei themselves have to be brought into collision. As both nuclei are positively charged there is a strong repulsive force between them, which can only be overcome if the reacting nuclei have very high kinetic energies. These high kinetic energies imply temperatures of the order of 10^{8} K. As the kinetic energy required increases with the nuclear charge (i.e. atomic number), reactions involving low atomic-number nuclei are the easiest to produce. At these elevated temperatures, however, fusion reactions are self-sustaining; the reactants at these temperatures are in the form of a *plasma (i.e. nuclei and free electrons) with the nuclei possessing sufficient energy to overcome electrostatic repulsion forces. The fusion bomb (*see* NUCLEAR WEAPONS) and the stars generate energy in this way. It is hoped that the method will be harnessed in the *thermonuclear reactor

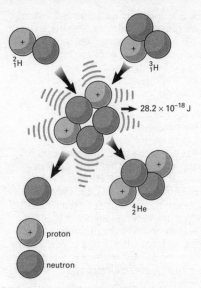

Nuclear fusion: tritium–deuterium reaction

as a source of energy for human use. A great deal of effort has been devoted to this but considerable difficulties remain.

Typical fusion reactions with the energy release in joules are:

$D + D = T + p + 6.4 \times 10^{-13}$ J

$T + D = {}^4He + n + 28.2 \times 10^{-13}$ J

${}^6Li + D = 2{}^4He + 35.8 \times 10^{-13}$ J

By comparison the formation of a water molecule from hydrogen and oxygen is accompanied by the release of 1.5×10^{-19} J.

A large amount of work is currently being done on **cold fusion**, i.e. fusion that can occur at lower temperatures than those necessary to overcome the electrostatic repulsion between nuclei. The most productive approach is **meson-catalysed fusion**, in which the deuterium atoms have their electrons replaced by negative muons to give 'muonic atoms' of deuterium. The muon is 207 times heavier than the electron, so the muonic deuterium atom is much smaller and is able to approach another deuterium atom more closely, allowing nuclear fusion to occur. The muon is released to form another muonic atom, and the process continues. The limiting factor is the short lifetime of the muon, which restricts the number of fusion reactions it can catalyse. **2.** The production of new *transactinide elements by bombarding nuclei of one element with nuclei of another element at an energy precisely chosen to allow the fusion reaction to occur.

nuclear isomerism A condition in which atomic nuclei with the same number of neutrons and protons have different lifetimes. This occurs when nuclei exist in different unstable quantum states, from which they decay to lower excited states or to the ground state, with the emission of gamma-ray photons. If the lifetime of a particular excited state is unusually long it is said to be isomeric, although there is no fixed limit separating isomeric decays from normal decays.

nuclear magnetic resonance (**NMR**) The absorption of electromagnetic radiation at a suitable precise frequency by a nucleus with a nonzero magnetic moment in an external magnetic field. The phenomenon occurs if the nucleus has nonzero *spin, in which case it behaves as a small magnet. In an external magnetic field, the nucleus's magnetic moment vector precesses about the field direction but only certain orientations are allowed by quantum rules. Thus, for hydrogen (spin of ½) there are two possible states in the presence of a field, each with a slightly different energy. Nuclear magnetic resonance is the absorption of radiation at a photon energy equal to the difference between these levels, causing a transition from a lower to a higher energy state. For practical purposes, the difference in energy levels is small and the radiation is in the radiofrequency region of the electromagnetic spectrum. It depends on the field strength.

NMR can be used for the accurate determination of nuclear moments. It can also be used in a sensitive form of magnetometer to measure magnetic fields. In medicine, **magnetic resonance imaging** (**MRI**) has been developed, in which images of tissue are produced by magnetic-resonance techniques. See Feature.

The main application of NMR is as a technique for chemical analysis and structure determination, known as **NMR spectroscopy**. It depends on the fact that the electrons in a molecule shield the nucleus to some extent from the field, causing different atoms to absorb at slightly different frequencies (or at slightly different fields for a fixed frequency). Such effects are known as **chemical shifts**. There are two methods of NMR spectroscopy. In **continuous wave** (**CW**) **NMR**, the sample is subjected to a strong field, which can be varied in a controlled way over a small region. It is irradiated with radiation at a fixed frequency, and a detector monitors the field at the sample. As the field changes, absorption corresponding to transitions occurs at certain values, and this causes oscillations in the field, which induce a signal in the detector. **Fourier transform** (**FT**) **NMR** uses a fixed magnetic field and the sample is subjected to a high-intensity pulse of radiation covering a range of frequencies. The signal produced is analysed mathematically to give the NMR spectrum. The most common nucleus studied is ^{1}H. For instance, an NMR spectrum of ethanol (CH_3CH_2OH) has three peaks in the ratio 3:2:1, corresponding to the three different hydrogen-atom environments. The peaks also have a fine structure caused by interaction between spins in the molecule. Other nuclei can also be used for NMR spectroscopy (e.g. ^{13}C, ^{14}N, ^{19}F) although these generally have lower magnetic moment and natural abundance

MAGNETIC RESONANCE IMAGING

A diagnostic imaging technique based on the phenomenon of *nuclear magnetic resonance (NMR). NMR is a process in which *protons interact with a strong magnetic field and with radio waves to generate electrical pulses that can be processed in a similar way to computerized *tomography. The medical application of NMR, began in the 1950s, but the first images of live patients were not produced until the late 1970s. Images produced by MRI are similar to those produced by computerized tomography using X-rays, but without the radiation hazard.

A major factor in the high costs of MRI is the need for a *superconducting magnet to produce the very strong magnetic fields (0.1 – 2 tesla). A niobium–titanium alloy, which becomes superconducting at –269°C, is used to construct the field coils. These need to be immersed in liquid helium. Superimposed on this large magnetic field are smaller fields, with known gradients in two directions. These gradient fields produce a unique value of the magnetic field strength at each point within the instrument (see illustration).

Some nuclei in the atoms of a patient's tissues have a *spin, which makes them behave as tiny nuclear magnets. The purpose of the large magnetic field is to align these nuclear magnets. Having achieved this alignment, the area under examination is subjected to pulses of radiofrequency (RF) radiation. At a resonant frequency of RF pulses the nuclei under examination undergo *Larmor precession. This phenomenon may be thought of as a 'tipping' of the nuclear magnets away from the strong field alignment. The nuclear magnets then precess or 'wobble', about the axis of the main field as the nuclei regain their alignment with that field.

The speed at which the nuclei return to the steady stage gives rise to two parameters, known as **relaxation times**. Because these relaxation times for nuclei depend on their atomic environment, they may be used to identify nuclei. Small changes in the magnetic field produced as the nuclei precess induce currents in a receiving coil. These signals are digitized before being stored in the memory of a computer.

The resulting set of RF pulse sizes and sequences identify a variety of resonance situations. By analysing these sequences and knowing the unique value of magnetic field strength within the volume under investigation, the resonance signals may be decoded to give estimates of the compositions of the patient's tissues. A three dimensional map of the composition can then be produced, using colour to indicate contrast between differing tissue compositions.

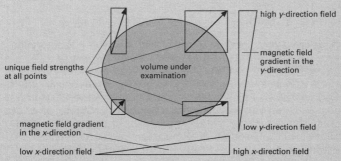

MRI. The way unique field strengths are produced at different points in a specimen.

these generally have lower magnetic moment and natural abundance than hydrogen. *See also* ELECTRON-SPIN RESONANCE.

nuclear models Because of the complexity of the many-body problem in the atomic nucleus, various models have been suggested to explain various aspects of the entity. In the *shell model, the nucleons are regarded as almost independent particles. In the *liquid-drop model all the nucleons in the nucleus are regarded as acting collectively in much the same way as the molecules in a liquid. As these models appear to be physically very different it is necessary to distinguish the circumstances under which they are applicable; moreover, to achieve a unified understanding of these different models would be desirable. Considerable insight is provided by considering the interactions between the *quasiparticles and *collective excitations in the many-nucleon problem. This enables more complicated models, which incorporate both the shell model and the liquid-drop model, to be constructed.

nuclear moment A property of atomic nuclei in which lack of spherical symmetry of the nuclear charge gives rise to electric moments and the intrinsic spin and rotational motion of the component nucleons give rise to magnetic moments.

nuclear physics The physics of atomic nuclei and their interactions, with particular reference to the generation of *nuclear energy.

nuclear power Electric power or motive power generated by a *nuclear reactor.

nuclear reaction Any reaction in which there is a change to an atomic nucleus. This may be a natural spontaneous disintegration or an artificial bombardment of a nucleus with an energetic particle, as in a *nuclear reactor. Nuclear reactions are commonly represented by enclosing within a bracket the symbols for the incoming and outgoing particles; the initial nuclide is shown before the bracket and the final nuclide after it. For example, the reaction:

$$^{12}C + ^{2}H \rightarrow ^{13}N + ^{1}n$$

is shown as $^{12}C(d,n)^{13}N$, where d is the symbol for a deuteron.

nuclear reactor A device in which a *nuclear fission *chain reaction is sustained and controlled in order to produce *nuclear energy, radioisotopes, or new nuclides. The fuels available for use in a fission reactor are uranium–235, uranium–233, and plutonium–239; only the first occurs in nature (as 1 part in 140 of natural uranium), the others have to be produced artificially (*see* NUCLEAR FUEL). When a uranium–235 nucleus is made to undergo fission by the impact of a neutron it breaks into two roughly equal fragments, which release either two or three very high-energy neutrons. These *fast neutrons need to be slowed down to increase the probability that they will cause further fissions of ^{235}U nuclei and thus sustain the chain reaction. This slowing down process occurs

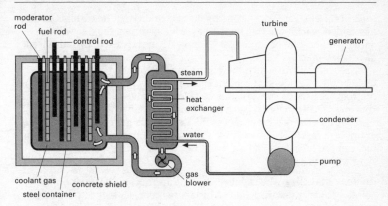

A schematic diagram of a gas-cooled reactor

naturally to a certain extent when the neutrons collide with other nuclei; unfortunately, however, the predominant uranium isotope, ^{238}U, absorbs fast neutrons to such an extent that in natural uranium the fission reaction is not self-sustaining. In order to create a controlled self-sustaining chain reaction it is necessary either to slow down the neutrons (using a *moderator in a **thermal reactor**) to greatly reduce the number absorbed by ^{238}U, or to reduce the predominance of ^{238}U in natural uranium by enriching it with more ^{235}U than it normally contains. In a **fast reactor** the fuel used is enriched uranium and no moderator is employed.

In thermal reactors, neutrons are slowed down by collisions with light moderator atoms (such as graphite, deuterium, or beryllium); they are then in thermal equilibrium with the surrounding material and are known as **thermal neutrons**. In a **heterogeneous thermal reactor** the fuel and moderator are in separate solid and liquid phases (e.g. solid uranium fuel and a heavy water moderator). In the **homogeneous thermal reactor** the fuel and moderator are mixed together, for example in a solution, molten dispersion, slurry, or suspension.

In the reactor **core** the **fuel elements** encase the fuel; in a heterogeneous reactor the fuel elements may fit into a lattice that also contains the moderator. The progress of the reaction is controlled by *control rods, which when lowered into the core absorb neutrons and so slow down or stop the chain reaction. The heat produced by the nuclear reaction in the core is used to generate electricity by the same means as in a conventional power station, i.e. by raising steam to drive a steam turbine that turns a generator. The heat is transferred to the steam-raising boiler or heat-exchanger by the *coolant. Water is frequently used as the coolant; in the case of the **boiling-water reactor** (**BWR**) and the **pressurized-water reactor** (**PWR**) water is both coolant and moderator. In the BWR the primary coolant drives the turbine; in the PWR the primary coolant raises steam in a secondary circuit for driving the turbine. In the **gas-cooled reactor** the coolant is a gas, usually carbon dioxide with an outlet temperature of about 350°C, or 600°C in the case of the **advanced gas-cooled reactor** (**AGR**).

In fast reactors, in which there is no moderator, the temperature is higher and a liquid-metal coolant is used, usually liquid sodium. Some fast reactors are used as converters or breeders. A **converter reactor** is one that converts *fertile material (such as ^{238}U) into *fissile material (such as ^{239}Pu). A **breeder reactor** produces the same fissile material as it uses. For example, a **fast breeder reactor** using uranium enriched with ^{239}Pu as the fuel can produce more ^{239}Pu than it uses by converting ^{238}U to ^{239}Pu. *See also* THERMONUCLEAR REACTOR.

nuclear stability The stability of an atomic nucleus depends on its *binding energy per nucleon, E. This energy represents the average energy that needs to be supplied to liberate a nucleon from the nucleus. One may visualize this qualitatively by imagining a well of potential energy, the depth of which represents E for a particular nucleus. Nuclei with deeper binding energy wells are more stable and occur at intermediate mass numbers. In the graph of E against mass number shown, $^{56}_{26}$Fe has a value of E of 8.8 MeV and is one of the most stable of nuclides. There are three other nuclides, $^{4}_{2}$He, $^{12}_{6}$C, and $^{16}_{8}$O lying significantly above the main curve. Though these are not the most stable of all nuclides, they are considerably more stable than those of other nuclides of adjacent mass numbers. Nuclides of intermediate masses, such as iron, have the greatest values of E. Therefore if the conditions are right, there will be a tendency for nuclides of mass lower than iron to undergo fusion to enhance stability by acquiring a deeper E well. When nuclei heavier than iron undergo fission, energy is released which is equivalent to the difference in E of the parent and daughter nuclei (*see also* LIQUID-DROP MODEL; Q-VALUE).

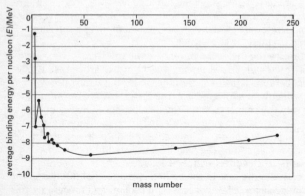

Nuclear stability

nuclear waste *See* RADIOACTIVE WASTE.

nuclear weapons Weapons in which an explosion is caused by
*nuclear fission, *nuclear fusion, or a combination of both. In the fission
bomb (**atomic bomb** or **A-bomb**) two subcritical masses (*see* CRITICAL
MASS) of a *fissile material (uranium–235 or plutonium–239) are brought
together by a chemical explosion to produce one supercritical mass.
The resulting nuclear explosion is typically in the kiloton range with
temperatures of the order 10^8 K being reached. The fusion bomb
(**thermonuclear weapon**, **hydrogen bomb**, or **H-bomb**) relies on a nuclear-
fusion reaction, which becomes self-sustaining at a critical temperature
of about 35×10^6 K. Hydrogen bombs consist of either two-phase fission-
fusion devices in which an inner fission bomb is surrounded by a
hydrogenous material, such as heavy hydrogen (deuterium) or lithium
deuteride, or a three-phase fission-fusion-fission device, which is even
more powerful. The megaton explosion produced by such a thermonuclear
reaction has not yet been used in war. A special type of fission-fusion
bomb is called a **neutron bomb**, in which most of the energy is released
as high-energy neutrons. This neutron radiation destroys people but
provides less of the shock waves and blast that destroy buildings.

nucleon A *proton or a *neutron.

nucleon emission A decay mechanism in which a particularly unstable
nuclide regains some stability by the emission of a nucleon (i.e. a proton
or neutron). Proton emitters have fewer neutrons than their stable
isotopes. Proton emitters are therefore found below the *Segrè plot
stability line. For example, ^{17}Ne (neon–17) has three fewer neutrons than
its most abundant stable isotope ^{20}Ne (neon–20). There are no naturally
occurring proton emitters. Neutron emitters have many more neutrons
than their stable isotopes. For this reason, emitters may be found above
the stability line on the Segrè plot and in most cases can also decay by
negative *beta decay. There are no naturally occurring neutron emitters.
They are usually produced in nuclear reactors by the negative beta decay
of fission products. An example of a neutron emitter is ^{99}Y (yttrium–99),
which has 10 more neutrons than the stable isotope ^{89}Y (yttrium–89).

nucleonics The technological aspects of *nuclear physics, including the
design of nuclear reactors, devices to produce and detect radiation, and
nuclear transport systems. It is also concerned with the technology of
*radioactive waste disposal and with radioisotopes.

nucleon number (**mass number**) Symbol A. The number of *nucleons
in an atomic nucleus of a particular nuclide.

nucleosynthesis The synthesis of chemical elements by nuclear
processes. There are several ways in which nucleosynthesis can take
place. **Primordial nucleosynthesis** took place very soon after the *big bang,
when the universe was extremely hot. This process was responsible for

the cosmic abundances observed for light elements, such as helium. Explosive nucleosynthesis can also occur during the explosion of a *supernova. Many of the elements that are heavier than iron are made in this way. However, **stellar nucleosynthesis**, which takes place in the centre of stars at very high temperatures, is the principal form of nucleosynthesis. The exact process occurring in stellar nucleosynthesis depends on the temperature, density, and chemical composition of the star. The synthesis of helium from protons and of carbon from helium can both occur in stellar nucleosynthesis. *See also* CARBON CYCLE; EARLY UNIVERSE; STELLAR EVOLUTION.

nucleus (of atom) The central core of an atom that contains most of its mass. Experiments performed in 1909 by Geiger and Marsden on the scattering of alpha particles (under the direction of *Rutherford) led to the discovery of a nuclear structure (*see* RUTHERFORD SCATTERING). The nucleus is positively charged and contains one or more nucleons (protons or neutrons). The positive charge of the nucleus is determined by the number of protons it contains (*see* ATOMIC NUMBER); in the neutral atom this positive charge is balanced by an equal number of negatively charged electrons orbiting the nucleus in a comparatively large region outside it.

The simplest nucleus is the hydrogen nucleus, consisting of a single proton. All other nuclei contain neutrons, which contribute to the atomic mass (*see* NUCLEON NUMBER) but not to the nuclear charge. Therefore nuclei with different numbers of neutrons but the same number of protons will possess the same number of electrons around the nuclei of their neutral atoms. Such atoms are called *isotopes. Since it is the electron configuration around the nucleus that dictates chemical properties, isotopes are chemically indistinguishable and separation is achieved by non-chemical methods (*see* ISOTOPE SEPARATION). Nuclei do not seem to have clearly defined perimeters. However, an estimate of a nuclear radius may be achieved by investigating the scattering of very high-energy alpha particles by the nucleus. The deflections of alpha particles in Rutherford scattering experiments agree very well with predicted data, except when the alpha particles have very high kinetic energies. The theoretical predictions made by Rutherford assumed the nucleus to be a point of positive charge. The incident alpha particle would behave as though the nucleus were a point as long as it did not begin to penetrate the outer regions of the nucleus itself. Deviation of the alpha particle scattering data from the point nucleus predictions, indicates the onset of nucleus penetration. Experiments of this kind, using a wide variety of target nuclei, reveal that the radius R of any nucleus can be represented by: $R = r_o A^{\frac{1}{3}}$, where r_o is a constant (= 1.414×10^{-15} m) and A is the nucleon number of the nucleus. Nuclear radii are commonly expressed in femtometres (1 fm = 10^{-15} m), which is sometimes called a fermi. Experiments using projectiles other than alpha particles (such as electrons, neutrons, etc.) give slightly different values of r_o.

The relationship between R and A has an interesting implication for the density of nuclear material. Since both mass and volume of nuclear material are proportional to A, an expression for the density of nuclear material is constant and therefore the same for all nuclei; that is, all nuclei have the same density, estimated to be about 2.3×10^{17} kg/m³. *See also* LIQUID-DROP MODEL.

nuclide A type of atom characterized by its *atomic number and its *neutron number. An *isotope refers to a member of a series of different atoms that have the same atomic number but different neutron numbers (e.g. uranium–238 and uranium–235 are isotopes of uranium), whereas a nuclide refers only to a particular nuclear species (e.g. the nuclides uranium–235 and plutonium–239 are fissile). The term is also used for a type of nucleus.

null method A method of making a measurement in which the quantity to be measured is balanced by another similar reading by adjusting the instrument to read zero (*see* WHEATSTONE BRIDGE).

numerical analysis The analysis of problems by means of calculations involving numbers rather than analytical formulae. Numerical analysis is used extensively for problems too complicated to be solved analytically (either exactly or approximately). Numerical analysis can be performed using either electronic calculators or computers. The use of numerical analysis has increased as the power of computers has increased.

nutation An irregular periodic oscillation of the earth's poles. It causes an irregularity of the precessional circle traced by the celestial poles and results from the varying distances and relative directions of the sun and the moon.

n

O

objective The *lens or system of lenses nearest to the object being examined through an optical instrument.

occlusion 1. The trapping of small pockets of liquid in a crystal during crystallization. **2.** The absorption of a gas by a solid such that atoms or molecules of the gas occupy interstitial positions in the solid lattice. Palladium, for example, can occlude hydrogen.

occultation The disappearance of a star, planet, or other celestial body behind the moon or another planet, while it is being observed. A solar *eclipse is a form of occultation.

octane number A number that provides a measure of the ability of a fuel to resist *knocking when it is burnt in a spark-ignition engine. It is the percentage by volume of iso-octane (C_8H_{18}; 2,2,4-trimethylpentane) in a blend with normal heptane (C_7H_{16}) that matches the knocking behaviour of the fuel being tested in a single cylinder four-stroke engine of standard design. *Compare* CETANE NUMBER.

octave The interval between two musical notes that have fundamental frequencies in the ratio 2:1; the word describes the interval in terms of the eight notes of the diatonic scale.

octet A stable group of eight electrons in the outer shell of an atom (as in an atom of a noble gas).

ocular *See* EYEPIECE.

odd–even nucleus An atomic nucleus containing an odd number of protons and an even number of neutrons.

odd–odd nucleus An atomic nucleus containing an odd number of protons and an odd number of neutrons. There are very few stable odd–odd nuclei.

oersted Symbol Oe. The c.g.s. unit of magnetic field strength in the. A field has a strength of one oersted if it exerts a force of one dyne on a unit magnetic pole placed in it. It is equivalent to $10^3/4\pi$ A m^{-1}. The unit is named after Hans Christian Oersted.

Oersted, Hans Christian (1777–1851) Danish physicist, who became a professor at Copenhagen in 1806. His best-known discovery came during a lecture in 1820, when he observed the deflection of a compass needle near a wire carrying an electric current. He had discovered electromagnetism.

ohm Symbol Ω. The derived *SI unit of electrical resistance, being the resistance between two points on a conductor when a constant potential difference of one volt, applied between these points, produces a current of one ampere in the conductor. The former **international ohm** (sometimes called the 'mercury ohm') was defined in terms of the resistance of a column of mercury. The unit is named after Georg Ohm.

Ohm, Georg Simon (1787–1854) German physicist, who taught in Cologne, Berlin, Nuremberg, and finally (1849) Munich. He is best known for formulating *Ohm's law in 1827. The unit of electrical resistance is named after him.

ohmmeter Any direct-reading instrument for measuring the value of a resistance in ohms. The instrument commonly used is a *multimeter capable of measuring also both currents and voltages. To measure resistance a dry cell and resistor are switched in series with the moving coil *galvanometer and the unknown resistance is connected across the instrument's terminals. The value of the resistance is then read off an ohms scale. Such instruments are increasingly being replaced by electronic digital multimeters.

Ohm's law The ratio of the potential difference between the ends of a conductor to the current flowing through it is constant. This constant is the *resistance of the conductor, i.e. $V = IR$, where V is the potential difference in volts, I is the current in amperes, and R is the resistance in ohms. The law was discovered in 1827 by Georg Ohm. Most materials do not obey this simple linear law; those that do are said to be **ohmic** but remain so only if physical conditions, such as temperature, remain constant. Metals are the most accurately ohmic conductors.

oil-immersion lens *See* IMMERSION OBJECTIVE.

Oklo reactors Naturally occurring nuclear fission reactors that are believed to have existed in uranium deposits at Oklo in Gabon, West Africa, about 2000 million years ago. In 1972, French scientists noticed a slight difference in the normal $^{235}U/^{238}U$ ratio in uranium ore from Oklo. Further detailed investigations showed that there had been 15 natural reactors in the ore deposits at Oklo, operating intermittently for about 1 million years. It is thought that the geology of the mine was an important factor in the creation of these reactors, in particular, the seepage of water through overlying rock, which functioned as a moderator. A similar natural reactor has been found at Bangombe, some miles south of Oklo, but no other comparable reactors have been found anywhere in the world. The Oklo reactors are of considerable interest. They involve basic nuclear processes occurring 2000 million years ago and might give insights into the time dependence of *fundamental constants. More practically, Oklo can be regarded as a 2000-million-year experiment in the containment of nuclear waste. The reactors shut down naturally when the proportion of ^{235}U decreased, and – for the same reason – natural reactors of this type could not occur today. The products of the reactor have, however, been localized because of the geology of the region, in particular, beds of granite underlying the ore deposits.

Olbers' paradox If the universe is infinite, uniform, and unchanging the sky at night would be bright, as in whatever direction one looked one would eventually see a star. The number of stars would increase in proportion to the square of the distance from the earth; the intensity of light reaching the earth from a given star is inversely proportional to the square of the distance. Consequently, the whole sky should be about as bright as the sun. The paradox, that this is not the case, was stated by Heinrich Olbers (1758–1840) in 1826. (It had been discussed earlier, in 1744, by J. P. L. Chesaux.) The paradox is resolved by the fact that, according to the *big-bang theory, the universe is not infinite, not uniform, and not unchanging. For instance, light from the most distant galaxies displays an extreme *redshift and ceases to be visible.

omega-minus particle A spin 3/2 *baryon made up three strange quarks (*see* STRANGENESS). The existence of the omega-minus particle, as well as its properties, were predicted by Murray *Gell-Mann in 1962 as part of a scheme to classify baryons, called the **eightfold way**. The omega-minus particle was subsequently discovered experimentally, thus demonstrating the validity of the eightfold way. This discovery was historically very important in the theoretical understanding of the *strong interactions. The mass of the omega mass particle is 1672.5 MeV and its average lifetime is 0.8×10^{-10} s. The omega minus particle has an electric charge of –1 (its *antiparticle has a charge of +1).

Oort cloud A cloud of comets around the sun a long way beyond the orbit of Pluto, extending to about half the distance to the nearest star. It consists of large numbers of objects and is the source of long-period *comets. The Oort cloud can be regarded as 'rubble' left over from the formation of the solar system. It is named after the Dutch astrophysicist Jan Hendrik Oort (1900–92).

opacity The extent to which a medium is opaque to electromagnetic radiation, especially to light. It is the reciprocal of the *transmittance. A medium that is opaque to X-rays and gamma rays is said to be **radiopaque**.

opalescence The milky appearance of certain transparent media when they are illuminated by visible polychromatic electromagnetic radiation. Opalescence occurs when particles are suspended in a solution. It also occurs in certain minerals, such as opal, in which there are thin surface films. Opalescence can occur in a system in which there are local fluctuations causing local fluctuations in the refractive index. **Critical opalescence** is opalescence that occurs at a critical point in various phase transitions, such as gas–liquid transitions due to density fluctuations.

open cluster (galactic cluster) *See* STAR CLUSTER.

open-hearth process A traditional method for manufacturing steel by heating together scrap, pig iron, hot metal, etc., in a refractory-lined shallow open furnace heated by burning producer gas in air.

opera glasses *See* BINOCULARS.

operator A mathematical symbol indicating that a specified operation should be carried out. For example, the operator $\sqrt{}$ in $\sqrt{x}$ indicates that the square root of x should be taken; the operator d/dx in dy/dx indicates that y should be differentiated with respect to x, etc.

Oppenheimer–Volkoff limit The maximum mass a neutron star can have before it undergoes gravitational collapse to a *black hole. It is more difficult to estimate this limit than the analogous *Chandrasekhar limit for white dwarf stars. It is thought that the Oppenheimer–Volkoff limit is between two and three times the mass of the sun. It was first calculated by Robert Oppenheimer and George Volkoff in 1939.

opposition The moment at which a planet having its orbit outside that of the earth is in a line with the earth and the sun. When a planet is in opposition it can be observed during the night and is near to its closest point to the earth; it is therefore a favourable opportunity for observation.

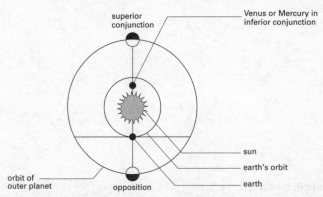

Opposition and conjunction

optical activity The ability of certain substances to rotate the plane of plane-polarized light as it passes through a crystal, liquid, or solution. It occurs when the molecules of the substance are asymmetric, so that they can exist in two different structural forms each being a mirror image of the other. The two forms are **optical isomers** or **enantiomers**. The existence of such forms is also known as **enantiomorphism** (the mirror images being **enantiomorphs**). One form will rotate the light in one direction and the other will rotate it by an equal amount in the other. The two possible forms are described as *dextrorotatory or *laevorotatory according to the direction of rotation, and prefixes *d*- and *l*-, respectively, are used to designate the isomer, as in *d*-tartaric and *l*-tartaric acids (see formulae). An equimolar mixture of the two forms is not optically active. It is called a **racemic mixture** (or **racemate**) and designated by *dl*-. Prefixes are used to designate the isomer: (+)- (dextrorotatory), (−)- (laevorotatory), and (±)- (racemic mixture) are now preferred to, and increasingly used for, the former *d*-, *l*-, and *dl*-, respectively. In addition, certain molecules can have a **meso form** in which one part of the molecule is a mirror image of the other. Such molecules are not optically active.

Molecules that show optical activity have no plane of symmetry. The commonest case of this is in organic compounds in which a carbon atom is linked to four different groups. An atom of this type is said to be a **chiral centre**. Asymmetric molecules showing optical activity can also occur in inorganic compounds. For example, an octahedral complex in which the central ion coordinates to eight different ligands would be optically active. Many naturally occurring compounds show optical isomerism and usually only one isomer occurs naturally. For instance, glucose is found in the dextrorotatory form. The other isomer, *l*- or (−)-glucose, can be synthesized in the laboratory, but cannot be synthesized by living organisms.

Optical activity is understood in terms of the quantum theory of the interaction between light and electrons in molecules.

Isomers of tartaric acid

optical axis (principal axis; optic axis) The line passing through the *optical centre and the centre of a curvature of a *lens or spherical *mirror.

optical centre The point at the geometrical centre of a *lens through which a ray of light entering the lens passes without deviation.

optical fibre A *wave guide through which light can be transmitted with very little leakage through the sidewalls. In the **step-index fibre** a pure glass core, with a diameter between 6 and 250 micrometres, is surrounded by a coaxial glass or plastic cladding of lower refractive index. The cladding is usually between 10 and 150 micrometres thick. The interface between core and cladding acts as a cylindrical mirror at which *total internal reflection of the transmitted light takes place. This structure enables a beam of light to travel through many kilometres of fibre. In the **graded-index fibre**, each layer of glass, from the fibre axis to its outer wall, has a slightly lower refractive index than the layer inside it. This arrangement also prevents light from escaping through the fibre walls by a combination of refraction and total internal reflection, and can be made to give the same transit time for rays at different angles.

Fibre-optic systems use optical fibres to transmit information, in the form of coded pulses or fragmented images (using bundles of fibres), from a source to a receiver. They are also used in medical instruments (**fibrescopes**) to examine internal body cavities, such as the stomach and bladder.

optical flat A flat glass disc having very accurately polished surfaces so that the deviation from perfect flatness does not exceed (usually) 50 nanometres. It is used to test the flatness of such plane surfaces as gauge anvils by means of the *interference patterns formed when parallel beams of light pass through the flat and are reflected by the surface being inspected.

Surfaces are said to be **optically flat** if the deviation from perfect flatness is smaller than the wavelength of light.

optical glass Glass used in the manufacture of lenses, prisms, and other optical parts. It must be homogeneous and free from bubbles and strain. Optical **crown glass** may contain potassium or barium in place of the sodium of ordinary crown glass and has a refractive index in the range 1.51 to 1.54. **Flint glass** contains lead oxide and has a refractive index between 1.58 and 1.72. Higher refractive indexes are obtained by adding lanthanoid oxides to glasses; these are now known as lanthanum crowns and flints.

optical isomers *See* OPTICAL ACTIVITY.

optical lever An experimental device used to measure angular rotation (e.g. in a *galvanometer or *torsion balance). Typically, a small mirror is attached to the rotating object, and a beam of light is directed onto the mirror and reflected onto a scale. The angle turned through by the beam is twice the angle turned through by the mirror.

optical maser *See* LASER.

optical microscope *See* MICROSCOPE.

optical molasses *See* LASER COOLING.

optical pumping *See* LASER.

optical pyrometer *See* PYROMETRY.

optical rotary dispersion (ORD) The effect in which the amount of rotation of plane-polarized light by an optically active compound depends on the wavelength. A graph of rotation against wavelength has a characteristic shape showing peaks or troughs. This phenomenon is explained in the same way as *optical activity.

optical rotation Rotation of plane-polarized light. *See* OPTICAL ACTIVITY.

optical telescope *See* TELESCOPE.

optical temperature *See* RADIATION TEMPERATURE.

optic axis 1. The direction in a doubly refracting crystal in which light is transmitted without double refraction. **2.** *See* OPTICAL AXIS.

optics The study of *light and the phenomena associated with its generation, transmission, and detection. In a broader sense, optics includes all the phenomena associated with infrared and ultraviolet radiation. **Geometrical optics** assumes that light travels in straight lines and is concerned with the laws controlling the reflection and refraction of rays of light. **Physical optics** deals with phenomena that depend on the wave nature of light, e.g. diffraction, interference, and polarization.

optoelectronics The branch of *electronics concerned with electronic devices (frequently solid-state devices) for generating, transmitting, modulating, and detecting *electromagnetic radiation, especially at optical, ultraviolet, and infrared frequencies. An **optoelectronic shutter** is a shutter in a camera that makes use of the *Kerr effect to control or shut off a beam of light. *Compare* PHOTONICS.

orbit 1. (in astronomy) The path through space of one celestial body about another. For one small body moving in the gravitational field of another the orbit is a *conic section. Most such orbits are elliptical and most planetary orbits in the solar system are nearly circular. The shape and size of an elliptical orbit is specified by its eccentricity, e, and the length of its semimajor axis, a (*see* ELLIPSE). **2.** (in physics) The path of an electron as it travels round the nucleus of an atom. *See* ORBITAL.

orbital A region in which an electron may be found in an atom or molecule. In the original *Bohr theory of the atom the electrons were assumed to move around the nucleus in circular orbits, but further advances in quantum mechanics led to the view that it is not possible to give a definite path for an electron. According to *wave mechanics, the electron has a certain probability of being in a given element of space. Thus for a hydrogen atom the electron can be anywhere from close to the nucleus to out in space but the maximum probability in spherical shells of equal thickness occurs in a spherical shell around the nucleus with a radius equal to the Bohr radius of the atom. The probabilities of finding an electron in different regions can be obtained by solving the Schrödinger wave equation to give the wave function ψ, and the probability of location per unit volume is then proportional to $|\psi|^2$. Thus the idea of electrons in fixed orbits has been replaced by that of a probability distribution around the nucleus – an **atomic orbital** (see illustration). Alternatively, the orbital can be thought of as an electric charge distribution (averaged over time). In representing orbitals it is convenient to take a surface enclosing the space in which the electron is likely to be found with a high probability.

The possible atomic orbitals correspond to subshells of the atom. Thus

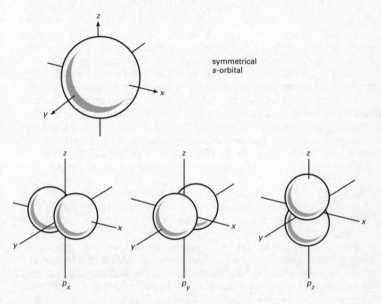

symmetrical
s-orbital

p_x

p_y

p_z

three equivalent *p*-orbitals, each having 2 lobes

Atomic orbitals

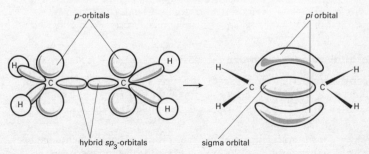

p-orbitals

pi orbital

hybrid sp_3-orbitals

sigma orbital

Molecular orbitals: formation of the double bond in ethene

Orbitals

there is one s-orbital for each shell (orbital quantum number $l = 0$). This is spherical. There are three p-orbitals (corresponding to the three values of l) and five d-orbitals. The shapes of orbitals depend on the value of l. For instance, p-orbitals each have two lobes; most d-orbitals have four lobes.

In molecules, the valence electrons move under the influence of two nuclei (in a bond involving two atoms) and there are corresponding **molecular orbitals** for electrons (see illustration). It is convenient in considering these to regard them as formed by overlap of atomic orbitals. In a hydrogen molecule the s-orbitals on the two atoms overlap and form a molecular orbital between the two nuclei. This is an example of a **sigma orbital**. In a double bond, as in ethene, one bond is produced by overlap along the line of axes to form a sigma orbital. The other is produced by sideways overlap of the lobes of the p-orbitals (see illustration). The resulting molecular orbital has two parts, one on each side of the sigma orbital – this is a **pi orbital**. It is also possible for a **delta orbital** to form by lateral overlap of two d-orbitals. In fact, the combination of two atomic orbitals produces two molecular orbitals with different energies. The one of lower energy is the **bonding orbital**, holding the atoms together; the other is the **antibonding orbital**, which would tend to push the atoms apart. In the case of valence electrons, only the lower (bonding) orbital is filled.

In considering the formation of molecular orbitals it is often useful to think in terms of **hybrid** atomic orbitals. For instance, carbon has in its outer shell one s-orbital and three p-orbitals. In forming methane (or other tetrahedral molecules) these can be regarded as combining to give four equivalent sp^3 hybrid orbitals, each with a lobe directed to a corner of a tetrahedron. It is these that overlap with the s-orbitals on the hydrogen atoms. In ethene, two p-orbitals combine with the s-orbital to give three sp^2 hybrids with lobes in a plane pointing to the corners of an equilateral triangle. These form the sigma orbitals in the C–H and C–C bonds. The remaining p-orbitals (one on each carbon) form the pi orbital. In ethyne, sp^2 hybridization occurs to give two hybrid orbitals on each atom with lobes pointing along the axis. The two remaining p-orbitals on each carbon form two pi orbitals. Hybrid atomic orbitals can also involve d-orbitals. For instance, square-planar complexes use sp^2d hybrids; octahedral complexes use sp^3d^2.

orbital quantum number *See* ATOM.

orbital velocity (orbital speed) The speed of a satellite, spacecraft, or other body travelling in an *orbit around the earth or around some other celestial body. If the orbit is elliptical, the orbital speed, v, is given by:

$$v = \sqrt{[gR^2(2/r - 1/a)]},$$

where g is the acceleration of free fall, R is the radius of the orbited body, a is the semimajor axis of the orbit, and r is the distance between the orbiting body and the centre of mass of the system. If the orbit is circular, $r = a$ and $v = \sqrt{(gR^2/r)}$.

OR circuit *See* LOGIC CIRCUITS.

order 1. The number of times a variable is differentiated. dy/dx represents a first-order derivative, d^2y/dx^2 a second-order derivative, etc. In a *differential equation the order of the highest derivative is the order of the equation. $d^2y/dx^2 + 2dy/dx = 0$ is a second-order equation of the first degree. *See also* DEGREE. **2.** (in physics) A category of *phase transition.

order of magnitude A value expressed to the nearest power of ten.

order parameter A quantity that characterizes the order of a phase of a system below its transition temperature. An order parameter has a nonzero value below the transition temperature and a zero value above the transition temperature. An example of an order parameter is magnetization (*see* MAGNETISM) in a ferromagnetic system. If the *phase transition is continuous (i.e. there is no *latent heat), the order parameter goes to zero continuously as the transition temperature is approached from below. **Disorder parameters** are quantities that are nonzero above the transition temperature and zero beneath it. Order parameters are associated with the *broken symmetry of a system.

ordinary ray *See* DOUBLE REFRACTION.

ordinate *See* CARTESIAN COORDINATES.

origin *See* CARTESIAN COORDINATES.

ortho- Prefix denoting the form of diatomic molecules in which nuclei have parallel spins, e.g. orthohydrogen. *Compare* PARA-.

orthopositronium *See* POSITRONIUM.

oscillation A *periodic motion about an equilibrium position. Examples of oscillations occur in the *pendulum, *simple harmonic motion, a *tuning fork, and vibrations in molecules and solids. An **electrical oscillation** occurs when an electric charge, current, or potential difference changes in a periodic way about an equilibrium value. One oscillation is one complete period of the periodic motion; the number of oscillations per second is called the *frequency of the oscillation.

oscillator An electronic device that produces an alternating output of known frequency. If the output voltage or current has the form of a sine wave with respect to time, the device is called a **sinusoidal** (or **harmonic**) **oscillator**. If the output voltage changes abruptly from one level to another (as in a *square wave or *sawtooth waveform) it is called a **relaxation oscillator**. A harmonic oscillator consists of a frequency-determining circuit or device, such as a *resonant circuit, maintained in oscillation by a source of power that by positive feedback also makes up for the resistive losses. In some relaxation oscillators the circuit is arranged so that in each cycle energy is stored in a reactive element (a capacitor or inductor) and subsequently discharged over a different time interval. *See also* MULTIVIBRATOR.

oscilloscope *See* CATHODE-RAY OSCILLOSCOPE.

osmiridium A hard white naturally occurring alloy consisting principally of osmium (17–48%) and iridium (49%). It also contains small quantities of platinum, rhodium, and ruthenium. It is used for making small items subject to wear, e.g. electrical contacts or the tips of pen nibs.

osmometer *See* OSMOSIS.

osmosis The passage of a solvent through a **semipermeable membrane** separating two solutions of different concentrations. A semipermeable membrane is one through which the molecules of a solvent can pass but the molecules of most solutes cannot. There is a thermodynamic tendency for solutions separated by such a membrane to become equal in concentration, the water (or other solvent) flowing from the weaker to the stronger solution. Osmosis will stop when the two solutions reach equal concentration, and can also be stopped by applying a pressure to the liquid on the stronger-solution side of the membrane. The pressure required to stop the flow from a pure solvent into a solution is a characteristic of the solution, and is called the **osmotic pressure** (symbol Π). Osmotic pressure depends only on the concentration of particles in the solution, not on their nature (i.e. it is a *colligative property). For a solution of n moles in volume V at thermodynamic temperature T, the osmotic pressure is given by $\Pi V = nRT$, where R is the gas constant. Osmotic-pressure measurements are used in finding the relative molecular masses of compounds, particularly macromolecules. A device used to measure osmotic pressure is called an **osmometer**.

   The distribution of water in living organisms is dependent to a large extent on osmosis, water entering the cells through their membranes. A cell membrane is not truly semipermeable as it allows the passage of certain solute molecules; it is described as **partially permeable**. Animals have evolved various means to counteract the effects of osmosis; in plant cells, excessive osmosis is prevented by the pressure exerted by the cell wall, which opposes the osmotic pressure.

osmotic pressure *See* OSMOSIS.

Ostwald ripening A process used in crystal growth in which a mixture of large and small crystals are both in contact with a solvent. The large crystals grow and the small crystals disappear. This occurs because there is a higher energy associated with the smaller crystals. When they dissolve the heat associated with this higher energy is released, enabling recrystallization to occur on the large crystals. Ostwald ripening is used in such applications as photography, requiring crystals with specific properties. The process was discovered by F. W. Ostwald (1853–1932).

Ostwald's dilution law An expression for the degree of dissociation of a weak electrolyte. For example, if a weak acid dissociates in water

$$HA \leftrightarrow H^+ + A^-$$

dissociation constant K_a is given by

$$K_a = \alpha^2 n/(1 - \alpha)V$$

where α is the degree of dissociation, n the initial amount of substance (before dissociation), and V the volume. If α is small compared with 1, then $\alpha^2 = KV/n$; i.e. the degree of dissociation is proportional to the square root of the dilution. The law was first put forward by F. W. Ostwald to account for electrical conductivities of electrolyte solutions.

Otto engine *See* INTERNAL-COMBUSTION ENGINE.

ounce 1. One sixteenth of a pound (avoirdupois), equal to 0.028 349 kg. **2.** Eight drachms (Troy), equal to 0.031 103 kg. **3. (fluid ounce)** Eight fluid drachms, equal to 0.028 413 dm^3.

overdamped *See* DAMPING.

overpotential A potential that must be applied in an electrolytic cell in addition to the theoretical potential required to liberate a given substance at an electrode. The value depends on the electrode material and on the current density. It occurs because of the significant activation energy for electron transfer at the electrodes, and is particularly important for the liberation of such gases as hydrogen and oxygen. For example, in the electrolysis of a solution of zinc ions, hydrogen ($E^\ominus$ = 0.00 V) would be expected to be liberated at the cathode in preference to zinc ($E^\ominus$ = −0.76 V). In fact, the high overpotential of hydrogen on zinc (about 1 V under suitable conditions) means that zinc can be deposited instead.

overtones *See* HARMONIC.

packing density **1.** The number of devices (such as *logic circuits) or integrated circuits per unit area of a *silicon chip. **2.** The quantity of information stored in a specified space of a storage system associated with a computer, e.g. *bits per inch of magnetic tape.

packing fraction The algebraic difference between the relative atomic mass of an isotope and its mass number, divided by the mass number.

pair distribution function The probability of finding a second particle as a function of distance from an initial particle; this probability is found from a statistical average over the system concerned (for example a liquid). A trivial case of the pair distribution function, denoted $g(x)$, is an *ideal gas consisting of point particles; in this case the function is independent of position, with $g(x) = 1$. The pair distribution function relates structure, interparticle interactions, and scattering experiments. Simplifications in the calculation of the pair distribution function arise when the system is homogeneous and translationally invariant. The function features strongly in the theory of liquids.

pair production The creation of an electron and a positron from a photon in a strong electric field, such as that surrounding an atomic nucleus. The electron and the positron each have a mass of about 9×10^{-31} kg, which is equivalent on the basis of the mass–energy equation ($E = mc^2$) to a total of 16×10^{-14} J. The frequency, ν, associated with a photon of this energy (according to $E = h\nu$) is 2.5×10^{20} Hz. Pair production thus requires photons of high quantum energy (Bremsstrahlung or gamma rays). Any excess energy is taken up as kinetic energy of the products.

palaeomagnetism The study of magnetism in rocks, which provides information on variations in the direction and intensity of the earth's magnetic field with time. During the formation of an igneous or sedimentary rock containing magnetic minerals the polarity of the earth's magnetic field at that time becomes 'frozen' into the rock. Studies of this fossil magnetism in samples of rocks have enabled the former positions of magnetic poles at various geological times to be located. It has also revealed that periodic reversals in the geomagnetic field have taken place (i.e. the N-pole becomes the S-pole and vice versa). This information has been important in plate tectonics in establishing the movements of lithospheric plates over the earth's surface. The magnetic reversals provided crucial evidence for the sea-floor spreading hypothesis proposed in the early 1960s.

para- Prefix denoting the form of diatomic molecules in which the nuclei have opposite spins, e.g. parahydrogen. *Compare* ORTHO-.

parabola A *conic with eccentricity $e = 1$. It is the locus of a point that moves so that its distance from the **focus** is equal to its perpendicular distance from the **directrix**. A chord through the focus, perpendicular to the axis, is called the **latus rectum**. For a parabola with its vertex at the origin, lying symmetrically about the x-axis, the equation is $y^2 = 4ax$, where a is the distance from the vertex to the focus. The directrix is the line $x = -a$, and the latus rectum is $4a$. See illustration.

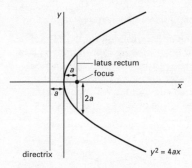

A parabola

parabolic reflector (paraboloidal reflector) A reflector having a section that is a parabola. A concave parabolic reflector will reflect a parallel beam of radiation through its focus and, conversely, will produce a parallel beam if the source of the radiation is placed at its focus. Parabolic mirrors are used in reflecting optical *telescopes to collect the light and in some light sources that require a parallel beam of light. In radio telescopes a dish aerial may also consist of a parabolic reflector.

paraboloid A solid formed by rotating a parabola about its axis of symmetry.

parallax **1.** An apparent displacement of a distant object (with respect to a more distant background) when viewed from two different positions. If such an object is viewed from two points at either end of a base line, the angle between the lines joining the object to the ends of the base line is the **angle of parallax**. If the base line is the distance between the two eyes of an observer the angle is called the **binocular parallax**. **2.** The angular displacement in the apparent position of a celestial body when observed from two different points. **Diurnal parallax** results from the earth's daily rotation, the celestial body being viewed from the surface of the earth rather than from its centre. **Annual parallax** is caused by the earth's motion round the sun, the celestial body being viewed from the earth rather than from the centre of the sun. **Secular parallax** is caused by the motion of the solar system relative to the fixed stars.

parallel circuits A circuit in which the circuit elements are connected so that the current divides between them. For resistors in parallel, the total resistance, R, is given by $1/R = 1/r_1 + 1/r_2 + 1/r_3 \ldots$, where r_1, r_2, and r_3 are the resistances of the individual elements. For capacitors in parallel, the total capacitance, C, is given by $C = c_1 + c_2 + c_3 \ldots$.

parallelepiped (parallelopiped) A solid with six faces, all of which are parallelograms.

parallelogram of forces *See* PARALLELOGRAM OF VECTORS.

parallelogram of vectors A method of determining the *resultant of two *vector quantities. The two vector quantities are represented by two adjacent sides of a parallelogram and the resultant is then the diagonal through their point of intersection. The magnitude and direction of the resultant is found by scale drawing or by trigonometry. The method is used for such vectors as forces (**parallelogram of forces**) and velocities (**parallelogram of velocities**). See illustration.

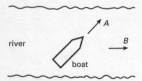

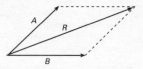

A is the velocity of the boat with respect to the water; *B* is the velocity of the water with respect to the bank

R is the resultant velocity of the boat with respect to the bank

Parallelogram of velocities

parallelogram of velocities *See* PARALLELOGRAM OF VECTORS.

parallel processing A technique that allows more than one process – stream of activity – to be running at any given moment in a computer system, hence processes can be executed in parallel. This means that two or more processors are active among a group of processes at any instant.

parallel spins Neighbouring spinning electrons in which the *spins, and hence the magnetic moments, of the electrons are aligned in the same direction. The interaction between the magnetic moments of electrons in atoms is dominated by *exchange forces. Under some circumstances the exchange interactions between magnetic moments favour parallel spins, while under other conditions they favour *antiparallel spins. The case of ferromagnetism (*see* MAGNETISM) is an example of a system with parallel spins.

parallel worlds *See* MANY-WORLDS INTERPRETATION.

paramagnetism *See* MAGNETISM.

parametric equation An equation of a curve expressed in the form of the parameters that locate points on the curve. The parametric equations of a straight line are $x = a + bt$, $y = c + dt$. For a circle, they are $x = a\cos\theta$, $y = a\sin\theta$.

parapositronium *See* POSITRONIUM.

parasitic capture The absorption of a neutron by a nuclide that does not result in either fission or a useful artificial element.

paraxial ray A ray of light that falls on a reflecting or refracting surface close to and almost parallel to the axis. It is for such rays that simple lens theory can be developed, by means of making small angle approximations.

parent *See* DAUGHTER.

parity Symbol P. The property of a *wave function that determines its behaviour when all its spatial coordinates are reversed in direction, i.e. when x,y,z are replaced by $-x,-y,-z$. If a wave function ψ satisfies the equation $\psi(x,y,z) = \psi(-x,-y,-z)$ it is said to have even parity, if it satisfies $\psi(x,y,z) = -\psi(-x,-y,-z)$ it has odd parity. In general, $\psi(x,y,z) = P\psi(-x,-y,-z)$, where P is a quantum number called parity that can have the value +1 or −1. The principle of **conservation of parity** (or **space-reflection symmetry**) would hold if all physical laws could be stated in a coordinate system independent of left- or right-handedness. If parity was conserved there would therefore be no fundamental way of distinguishing between left and right. In electromagnetic and strong interactions, parity is, in fact, conserved. In 1956, however, it was shown that parity is not conserved in weak interactions. In the beta decay of cobalt-60, for example, the electrons from the decay are emitted preferentially in a direction opposite to that of the cobalt spin. This experiment provides a fundamental distinction between left and right. It is not known why parity is violated in the weak interactions but not in the strong or electromagnetic reactions.

parking orbit *See* SYNCHRONOUS ORBIT.

parsec A unit of length used to express astronomical distance. The distance at which the mean radius of the earth's orbit subtends an angle of one second of arc. One parsec is equal to 3.0857×10^{16} metres or 3.2616 light years.

partial A simple component of a complex tone. When a musical instrument produces a note, say, middle C, it will produce a complex tone in which the fundamental frequency is mixed with a number of partials. Some of these partials, for example, if the note is produced by bowing a taut string, will be *harmonics, i.e. integral multiples of the fundamental. If the string is struck, however, some of the partials can be inexact multiples of the fundamental. Partials are not therefore identical with harmonics.

partial derivative The infinitesimal change in a function consisting of two or more variables when one of the variable changes and the others remain constant. If $z = f(x,y)$, $\partial z/\partial x$ is the partial derivative of z with respect to x, while y remains unchanged. A **partial differential equation**, such as the *Laplace's equation, is an equation containing partial derivatives of a function.

partial eclipse *See* ECLIPSE.

partial pressure *See* DALTON'S LAW.

particle 1. (in physics) One of the fundamental components of matter. *See* ELEMENTARY PARTICLES. **2.** (in mechanics) A hypothetical body that has mass but no physical extension. As it is regarded as having no volume, a particle is incapable of rotation and therefore can only have translational motion. Thus a real body may often, for translational purposes, be regarded as a particle located at the body's centre of mass and having a mass equal to that of the whole body.

particle-beam experiments There are essentially two arrangements for particle-beam experiments in *elementary-particle physics. Before the early 1970s, **fixed-target experiments** were the only means of probing the structures of subatomic particles. In a fixed-target experiment a beam of accelerated particles is directed at a stationary target. An example of such an experimental arrangement is the *Bevatron at Berkeley, built to produce antiprotons by striking a proton at rest with a high-energy proton: $p + p \rightarrow p + p + p + \bar{p}$. According to the mass–energy equation (*see* MASS) some of the kinetic energy of the accelerated proton becomes the mass energy of the proton-antiproton pair. Since this newly formed mass has only the mass equivalence of two proton rest masses, it is remarkable that in practice this fixed target experiment requires an incident kinetic energy equivalent to six times the proton rest mass to create a single proton–antiproton pair. This example illustrates the inefficiency of scattering using a stationary target; conservation of momentum (*see* CONSERVATION LAW) means that the products of a fixed target experiment must possess some residual kinetic energy which therefore cannot be involved in forming the mass-energy of collision products. This problem does not arise in **colliding-beam experiments** introduced in the early 1970s. In these, the beams of particles are caused to collide with their antiparticle counterparts moving in the opposite direction at the same speed. The total momentum of the colliding particles is therefore equal to zero, and their total energy is available to form the mass energy of the collision products. Colliding-beam experiments are known to have lower collision rates than their fixed target counterparts. However, the likelihood of collision is increased by using finely focused beams and by allowing particles to accumulate in storage rings before they are made to collide. New machines in high-energy physics are therefore based on a colliding-beam arrangement.

particle-in-a-box *See* ELECTRICAL CONDUCTIVITY IN METALS; FREE ELECTRON THEORY.

particle physics The study of *elementary particles.

partition If a substance is in contact with two different phases then, in general, it will have a different affinity for each phase. Part of the substance will be absorbed or dissolved by one and part by the other, the relative amounts depending on the relative affinities. The substance is said to be **partitioned** between the two phases. For example, if two immiscible liquids are taken and a third compound is shaken up with them, then an equilibrium is reached in which the concentration in one solvent differs from that in the other. The ratio of the concentrations is the **partition coefficient** of the system. The **partition law** states that this ratio is a constant for given liquids.

partition coefficient *See* PARTITION.

partition function The quantity Z defined by $Z = {}_i\Sigma\exp(-E_i/kT)$, where the sum is taken over all states i of the system. E_i is the energy of the ith state, k is the *Boltzmann constant, and T is the *thermodynamic temperature. Z is a quantity of fundamental importance in equilibrium *statistical mechanics. For a system in which there are nontrivial interactions, it is very difficult to calculate the partition function exactly. For such systems it is necessary to use *approximation techniques and/or *model systems. The partition function links results at the atomic level to *thermodynamics, since Z is related to the Helmholtz *free energy F by $F = kT\ln Z$.

parton A pointlike, almost free, particle postulated as a component of nucleons. The parton model enabled the results of very high-energy experiments on nucleons to be understood. *See* QUANTUM CHROMODYNAMICS.

pascal The *SI unit of pressure equal to one newton per square metre. It is named after Blaise Pascal.

Pascal, Blaise (1623–62) French mathematician and physicist. An infant prodigy, he had already made a mechanical calculating machine by 1642. In physics he formulated *Pascal's law concerning fluid pressure and the principle behind the hydraulic press. The SI unit of pressure is named after him.

Pascal's law In a confined fluid, externally applied pressure is transmitted uniformly in all directions. In a static fluid, force is transmitted at the speed of sound throughout the fluid and acts at right angles to any surface in or bounding the fluid. This principle is made use of in the hydraulic jack, the pneumatic tyre, and similar devices. The law was discovered in 1647 by Blaise Pascal.

Paschen–Back effect An effect in atomic line *spectra that occurs when the atoms are placed in a strong magnetic field. Spectral lines that give the anomalous *Zeeman effect when the atoms are placed in a weaker magnetic field have a different splitting pattern in a very strong magnetic field in which the spectral lines go back to the pattern of the normal Zeeman effect but are split into a set of components that are close in energy. The Paschen–Back effect is named after the German physicists Louis Carl Heinrich Friedrich Paschen (1865–1947) and Ernest E. A. Back (1881–1959), who discovered it in 1912. In the quantum theory of atoms the Paschen–Back effect is explained by the fact that the energies of precession of the electron orbital angular momentum l and the electron spin angular momentum s about the direction of the magnetic field H are greater than the energies of coupling between l and s. In the Paschen–Back effect the orbital magnetic moment and the spin magnetic moment precess independently about the direction of H.

Paschen series *See* HYDROGEN SPECTRUM.

passive device **1.** An electronic component, such as a capacitor or resistor, that is incapable of amplification. **2.** An artificial *satellite that reflects an incoming signal without amplification. **3.** A solar-power device that makes use of an existing structure to collect and utilize solar energy without the use of pumps, fans, etc. **4.** A radar device that provides information for navigation, guidance, surveillance, etc., by receiving the microwave radiation. Such a passive device emits no microwave energy itself and therefore does not disclose its position. **5.** A system that detects an object by the radiation that it emits, rather than by reflecting radiation off it, as in a passive infrared detector (**PIR detector**). *Compare* ACTIVE DEVICE.

path integral formulation A formulation of quantum mechanics put forward by Richard Feynman in 1942 in which all the possible paths a particle in a quantum mechanical system can take, weighted by the probability of each path occurring, are added up. Path integrals have been used extensively, both in analysing the foundations of quantum mechanics and in solving certain types of problem.

Pauli, Wolfgang Ernst (1900–58) Austrian-born Swiss physicist. After studying with Niels *Bohr and Max *Born, he taught at Heidelberg and, finally Zurich. His formulation in 1925 of the *Pauli exclusion principle explained the electronic make-up of atoms. For this work he was awarded the 1945 Nobel Prize for physics. In 1930 he predicted the existence of the *neutrino, which was finally discovered in 1956 by Clyde Cowan (1919–) and Frederick Reines (1918–). He also made many other contributions to quantum field theory, including the *spin–statistics theorem.

Pauli exclusion principle The quantum-mechanical principle, applying to fermions but not to bosons, that no two identical particles in a system, such as electrons in an atom or quarks in a hadron, can possess an identical set of quantum numbers. It was first formulated by Wolfgang Pauli in 1925. The origin of the Pauli exclusion principle lies in the *spin–statistics theorem of relativistic quantum field theory.

PC *See* PERSONAL COMPUTER.

p.d. (potential difference) *See* ELECTRIC POTENTIAL.

pearlite *See* STEEL.

pebi- *See* BINARY PREFIXES.

peculiar motion Motion of a galaxy that is a departure from the expansion described by Hubble's law (*see* HUBBLE CONSTANT). Peculiar motions occur because of the existence of large-scale structure in the universe.

Peierls instability A periodic distortion in a one-dimensional metal; i.e. a chain of atoms with free electrons. A system of this type has a lower energy if there is a periodic lattice distortion, giving alternate longer and shorter bonds. It is observed in organic compounds containing alternating double and single bonds. It was identified by the German-born British physicist Sir Rudolph Peierls (1907–95) in 1955.

pellet fusion *See* THERMONUCLEAR REACTOR.

Peltier effect The change in temperature produced at a junction between two dissimilar metals or semiconductors when an electric current passes through the junction. The direction of the current determines whether the temperature rises or falls. The first metals to be investigated were bismuth and copper; if the current flows from bismuth to copper the temperature rises. If the current is reversed the temperature falls. The effect was discovered in 1834 by J. C. A. Peltier (1785–1845) and has been used recently for small-scale refrigeration. *Compare* SEEBECK EFFECT.

pen drive *See* USB DRIVE.

pendulum Any rigid body that swings about a fixed point. The **ideal simple pendulum** consists of a bob of small mass oscillating back and forth through a small angle at the end of a string or wire of negligible mass. Such a device has a period $2\pi\sqrt{(l/g)}$, where l is the length of the string or wire and g is the *acceleration of free fall. This type of pendulum moves with *simple harmonic motion.

The **compound pendulum** consists of a rigid body swinging about a point within it. The period of such a pendulum is given by $T = 2\pi\sqrt{[(h^2 + k^2)/hg]}$, where k is the radius of gyration about an axis through the centre of mass and h is the distance from the pivot to the centre of mass. *See also* KATER'S PENDULUM.

penetration depth Symbol λ_L. The depth in a superconductor that a magnetic field can penetrate without being expelled (assuming that the field strength is not too high to destroy the superconducting state). This thin surface layer usually has a depth of about 10^{-5} cm and is temperature dependent.

Penrose, Sir Roger *See* HAWKING, STEPHEN WILLIAM.

Penrose process A process by which the rotational energy of a rotating black hole can be extracted. An object close to the event horizon may split into two particles. One, with negative energy, falls into the black hole, causing the rotation rate to decrease. The other, with positive rotation, moves away. The result is that energy is extracted at the expense of rotational energy of the black hole. This process was suggested by Sir Roger Penrose in 1969. *See also* BLANDFORD–ZNAJEK PROCESS.

pentaquark A long-lived particle consisting of five quarks with a mass of just over 1500 MeV, which has been predicted to exist. There is some evidence for the existence of this particle but, at present, this evidence is not conclusive.

pentode A *thermionic valve with a **suppressor grid** between the anode and the screen grid of a tetrode. Its purpose is to suppress the loss of electrons from the anode as a result of secondary emission. The suppressor grid is maintained at a negative potential relative to the anode and to the screen grid.

penumbra *See* SHADOW.

percentile For a random variable in *statistics, any of the 99 values that divide its distribution such that an integral percentage of the collection lies below that value. For example, the 85th percentile is the value of a variable that has 85% of the collection below that value. The 25th percentile is called the lower **quartile**, the 50th percentile is the **median**, and the 75th percentile is the upper quartile.

perfect gas *See* IDEAL GAS; GAS.

perfect pitch *See* ABSOLUTE PITCH.

perfect solution *See* RAOULT'S LAW.

pericynthion The point in the orbit around the moon of a satellite launched from the earth that is nearest to the moon. For a satellite launched from the moon the equivalent point is the **perilune**. *Compare* APOCYNTHION.

perigee *See* APOGEE.

perihelion The point in the orbit of a planet, comet, or artificial satellite in solar orbit at which it is nearest to the sun. The earth is at perihelion on about 3 January. *Compare* APHELION.

period *See* PERIODIC MOTION.

period doubling A mechanism for describing the transition to *chaos in certain dynamical systems. If the force on a body produces a regular orbit with a specific period (*see* PERIODIC MOTION) a sudden increase in the force can suddenly double the period of the orbit and the motion becomes more complex. The original simple motion is called a **one-cycle**, while the more complicated motion after the period doubling is called a **two-cycle**. The process of period doubling can continue until a motion called an **n-cycle** is produced. As *n* increases to infinity the motion becomes nonperiodic. The period-doubling route to chaos occurs in many systems involving nonlinearity, including lasers and certain chaotic chemical reactions. The period-doubling route to chaos was postulated and investigated by the US physicist Mitchell Feigenbaum in the early 1980s. Routes to chaos other than period doubling also exist.

periodic law The principle that the physical and chemical properties of elements are a periodic function of their proton number. The concept was first proposed in 1869 by Dimitri *Mendeleev, using relative atomic mass rather than proton number, as a culmination of efforts to rationalize chemical properties by J. W. Döbereiner (1817), J. A. R. Newlands (1863), and Lothar Meyer (1864). One of the major successes of the periodic law was its ability to predict chemical and physical properties of undiscovered elements and unknown compounds that were later confirmed experimentally.

periodic motion Any motion of a system that is continuously and identically repeated. The time T that it takes to complete one cycle of an oscillation or wave motion is called the **period**, which is the reciprocal of the *frequency. *See* PENDULUM; SIMPLE HARMONIC MOTION.

peripheral device Any device, such as an input or output device, connected to the central processing unit of a *computer. Backing store is also usually regarded as a peripheral.

periscope An optical device that enables an observer to see over or around opaque objects. The simplest type consists of a long tube with mirrors at each end set at 45° to the direction to be viewed. A better type uses internally reflecting prisms instead of plane mirrors. Periscopes are used in tanks (to enable the observer to see over obstacles without being shot at) and in submarines (when the vessel is submerged). Such periscopes are usually quite complicated instruments and include telescopes.

Permalloys A group of alloys of high magnetic permeability consisting of iron and nickel (usually 40–80%) often with small amounts of other elements (e.g. 3–5% molybdenum, copper, chromium, or tungsten). They are used in thin foils in electronic transformers, for magnetic shielding, and in computer memories.

permanent gas A gas, such as oxygen or nitrogen, that was formerly thought to be impossible to liquefy. A permanent gas is now regarded as one that cannot be liquefied by pressure alone at normal temperatures (i.e. a gas that has a critical temperature below room temperature).

permanent magnet *See* MAGNET.

permeability (magnetic permeability) Symbol μ. The ratio of the magnetic flux density, B, in a substance to the external field strength, H; i.e. $\mu = B/H$. The permeability of free space, μ_0, is also called the **magnetic constant** and has the value $4\pi \times 10^{-7}$ H m^{-1} in *SI units. The relative permeability of a substance, μ_r, is given by μ/μ_0 and is therefore dimensionless. *See* MAGNETISM.

permittivity Symbol ε. The ratio of the *electric displacement in a medium to the intensity of the electric field producing it. It is important for electrical insulators used as *dielectrics.

If two charges Q_1 and Q_2 are separated by a distance r in a vacuum, the force F between the charges is given by:

$$F = Q_1 Q_2 / r^2 4\pi\varepsilon_0$$

In this statement of *Coulomb's law using *SI units, ε_0 is called the absolute permittivity of free space, which is now known as the **electric constant**. It has the value 8.854×10^{-12} F m^{-1}.

If the medium between the charges is anything other than a vacuum the equation becomes:

$$F = Q_1 Q_2 / r^2 4\pi\varepsilon$$

and the force between the charges is reduced. ε is the **absolute permittivity** of the new medium. The **relative permittivity** (ε_r) of a medium, formerly called the **dielectric constant**, is given by $\varepsilon_r = \varepsilon/\varepsilon_0$.

permutations and combinations A combination is any subset of a particular set of objects, regardless of the order of selection. If the set consists of n objects, r objects can be selected giving $n!/r!(n-r)!$ different combinations. This can be written $_nC_r$.

A permutation is an ordered subset (i.e. attention is paid to the order of selection or arrangement) of a particular set of objects. If the set consists of n objects, r such objects can be selected to give $n!/(n-r)!$ permutations. This is written $_nP_r$.

perpetual motion 1. Perpetual motion of the first kind. Motion in which a mechanism, once started, would continue indefinitely to perform useful work without being supplied with energy from an outside source. Such a device would contravene the first law of *thermodynamics and is therefore not feasible. Many historical attempts, exercising great ingenuity, were constructed before the concept of energy and its conservation were understood. Some attempts have been made, since the first law of thermodynamics became generally accepted, by inventors seeking to establish loopholes in the laws of nature. **2.** Perpetual motion of the second kind. Motion in which a mechanism extracts heat from a source and converts all of it into some other form of energy. An example of such a mechanism would be a ship that utilized the internal energy of the oceans for propulsion. Such a device does not contravene the first law of thermodynamics but it does contravene the second law. In the case of the ship, the sea would have to be at a higher temperature than the ship to establish a useful flow of heat. This could not occur without an external energy source. **3.** Perpetual motion of the third kind. A form of motion that continues indefinitely but without doing any useful work. An example is the random molecular motion in a substance. This type postulates the complete elimination of friction. A mechanism consisting of frictionless bearings maintained in a vacuum could turn indefinitely, once started, without contravening the first or second laws of thermodynamics, provided it did no external work. Experience indicates that on the macroscopic scale such a condition cannot be achieved. On the microscopic scale, however, a superconducting ring of wire will apparently sustain a perpetual current flow without the application of an external force. This could be considered a form of perpetual motion of the third kind, if the energy required to cool the wire to superconducting temperatures is ignored.

personal computer A general-purpose *microcomputer designed for use by one person at a time. The original Personal Computer (or PC) was a highly successful product from IBM. An **IBM-compatible** computer is functionally identical to an IBM PC and able to accept all hardware and software intended for it. The abbreviation **PC** is now most often used to mean an IBM-compatible computer as opposed to other systems. Personal computers range widely in capability and cost. They may take the form of desktop computers or be portable versions, such as laptop, notebook, or subnotebook computers.

perturbation A departure by a celestial body from the trajectory or orbit it would follow if it moved only under the influence of a single central force. According to *Kepler's law, for example, a single planet orbiting the sun would move in an elliptical orbit. In fact, planets are perturbed from elliptical orbits by the gravitational forces exerted on them by other planets. Similarly, the moon's orbit round the earth is perturbed by the gravitational effect of the sun and the trajectories of comets are perturbed when they pass close to planets.

perturbation theory A method used in calculations in both classical physics (e.g. planetary orbits) and quantum mechanics (e.g. atomic structure), in which the system is divided into a part that is exactly calculable and a small term, which prevents the whole system from being exactly calculable. The technique of perturbation theory enables the effects of the small term to be calculated by an infinite series (which in general is an asymptotic series). Each term in the series is a 'correction term' to the solutions of the exactly calculable system. In classical physics, perturbation theory can be used for calculating planetary orbits. In quantum mechanics, it can be used to calculate the energy levels in molecules. In the many-body problem in quantum mechanics and in relativistic quantum field theory, the terms in perturbation theory may be represented pictorially by Feynman diagrams (*see* QUANTUM ELECTRODYNAMICS).

perversion *See* LATERAL INVERSION.

PET *See* POSITRON EMISSION TOMOGRAPHY.

peta- Symbol P. A prefix used in the metric system to denote one thousand million million times. For example, 10^{15} metres = 1 petametre (Pm).

pewter An alloy of lead and tin. It usually contains 63% tin; pewter tankards and food containers should have less than 35% of lead so that the lead remains in solid solution with the tin in the presence of weak acids in the food and drink. Copper is sometimes added to increase ductility and antimony is added if a hard alloy is required.

Pfund series *See* HYDROGEN SPECTRUM.

pH *See* pH SCALE.

phase 1. A homogeneous part of a heterogeneous system that is separated from other parts by a distinguishable boundary. A mixture of ice and water is a two-phase system. A solution of salt in water is a single-phase system. **2.** A description of the stage that a periodic motion has reached, usually by comparison with another such motion of the same frequency. Two varying quantities are said to be **in phase** if their maximum and minimum values occur at the same instants; otherwise, there is said to be a **phase difference**. *See also* PHASE ANGLE. **3.** One of the circuits in an electrical system or device in which there are two or more alternating currents that are not in phase with each other. In a three-phase system the displacement between the currents is one third of a period. **4.** *See* PHASES OF THE MOON.

phase angle The difference in *phase between two sinusoidally varying quantities. The displacement x_1 of one quantity at time t is given by $x_1 = a\sin\omega t$, where ω is the angular frequency and a is the amplitude. The displacement x_2 of a similar wave that reaches the end of its period T, a fraction β of the period before the first is said to **lead** the first quantity by a time βT; if it reaches the end of its period, a fraction β of the period after the first quantity it **lags** by a time βT. The value of x_2 is then given by $x_2 = a\sin(\omega t + \phi)$. ϕ is called the phase angle and it is equal to $2\pi\beta$.

phase-contrast microscope A type of *microscope that is widely used for examining such specimens as biological cells and tissues. It makes visible the changes in phase that occur when nonuniformly transparent specimens are illuminated. In passing through an object the light is slowed down and becomes out of phase with the original light. With transparent specimens having some structure *diffraction occurs, causing a larger phase change in light outside the central maximum of the pattern. The phase-contrast microscope provides a means of combining this light with that of the central maximum by means of an annular diaphram and a **phase-contrast plate**, which produces a matching phase change in the light of the central maximum only. This gives greater contrast to the final image, due to constructive interference between the two sets of light waves. This is **bright contrast**; in **dark contrast** a different phase-contrast plate is used to make the same structure appear dark, by destructive interference of the same waves. See illustration.

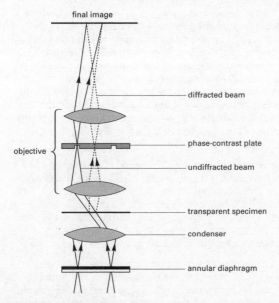

Phase-contrast microscope

phase diagram A graph showing the relationship between solid, liquid, and gaseous *phases over a range of conditions (e.g. temperature and pressure).

phase modulation *See* MODULATION.

phase rule For any system at equilibrium, the relationship $P + F = C + 2$ holds, where P is the number of distinct phases, C the number of components, and F the number of degrees of freedom of the system. The rule was discovered by Josiah Willard *Gibbs in the 1870s.

phases of the moon The shapes of the illuminated surface of the moon as seen from the earth. The shape changes as a result of the relative positions of the earth, sun, and moon.

New moon occurs when the nearside is totally unilluminated by the sun. As the moon moves eastwards in its orbit the sunrise *terminator crosses the nearside from east to west producing a **crescent moon**. The moon is half illuminated at **first quarter**. When it is more than half-phase but less than full phase it is said to be a **gibbous moon**. When the moon is at *opposition the nearside is fully illuminated producing a **full moon**. The sunset terminator then follows to produce a waning gibbous moon, **last quarter**, a waning crescent moon, and eventually the next new moon.

phase space For a system with n degrees of freedom, the $2n$-dimensional space with coordinates $(q_1, q_2, ..., q_n, p_1, p_2, ..., p_n)$, where the qs describe the degrees of freedom of the system and the ps are the corresponding momenta. Each point represents a state of the system. In a gas of N point particles, each particle has three positional coordinates and three corresponding momentum coordinates, so that the phase space has $6N$-dimensions. If the particles have internal degrees of freedom, such as the vibrations and rotations of molecules, then these must be included in the phase space, which is consequently of higher dimension than that for point particles. As the system changes with time the representative points trace out a curve in phase space known as a **trajectory**. *See also* ATTRACTOR; CONFIGURATION SPACE; STATISTICAL MECHANICS.

phase speed (**phase velocity**) Symbol V_p. The speed of propagation of a pure sine wave. $V_p = \lambda f$, where λ is the wavelength and f is the frequency. The value of the phase speed depends on the nature of the medium through which it is travelling and may also depend on the mode of propagation. For electromagnetic waves travelling through space the phase speed c is given by $c^2 = 1/\varepsilon_0\mu_0$, where ε_0 and μ_0 are the *electric constant and the *magnetic constant respectively.

phase transition A change in a feature that characterizes a system. Examples of phase transitions are changes from solid to liquid, liquid to gas, and the reverse changes. Other examples of phase transitions include the transition from a paramagnet to a ferromagnet (*see* MAGNETISM) and the transition from a normally conducting metal to a superconductor. Phase transitions can occur by altering such variables as temperature and pressure.

Phase transitions can be classified by their **order**. If there is nonzero *latent heat at the transition it is said to be a **first-order transition**. If the latent heat is zero it is said to be a **second-order transition**.

Some *models describing phase transitions, particularly in *low-dimensional systems, are amenable to exact mathematical solutions. An effective technique for understanding phase transitions is the *renormalization group since it can deal with problems involving different length-scales, including the feature of **universality**, in which very different physical systems behave in the same way near a phase transition. *See also* RENORMALIZATION GROUP; TRANSITION POINT; BROKEN SYMMETRY; EARLY UNIVERSE.

phasor A rotating *vector that represents a sinusoidally varying quantity. Its length represents the amplitude of the quantity and it is imagined to rotate with angular velocity equal to the angular frequency of the quantity, so that the instantaneous value of the quantity is represented by its projection upon a fixed axis. The concept is convenient for representing the *phase angle between two quantities; it is shown on a diagram as the angle between their phasors.

phlogiston theory A former theory of combustion in which all flammable objects were supposed to contain a substance called **phlogiston**, which was released when the object burned. The existence of this hypothetical substance was proposed in 1669 by the German chemist Johann Becher (1635–81), who called it 'combustible earth' (*terra pinguis*: literally 'fat earth'). For example, according to Becher, the conversion of wood to ashes by burning was explained on the assumption that the original wood consisted of ash and *terra pinguis*, which was released on burning. In the early 18th century Georg Stahl renamed the substance phlogiston (from the Greek for 'burned') and extended the theory to include the calcination (and corrosion) of metals. Thus, metals were thought to be composed of **calx** (a powdery residue) and phlogiston; when a metal was heated, phlogiston was set free and the calx remained. The process could be reversed by heating the metal over charcoal (a substance believed to be rich in phlogiston, because combustion almost totally consumed it). The calx would absorb the phlogiston released by the burning charcoal and become metallic again.

The theory was finally demolished by Antoine *Lavoisier, who showed by careful experiments with reactions in closed containers that there was no *absolute* gain in mass – the gain in mass of the substance was matched by a corresponding loss in mass of the air used in combustion. After experiments with Priestley's dephlogisticated air, Lavoisier realized that this gas, which he named oxygen, was taken up to form a calx (now called an oxide). The role of oxygen in the new theory was almost exactly the opposite of phlogiston's role in the old. In combustion and corrosion phlogiston was released; in the modern theory, oxygen is taken up to form an oxide.

phon A unit of loudness of sound that measures the intensity of a sound relative to a reference tone of defined intensity and frequency. The reference tone usually used has a frequency of 1 kilohertz and a root-mean-square sound pressure of 2×10^{-5} pascal. The observer listens with both ears to the reference tone and the sound to be measured alternately. The reference tone is then increased until the observer judges it to be of equal intensity to the sound to be measured. If the intensity of the reference tone has been increased by n *decibels to achieve this, the sound being measured is said to have an intensity of n phons. The decibel and phon scales are not identical as the phon scale is subjective and relies on the sensitivity of the ear to detect changes of intensity with frequency.

phonon A quantum of *crystal lattice vibrational energy having a magnitude hf, where h is the *Planck constant and f is the frequency of the vibration. Phonons are analogous to the quanta of light, i.e. *photons. The concept of phonons is useful in the treatment of the thermal conductivity of nonmetallic solids and, through consideration of electron–phonon interactions, the temperature dependence of the electrical conductivity of metals and understanding the phenomenon of superconductivity.

phosphor bronze An alloy of copper containing 4% to 10% of tin and 0.05% to 1% of phosphorus as a deoxidizing agent. It is used particularly for marine purposes and where it is exposed to heavy wear, as in gear wheels. *See also* BRONZE.

phosphorescence *See* LUMINESCENCE.

phot A unit of illuminance equal to 10^4 lux or one lumen per square centimetre.

photino *See* SUPERSYMMETRY.

photocathode A *cathode that emits electrons when light falls upon it, as a result of the *photoelectric effect. *See* PHOTOELECTRIC CELL.

photocell *See* PHOTOELECTRIC CELL.

photochemical reaction A chemical reaction caused by light or ultraviolet radiation. The incident photons are absorbed by reactant molecules to give excited molecules or free radicals, which undergo further reaction.

photochemistry The branch of chemistry concerned with *photochemical reactions.

photochromism A change of colour occurring in certain substances when exposed to light. Photochromic materials are used in sunglasses that darken in bright sunlight.

photoconductive effect *See* PHOTOELECTRIC EFFECT.

photodiode A *semiconductor diode used to detect the presence of light or to measure its intensity. It usually consists of a p–n junction device in a container that focuses any light in the environment close to the junction. The device is usually biased in reverse so that in the dark the current is small; when it is illuminated the current is proportional to the

amount of light falling on it. *See* PHOTOELECTRIC CELL; PHOTOELECTRIC
EFFECT.

photodisintegration The decay of a nuclide as a result of the
absorption of a gamma-ray photon.

photoelasticity An effect in which certain materials exhibit double
refraction when subjected to stress. It is used in a technique for detecting
strains in transparent materials (e.g. Perspex, celluloid, and glass). When
polarized white light is passed through a stressed sample, the
birefringence causes coloured patterns to be seen on the viewing screen
of a suitable *polariscope. If monochromatic polarized light is used, a
complex pattern of light and dark fringes is produced.

photoelectric cell (photocell) Any of several devices that produce an
electric signal in response to exposure to electromagnetic radiation. The
original photocells utilized photoemission from a photosensitive cathode
(**photocathode**). The electrons emitted are attracted to an anode. A
positive potential on the anode enables a current to flow through an
external circuit, the current being proportional to the intensity of the
illumination on the cathode. The electrodes are enclosed in an evacuated
glass tube (*see also* PHOTOMULTIPLIER).

  More modern light-sensitive devices utilize the photoconductive effect
and the photovoltaic effect (*see* PHOTOELECTRIC EFFECT; PHOTODIODE;
PHOTOTRANSISTOR; SOLAR CELL). See illustration.

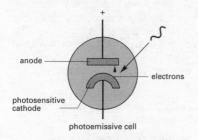

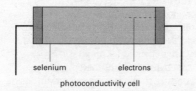

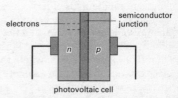

Photoelectric cells

photoelectric effect The liberation of electrons (*see* PHOTOELECTRON)
from a substance exposed to electromagnetic radiation. The number of

electrons emitted depends on the intensity of the radiation. The kinetic energy of the electrons emitted depends on the frequency of the radiation. The effect is a quantum process in which the radiation is regarded as a stream of *photons, each having an energy hf, where h is the Planck constant and f is the frequency of the radiation. A photon can only eject an electron if the photon energy exceeds the *work function, ϕ, of the solid, i.e. if $hf_0 = \phi$ an electron will be ejected; f_0 is the minimum frequency (or **threshold frequency**) at which ejection will occur. For many solids the photoelectric effect occurs at ultraviolet frequencies or above, but for some materials (having low work functions) it occurs with light. The maximum kinetic energy, E_m, of the photoelectron is given by *Einstein's equation: $E_m = hf - \phi$ (*see also* PHOTOIONIZATION).

Apart from the liberation of electrons from atoms, other phenomena are also referred to as photoelectric effects. These are the **photoconductive effect** and the **photovoltaic effect**. In the photoconductive effect, an increase in the electrical conductivity of a semiconductor is caused by radiation as a result of the excitation of additional free charge carriers by the incident photons. **Photoconductive cells**, using such photosensitive materials as cadmium sulphide, are widely used as radiation detectors and light switches (e.g. to switch on street lighting).

In the photovoltaic effect, an e.m.f. is produced between two layers of different materials as a result of irradiation. The effect is made use of in **photovoltaic cells**, most of which consist of p–n *semiconductor junctions (*see also* PHOTODIODE; PHOTOTRANSISTOR). When photons are absorbed near a p–n junction new free charge carriers are produced (as in photoconductivity); however, in the photovoltaic effect the electric field in the junction region causes the new charge carriers to move, creating a flow of current in an external circuit without the need for a battery. *See also* PHOTOELECTRIC CELL.

photoelectron An electron emitted from a substance by irradiation as a result of the *photoelectric effect or *photoionization.

photoelectron spectroscopy A technique for determining the *ionization potentials of molecules. The sample is a gas or vapour irradiated with a narrow beam of ultraviolet radiation (usually from a helium source at 58.4 nm, 21.21 eV photon energy). The photoelectrons produced in accordance with *Einstein's equation are passed through a slit into a vacuum region, where they are deflected by magnetic or electrostatic fields to give an energy spectrum. The photoelectron spectrum obtained has peaks corresponding to the ionization potentials of the molecule (and hence the orbital energies). The technique also gives information on the vibrational energy levels of the ions formed. **ESCA** (electron spectroscopy for chemical analysis) is a similar analytical technique in which a beam of X-rays is used. In this case, the electrons ejected are from the inner shells of the atoms. Peaks in the electron spectrum for a particular element show characteristic chemical shifts, which depend on the presence of other atoms in the molecule.

photoemission The process in which electrons are emitted by a substance as a result of irradiation. *See* PHOTOELECTRIC EFFECT; PHOTOIONIZATION.

photofission A *nuclear fission that is caused by a gamma-ray photon.

photographic density A measure of the opacity of a photographic emulsion (negative or transparency). *See* DENSITOMETER.

photography The process of forming a permanent record of an image on specially treated film or paper. In normal black-and-white photography a camera is used to expose a film or plate to a focused image of the scene for a specified time. The film or plate is coated with an emulsion containing silver salts and the exposure to light causes the silver salts to break down into silver atoms; where the light is bright dark areas of silver are formed on the film after development (by a mild reducing agent) and fixing. The negative so formed is printed, either by a contact process or by projection. In either case light passing through the negative film falls on a sheet of paper also coated with emulsion. Where the negative is dark, less light passes through and the resulting positive is light in this area, corresponding with a light area in the original scene. As photographic emulsions are sensitive to ultraviolet and X-rays, they are widely used in studies involving these forms of electromagnetic radiation. *See also* COLOUR PHOTOGRAPHY.

photoionization The *ionization of an atom or molecule as a result of irradiation by electromagnetic radiation. For a photoionization to occur the incident photon of the radiation must have an energy in excess of the *ionization potential of the species being irradiated. The ejected photoelectron will have an energy, E, given by $E = hf - I$, where h is the Planck constant, f is the frequency of the incident radiation, and I is the ionization potential of the irradiated species.

photolithography A technique used in the manufacture of semiconductor components, integrated circuits, etc. It depends on the principle of masking selected areas of a surface and exposing the unmasked areas to such processes as the introduction of impurities, deposition of thin films, removal of material by etching, etc. The technique has been developed for use on tiny structures (typically measured in micrometres), which can only be examined by means of an electron microscope.

photoluminescence *See* LUMINESCENCE.

photolysis A chemical reaction produced by exposure to light or ultraviolet radiation. Photolytic reactions often involve free radicals, the first step being homolytic fission of a chemical bond. (*See* FLASH

PHOTOLYSIS.) The photolysis of water, using energy from sunlight absorbed by chlorophyll, produces gaseous oxygen, electrons, and hydrogen ions and is a key reaction in photosynthesis.

photometer An instrument used to measure *luminous intensity, illumination, and other photometric quantities. The older types rely on visual techniques to compare a source of light with a standard source. More modern photometers use *photoelectric cells based on the photoconductive, photoemissive, or photovoltaic effect. The photovoltaic types do not require an external power source and are therefore very convenient to use but are relatively insensitive. The photoemissive type usually incorporates a *photomultiplier, especially for use in astronomy and with other weak sources. Photoconductive units require only low-voltage supplies, which makes them convenient for commercial illumination meters and photographers' exposure meters.

photometric brightness *See* LUMINANCE.

photometry The study of visual radiation, especially the calculations and measurements of *luminous intensity, *luminous flux, etc. In some cases photometric calculations and measurements extend into the near infrared and the near ultraviolet.

In photometry, two types of measurement are used: those that measure **luminous** quantities rely on the use of the human eye (for example, to compare the illuminance of two surfaces); those called **radiant** quantities rely on the use of photoelectric devices to measure electromagnetic energy. *See also* PHOTOMETER.

photomicrography The use of photography to obtain a permanent record (a **photomicrograph**) of the image of an object as viewed through a microscope.

photomultiplier A sensitive type of *photoelectric cell in which electrons emitted from a photocathode are accelerated to a second electrode where several electrons are liberated by each original photoelectron, as a result of *secondary emission. The whole process is repeated as many times as necessary to produce a useful electric current by secondary emission from the last electrode. A photomultiplier is thus a photocathode with the output amplified by an electron multiplier. The initial photocurrent can be amplified by a factor of 10^8. Photomultipliers are thus useful when it is necessary to detect low intensities of light, as in stellar photometry, star and planet tracking in guidance systems, and more mundanely in process control.

photon A particle with zero *rest mass consisting of a *quantum of electromagnetic radiation. The photon may also be regarded as a unit of energy equal to hf, where h is the *Planck constant and f is the frequency of the radiation in hertz. Photons travel at the speed of light. They are required to explain the photoelectric effect and other phenomena that require *light to have particle character.

photoneutron A neutron emitted by an atomic nucleus undergoing a *photonuclear reaction.

photonic crystal A crystal lattice in which diffraction of electromagnetic radiation occurs with visible light, as with *X-ray diffraction. Because the wavelength of visible light is much longer than the wavelength of X-rays the distance between atoms is much greater in a photonic crystal than an ordinary crystal.

photonics The study of devices analogous to those used in electronics, but with the electrons replaced by photons. Thus, photonics is concerned with devices involving the transmission, modulation, reflection, refraction, amplification, detection, and guidance of light. Examples are *lasers and *optical fibres. Photonics is used extensively in *telecommunications. *Compare* OPTOELECTRONICS.

photonuclear reaction A *nuclear reaction that is initiated by a (gamma-ray) photon.

photopic vision The type of vision that occurs when the cones in the eye are the principal receptors, i.e. when the level of illumination is high. Colours can be identified with photopic vision. *Compare* SCOTOPIC VISION.

photoreceptor A sensory cell or group of cells that reacts to the presence of light. It usually contains a pigment that undergoes a chemical change when light is absorbed, thus stimulating a nerve. *See* EYE.

photosensitive substance 1. Any substance that when exposed to electromagnetic radiation produces a photoconductive, photoelectric, or photovoltaic effect. **2.** Any substance, such as the emulsion of a photographic film, in which electromagnetic radiation produces a chemical change.

photosphere The visible surface of the *sun or other star and the source of its continuous spectrum. It is a gaseous layer several hundreds of kilometres thick with an average temperature of 5780 K. Where the photosphere merges with the *chromosphere the temperature is 4000 K.

phototransistor A junction *transistor that is photosensitive. When radiation falls on the emitter-base junction, new free charge carriers are created in the base region and the collector current is increased. Phototransistors are similar to *photodiodes except that the primary photoelectric current is amplified internally and it is therefore more sensitive to light than the photodiode. Some types can be used as switching or bistable devices, a small intensity of radiation switching them from a low to high current state.

photovoltaic effect *See* PHOTOELECTRIC EFFECT.

pH scale A logarithmic scale for expressing the acidity or alkalinity of a solution. To a first approximation, the pH of a solution can be defined as $-\log_{10}c$, where c is the concentration of hydrogen ions in moles per cubic decimetre. A neutral solution at 25°C has a hydrogen-ion concentration of 10^{-7} mol dm^{-3}, so the pH is 7. A pH below 7 indicates an acid solution; one above 7 indicates an alkaline solution. More accurately, the pH depends not on the concentration of hydrogen ions but on their *activity, which cannot be measured experimentally. For practical purposes, the pH scale is defined by using a hydrogen electrode in the solution of interest as one half of a cell, with a reference electrode (e.g. a calomel electrode) as the other half cell. The pH is then given by $(E - E_R)F/2.303RT$, where E is the e.m.f. of the cell, E_R the standard electrode potential of the reference electrode, and F the Faraday constant. In practice, a glass electrode is more convenient than a hydrogen electrode. pH stands for 'potential of hydrogen'. The scale was introduced by S. P. Sørensen in 1909.

physical chemistry The branch of chemistry concerned with the effect of chemical structure on physical properties. It includes chemical thermodynamics and electrochemistry.

physics The study of the laws that determine the structure of the universe with reference to the matter and energy of which it consists. It is concerned not with chemical changes that occur but with the forces that exist between objects and the interrelationship between matter and energy. Traditionally, the study was divided into separate fields: heat, light, sound, electricity and magnetism, and mechanics (*see* CLASSICAL PHYSICS). Since the turn of the century, however, quantum mechanics and relativistic physics have become increasingly important; the growth of modern physics has been accompanied by the studies of atomic physics, nuclear physics, and particle physics. The physics of astronomical bodies and their interactions is known as **astrophysics**, the physics of the earth is known as **geophysics**, and the study of the physical aspects of biology is called **biophysics**. *See also* THEORETICAL PHYSICS.

physisorption *See* ADSORPTION.

pi Symbol π. The ratio of the circumference of any circle to its diameter. It is a *transcendental number with the value 3.141 592....

pico- Symbol p. A prefix used in the metric system to denote 10^{-12}. For example, 10^{-12} farad = 1 picofarad (pF).

pie chart A diagram in which percentages are shown as sectors of a circle. If x percent of the electorate vote for party X, y percent for party Y, and z percent for party Z, a pie chart would show three sectors having central angles $3.6x°$, $3.6y°$, and $3.6z°$.

pi electron An electron in a pi orbital. *See* ORBITAL.

piezoelectric crystal *See* PIEZOELECTRIC EFFECT.

piezoelectric effect The generation of a potential difference across the opposite faces of certain nonconducting crystals (**piezoelectric crystals**) as a result of the application of mechanical stress between these faces. The electric polarization (see DIELECTRIC) produced is proportional to the stress and the direction of the polarization reverses if the stress changes from compression to tension. The **reverse piezoelectric effect** is the opposite phenomenon: if the opposite faces of a piezoelectric crystal are subjected to a potential difference, the crystal changes its shape. Rochelle salt and quartz are the most frequently used piezoelectric materials. While Rochelle salt produces the greater polarization for a given stress, quartz is more widely used as its crystals have greater strength and are stable at temperatures in excess of 100°C.

If a quartz plate is subjected to an alternating electric field, the reverse piezoelectric effect causes it to expand and contract at the field frequency. If this field frequency is made to coincide with the natural elastic frequency of the crystal, the plate resonates; the direct piezoelectric effect then augments the applied electric field. This is the basis of the *crystal oscillator and the *quartz clock. See also CRYSTAL MICROPHONE; CRYSTAL PICK-UP.

piezoelectric oscillator See PIEZOELECTRIC EFFECT.

pig iron The impure form of iron produced by a blast furnace, which is cast into pigs (blocks) for converting at a later date into cast iron, steel, etc. The composition depends on the ores used, the smelting procedure, and the use to which the pigs will later be put.

pi-meson See PION.

pinch effect A magnetic attraction between parallel conductors carrying currents flowing in the same direction. The force was noticed in early induction furnaces. Since the late 1940s it has been widely studied as a means of confining the hot plasma in a *thermonuclear reactor. In an experimental toroidal thermonuclear reactor a large electric current is induced in the plasma by electromagnetic induction; this current both heats the plasma and draws it away from the walls of the tube as a result of the pinch effect.

pion (pi-meson) An *elementary particle classified as a *meson. It exists in three forms: neutral, positively charged, and negatively charged. The charged pions decay into muons and neutrinos; the neutral pion decays into two gamma-ray photons. Pions consist of a quark and an anti-quark.

Pirani gauge An instrument devised in 1906 by Marcello Pirani to measure low pressures (1–10^{-4} torr; 100–0.01 Pa). It consists of an electrically heated filament, which is exposed to the gas whose pressure is to be measured. The extent to which heat is conducted away from the filament depends on the gas pressure, which thus controls its equilibrium temperature. Since the resistance of the filament is dependent on its temperature, the pressure is related to the resistance of the filament. The filament is arranged to be part of a *Wheatstone bridge circuit and the pressure is read from a microammeter calibrated in pressure units. As the effect depends on the thermal conductivity of the gas, the calibration has to be made each time the pressure of a different gas is measured.

PIR detector *See* PASSIVE DEVICE.

pitch 1. (in physics) The property of a sound that characterizes its highness or lowness to an observer. It is related to, but not identical with, frequency. Below about 1000 Hz the pitch is slightly higher than the frequency and above 1000 the position is reversed. The loudness of a sound also affects the pitch. Up to 1000 Hz an increase in loudness causes a decrease in pitch. From about 1000 to 3000 Hz the pitch is independent of loudness, while above 3000 Hz an increase in loudness seems to cause a raising of pitch. Pitch is usually measured in mels; a note of 1000 Hz frequency with a loudness of 40 decibels above the absolute threshold of hearing has a pitch of 1000 mels. **2.** (in mechanics) *See* SCREW.

Pitot tube A device for measuring the speed of a fluid. It consists of two tubes, one with an opening facing the moving fluid and the other with an opening at 90° to the direction of the flow. The two tubes are connected to the opposite sides of a manometer so that the difference between the dynamic pressure in the first tube and the static pressure in the second tube can be measured. The speed v of the flow of an incompressible fluid is then given by: $v^2 = 2(P_2 - P_1)/\rho$, where P_2 is the dynamic pressure, P_1 is the static pressure, and ρ is the density of the fluid. The device has a wide variety of applications. It was devised by Henri Pitot (1695–1771).

pixel One of the tiny dots that make up an image on the screen of a computer's *visual-display unit (VDU) or on some types of television receiver; it is short for picture element. Screen resolution is determined by the number of pixels (the more pixels, the better the resolution), and each pixel is given a brightness and colour. A typical high-resolution colour VDU screen has a 1024×768 array of pixels.

Planck, Max Karl Ernst Ludwig (1858–1947) German physicist, who became a professor at Berlin University in 1892. Here he formulated the *quantum theory, which had its basis in a paper of 1900. (*See also* PLANCK CONSTANT; PLANCK'S RADIATION LAW.) One of the most important scientific discoveries of the century, this theory earned him the 1918 Nobel Prize for physics.

Planck constant Symbol h. The fundamental constant equal to the ratio of the energy E of a quantum of energy to its frequency v: $E = hv$. It has the value $6.626\ 176 \times 10^{-34}$ J s. It is named after Max Planck. In quantum-mechanical calculations (especially particle physics) the **rationalized Planck constant** $= h/2\pi = 1.054\ 589 \times 10^{-34}$ J s is frequently used. It is also known as the **Dirac constant**.

Planck length The length scale at which a classical description of gravity ceases to be valid, and *quantum mechanics must be taken into account. It is given by $L_P = \sqrt{(G\hbar/c^3)}$, where G is the gravitational constant, is the rationalized Planck constant, and c is the speed of light. The value of the Planck length is of order 10^{-35} m (twenty orders of magnitude smaller than the size of a proton, 10^{-15} m).

Planck mass The mass of a particle whose Compton wavelength is equal to the *Planck length. It is given by $m_P = \sqrt{(c/G)}$, where is the rationalized Planck constant, c is the speed of light, and G is the gravitational constant. The description of an elementary particle of this mass, or particles interacting with energies per particle equivalent to it (through $E = mc^2$), requires a *quantum theory of gravity. Since the Planck mass is of order 10^{-8} kg (equivalent energy 10^{19} GeV), and, for example, the proton mass is of order 10^{-27} kg and the highest energies attainable in present-day particle accelerators are of order 10^3 GeV, quantum-gravitational effects do not arise in laboratory particle physics. However, energies equivalent to the Planck mass did occur in the early universe according to *big-bang theory, and a quantum theory of gravity is important for discussing conditions there (*see* PLANCK TIME).

Planck's radiation law A law stated by Max Planck giving the distribution of energy radiated by a *black body. It introduced into physics the novel concept of energy as a quantity that is radiated by a body in small discrete packets rather than as a continuous emission. These small packets became known as quanta and the law formed the basis of *quantum theory. The **Planck formula** gives the energy radiated per unit time at frequency ν per unit frequency interval per unit solid angle into an infinitesimal cone from an element of the black-body surface that is of unit area in projection perpendicular to the cone's axis. The expression for this **monochromatic specific intensity** I_ν is:

$$I_\nu = 2hc^{-2}\nu^3/[\exp(h\nu/kT - 1)],$$

where h is the *Planck constant, c is the *speed of light, k is the *Boltzmann constant, and T is the thermodynamic temperature of the black body. I_ν has units of watts per square metre per steradian per hertz (W m^{-2} sr^{-1} Hz^{-1}). The monochromatic specific intensity can also be expressed in terms of the energy radiated at wavelength λ per unit wavelength interval; it is then written as I_λ, and the Planck formula is:

$$I_\lambda = 2hc^2\lambda^{-5}/[\exp(hc/\lambda kT) - 1].$$

There are two important limiting cases of the Planck formula. For low frequencies ν << kT/h (equivalently, long wavelengths λ >> hc/kT) the **Rayleigh–Jeans formula** is valid:

$$I_\nu = 2c^{-2}\nu^2 kT,$$

or

$$I_\lambda = 2c\lambda^{-4}kT.$$

Note that these expressions do not involve the Planck constant. They can be derived classically and do not apply at high frequencies, i.e. high energies, when the quantum nature of *photons must be taken into account. The second limiting case is the **Wien formula**, which applies at high frequencies $\nu \gg kT/h$ (equivalently, short wavelengths $\lambda \ll hc/kT$):

$$I_\nu = 2hc^{-2}\nu^3 \exp(-h\nu/kT),$$

or

$$I_\lambda = 2hc^2\lambda^{-5} \exp(-hc/\lambda kT).$$

See also WIEN'S DISPLACEMENT LAW.

Planck time The time taken for a photon (travelling at the speed of light c) to move through a distance equal to the *Planck length. It is given by $t_P = \sqrt{(G\hbar/c^5)}$, where G is the gravitational constant and $\hbar$ is the rationalized Planck constant. The value of the Planck time is of order 10^{-43} s. In the *big-bang cosmology, up until a time t_P after the initial instant, it is necessary to use a *quantum theory of gravity to describe the evolution of the universe.

Planck units A system of *units, used principally in discussions of *quantum theories of gravity, in which length, mass, and time are expressed as multiples of the *Planck length, mass, and time respectively. This is equivalent to setting the gravitational constant, the speed of light, and the reduced Planck constant all equal to unity. All quantities that ordinarily have dimensions involving length, mass, and time become dimensionless in Planck units. Since, in the subject area where Planck units are used, it is normal to employ *Gaussian units or *Heaviside–Lorentz units for electromagnetic quantities, these then become dimensionless. *See also* GEOMETRIZED UNITS; NATURAL UNITS.

plane A flat surface defined by the condition that any two points in the plane are joined by a straight line that lies entirely in the surface.

plane-polarized light *See* POLARIZATION OF LIGHT.

planet A body that revolves around a central astronomical body, especially one of the nine bodies (Mercury; Venus; earth; Mars; Jupiter; Saturn; Uranus; Neptune; Pluto) that revolve in elliptical orbits around the sun. *See* SOLAR SYSTEM.

planimeter An instrument used to measure the area of a closed curve. The outline of the curve is followed by a pointer on the instrument and the area is given on a graduated disc.

plano-concave lens *See* CONCAVE; LENS.

plano-convex lens *See* CONVEX; LENS.

plasma A highly ionized gas in which the number of free electrons is approximately equal to the number of positive ions. Sometimes described as the fourth state of matter, plasmas occur in interstellar space, in the atmospheres of stars (including the sun), in discharge tubes, and in experimental thermonuclear reactors.

Because the particles in a plasma are charged, its behaviour differs in some respects from that of a gas. Plasmas can be created in the laboratory by heating a low-pressure gas until the mean kinetic energy of the gas particles is comparable to the *ionization potential of the gas atoms or molecules. At very high temperatures, from about 50 000 K upwards, collisions between gas particles cause cascading ionization of the gas. However, in some cases, such as a fluorescent lamp, the temperature remains quite low as the plasma particles are continually colliding with the walls of the container, causing cooling and recombination. In such cases ionization is only partial and requires a large energy input. In *thermonuclear reactors an enormous plasma temperature is maintained by confining the plasma away from the container walls using electromagnetic fields (*see* PINCH EFFECT). The study of plasmas is known as **plasma physics**.

plasma oscillation A *collective oscillation that occurs in a *plasma. To ensure that the plasma is electrically neutral there has to be a positive background to the free electrons. This causes a charge disturbance in the electron density, which the electrons move to screen. The electrons tend to overshoot in doing so and are pulled back, overshoot again, etc., with the result that the charge density is described by a *simple harmonic motion, which has a frequency called the *Langmuir frequency, named after the US chemist Irving Langmuir (1881–1957), who analysed plasma oscillations in 1928. In plasmas described by *quantum mechanics, such as the *electron gas and the electrons in a metal plasma, oscillations become *plasmons. *See also* LANDAU DAMPING.

plasmon A *collective excitation for quantized oscillations of the electrons in a metal.

plasticity The property of solids that causes them to change permanently in size or shape as a result of the application of a stress in excess of a certain value, called the **yield point**.

pleochroic Denoting a crystal that appears to be of different colours, depending on the direction from which it is viewed. It is caused by polarization of light as it passes through an anisotropic medium.

plutonium Symbol Pu. A dense silvery radioactive metallic transuranic element belonging to the actinoids; a.n. 94; mass number of most stable isotope 244 (half-life 7.6×10^7 years); r.d. 19.84; m.p. 641°C; b.p. 3232°C. Thirteen isotopes are known, by far the most important being plutonium–239 (half-life 2.44×10^4 years), which undergoes *nuclear fission with slow neutrons and is therefore a vital power source for *nuclear weapons and some *nuclear reactors. About 20 tonnes of plutonium are produced annually by the world's nuclear reactors, a detailed inventory of every gram of which is kept in order to prevent its military misuse. The element was first produced by Seaborg, McMillan, Kennedy, and Wahl in 1940.

Poincaré group *See* LORENTZ GROUP.

Poincaré sphere A geometrical representation of the state of polarization of a wave of *monochromatic radiation. The *Stokes' parameters s_1, s_2, s_3 can be regarded as the Cartesian coordinates of a point P on a sphere S, which has a radius s_0. This representation means that every possible state of polarization for a plane monochromatic wave with a given intensity (meaning that $s_0 = c$, where c is a constant) corresponds to one point on S and vice versa. This representation was introduced in 1892 by the French physicist Henri Poincaré (1854–1912) with S being called the Poincaré sphere. Applications of the Poincaré sphere include the polarization of light in crystals.

Poincaré stresses Nonelectric forces postulated to give stability to a model of the *electron. Because of the difficulties in regarding an electron as a point charge it is possible to postulate that the electron is a charge distribution with a nonzero radius. However, an electric charge distribution alone is unstable. In 1906 Henri Poincaré postulated unknown nonelectric forces, now called Poincaré stresses, to give stability to the electron. Considerations such as these are now thought to be irrelevant, as it is accepted that an electron should be described by *quantum electrodynamics rather than *classical field theory.

point-contact transistor *See* TRANSISTOR.

point defect *See* DEFECT.

point discharge *See* CORONA.

point group The group formed by all the symmetry operations applied to a pattern arranged around a fixed point. The symmetry operations of molecules form point groups. There are 32 point groups, called **crystallographic point groups**, that have symmetries that are also compatible with the translational symmetries of crystals.

poise A c.g.s. unit of viscosity equal to the tangential force in dynes per square centimetre required to maintain a difference in velocity of one centimetre per second between two parallel planes of a fluid separated by one centimetre. 1 poise is equal to 10^{-1} N s m^{-2}.

Poiseuille's equation An equation relating the volume flow rate, V, of a fluid through a cylindrical tube to the pressure difference, p, between the ends of the tube: $V = \pi p r^4/8l\eta$, where r is the radius and l the length of the tube; η is the viscosity of the fluid. It applies if the Reynolds number is less than 2000 and was first stated by Jean Louis Poiseuille (1799–1869).

poison A substance that absorbs neutrons in a nuclear reactor and therefore slows down the reaction. It may be added intentionally for this purpose or may be formed as a fission product and need to be periodically removed.

Poisson's ratio The ratio of the lateral strain to the longitudinal strain in a stretched rod. If the original diameter of the rod is d and the contraction of the diameter under stress is Δd, the lateral strain $\Delta d/d = s_d$; if the original length is l and the extension under stress Δl, the longitudinal strain is $\Delta l/l = s_l$. Poisson's ratio is then s_d/s_l. For steels the value is between 0.28 and 0.30 and for aluminium alloys it is about 0.33. It was first introduced by Siméon-Denis Poisson (1781–1840).

Poisson's spot A bright spot at the centre of the shadow of a circular opaque obstacle. It was predicted to occur by Siméon-Denis Poisson as a consequence of the wave theory of light. The effect was observed by Dominique François Jean Arago (1786–1853), and is sometimes known as **Arago's spot**.

polar coordinates A system used in analytical geometry to locate a point P, with reference to two or three axes. The distance of P from the

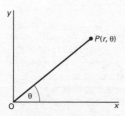

two-dimensional coordinates

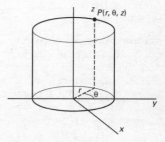

cylindrical polar coordinates

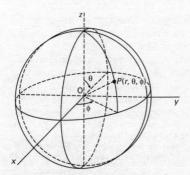

spherical polar coordinates

Polar coordinates

origin is r, and the angle between the x-axis and the **radius vector** OP is θ, thus in two-dimensional polar coordinates the coordinates of P are $(r,θ)$. If the *Cartesian coordinates of P are (x,y) then $x = r\cosθ$ and $y = r\sinθ$.

In three dimensions the point P may be regarded as lying on the surface of a cylinder, giving **cylindrical polar coordinates**, or on the surface of a sphere, giving **spherical polar coordinates**. In the former the coordinates of P would be $(r,θ,z)$; in the latter they would be $(r,θ,φ)$ (see illustration).

polarimeter (polariscope) An instrument used to determine the angle through which the plane of polarization of plane-polarized light is rotated on passing through an optically active substance. Essentially, a polarimeter consists of a light source, a **polarizer** (e.g. a sheet of Polaroid) for producing plane-polarized light, a transparent cell containing the sample, and an **analyser**. The analyser is a polarizing material that can be rotated. Light from the source is plane-polarized by the polarizer and passes through the sample, then through the analyser into the eye or onto a light-detector. The angle of polarization is determined by rotating the analyser until the maximum transmission of light occurs. The angle of rotation is read off a scale. Simple portable polarimeters are used for estimating the concentrations of sugar solutions in confectionary manufacture. *See also* PHOTOELASTICITY.

polarizability Symbol α. A measure of the response of a molecule to an external electric field. When a molecule is placed in an external electric field, the displacement of electric charge induces a dipole in the molecule (*see* ELECTRIC DISPLACEMENT). If the *electric field strength is denoted E and the electrical dipole moment induced by this electric field p, the polarizability α is defined by $p = αE$. To calculate the polarizability from first principles it is necessary to use the quantum mechanics of molecules. However, if regarded as a parameter, the polarizability α provides a link between microscopic and macroscopic theories as in the *Clausius–Mossotti equation and the *Lorentz–Lorenz equation.

polarization 1. The process of confining the vibrations of the vector constituting a transverse wave to one direction. In unpolarized radiation the vector oscillates in all directions perpendicular to the direction of propagation. *See* POLARIZATION OF LIGHT. **2.** The formation of products of the chemical reaction in a *voltaic cell in the vicinity of the electrodes resulting in increased resistance to current flow and, frequently, to a reduction in the e.m.f. of the cell. *See also* DEPOLARIZATION. **3.** The partial separation of electric charges in an insulator subjected to an electric field.

polarization of light The process of confining the vibrations of the electric vector of light waves to one direction. In unpolarized light the electric field vibrates in all directions perpendicular to the direction of propagation. After reflection or transmission through certain substances (*see* POLAROID) the electric field is confined to one direction and the radiation is said to be **plane-polarized light**. The plane of plane-polarized light can be rotated when it passes through certain substances (*see* OPTICAL ACTIVITY).

In **circularly polarized light**, the tip of the electric vector describes a circular helix about the direction of propagation with a frequency equal to the frequency of the light. The magnitude of the vector remains constant. In **elliptically polarized light**, the vector also rotates about the direction of propagation but the amplitude changes; a projection of the vector on a plane at right angles to the direction of propagation describes an ellipse. Circularly and elliptically polarized light are produced using a *retardation plate.

polarizer A device used to plane-polarize light (*see* POLARIZATION OF LIGHT). *Nicol prisms or *Polaroid can be used as polarizers. If a polarizer is placed in front of a source of unpolarized light, the transmitted light is plane-polarized in a specific direction. As the human eye is unable to detect that light is polarized, it is necessary to use an *analyser to detect the direction of polarization. **Crossing** a polarizer and analyser causes extinction of the light, i.e. if the plane of polarization of the polarizer and the plane of the analyser are perpendicular, no light is transmitted when the polarizer and analyser are combined. Both a polarizer and an analyser are components of a *polarimeter.

polarizing angle *See* BREWSTER'S LAW.

polar molecule A molecule that has a dipole moment; i.e. one in which there is some separation of charge in the chemical bonds, so that one part of the molecule has a positive charge and the other a negative charge.

Polaroid A doubly refracting material that plane-polarizes unpolarized light passed through it. It consists of a plastic sheet strained in a manner that makes it birefringent by aligning its molecules. Sunglasses incorporating a Polaroid material absorb light that is vibrating horizontally – produced by reflection from horizontal surfaces – and thus reduce glare.

polaron A coupled electron–ion system that arises when an electron is introduced into the conduction band of a perfect ionic crystal and induces lattice polarization around itself.

polar vector A *vector that reverses its sign when the coordinate system is changed to a new system by a reflection in the origin (i.e. $x'_i = -x_i$). *Compare* AXIAL VECTOR.

pole 1. *See* MAGNETIC POLES; MAGNETIC MONOPOLE. **2.** The *optical centre of a curved mirror.

pollution An undesirable change in the physical, chemical, or biological characteristics of the natural environment, brought about by man's activities. It may be harmful to human or nonhuman life. Pollution may affect the soil, rivers, seas, or the atmosphere. There are two main classes of pollutants: those that are **biodegradable** (e.g. sewage), i.e. can be rendered harmless by natural processes and need therefore cause no permanent harm if adequately dispersed or treated; and those that are **nonbiodegradable** (e.g. heavy metals such as lead in industrial effluents and DDT and other chlorinated hydrocarbons used as pesticides), which eventually accumulate in the environment and may be concentrated in food chains. Other forms of pollution in the environment include noise (e.g. from jet aircraft, traffic, and industrial processes) and thermal pollution (e.g. the release of excessive waste heat into lakes or rivers causing harm to wildlife). Recent pollution problems include the disposal of *radioactive waste; acid rain; photochemical smog; increasing levels of human waste; high levels of carbon dioxide and other greenhouse gases in the atmosphere (*see* GREENHOUSE EFFECT); damage to the ozone layer by nitrogen oxides, chlorofluorocarbons (CFCs), and halons; and pollution of inland waters by agricultural fertilizers and sewage effluent, causing eutrophication. Attempts to contain or prevent pollution include strict regulations concerning factory emissions, the use of smokeless fuels, the banning of certain pesticides, the increasing use of lead-free petrol, restrictions on the use of chlorofluorocarbons, and the introduction, of catalytic converters to cut pollutants in vehicle exhausts.

polonium Symbol Po. A rare radioactive metallic element of group 16 (formerly VIB) of the periodic table; a.n. 84; r.a.m. 210; r.d. 9.32; m.p. 254°C; b.p. 962°C. The element occurs in uranium ores to an extent of about 100 micrograms per 1000 kilograms. It has over 30 isotopes, more than any other element. The longest-lived isotope is polonium–209 (half-life 103 years). Polonium has attracted attention as a possible heat source for spacecraft as the energy released as it decays is 1.4×10^5 J kg^{-1} s^{-1}. It was discovered by Marie *Curie in 1898 in a sample of pitchblende.

polychromatic radiation Electromagnetic radiation that consists of a mixture of different wavelengths. This need not refer only to visible radiation. *Compare* MONOCHROMATIC RADIATION.

polygon A plane figure with a number of sides. In a **regular polygon** all the sides and internal angles are equal. For such a polygon with n sides, the interior angle is $(180 - 360/n)$ degrees and the sum of the interior angles is $(180n - 360)$ degrees.

polygon of forces A polygon in which the sides represent, in magnitude and direction, all forces acting on a rigid body. The side required to close the polygon represents the resultant of a system of forces.

polyhedron A solid bounded by polygonal faces. In a **regular polyhedron** all the faces are congruent regular polygons. The cube is one of five possible regular polyhedrons. The others are the **tetrahedron** (four triangular faces), the **octahedron** (eight triangular faces), the **dodecahedron** (twelve pentagonal faces), and the **icosahedron** (twenty triangular faces).

polynomial A mathematical expression containing three or more terms. It has the general form $a_0x^n + a_1x^{n-1} + a_2x^{n-2} + \ldots + a_n$, where a_0, a_1, etc., are constants and n is the highest power of the variable, called the **degree** of the polynomial.

population inversion *See* LASER.

population type A method of classifying stars as either population I or population II bodies, devised in 1944 by Wilhelm Baade (1893–1960). Population I stars are the young metal-rich highly luminous stars found in the spiral arms of galaxies. Population II stars are older metal-deficient stars that occur in the centres of galaxies.

positive charge *See* CHARGE.

positive feedback *See* FEEDBACK.

positron The antiparticle of the *electron. *See also* ANNIHILATION; ELEMENTARY PARTICLES; PAIR PRODUCTION.

positron emission tomography (PET) A type of *tomography in which a natural biochemical substance containing positron-emitting isotopes is introduced into human tissues. This causes the emission of *gamma radiation which is detected using a *gamma camera. The sources of the gamma radiation are calculated using a computer. The resulting tomogram provides evidence for the positions of local concentrations of the substance containing the isotope. PET is used in metabolic and physiological studies *in vivo*.

positronium A bound state consisting of an electron and a positron. There are two types of positronium: **orthopositronium**, in which the spins of the two constituents are parallel, and **parapositronium**, in which the spins are anti-parallel. Both forms have brief existences, with orthopositronium decaying into three photons in about 1.5×10^{-7} s and parapositronium decaying into two photons in the even shorter time of about 10^{-10} s. Positronium has a hydrogen-like spectrum, but with different values of the frequencies since a positron is much lighter than a proton.

potassium–argon dating A *dating technique for certain rocks that depends on the decay of the radioisotope potassium–40 to argon–40, a process with a half-life of about 1.27×10^{10} years. It assumes that all the argon–40 formed in the potassium-bearing mineral accumulates within it and that all the argon present is formed by the decay of potassium–40. The mass of argon–40 and potassium–40 in the sample is estimated and the sample is then dated from the equation:

$$^{40}\text{Ar} = 0.1102\ ^{40}\text{K}(e^{\lambda t} - 1),$$

where λ is the decay constant and t is the time in years since the mineral cooled to about 300°C, when the ^{40}Ar became trapped in the crystal lattice. The method is effective for micas, feldspar, and some other minerals.

potential A scalar quantity associated with a field. *See* ELECTRIC POTENTIAL; GRAVITATIONAL POTENTIAL.

potential barrier A region containing a maximum of potential that prevents a particle on one side of it from passing to the other side. According to classical theory a particle must possess energy in excess of the height of the potential barrier to pass it. However, in quantum theory there is a finite probability that a particle with less energy will pass through the barrier (*see* TUNNEL EFFECT). A potential barrier surrounds the atomic nucleus and is important in nuclear physics; a similar but much lower barrier exists at the interface between *semiconductors and metals and between differently doped semiconductors. These barriers are important in the design of electronic devices.

potential difference *See* ELECTRIC POTENTIAL.

potential divider *See* VOLTAGE DIVIDER.

potential energy *See* ENERGY.

potentiometer 1. *See* VOLTAGE DIVIDER. **2.** An instrument for measuring, comparing, or dividing small potential differences. A typical example of its use is the measurement of the e.m.f. (E_1) of a cell by comparing it with the e.m.f. (E_2) of a standard cell. In this case a circuit is set up as illustrated, in which AB is a wire of uniform resistance and S is a sliding contact onto this wire. An accumulator X maintains a steady current through the wire. To measure the e.m.f. of a cell C, it is connected up as shown in the diagram and the sliding contact moved until the e.m.f. of C exactly balances the p.d. from the accumulator, as indicated by a zero reading on the galvanometer G. If the length AS is then l_1, the value of E_1 is given by $E_1/E_2 = l_1/l_2$, where l_2 is the length AS when the standard cell is used as the cell C.

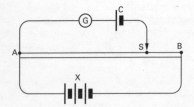

Potentiometer

pound The unit of mass in *f.p.s. units defined as 0.453 592 37 kilogram. Before 1963 it was defined in terms of a platinum cylinder called the Imperial Standard Pound.

poundal The unit of force in *f.p.s. units equal to the force required to impart to a mass of one pound an acceleration of one foot per second per second.

powder metallurgy A process in which powdered metals or alloys are pressed into a variety of shapes at high temperatures. The process started with the pressing of powdered tungsten into incandescent lamp filaments in the first decade of this century and is now widely used for making self-lubricating bearings and cemented tungsten carbide cutting tools.

The powders are produced by atomization of molten metals, chemical decomposition of a compound of the metal, or crushing and grinding of the metal or alloy. The parts are pressed into moulds at pressures ranging from 140×10^6 Pa to 830×10^6 Pa after which they are heated in a controlled atmosphere to bond the particles together (*see* SINTERING).

powder method *See* DEBYE–SCHERRER METHOD.

power **1.** (in physics) Symbol P. The rate at which work is done or energy is transferred. In SI units it is measured in watts (joules per second). *See also* HORSEPOWER. **2.** (in mathematics) The number of times a quantity is multiplied; e.g. x^5 is the fifth power of x. A **power series** is one in which the power of the variable increases with each term, e.g. $a_0 + a_1x + a_2x^2 + a_3x^3 + \ldots + a_nx^n$.

power factor *See* ELECTRIC POWER.

power reactor A *nuclear reactor designed to produce electrical power.

Poynting vector Symbol S. A vector that gives a measure of the flow of energy in an electromagnetic field. It is given by the vector product $S = E \times H$, where E is the electric field strength and H is the magnetic field strength. It was discovered by the British physicist John Henry Poynting (1852–1914) in 1884.

preamplifier An *amplifier in a radio, record player, etc., providing a first stage of amplification. It is usually located close to the signal source (i.e. the aerial or pick-up) and the signal is then transmitted by cable to the main amplifier. Preamplification at this early stage improves the signal-to-noise ratio of the whole system.

precessional motion A form of motion that occurs when a torque is applied to a rotating body in such a way that it tends to change the direction of its axis of rotation. It arises because the resultant of the angular velocity of rotation and the increment of angular velocity produced by the torque is an angular velocity about a new direction; this commonly changes the axis of the applied torque and leads to sustained rotation of the original axis of rotation.

A spinning top, the axis of which is not exactly vertical, has a torque acting on it as a result of gravity. Instead of falling over, the top precesses about a vertical line through the pivot. The earth also experiences a torque and undergoes a slow precession, primarily as a result of the gravitational attraction of the sun and the moon on its equatorial bulge (*see* PRECESSION OF THE EQUINOXES).

precession of the equinoxes The slow westward motion of the
*equinoxes about the ecliptic as a result of the earth's *precessional
motion. The equinoxes move round the ecliptic with a period of 25 800
years.

precipitation Any of the particles of water that fall from the
atmosphere and reach the ground, e.g. rain and snow. Rain is an example
of **liquid precipitation**, while snow is an example of **solid precipitation**.
Precipitation is distinguished from other forms of *hydrometeor, such as
*dew, fog, and hoarfrost, since they do not involve the water particles
falling to the ground.

preons Hypothetical entities that are postulated as being 'building
blocks' of quarks and leptons. There is no experimental evidence for
preons but the idea has considerable theoretical appeal. It is expected that
evidence for their existence would only be obtained at much higher
energies than are available from present accelerators.

presbyopia A loss of accommodation that normally develops in human
eyes over the age of 45–50 years. Vision of distant objects remains
unchanged but accommodation of the eye to near objects is reduced as a
result of loss of elasticity in the lens of the eye. The defect is corrected by
reading glasses using weak converging lenses.

pressure The force acting normally on unit area of a surface or the ratio
of force to area. It is measured in *pascals in SI units. **Absolute pressure** is
pressure measured on a gauge that reads zero at zero pressure rather than
at atmospheric pressure. **Gauge pressure** is measured on a gauge that
reads zero at atmospheric pressure.

p

pressure gauge Any device used to measure pressure. Three basic types
are in use: the liquid-column gauge (e.g. the mercury *barometer and the
*manometer), the expanding-element gauge (e.g. the *Bourdon gauge and
the aneroid *barometer), and the electrical transducer. In the last category
the *strain gauge is an example. Capacitor pressure gauges also come into
this category. In these devices, the pressure to be measured displaces one
plate of a capacitor and thus alters its capacitance.

pressurized-water reactor *See* NUCLEAR REACTOR.

Prévost's theory of exchanges A body emits and absorbs radiant
energy at equal rates when it is in equilibrium with its surroundings. Its
temperature then remains constant. If the body is not at the same
temperature as its surroundings there is a net flow of energy between the
surroundings and the body because of unequal emission and absorption.
The theory was proposed by Pierre Prévost (1751–1839) in 1791.

primary cell A *voltaic cell in which the chemical reaction producing
the e.m.f. is not satisfactorily reversible and the cell cannot therefore be

recharged by the application of a charging current. *See* DANIELL CELL; LECLANCHÉ CELL; WESTON CELL; MERCURY CELL. *Compare* SECONDARY CELL.

primary colour Any one of a set of three coloured lights that can be mixed together to give the sensation of white light as well as approximating all the other colours of the spectrum. An infinite number of such sets exists, the condition being that none of the individual colours of a set should be able to be matched by mixing the other two; however, unless the colours are both intense and very different the range that they can match well will be limited. The set of primary colours most frequently used is red, green, and blue. *See also* COLOUR.

primary winding The winding on the input side of a *transformer or *induction coil. *Compare* SECONDARY WINDING.

principal axis *See* OPTICAL AXIS.

principal focus A point through which rays close to and parallel to the axis of a lens or spherical mirror pass, or appear to pass, after refraction or reflection. A mirror has one principal focus, a lens has a principal focus on both sides.

principal plane The plane that is perpendicular to the optical axis of a lens and that passes through the optical centre. A thick lens has two principal planes, each passing through a *principal point.

principal point Either of two points on the principal axis of a thick lens from which simply related distances can be measured, as from the optical centre of a thin lens.

principle of least action A *variational principle that can be regarded as the basis for Newtonian mechanics. It states that the integral of the kinetic energy minus the potential energy with respect to time is always smaller for the actual motion of a body than for any other motion.

principle of superposition *See* SUPERPOSITION PRINCIPLE.

printed circuit An electronic circuit consisting of a conducting material deposited (printed) onto the surface of an insulating sheet. These devices are now common in all types of electronic equipment, facilitating batch production and eliminating the unreliability of the hand-soldered joint.

printer A device for producing a printed version of text (and, sometimes, pictures) from a computer. There are several types. **Impact printers** work on the same principle as a typewriter, in which a ribbon is hit by a surface embossed with the character. A **line printer** produces a whole line of text at a time. In this device the characters are held on a row of spinning cylinders. In a **daisywheel printer**, the characters are held on the ends of a series of arms radiating from the centre of a wheel. A **dot-matrix printer** forms the image of each character from a rectangular matrix of dots. An **inkjet printer** works by spraying fine jets of quick-drying ink onto the paper. A **laser printer** uses a xerographic technique in which a photosensitive plate is scanned by a low-power laser.

prism 1. (in mathematics) A polyhedron with two parallel congruent polygons as bases and parallelograms for all other faces. A **triangular prism** has triangular bases. **2.** (in optics) A block of glass or other transparent material, usually having triangular bases. Prisms have several uses in optical systems: they can be used to deviate a ray, to disperse white light into the visible spectrum, or to erect an inverted image (*see* BINOCULARS). Prisms of other materials are used for different kinds of radiation. *See also* NICOL PRISM; WOLLASTON PRISM.

probability The likelihood of a particular event occurring. If there are n equally likely outcomes of some experiment, and a ways in which event E could occur, then the probability of event E is a/n. For instance, if a die is thrown there are 6 possible outcomes and 3 ways in which an odd number may occur. The probability of throwing an odd number is $3/6 = 1/2$.

program *See* COMPUTER.

progressive wave *See* WAVE.

projectile Any body that is thrown or projected. If the projectile is discharged on the surface of the earth at an angle θ to the horizontal it will describe a parabolic flight path (if θ < 90° and the initial velocity < the *escape velocity). Neglecting air resistance, the maximum height of this flight path will be $(v^2\sin^2θ)/2g$, where v is the velocity of discharge and g is the acceleration of free fall. The horizontal distance covered will be $(v^2\sin2θ)/g$ and the time of the flight will be $(2v\sinθ)/g$. This means that, if air resistance is neglected, the launch angle needed for the furthest possible horizontal distance is 45°.

projector An optical device for throwing a large image of a two-dimensional object onto a screen. In an **episcope**, light is reflected from the surface of an opaque two-dimensional object (such as a diagram or photographic print) and an enlarged image is thrown onto a distant screen by means of a system of mirrors and lenses. The **diascope** passes light through the two-dimensional object (such as a photographic transparency, slide, or film) and uses a converging projection lens to form an enlarged image on a distant screen. An **epidiascope** is a device that can be used as both episcope and diascope. An **overhead projector** is a form of diascope that throws its image on a wall or screen behind and above the operator. In a **motion-picture projector** (or **ciné projector**) the film, consisting of a long sequence of transparent pictures, is driven by a motor past the light source in such a way that each picture comes to rest for a brief period in front of the light source. The illusion of motion is created as each image on the screen is replaced by the next; during the picture change the light is interrupted.

PROM (programmable read-only memory) An *integrated circuit with a memory that can be programmed once and cannot be changed thereafter. *See also* ROM.

prompt neutrons The neutrons emitted during a nuclear fission process within less than a microsecond of fission. *Compare* DELAYED NEUTRONS.

proper motion The apparent angular motion of a star on the *celestial sphere. This is motion in a direction that is perpendicular to the line of sight.

proportional counter A type of detector for *ionizing radiation in which the size of the output pulse is proportional to the number of ions formed in the initial ionizing event. It operates in a voltage region, called the **proportional region**, intermediate between that of an *ionization chamber and a *Geiger counter, avalanche ionization being limited to the immediate vicinity of the primary ionization rather than the entire length of the central wire electrode.

proportional limit *See* ELASTICITY.

protactinium Symbol Pa. A radioactive metallic element belonging to the actinoids; a.n. 91; r.a.m. 231.036; r.d. 15.37 (calculated); m.p. <1600°C (estimated). The most stable isotope, protactinium–231, has a half-life of 3.43×10^4 years; at least ten other radioisotopes are known. Protactinium–231 occurs in all uranium ores as it is derived from uranium–235. Protactinium has no practical applications; it was discovered by Lise *Meitner and Otto *Hahn in 1917.

proton An *elementary particle that is stable, bears a positive charge equal in magnitude to that of the *electron, and has a mass of $1.672\ 614 \times 10^{-27}$ kg, which is 1836.12 times that of the electron. The proton occurs in all atomic nuclei (the hydrogen nucleus consists of a single proton).

proton decay A process of the type

$$p \to e^+ + \pi^0$$

where a proton decays into a positron and a pion, predicted to occur in *grand unified theories (GUTs) because baryon number is no longer conserved. The lifetime depends on the theory used and is typically 10^{35} years, but a combination of GUTs and *supersymmetry gives a lifetime of about 10^{45} years. Considerable experimental effort has been spent in looking for proton decay, so far with no success. This negative result means that GUTs that are not combined with supersymmetry can be ruled out.

proton number *See* ATOMIC NUMBER.

protostar *See* STELLAR EVOLUTION.

pseudo-scalar A *scalar quantity that changes sign when the coordinate system is changed to a new system by a reflection in the origin (i.e. $x'_i = -x_i$). It is the *scalar product of an *axial vector and a *polar vector.

pseudo-vector *See* AXIAL VECTOR.

psi particle (J particle) A *meson discovered in 1974, which led to the extension of the quark model and the hypothesis that a fourth quark existed with the property of charm (*see* ELEMENTARY PARTICLES). The psi particle is believed to consist of a charmed quark and its antiquark.

psychrometer *See* HYGROMETER.

Ptolemaic astronomy The system of astronomy originally proposed by Apollonius of Perga in the third century BC and completed by Claudius Ptolemaeus of Alexandria (100–178 AD). It assumed that the earth was at the centre of the universe and that each known planet, the moon, and the sun moved round it in a circular orbit, called the deferent. In addition to this motion the orbiting bodies also described epicycles, small circles about points on the deferent. The system gave moderately good predictions, but was completely replaced by the heliocentric astronomy of Copernicus in the 16th century. The Ptolemaic system was published by Ptolemaeus in the work known by its Arabic name, the *Almagest*.

***p*-type conductivity** *See* SEMICONDUCTOR; TRANSISTOR.

pulley A simple machine consisting of a wheel with a flat, crowned, or grooved rim to take a belt, rope, or chain with which a load can be raised. See illustration.

 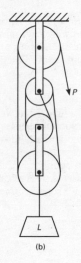

(a) (b)

Pulleys

In fig (a), assuming the system is frictionless, the force P in any part of the rope is constant; therefore $2P = L$, where L is the load. In general, $nP = L$, where n is the number of supporting ropes. In fig (b), the number of supporting ropes is 4. The mechanical advantage of a pulley system is the ratio of the load, L, to the effort applied to the free end of the rope, P; i.e. mechanical advantage = $L/P = L(L/n)^{-1} = n$. Thus in fig (b) the mechanical advantage is 4. A combination of ropes and pulleys as in fig (b) is called a **block and tackle**.

pulsar A celestial source of radiation emitted in brief (0.03 second to 4 seconds) regular pulses. First discovered in 1968, pulsars are believed to be rotating *neutron stars. The strong magnetic field of the neutron star concentrates charged particles in two regions and the radiation is emitted in two directional beams. The pulsing effect occurs as the beams rotate. Most pulsars are radio sources (emit electromagnetic radiation of radio frequencies) but a few that emit light or X-rays have been detected. Over 300 pulsars are now known in our Galaxy.

pulsatance *See* ANGULAR FREQUENCY.

pulse **1.** A brief variation in a quantity, usually for a finite time, especially in a quantity that is normally constant. **2.** A series of such variations having a regular waveform in which the variable quantity rises sharply from a base value to a maximum value and then falls back to the base value in a relatively short time.

pulse jet A type of ramjet (*see* JET PROPULSION) in which a louvred valve at the front of the projectile is blown open by the ram effect of the moving projectile and remains open until pressure has built up in the combustion chamber. Fuel is then admitted and the mixture exploded by spark ignition. This closes the louvred valve and produces thrust at the open rear end of the projectile. The German flying bombs of World War II were powered by pulse jets.

pulse modulation *See* MODULATION.

pump A device that imparts energy to a fluid in order to move it from one place or level to another or to raise its pressure (*compare* VACUUM PUMP). **Centrifugal pumps** and *turbines have rotating impellers, which increase the velocity of the fluid, part of the energy so acquired by the fluid then being converted to pressure energy. Displacement pumps act directly on the fluid, forcing it to flow against a pressure. They include piston, plunger, gear, screw, and cam pumps. *See also* ELECTROMAGNETIC PUMP.

pyramid A solid having a polygonal base with n sides, each side forming the base of a triangle. The n triangles so formed have a common vertex. The **axis** of the pyramid is a line joining the vertex to the centre of symmetry of the base. If the axis is perpendicular to the base the solid is a **right pyramid**. A **square pyramid** has a square base and a **triangular**

pyramid has a triangular base (*see* TETRAHEDRON). The volume of a pyramid is one third of the base area multiplied by the height.

pyroelectricity The property of certain crystals, such as tourmaline, of acquiring opposite electrical charges on opposite faces when heated. In tourmaline a rise in temperature of 1 K at room temperature produces a polarization of some 10^{-5} C m^{-2}.

pyrometric cones *See* SEGER CONES.

pyrometry The measurement of high temperatures from the amount of radiation emitted, using a **pyrometer**. Modern **narrow-band** or **spectral pyrometers** use infrared-sensitive *photoelectric cells behind filters that exclude visible light. In the **optical pyrometer** (or disappearing filament pyrometer) the image of the incandescent source is focused in the plane of a tungsten filament that is heated electrically. A variable resistor is used to adjust the current through the filament until it blends into the image of the source, when viewed through a red filter and an eyepiece. The temperature is then read from a calibrated ammeter or a calibrated dial on the variable resistor. In the **total-radiation pyrometer** radiation emitted by the source is focused by a concave mirror onto a blackened foil to which a thermopile is attached. From the e.m.f. produced by the thermopile the temperature of the source can be calculated.

pyrophoric Igniting spontaneously in air. **Pyrophoric alloys** are alloys that give sparks when struck. *See* MISCH METAL.

p

QCD *See* QUANTUM CHROMODYNAMICS.

QED *See* QUANTUM ELECTRODYNAMICS.

QFD Quantum flavourdynamics. *See* ELECTROWEAK THEORY.

QSG *See* QUASARS.

QSO *See* QUASARS.

QSS *See* QUASARS.

quadratic equation An equation of the second degree having the form $ax^2 + bx + c = 0$. Its roots are:

$$x = [-b \pm \sqrt{(b^2 - 4ac)}]/2a.$$

quadrature The position of the moon or an outer planet when the line joining it to the earth makes a right angle with a line joining the earth to the sun.

quality of sound (timbre) The quality a musical note has as a result of the presence of *harmonics. A pure note consists only of the *fundamental; however, a note from a musical instrument will have several harmonics present, depending on the type of instrument and the way in which it is played. For example, a plucked string (as in a guitar) produces a series of harmonics of diminishing intensity, whereas a struck string (as in a piano) produces a series of harmonics of more nearly equal intensity.

quantization The process of constructing a quantum theory for a system, using the original classical theory as a basis. The starting point for such a process is to write the *Lagrangian or *Hamiltonian of the classical system. The formulation of the quantum theory for the system can be performed using a formalism, such as *matrix mechanics or *wave mechanics. The application of these methods leads to the conclusion that energy levels in systems, such as atoms, are discrete (**quantized**) rather than continuous. Before the discovery of quantum mechanics in the mid 1920s, quantization involved a series of ad hoc postulates for atomic systems, such as the *Bohr theory and its extensions.

quantum The minimum amount by which certain properties, such as energy or angular momentum, of a system can change. Such properties do not, therefore, vary continuously, but in integral multiples of the relevant quantum. This concept forms the basis of the *quantum theory.

In waves and fields the quantum can be regarded as an excitation, giving a particle-like interpretation to the wave or field. Thus, the quantum of the electromagnetic field is the *photon and the *graviton is the quantum of the gravitational field. *See* QUANTUM MECHANICS.

quantum chaos The *quantum mechanics of systems for which the corresponding classical system can exhibit *chaos. This subject was initiated by Einstein in 1917, who showed that the quantization conditions associated with the *Bohr theory need to be modified for systems that show chaos in classical mechanics. The subject of quantum chaos is an active field of research in which many basic issues still require clarification. It appears that systems exhibiting chaos in classical mechanics do not necessarily exhibit chaos in quantum mechanics. In quantum mechanics, chaos can be investigated in terms of randomness, either of the evolution in time of the *Schrödinger equation or its *eigenfunctions and eigenvalues.

quantum chromodynamics (QCD) A *gauge theory that describes the strong interaction in terms of quarks and antiquarks and the exchange of massless gluons between them (*see also* ELEMENTARY PARTICLES). Quantum chromodynamics is similar to quantum electrodynamics (QED), with colour being analogous to electric charge and the gluon being the analogue of the photon. The gauge group of QCD is non-Abelian and the theory is much more complicated than quantum electrodynamics; the gauge symmetry in QCD is not a *broken symmetry.

QCD has the important property of *asymptotic freedom – that at very high energies (and, hence, short distances) the interactions between quarks tend to zero as the distance between them tends to zero. Because of asymptotic freedom, perturbation theory may be used to calculate the high energy aspects of strong interactions, such as those described by the *parton model.

quantum computing The design and theory of computer systems that depend on quantum effects for their operation. On one level, this can be the use of small components, at the atomic or molecular level, to store or process information. An example would be a storage system that used two different spin states of atoms to store bits of information, or a logic gate that depends on the movement or spin of a single electron. Systems of this type are studied in **nanocomputing** (*see also* SPINTRONICS). At a more fundamental level, the term 'quantum computing' implies the use of quantum effects that have no classical analogue to process information. In a 'classical' computer information is held in bits, which can have two alternative values (0 and 1). In a quantum computer the 0 and 1 values are held simultaneously in an entangled state (*see* QUANTUM ENTANGLEMENT). This unit of information is a quantum bit (or **qubit**). Much more information can be held in this way and, in principle, it is possible to do parallel processing of the information. Quantum computers would be much faster than conventional machines and capable of performing calculations that could not realistically be done otherwise. Ion traps and spin measurements have been used in research in this area.

quantum cryptography Cryptography that is based on quantum mechanics. Since any attempt to measure the quantum state of a system alters that state, the basic idea of quantum cryptography is that if a message exists as a quantum state then any eavesdropper would change the state. At present, much work is being undertaken to make quantum cryptography a practical possibility.

quantum electrodynamics (QED) The study of the properties of electromagnetic radiation and the way in which it interacts with charged matter in terms of *quantum mechanics. The collision of a moving electron with a proton, in this theory, can be visualized by a space–time diagram (**Feynman diagram**) in which photons are exchanged (see illustration).

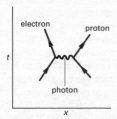

An electron–proton collision

  *Perturbation-theory calculations using Feynman diagrams enable an agreement between theory and experiment to a greater accuracy than one part in 10^9 to be obtained. Because of this, QED is the most accurate theory known in physical science. Although many of the effects calculated in QED are very small (about 4×10^{-6} eV), such as *energy level splitting in the spectra of *atoms, they are of great significance for demonstrating the physical reality of fluctuations and polarization in the vacuum state.

  QED is a *gauge theory for which the gauge group is Abelian (*see* GROUP THEORY).

quantum electronics The application of *quantum optics and the specifically quantum-mechanical properties of electrons to the design of electronic devices.

quantum entanglement A phenomenon in quantum mechanics in which a particle or system does not have a definite state but exists as an intermediate form of two 'entangled' states. One of these states is realized when a 'measurement' is made. *See* BELL'S THEOREM.

quantum field theory A quantum-mechanical theory applied to systems that have an infinite number of degrees of freedom. In quantum field theories, particles are represented by fields that have quantized normal modes of oscillation. For instance, *quantum electrodynamics is a quantum field theory in which the photon is emitted or absorbed by particles; the photon is the quantum of the electromagnetic field. **Relativistic quantum field theories** are used to describe fundamental interactions between elementary particles. They predict the existence of *antiparticles and also show the connection between spin and statistics that leads to the Pauli exclusion principle (*see* SPIN–STATISTICS THEOREM). In spite of their success, it is not clear whether a quantum field theory can give a completely unified description of all interactions (including the gravitational interaction).

quantum flavourdynamics (QFD) *See* ELECTROWEAK THEORY.

quantum gravity An aspect of *quantum theory that attempts to incorporate the *gravitational field as described by the general theory of *relativity; no such theory has yet been accepted, however. Unlike the *quantum field theories for the other three *fundamental interactions, the procedure of *renormalization does not work for quantum gravity, although there is some evidence that *superstring theory can provide a quantum theory of gravity free of infinities. An approximation to quantum gravity is given by **quantum field theory in curved space–time**, in which the gravitational interactions are treated classically, while all other interactions are treated by *quantum mechanics. An important aspect of quantum field theory in curved space–time is its description of the *Hawking process. It is necessary to consider quantum gravity in the very *early universe, just after the *big bang, and the singularities associated with *black holes can also be interpreted as requiring a quantum theory of gravity.

quantum Hall effect A quantum mechanical version of the *Hall effect found at very low temperatures, in which the Hall coefficient R_H is proportional to h/e^2, where h is the Planck constant and e is the charge of the electron. Thus, the Hall coefficient is quantized. There are two types of quantum Hall effect. The **integer quantum Hall effect** has R_H given as an integer with great precision. It can be used for precision measurements of constants such as e and h. In the **fractional quantum Hall effect**, R_H has fractional values.

The integer quantum Hall effect can be understood in terms of noninteracting electrons, whereas the fractional effect is thought to result from many-electron interactions in two-dimensional systems, and be an example of anyons (*see* QUANTUM STATISTICS).

quantum jump A change in a system (e.g. an atom or molecule) from one quantum state to another.

quantum mechanics A system of mechanics based on *quantum theory, which arose out of the failure of classical mechanics and electromagnetic theory to provide a consistent explanation of both *electromagnetic waves and atomic structure. Many phenomena at the atomic level puzzled physicists at the beginning of the 20th century because there seemed to be no way of explaining them without making use of incompatible concepts. One such phenomenon was the emission of *electrons from the surface of a metal illuminated by light. Einstein realized that the classical description of light as a wave on an electromagnetic field could not explain this *photoelectric effect, as it is called. Experiments showed that electrons would be emitted only if the incident light was of a sufficiently short wavelength, while the intensity of the light appeared not to be relevant. It seemed not to make sense that small ripples of short wavelength could easily knock electrons out of the metal, but a huge tidal wave of long wavelength could not. In 1905 Einstein abandoned classical mechanics and sought an explanation of this photoelectric effect in Planck's work on thermal radiation (*see* PLANCK'S RADIATION LAW). In this work light energy is regarded as being imparted to matter in discrete packets rather than continuously, as one might expect from a wave. Einstein assumed that in the photoelectric effect, light behaves as a shower of particles, each with energy E given by Planck's expression:

$$E = hf,$$

where f is the light's frequency and h is the *Planck constant. Each particle of light, which Einstein called a *photon, would impart its energy to a single electron in the metal. The electron would be liberated only if the photon could impart at least the required amount of energy. However many photons were falling on the surface of the metal, no electrons would be liberated unless individual photons had the required energy (hf) to break the attractive forces holding the electrons in the metal. This elegantly quantified reversion to Newton's corpuscular theory of light by Einstein was one of the milestones in the development of quantum mechanics.

Further verification that light could indeed behave as a shower of particles came from the *Compton effect. In Compton scattering, X-radiation is scattered off an electron in a manner that resembles a particle collision. The momentum imparted to the electron can be predicted by assuming that the X-ray possesses the momentum of a photon. An expression for photon momentum is suggested by the classical theory of radiation pressure. It is known that if energy is transported by an electromagnetic wave at a rate W joules per unit area per second, the wave exerts a radiation pressure W/c, where c is the speed of light. Planck's expression for the energy of photons therefore led to an equivalent expression for the momentum p of these photons:

$$p = h/\lambda,$$

where λ is the wavelength of the light. Experimental studies of the Compton effect produce results in good agreement with this expression.

Both the photoelectric effect and the Compton effect imply that light imparts energy and momentum to matter in the form of packets. It is as if energy and momentum are fundamental 'currencies' of physical interaction, and that these currencies exist in denominations that are all multiples of the Planck constant. These quantities are said to be 'quantized' and a packet of energy or momentum is called a *quantum. Quantum mechanics is essentially concerned with the exchange of these quanta of energy and momentum. For more than a century before the birth of quantum mechanics, experiments had indicated that light behaves as a wave. The successful explanation of the photoelectric and Compton effects demonstrated that in some situations light interacts with matter as if it is a shower of particles. The principle that two models are required to explain the nature of light was called by Niels Bohr *complementarity. This principle was extended by the French aristocrat, Louis de Broglie, who suggested in 1923 that particles of matter might also behave as waves in certain circumstances.

Louis de Broglie received the Nobel Prize for this idea in 1929 after the successful measurement of the *de Broglie wavelength of an electron in 1927 by Clinton Davisson and Lester Germer, who had observed the *diffraction of electrons by single crystals of nickel. The behaviour of individual electrons seemed random and unpredictable, but when a large number had passed through the crystal a typical diffraction pattern emerged. This provided proof that the electron, which until then had been thought of simply as a particle of matter, could under the right circumstances exhibit wavelike properties. Classical mechanics and electromagnetism were based on two kinds of entity: matter and fields. In classical physics, matter consists of particles and waves are oscillations on a field. Quantum mechanics blurs the distinction between matter and field. Modern physicists are forced to concede that the universe is made up of entities that exhibit a *wave–particle duality.

A new representation of matter and field is needed to fully appreciate this wave–particle duality. In quantum mechanics, an electron is represented by a *complex number called a *wave function that depends on time and space coordinates. The wave function behaves like a classical wave displacement on a medium (e.g. on a string), exhibiting wave behaviour, interference, diffraction, etc. However, unlike a classical wave displacement, the wave function is essentially a complex quantity. Since an electron's observable properties do not involve complex numbers, it follows that an electron's wave function cannot itself be identified with a single physical property of the electron. The diffraction of electrons observed by Davisson and Germer revealed that, although the behaviour of the individual electrons is random and unpredictable, when a large number have passed through the apparatus, a diffraction pattern is formed whose intensity distribution is proportional to the intensity of the associated wave function. The intensity of the wave function, Ψ, is the square of its *absolute value $|\Psi|^2$. Therefore, although an electron's wave function itself has no physical significance, the square of its absolute value at a point turns out to be proportional to the probability of finding an electron at that point.

The electron wave function must satisfy a wave equation based on the conservation of energy and momentum for the electron. There are two main ways of treating this wave equation: a classical or a relativistic treatment. The resulting wave equations are called **eigenvalue equations** because they have the same form as equations in a branch of mathematics called **eigenvalue problems**, i.e.

$$\Omega\Psi = \omega\Psi,$$

where Ω is some mathematical operation (multiplication by some number, differentiation, etc.) on the wave function Ψ and ω, called the **eigenvalue** in quantum mechanics, is always a real number. In such an equation the wave function is often called the **eigenfunction**. This form of treatment of quantum mechanics is known as **wave mechanics** (*see also* SCHRÖDINGER EQUATION).

The energy E and momentum p of an electron are associated with the frequency f and wavelength λ of the electron's wave using the expressions $E = hf$ and $p = h/\lambda$. While the wave equation expresses the behaviour of the wavelike properties of a particle, it does not define the physical attributes it has as a particle. As a particle, the electron has an easily defined spatial and temporal position, not possessed by an oscillation of some kind of field, whose influence extends over a region of space and time. The incompatibility of these two views of the electron leads to the Heisenberg *uncertainty principle. Heisenberg recognized that if matter had wavelike properties, the physical attributes normally associated with matter (position, momentum, kinetic energy, etc.) would have to be expressed in a statistical, rather than a deterministic, manner. This is illustrated by the electron diffraction patterns of Davisson and Germer. Individual electrons somehow fell onto the apparatus to form a pattern statistically consistent with the intensity of a wave function. It is as if the final wave function were made up of a superposition of all the possible positions that the electrons could fall onto, the waves of electrons constructively and destructively interfering to form the final diffraction pattern.

It is known that a clever superposition of waves of different wavelengths can lead to a construction of a wave packet of finite extension (*see* FOURIER ANALYSIS). However, to produce a packet that exists at a point of zero width requires an infinite number of waves to superimpose. Heisenberg realized that these packets of waves must account for the way in which matter particles, such as electrons, could retain some semblance of their particle-like qualities. However, this must also mean that there is an inherent uncertainty in position and momentum associated with electrons and indeed all particles of matter (*see* UNCERTAINTY PRINCIPLE). Since waves of different wavelength correspond to different possible momentum values of an electron, a superposition of such waves to form a particle at a point would correspond to an infinite uncertainty in the momentum of the electron. Therefore the more one knows about the position of an electron the less one will know about its momentum and vice versa. A similar uncertainty between the energy of the electron and its temporal position also exists.

Quantities that are related by such an uncertainty principle are said to be **incompatible**.

An alternative to the wave mechanical treatment of quantum mechanics is an equivalent formalism called *matrix mechanics, which is based on mathematical operators. *See also* BELL'S THEOREM; HIDDEN-VARIABLES THEORY.

quantum number *See* ATOM; SPIN.

quantum optics The study of those aspects of light that rely on *quantum mechanics. Quantum optics makes use of the *quantum theory of radiation to describe photons, *coherent radiation, and the interaction of photons and atoms. The study of *lasers is an important branch of quantum optics; other applications include *photonics and *quantum electronics.

quantum simulation The mathematical modelling of systems of large numbers of atoms or molecules by computer studies of relatively small clusters. It is possible to obtain information about solids and liquids in this way and to study surface properties and reactions.

quantum state The state of a quantized system as described by its quantum numbers. For instance, the state of a hydrogen *atom is described by the four quantum numbers n, l, m_l, m_s. In the ground state they have values 1, 0, 0, and ½ respectively.

quantum statistics A statistical description of a system of particles that obeys the rules of *quantum mechanics rather than classical mechanics. In quantum statistics, energy states are considered to be quantized. **Bose–Einstein statistics** apply if any number of particles can occupy a given quantum state. Such particles are called **bosons**. Bosons have an angular momentum $nh/2\pi$, where n is zero or an integer and h is the Planck constant. For identical bosons the *wave function is always symmetric. If only one particle may occupy each quantum state, **Fermi–Dirac statistics** apply and the particles are called **fermions**. Fermions have a total angular momentum $(n + \frac{1}{2})h$ and any wave function that involves identical fermions is always antisymmetric.

The relation between the spin and statistics of particles is given by the *spin–statistics theorem.

In two-space dimensions, it is possible that there are particles (or *quasiparticles) that have statistics intermediate between bosons and fermions. These particles are known as **anyons**; for identical anyons the wave function is not symmetric (a phase sign of +1) or antisymmetric (a phase sign of –1), but interpolates continuously between +1 and –1. Anyons may be involved in the fractional *quantum Hall effect.

quantum teleportation The teleportation of quantum states, i.e. sending the states across space. This cannot be done directly because of the *no-cloning theorem. However, it can be achieved by using a system in which there is *quantum entanglement. Quantum teleportation has been performed both for photons and for systems with electrons.

quantum theory The theory devised by Max Planck in 1900 to account for the emission of the black-body radiation from hot bodies. According to this theory energy is emitted in quanta (*see* QUANTUM), each of which has an energy equal to $h\nu$, where h is the *Planck constant and ν is the frequency of the radiation. This theory led to the modern theory of the interaction between matter and radiation known as *quantum mechanics, which generalizes and replaces classical mechanics and Maxwell's electromagnetic theory. In **nonrelativistic quantum theory** particles are assumed to be neither created nor destroyed, to move slowly relative to the speed of light, and to have a mass that does not change with velocity. These assumptions apply to atomic and molecular phenomena and to some aspects of nuclear physics. **Relativistic quantum theory** applies to particles that travel at or near the speed of light.

quantum theory of radiation The study of the emission and absorption of *photons of electromagnetic radiation by atomic systems using *quantum mechanics. Photons are emitted by atoms when a transition occurs from an excited state to the ground state. If an atom is exposed to external electromagnetic radiation a transition can occur from the ground state to an excited state by absorption of a photon. An excited atom can lose the energy it has gained by stimulated emission (*see* LASER). However, an atom can also emit a photon in the absence of external electromagnetic radiation, a process called *spontaneous emission. The quantum theory of radiation was initiated by Einstein in 1916–17, as an extension of *Planck's radiation law. The theory is quantified by the *Einstein coefficients. The quantum theory of radiation is the basis of the theory behind the operation of *lasers and *masers.

quark *See* ELEMENTARY PARTICLES.

quark confinement The hypothesis that free quarks can never be seen in isolation. It is a result of *quantum chromodynamics, in which the property of *asymptotic freedom means that the interactions between quarks get weaker as the distance between them gets smaller, and tends to zero as the distance between them tends to zero. Conversely, the attractive interactions between quarks get stronger as the distance between them gets greater, and the quark-confinement hypothesis is that the quarks cannot escape from one another. It is possible that at very high temperatures, such as those in the *early universe, quarks may become free. The temperature at which this occurs is called the **deconfinement temperature**. The hypothesis of quark confinement has not been proved theoretically in a conclusive way although there is much evidence for it.

quark–gluon plasma A state of matter in which quarks and gluons are not confined into baryons and mesons but exist as a hot plasma. It is thought that this state existed in the early universe until there was a phase transition about 10^{-5} s after the big bang, when the universe had cooled to a sufficiently low temperature that *quark confinement into hadrons occurred.

quarter-wave plate *See* RETARDATION PLATE.

quartile *See* PERCENTILE.

quartz The most abundant and common mineral, consisting of crystalline silica (silicon dioxide, SiO_2), crystallizing in the trigonal system. It has a hardness of 7 on the Mohs' scale. The mineral has the property of being piezoelectric and hence is used to make oscillators for clocks (*see* QUARTZ CLOCK), radios, and radar instruments. It is also used in optical instruments and in glass, glaze, and abrasives.

quartz clock A clock based on a piezoelectric crystal of quartz (*see* PIEZOELECTRIC EFFECT). Each quartz crystal has a natural frequency of vibration, which depends on its size and shape. If such a crystal is introduced into an oscillating electronic circuit that resonates at a frequency very close to that of the natural frequency of the crystal, the whole circuit (including the crystal) will oscillate at the crystal's natural frequency and the frequency will remain constant over considerable periods (a good crystal will maintain oscillation for a year with an accumulated error of less than 0.1 second). In a quartz clock or watch the alternating current from the oscillating circuit containing such a crystal is amplified and the frequency subdivided until it is suitable to drive a synchronous motor, which in turn drives a gear train to operate hands. Alternatively it is used to activate a digital display.

quasars A class of astronomical objects that appear on optical photographs as starlike but have large *redshifts quite unlike those of stars. They were first observed in 1961 when it was found that strong radio emission was emanating from many of these starlike bodies. Over 600 such objects are now known and their redshifts can be as high as 4. The redshifts are characteristic of the *expansion of the universe. This **cosmological redshift** is the explanation of the high observed redshifts of quasars favoured by most astronomers. (A few, however, maintain that the redshift could be a local Doppler effect, characteristic of movement relative to the earth and sun of nearby objects in the Galaxy, or a gravitational effect.) If the redshifts are cosmological, quasars are the most distant objects in the universe, some being up to 10^{10} light-years away. The exact nature of quasars is unknown but it is believed that they are the nuclei of galaxies in which there is violent activity. The luminosity of the nucleus is so much greater than that of the rest of the galaxy that the source appears pointlike. It has been proposed that the power source in a quasar is a supermassive *black hole accreting material from the stars and gas in the surrounding galaxy. There is a considerable amount of evidence in support of this point of view.

The name **quasar** is a contraction of quasistellar object (**QSO**) or

quasistellar galaxy (**QSG**). Quasars that are also radio sources are
sometimes called quasistellar radio sources (**QSS**).

quasicrystal A solid structure in which there is (1) long-range
incommensurate translational order (*see* INCOMMENSURATE LATTICE) and (2)
long-range orientational order with a point group, which is not allowed in
*crystallography. Condition (1) is called **quasiperiodicity**. In two
dimensions, the fivefold symmetry of a pentagon is an example of a point
symmetry, which is not allowed in crystallography, but for which
quasicrystals exist. In three dimensions the symmetry of an icosahedron
is not allowed in crystallography, but quasicrystals with this symmetry
exist (e.g. AlMn). Diffraction patterns for quasicrystals have Bragg peaks,
with the density of Bragg peaks in each plane being higher than would be
expected for a perfect periodic crystal.

quasiparticle A long-lived single-particle *excitation in the quantum
theory of many-body systems, in which the excitations of the individual
particles are modified by their interactions with the surrounding medium.

quasiperiodicity *See* QUASICRYSTAL.

qubit *See* QUANTUM COMPUTING.

quenching 1. The rapid cooling of a metal by immersing it in a bath of
liquid in order to improve its properties. Steels are quenched to make
them harder but some nonferrous metals are quenched for other reasons
(copper, for example, is made softer by quenching). **2.** The process of
inhibiting a continuous discharge in a *Geiger counter so that the
incidence of further ionizing radiation can cause a new discharge. This is
achieved by introducing a quenching vapour, such as methane mixed
with argon or neon into the tube.

Q-value The total energy released in a *decay process. The *binding
energy curve (see diagram 1) shows the fission of a nuclide A into its
primary fission fragments B and C. B and C have greater binding energies
per nucleon. The fragments gain stability (*see* NUCLEAR STABILITY) because
there is no change in the total number of nucleons. The effective increase
in binding energy per nucleon corresponds to a release of stored potential
energy, which subsequently appears as either the kinetic energy of the
primary fragments or as gamma radiation. The Q-value of a decay process
is therefore equal to the increase in binding energy of the decay products.
For example, ^{236}U (uranium–236) undergoes a decay process in which two
fission fragments are formed, ^{146}La (lanthanum–146) and ^{87}Br (bromine–
87):

$$^{236}U \rightarrow\, ^{146}La + ^{87}Br + 3\,^{1}_{0}n.$$

The total binding energy of the products is 1975.19 MeV, whereas the
binding energy of the uranium nucleus is 1791.24 MeV. The increase in
binding energy is therefore 183.95 MeV; the Q-value for this fission

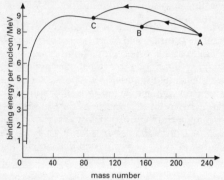

(1) The fission of nuclide A into fragments B and C.

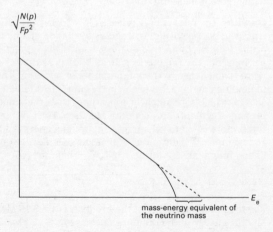

(2) The energy spectrum for electrons in beta decay is continuous; the Q-value is the upper limit.

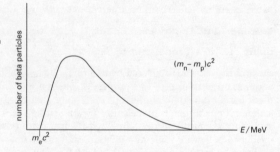

(3) Lymbimov's Curie plot variable is made up of the number of beta particles per unit, momentum range, $N(p)$, the momentum of these beta particles, p, and a correction factor, F.

Q-value

process is said to be 184 MeV. The possible existence of neutrino mass may lead to a modification of published Q-value tables of beta particle energies. As the energy spectrum for electrons emitted in beta decays is continuous, the upper limit of the spectrum is the Q-value (see diagram 2). If an emitted electron were to carry away this maximum energy there would be no more energy to make up the mass of the neutrino, however small that may be. In 1985, a Russian team led by Lymbimov investigated the beta decay of tritium:

$^{3}H \rightarrow\ ^{3}He + e^{-} + \gamma_{e}.$

They reported a nonzero neutrino mass calculated by measuring the shortfall in energy of the decay products. Lymbimov applied weak-interaction theory to obtain a relationship between a number of experimentally accessible variables. His team analysed their data by plotting these variables in a Curie plot, which according to theory should have yielded a straight line. The straight line predicted by theory would intersect the energy axis at the Q-value, suggesting a zero neutrino mass (see diagram 3). However, Lymbimov's data showed a small deviation from the expected line, and a value of the neutrino mass was calculated to be $(26 \pm 5)\ eV/c^{2}$.

q

racemic mixture (racemate) A mixture of equal quantities of the (+)- or *d*- and (–)- or *l*- forms of an optically active compound. Racemic mixtures are denoted by the prefix (±)- or *dl*- (e.g. (±)-lactic acid). A racemic mixture shows no *optical activity.

rad *See* RADIATION UNITS.

radar (radio detection and ranging) A method of detecting the presence, position, and direction of motion of distant objects (such as ships and aircraft) by means of their ability to reflect a beam of electromagnetic radiation of centimetric wavelengths. It is also used for navigation and guidance. It consists of a transmitter producing radio-frequency radiation, often pulsed, which is fed to a movable aerial from which it is transmitted as a beam. If the beam is interrupted by a solid object, a part of the energy of the radiation is reflected back to the aerial. Signals received by the aerial are passed to the receiver, where they are amplified and detected. An echo from a reflection of a solid object is indicated by a sudden rise in the detector output. The time taken for a pulse to reach the object and be reflected back (t) enables the distance away (d) of the target to be calculated from the equation $d = ct/2$, where c is the speed of light. In some systems the speed of the object can be measured using the *Doppler effect. The output of the detector is usually displayed on a cathode-ray tube in a variety of different formats (see illustration).

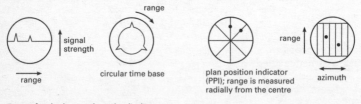

Types of cathode-ray tube radar display

radial field A field in which the field lines are radii that radiate from a centre. Point charges and masses and spheres have radial fields. The point sources of the fields and the centres of the spheres are the centres from which the field lines radiate.

radial velocity *See* LINE-OF-SIGHT VELOCITY.

radian *See* CIRCULAR MEASURE.

radiance Symbol L_e. The radiant intensity per unit transverse area, in a given direction, of a source of radiation. It is measured in W sr^{-1} m^{-2}.

radiant energy Energy transmitted as electromagnetic radiation.

radiant exitance *See* EXITANCE.

radiant flux Symbol Φ_e. The total power emitted, received, or passing in the form of electromagnetic radiation. It is measured in watts.

radiant intensity Symbol I_e. The *radiant flux per unit solid angle emitted by a point source. It is measured in watts per steradian.

radiation 1. Energy travelling in the form of electromagnetic waves or photons. **2.** A stream of particles, especially alpha- or beta-particles from a radioactive source or neutrons from a nuclear reactor.

radiation belts *See* VAN ALLEN BELTS.

radiation damage Harmful changes that occur to inanimate materials and living organisms as a result of exposure to energetic electrons, nucleons, fission fragments, or high-energy electromagnetic radiation. In inanimate materials the damage may be caused by electronic excitation, ionization, transmutation, or displacement of atoms. In organisms, these mechanisms can cause changes to cells that alter their genetic structure, interfere with their division, or kill them. In humans, these changes can lead to **radiation sickness**, **radiation burns** (from large doses of radiation), or to long-term damage of several kinds, the most serious of which result in various forms of cancer (especially leukaemia).

radiationless decay Decay of an atom or molecule from an excited state to a lower energy state without the emission of electromagnetic radiation. A common example of a radiationless process is the *Auger effect, in which an electron rather than a photon is emitted as a result of decay.

radiation pressure The pressure exerted on a surface by electromagnetic radiation. As radiation carries momentum as well as energy it exerts a force when it meets a surface, i.e. the *photons transfer momentum when they strike the surface. The pressure is usually negligible on large bodies, for example, the pressure of radiation from the sun on the surface of the earth is of the order of 10^{-5} pascal, but on small bodies it can have a considerable effect, driving them away from the radiation source. Radiation pressure is also important in the interiors of stars of very high mass.

radiation temperature The surface temperature of a celestial body as calculated by *Stefan's law, assuming that the body behaves as a *black body. The radiation temperature is usually measured over a narrow portion of the electromagnetic spectrum, such as the visible range (which gives the **optical temperature**).

radiation units Units of measurement used to express the *activity of a radionuclide and the *dose of ionizing radiation. The units **curie**, **roentgen**, **rad**, and **rem** are not coherent with SI units but their temporary use with SI units has been approved while the derived SI units **becquerel**, **gray**, and **sievert** become familiar.

The becquerel (Bq), the SI unit of activity, is the activity of a radionuclide decaying at a rate, on average, of one spontaneous nuclear transition per second. Thus $1\ Bq = 1\ s^{-1}$. The former unit, the curie (Ci), is equal to 3.7×10^{10} Bq. The curie was originally chosen to approximate the activity of 1 gram of radium–226.

The gray (Gy), the SI unit of absorbed dose, is the absorbed dose when the energy per unit mass imparted to matter by ionizing radiation is 1 joule per kilogram. The former unit, the rad (rd), is equal to 10^{-2} Gy.

The sievert (Sv), the SI unit of dose equivalent, is the dose equivalent when the absorbed dose of ionizing radiation multiplied by the stipulated dimensionless factors is $1\ J\ kg^{-1}$. As different types of radiation cause different effects in biological tissue a weighted absorbed dose, called the **dose equivalent**, is used in which the absorbed dose is modified by multiplying it by dimensionless factors stipulated by the International Commission on Radiological Protection. The former unit of dose equivalent, the rem (originally an acronym for *r*oentgen *e*quivalent *m*an), is equal to 10^{-2} Sv.

In SI units, exposure to ionizing radiation is expressed in coulombs per kilogram, the quantity of X- or gamma-radiation that produces ion pairs carrying 1 coulomb of charge of either sign in 1 kilogram of pure dry air. The former unit, the roentgen (R), is equal to $2.58 \times 10^{-4}\ C\ kg^{-1}$.

radiative collision A collision between charged particles in which part of the kinetic energy is radiated in the form of photons.

radical A root of a number or a quantity. The symbol $\sqrt{}$ is called the **radical sign**.

radio A means of transmitting information in which the transmission medium consists of electromagnetic radiation. Information is transmitted by means of the *modulation of a *carrier wave in a transmitter; the modulated carrier wave is fed to a transmitting aerial from which it is broadcast through the atmosphere or through space. A receiving aerial forms part of a *resonant circuit, which can be tuned to the frequency of the carrier wave, enabling the receiver that it feeds selectively to amplify and then to demodulate the transmitted signal. A replica of the original information is thus produced by the receiver. *See also* RADIO TRANSMISSION.

radioactive age The age of an archaeological or geological specimen as determined by a process that depends on a radioactive decay. *See* CARBON DATING; FISSION-TRACK DATING; POTASSIUM–ARGON DATING; RUBIDIUM–STRONTIUM DATING; URANIUM–LEAD DATING.

radioactive equilibrium The equilibrium reached by a *radioactive series in which the rate of decay of each nuclide is equal to its rate of production. It follows that all the rates of decay of the different nuclides within the sample are equal when radioactive equilibrium is achieved. For example, in the **uranium series**, uranium–238 decays to thorium–234. Initially the rate of production of thorium will exceed the rate at which it is decaying, and the thorium content of the sample will rise. As the amount of thorium increases, its activity increases; eventually a situation is reached in which the rate of production of thorium is equal to its rate of decay. The proportion of thorium in the sample will then remain constant. Thorium decays to produce protactinium–234; some time after the stabilization of the thorium content the protactinium content will also stabilize. When the whole radioactive series attains stabilization, the sample is said to be in radioactive equilibrium. There is a simple relationship between the numbers of nuclides of a specific type and their associated half-lives: $N_1/T_1 = N_2/T_2 = N_3/T_3 = \ldots$, where N_1, N_2, … are the numbers of nuclides of type 1, 2, etc. T_1, T_2, … are their respective half-lives.

radioactive series A series of radioactive nuclides in which each member of the series is formed by the decay of the nuclide before it. The series ends with a stable nuclide. Three radioactive series occur naturally, those headed by thorium–232 (**thorium series**), uranium–235 (**actinium series**), and uranium–238 (**uranium series**). All three series end with an isotope of lead. The neptunium series starts with the artificial isotope plutonium–241, which decays to neptunium–237, and ends with bismuth–209.

radioactive tracing *See* LABELLING.

radioactive waste (**nuclear waste**) Any solid, liquid, or gaseous waste material that contains *radionuclides. These wastes are produced in the mining and processing of radioactive ores, the normal running of nuclear power stations and other reactors, the manufacture of nuclear weapons, and in hospitals and research laboratories. Because high-level radioactive wastes can be extremely dangerous to all living matter and because they may contain radionuclides having half-lives of many thousands of years, their disposal has to be controlled with great stringency.

High-level waste (e.g. spent nuclear fuel) requires to be cooled artificially and is therefore stored for several decades by its producers before it can be disposed of. Intermediate-level waste (e.g. processing plant sludge and reactor components) is solidified, mixed with concrete, packed in steel drums, and stored in special sites at power stations before being buried in concrete chambers in deep mines or below the seabed. Low-level waste (e.g. solids or liquids lightly contaminated by radioactive substances) is disposed of in steel drums in special sites in concrete-lined trenches. In the UK, a company (Nirex Ltd) was set up by the nuclear industry and the government in 1988 to handle the disposal of nuclear waste. There are also nuclear reprocessing plants at Dounreay and Sellafield (Thermal

Oxide Reprocessing Plant). Other countries have similar arrangements. Since 1983 no wastes have been disposed of in steel drums cast in concrete in the Atlantic deeps.

radioactivity The spontaneous disintegration of certain atomic nuclei accompanied by the emission of alpha-particles (helium nuclei), beta-particles (electrons or positrons), or gamma radiation (short-wavelength electromagnetic waves).

Natural radioactivity is the result of the spontaneous disintegration of naturally occurring radioisotopes. Many radioisotopes can be arranged in three *radioactive series. The rate of disintegration is uninfluenced by chemical changes or any normal changes in their environment. However, radioactivity can be induced in many nuclides by bombarding them with neutrons or other particles. *See also* HALF-LIFE; IONIZING RADIATION; RADIATION UNITS.

radio astronomy The study of the radio-frequency radiation emitted by celestial bodies. This branch of astronomy began in 1932 when a US engineer, Karl Jansky (1905–40), first detected radio waves from outside the earth's atmosphere. *See* RADIO SOURCE; RADIO TELESCOPE; RADIO WINDOW.

radiobiology The branch of biology concerned with the effects of radioactive substances on living organisms and the use of radioactive tracers to study metabolic processes (*see* LABELLING).

radiocarbon dating *See* CARBON DATING.

radiochemistry The branch of chemistry concerned with radioactive compounds and with ionization. It includes the study of compounds of radioactive elements and the preparation and use of compounds containing radioactive atoms. *See* LABELLING; RADIOLYSIS.

radio frequencies The range of frequencies, between about 3 kilohertz and 300 gigahertz, over which electromagnetic radiation is used in radio transmission. It is subdivided into eight equal bands, known as very low frequency, low frequency, medium frequency, high frequency, very high frequency, and extremely high frequency.

radio galaxies A *radio source outside the Galaxy that has been identified with an optically visible galaxy. These radio galaxies are distinguished from normal galaxies by having a radio power output some 10^6 times greater (i.e. up to 10^{38} watts rather than 10^{32} W). The source of the radio-frequency energy is associated with violent activity, involving the ejection of relativistic jets of particles from the nucleus of the galaxy. It has been suggested that the radio sources are powered by supermassive *black holes in the nucleus.

radiogenic Denoting a material produced by radioactive decay, e.g. the lead formed from uranium. **Radiogenic dating** is a technique for calculating the age of an object by determining the ratio of the concentration of a particular *radioisotope in the object to the concentration of its stable isotope. For example, in archaeology the age of objects, such as bones, can be estimated by finding the ratio of carbon–14 to carbon–12. *See also* RADIOACTIVE AGE.

radiography The process or technique of producing images of an opaque object on photographic film or on a fluorescent screen by means of radiation (either particles or electromagnetic waves of short wavelength, such as X-rays and gamma-rays). The photograph produced is called a **radiograph**.

In a medical radiograph part of a patient's body is exposed to X-rays so that a shadow image of that body part is produced on a film placed behind the body. The image is formed by the varying degrees of attenuation offered by the various media within the body. The degree of attenuation depends on three main factors:
1. The attenuating character of the medium. This is usually characterized by an **attenuation coefficient**.
2. The density of the medium.
3. The thickness of the structure.

Bone offers the greatest degree of attenuation, followed by soft tissue, and then pockets of air. Conventional radiography cannot easily distinguish between different types of soft tissue (e.g. liver and kidney tissues) owing to their similar attenuation characters and densities. Bone, however, is very easily distinguished from neighbouring soft tissue, such as muscle.

If better contrast is required, a contrast medium, such as barium, can be used, which preferentially absorbs X-rays and enables otherwise difficult anatomical features to be observed. Kidney, arterial, and intestinal functions are examined by this method, using various contrast media.

When an X-ray interacts with an atom in the medium, there are several possible mechanisms that may attenuate the X-ray. The attenuation mechanism that the X-ray undergoes depends on its energy. The main attenuation mechanisms are:

Simple scattering: this occurs when the X-ray has insufficient energy to cause an ionization of the atom. The X-ray is simply deflected without significant change of energy.

Photoelectric effect: the X-ray transfers all of its energy to an electron in the atom and therefore ionizes the atom (*see* PHOTOELECTRIC EFFECT).

Compton scattering: the high-energy X-ray collides with an electron, transferring both energy and momentum to it. The X-ray recoils with a loss of energy (*see* COMPTON EFFECT).

Pair production: this mechanism occurs at very high energies. Within the field of the interacting atomic nucleus, the X-ray spontaneously creates a positron–electron pair.

The table overleaf shows the thresholds for the various attenuation mechanisms and their variation with the atomic number Z of the medium material and the energy E of the X-ray.

Mechanism	Variation of the attenuation with the energy (E) of the X-ray	Variation of the attenuation with the atomic number (Z) of the medium substance	Energy range in which this kind of attenuation is dominant
simple scattering	$\propto 1/E$	$\propto Z^2$	1–20 keV
photoelectric effect	$\propto 1/E^3$	$\propto Z^3$	1–30 keV
Compton scattering	falls gradually with E	independent	30 keV – 20 MeV
pair production	rises slowly with E	$\propto Z^2$	above 20 MeV

The optimum X-ray energy for diagnostic radiography is in the range of the photoelectric effect. At about 30 keV, the photoelectric effect dominates the attenuation of the X-radiation and leads to a better contrast between media of different atomic numbers. The Z^3 variation of the attenuation within the photoelectric-dominated regime means that the distinction between media of different Z is more pronounced.

Air has a higher average Z than soft tissue, which consists of mostly carbon ($_6$C) and hydrogen ($_1$H). However, the low average density of air means that it does not attenuate X-rays strongly. Barium, on the other hand, has an atomic number of 56, and thus attenuates very strongly, which is why it is used as a contrast medium in barium meals and barium enemas.

Diagnostic radiography requires high-quality images of internal structures. This means that a high contrast and low blurring must be achieved. It is therefore important to prevent X-rays falling onto the film, which have scattered off many different parts of the body. This is made possible by the introduction of a grid between the patient and the film as shown in the diagram.

The grid consists of narrow lead strips that allow passage of the main unabsorbed beam, but intercept randomly scattered X-rays. The entire grid is oscillated during the formation of the radiograph in order to prevent the formation of its own well-defined X-ray shadow. Before the X-ray beam interacts with the patient's body it is controlled by a diaphragm to limit the beam width. This gives a relatively well-defined beam that can be directed onto the desired area. Wide beams are undesirable because they generate unwanted scattering and deliver an

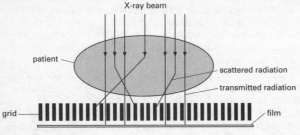

Radiography (medical)

unnecessarily high dose to the patient. Longer exposure at low beam intensity can, in principle, improve the contrast. However, blurring becomes more likely owing to the involuntary movements of the patient during the scan.

A natural magnification of the subject structure occurs during an X-ray scan because the X-radiation source acts as a point source. The area of the X-ray shadow is greater than that occupied by the two-dimensional projection of the structure under examination. In practice the X-ray source is not a point and the image is surrounded by a penumbra, which constitutes a relatively blurred region. The more point-like the source, the smaller the penumbra on the image.

radio interferometry *See* RADIO TELESCOPE.

radioisotope (**radioactive isotope**) An isotope of an element that is radioactive. *See* LABELLING.

radioisotope imaging The formation of images to provide information about the function of various organs in the body, using internally administered *radioisotopes as a radiation source. The technique is widely used in medicine to locate tumours or cancers and to examine the flow patterns of body fluids.

Alpha and beta radiation is not suitable for imaging procedures as they are both absorbed by tissue, subjecting healthy cells to an unnecessary dose of radiation. Gamma radiation, however, is not easily absorbed by tissue and can therefore be conveniently detected outside the body. The gamma radiation emitted from the area under examination is detected either by a **rectilinear scanner** or a **gamma camera** (see illustrations).

In the rectilinear scanner, the radiation detector is mounted onto an armature that moves in straight lines as shown in the diagram. The intensity of the radiation from the radioisotope within the patient's body is picked up and transformed by electronics into a signal that controls the intensity of light produced by a light source. The light source mimics the movement of the detector exactly, enabling an image of the distribution of the radioisotope to be produced.

In the gamma camera, the gamma radiation emitted from the radioisotope distributed in the patient's body is detected by a series of sodium iodide crystals set behind a collimator. The sodium iodide crystals produce a flash for each detected gamma ray. The light from the flash is amplified in photomultiplier tubes, which are arranged in an array to enable different tubes to correspond to different regions of the body under examination. Electronic circuits convert the output of the photomultiplier array into electrical potential differences, which are applied across the X and Y plates of a cathode-ray oscilloscope to form a two dimensional image of the distribution of the radioisotope.

Nuclides are often injected into the patient in the form of a dissolved compound, or sometimes attached to antibodies. Pure gamma-emitters of energy 60–40 keV are usually chosen as radioisotopes for these procedures. The nuclide should also have a suitable half-life, typically a few hours, so that the overall radiation dose is kept to a minimum. However, the activity must remain high enough for a clinical examination to take place. It is also important that the nuclide and its carrier fluid are non-toxic and sterile. Two isotopes commonly used for this procedure are iodine–131 (^{131}I) and technecium–99 (^{99}Tc). ^{131}I is produced in a nuclear

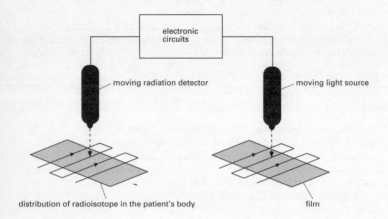

Rectilinear scanner

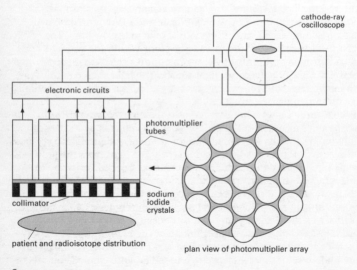

Gamma camera

reactor, when neutrons are captured by tellurium ($^{130}_{52}$Te). The isotope formed then decays by the emission of a beta particle into ^{131}I. The iodine is then separated chemically from the tellurium:

$$^{130}_{52}\text{Te} + ^{1}_{0}\text{n} \rightarrow ^{131}_{52}\text{Te} \rightarrow ^{131}_{53}\text{I} + ^{0}_{-1}\beta.$$

^{131}I has a half-life of 8 days, and itself decays by beta emission, enabling it to be sometimes used to treat thyroid cancer (*see* RADIOLOGY). However, it also emits a gamma ray of suitable energy for detection (360 keV):

$$^{131}_{53}\text{I} \rightarrow ^{131}_{54}\text{Xe} + ^{0}_{-1}\beta + ^{0}_{0}\gamma.$$

The 8-day half-life is ideal for clinical purposes. ^{131}I is commonly used for thyroid studies, since the thyroid gland readily takes up iodine from the bloodstream. ^{99}Tc used in the imaging procedure is a metastable state; that is, it is a nuclide remaining in an excited state over a long half-life. The metastable ^{99}Tc (usually labelled ^{99}Tc*) decays to a ground state by gamma emission. The ^{99}Tc* itself is a product of the beta decay of a molybdenum radionuclide:

$$^{99}_{42}\text{Mo} \rightarrow ^{99}_{43}\text{Tc*} + ^{0}_{-1}\beta$$

$$^{99}_{43}\text{Tc*} \rightarrow ^{99}_{43}\text{Te} + ^{0}_{0}\gamma.$$

The second process in this chain has a half-life of 6 hours, which is ideal for relatively short clinical studies. The gamma ray emission is also at an energy of 160 keV, which is easily detectable. No harmful beta particles are produced.

^{99}Tc* is obtained from a device, sometimes called a 'Mo-cow' or 'molybdenum cow', on account of the way in which the isotope is produced and tapped off the apparatus. The device operates by elution of a specially prepared column of ammonium molybdenate. The molybdenum in this column is originally obtained from the fission of uranium in a nuclear reactor. The metastable technetium occurs in the column in the form of a pertechnate ion, which can be exchanged with chloride ions when sodium chloride solution is passed through the column. In this way pertechnate ions are removed from the column, leaving behind the molybdate that has not yet decayed. The fluid that one 'milks' from this apparatus is sterile and compatible with blood. ^{99}Tc* has many applications in imaging the brain or tumours, which have rich or poor blood supplies.

radiolocation The location of distant objects by means of *radar.

radiology The study and use of X-rays, radioactive materials, and other ionizing radiations for medical purposes, especially for diagnosis (**diagnostic radiology**) and the treatment of cancer and allied diseases (*see* RADIOTHERAPY).

radiolysis The use of ionizing radiation to produce chemical reactions. The radiation used includes alpha particles, electrons, neutrons, X-rays, and gamma rays from radioactive materials or from accelerators. Energy transfer produces ions and excited species, which undergo further reaction. A particular feature of radiolysis is the formation of short-lived solvated electrons in water and other polar solvents.

radiometric dating (radioactive dating) *See* DATING TECHNIQUES; RADIOACTIVE AGE.

radionuclide (radioactive nuclide) A *nuclide that is radioactive.

radiopaque *See* OPACITY.

radiosonde A meteorological instrument that measures temperature, pressure, humidity, and winds in the upper atmosphere. It consists of a package of instruments and a radio transmitter attached to a balloon. The data is relayed back to earth by the transmitter. The position of the balloon can be found by radar and from its changes in position the wind velocities can be calculated.

radio source An astronomical object that has been observed with a *radio telescope to emit radio-frequency electromagnetic radiation. Radio sources within the Galaxy include Jupiter, the sun, pulsars, and background radiation arising from *synchrotron radiation. Sources outside the Galaxy include spiral galaxies, *radio galaxies, and *quasars. Radio sources were formerly known as **radio stars**.

radio star *See* RADIO SOURCE.

radio telescope An instrument for detecting and measuring electromagnetic radiation of radio frequencies that have passed through the *radio window in the earth's atmosphere and reached the surface of the earth. There are a great diversity of *radio sources within the universe and radio telescopes are required to detect both continuous emissions and specific spectral lines. They therefore require the highest possible angular resolution so that the details of radio sources can be studied and they should be able to pick up weak signals. The simplest radio telescope consists of a paraboloidal steerable-dish aerial together with ancillary amplifiers. The paraboloidal reflecting dish surface reflects the incoming signal to the principal focus of the reflector. At this point the radio-frequency signals are amplified up to 1000 times and converted to a lower, intermediate, frequency before transmission by cable to the control building. Here the intermediate frequency is amplified again and passed to the detector and display unit. As the radio waves arriving from the surface of the reflector at the focus must be in phase, the surface of the dish must be very accurately constructed; for example, a 100-metre-diameter dish must be accurate to the nearest millimetre, when receiving radiation of 1 cm wavelength. To overcome the problem of constructing large dishes to such a high accuracy, the technique of **radio interferometry** has been developed. In this technique an array of small aerials connected by cable is used to simulate a large dish aerial. In earth-rotation **aperture synthesis**, the individual positions and displacements of an array of only a few such small aerials can be made to simulate an enormous dish aerial as the earth rotates. All but the smallest steerable dishes are constructed from metal mesh so that wind can pass through them. A few very large fixed dishes have been built into the earth's surface.

radiotherapy The use of *ionizing radiation in the treatment of cancer and some diseases of bone marrow and blood. *X-rays and *gamma radiation used in diagnostic *radiology have low energy so that they do not damage the tissues through which they pass. Radiotherapy, however, makes use of the property of ionizing radiation that diagnostic radiology tries to avoid; that is, it has to have sufficient energy to kill the cells on which it falls.

Ionizing radiation kills cells indiscriminately. Therefore when a cancerous tumour is irradiated, the radiation will kill a number of normal tissue cells in the vicinity of the tumour. Since high doses are usually required to kill tumours, steps must be taken to limit the damage to normal tissue. This is achieved by a method called **fractionation**, in which the necessary radiation dose is divided into a number of small fractions administered over a period. This allows the normal tissue partially to recover between treatments. Since, in some cases, cancer cells are more vulnerable to radiation than normal cells, normal tissue recovery is more rapid, enabling fractionation to increase the ability of the patient to tolerate the therapy.

A radiation dose may be administered to the patient either from outside the body or inside the body. **Teletherapy** is the use of radiation beams external to the body, and **brachytherapy** is the use of small-volume sealed radionuclide sources implanted near the tumour.

radio transmission The transmission of radio waves from a transmitting aerial to a receiving aerial. The radiation may take several

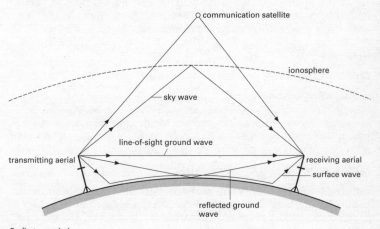

Radio transmission

paths (see illustration). The sum of the line-of-sight ground wave, the reflected ground wave, and the surface wave is called the **ground wave** (or **tropospheric wave**). **Sky waves** (or **ionospheric waves**) are reflected by the ionosphere (*see* EARTH'S ATMOSPHERE) and enable long-distance transmissions to be made. The ionization of atoms and molecules in the ionosphere is caused largely by solar ultraviolet and X-radiation and therefore conditions differ between night and day. Ionization in the lower E-region of the ionosphere falls off at night in the absence of sunlight and ions and electrons tend to recombine. However, in the less dense (higher) F-region there are fewer collisions between ions and electrons and therefore fewer recombinations at night. The F-region is therefore a more effective reflector at night.

The UHF and VHF waves used in television broadcasting penetrate the ionosphere with little reflection. Therefore TV broadcasts can only be made over long distances by means of artificial satellites. *See also* RADIO.

radiotransparent Transparent to radiation, especially to X-rays and gamma-rays.

radio window A region of the electromagnetic spectrum in the radio-frequency band within which radio waves can be transmitted through the *earth's atmosphere without significant reflection or attenuation by constituents of the atmosphere. It extends from about 10 megahertz to 100 gigahertz and enables radiation in this range from celestial radio sources to be picked up by *radio telescopes on the earth's surface. Below 100 MHz incoming radio waves are reflected by the ionosphere and those above 100 GHz are increasingly affected by molecular absorption.

radium Symbol Ra. A radioactive metallic element belonging to group 2 (formerly IIA) of the periodic table; a.n. 88; r.a.m. 226.0254; r.d. ~5; m.p. 700°C; b.p. 1140°C. It occurs in uranium ores (e.g. pitchblende). The most stable isotope is radium–226 (half-life 1602 years), which decays to radon. It is used as a radioactive source in research and, to some extent, in radiotherapy. The element was isolated from pitchblende in 1898 by Marie and Pierre *Curie.

radius of curvature *See* CENTRE OF CURVATURE.

radius of gyration Symbol k. The square root of the ratio of the *moment of inertia of a rigid body about an axis to the body's mass, i.e. $k^2 = I/m$, where I is the body's moment of inertia and m is its mass. If a rigid body has a moment of inertia I about an axis and mass m it behaves as if all its mass is rotating at a distance k from the axis.

radius vector *See* POLAR COORDINATES.

radon Symbol Rn. A colourless radioactive gaseous element belonging to group 18 of the periodic table (the noble gases); a.n. 86; r.a.m. 222; d. 9.73 g dm^{-3}; m.p. –71°C; b.p. –61.8°C. At least 20 isotopes are known, the most stable being radon–222 (half-life 3.8 days). It is formed by decay of radium–226 and undergoes alpha decay. It is used in radiotherapy. Radon occurs naturally, particularly in areas underlain by granite, where it is thought to be a health hazard. As a noble gas, radon is practically inert, although a few compounds, e.g. radon fluoride, can be made. It was first isolated by William Ramsey and Robert Whytlaw-Gray (1877–1958) in 1908.

rainbow An optical phenomenon that appears as an arc of the colours of the spectrum across the sky when falling water droplets are illuminated by sunlight from behind the observer. The colours are produced by the refraction and internal reflection of the sunlight by the water drops. Two bows may be visible, the inner ring being known as the primary bow and the outer, in which the colours are reversed, as the secondary bow (see illustration).

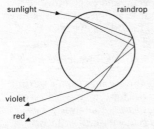

Primary rainbow (one internal reflection)

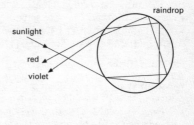

Secondary rainbow (two internal reflections)

RAM Random-access memory. The main memory of a computer, fabricated from *integrated circuits, in which data can only be stored temporarily – until the power supply is turned off. RAM consists of arrays of 'cells', each capable of holding one *bit of data. Cells are completely independent so that the access time to any location is fixed (and extremely rapid) – hence the term random access.

r.a.m. *See* RELATIVE ATOMIC MASS.

Raman effect A type of *scattering of electromagnetic radiation in which light suffers a change in frequency and a change in phase as it passes through a material medium. The intensity of Raman scattering is about one-thousandth of that in Rayleigh scattering in liquids; for this reason it was not discovered until 1928. However, it was not until the development of the laser that the effect was put to use.

In **Raman spectroscopy** light from a laser is passed through a substance and the scattering is analysed spectroscopically. The new frequencies in the **Raman spectrum** of monochromatic light scattered by a substance are characteristic of the substance. Both inelastic and superelastic scattering occurs. The technique is used as a means of determining molecular structure and as a tool in chemical analysis. The effect was discovered by the Indian scientist Sir C. V. Raman (1888–1970). It was predicted theoretically by H. A. Kramers and Werner Heisenberg in 1925.

ramjet *See* JET PROPULSION.

Ramsay, Sir William (1852–1916) British chemist, born in Glasgow. After working under Robert *Bunsen, he returned to Glasgow before taking up professorships at Bristol (1880–87) and London (1887–1912). In the early 1890s he worked with Lord *Rayleigh on the gases in air and in 1894 they discovered argon. In 1898, with Morris Travers (1872–1961), he discovered neon, krypton, and xenon. Six years later he discovered the last of the noble gases, *radon. He was awarded the Nobel Prize for chemistry in 1904, the year in which Rayleigh received the physics prize.

Ramsden eyepiece An eyepiece for optical instruments consisting of two identical plano-convex lenses with their convex faces pointing towards each other. They are separated by a distance of two thirds of the focal length of either lens. It was invented by the British optical instrument maker Jesse Ramsden (1735–1800).

random alloy *See* DISORDERED SOLID.

random walk The problem of determining the distance from a starting position made by a walker, who can either move forward (toward +x) or backwards (toward –x) with the choice being made randomly (e.g. by tossing a coin). The progress of the walker is characterized by the net distance D_N travelled in N steps. After N steps the *root-mean-square value D_{rms}, which is the average distance away from the starting position, is given by $D_{rms} = \sqrt{N}$. Physical applications of the random walk include diffusion and the related problem of Brownian motion as well as problems involving the structures of polymers and disordered solids.

Rankine cycle A cycle of operations in a heat engine. The Rankine cycle more closely approximates to the cycle of a real steam engine that does the *Carnot cycle. It therefore predicts a lower ideal thermal efficiency than the Carnot cycle. In the Rankine cycle (see illustration), heat is added at constant pressure p_1, at which water is converted in a boiler to

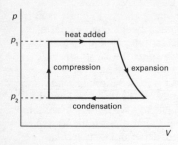

Rankine cycle

superheated steam; the steam expands at constant entropy to a pressure p_2 in the cylinder; heat is rejected at constant pressure p_2 in a condenser; the water so formed is compressed at constant entropy to pressure p_1 by a feed pump. The cycle was devised by the Scottish engineer W. J. M. Rankine (1820–70).

Rankine temperature scale An absolute temperature scale based on the Fahrenheit scale. Absolute zero on this scale, 0°R, is equivalent to –459.67°F and the melting point of ice (32°F) is therefore (459.67 + 32 = 491.67)°R. The scale was devised by W. J. M. Rankine.

Raoult's law The partial vapour pressure of a solvent is proportional to its mole fraction. If p is the vapour pressure of the solvent (with a substance dissolved in it) and X the mole fraction of solvent (number of moles of solvent divided by total number of moles) then $p = p_0X$, where p_0 is the vapour pressure of the pure solvent. A solution that obeys Raoult's law is said to be an **ideal solution**. In general the law holds only for dilute solutions, although some mixtures of liquids obey it over a whole range of concentrations. Such solutions are **perfect solutions** and occur when the intermolecular forces between molecules of the pure substances are similar to the forces between molecules of one and molecules of the other. Deviations in Raoult's law for mixtures of liquids cause the formation of *azeotropes. The law was discovered by the French chemist François Raoult (1830–1901).

rarefaction A reduction in the pressure of a fluid and therefore of its density.

raster The pattern of scanning lines on the screen of the cathode-ray tube in a television receiver or other device that provides a visual display.

rationalized Planck constant *See* PLANCK CONSTANT.

rationalized units A system of units in which the defining equations have been made to conform to the geometry of the system in a logical way. Thus equations that involve circular symmetry contain the factor 2π, while those involving spherical symmetry contain the factor 4π. *SI units and *Heaviside–Lorentz units are rationalized; *Gaussian units are unrationalized.

rational number Any number that can be expressed as the ratio of two integers. For example, 0.3333… is rational because it can be written as 1/3. $\sqrt{2}$, however, is an *irrational number.

ray A narrow beam of radiation or an idealized representation of such a beam on a **ray diagram**, which can be used to indicate the positions of the object and image in a system of lenses or mirrors.

Rayleigh, Lord (John William Strutt; 1842–1919) British physicist, who built a private laboratory after working at Cambridge University. His work in this laboratory included the discovery of Rayleigh scattering (*see*

SCATTERING OF ELECTROMAGNETIC RADIATION). He also worked in acoustics, electricity, and optics, as well as collaborating with William *Ramsay on the discovery of argon. He was awarded the 1904 Nobel Prize for physics.

Rayleigh criterion *See* RESOLVING POWER.

Rayleigh–Jeans formula *See* PLANCK'S RADIATION LAW.

Rayleigh scattering *See* SCATTERING OF ELECTROMAGNETIC RADIATION.

reactance Symbol X. A property of a circuit containing inductance or capacitance that together with any resistance makes up its *impedance. The impedance Z is given by $Z^2 = R^2 + X^2$, where R is the resistance. For a pure capacitance C, the reactance is given by $X_C = 1/2\pi fC$, where f is the frequency of the *alternating current; for a pure inductance L, $X_L = 2\pi fL$. If the resistance, inductance, and capacitance are in series the impedance $Z = \sqrt{[R^2 + (X_L - X_C)^2]}$. Reactance is measured in ohms.

reaction A force that is equal in magnitude but opposite in direction to some other force, in accordance with Newton's third law of motion. If a body A exerts a force on body B, then B exerts an equal and opposite force on A. Thus, every force could be described as 'a reaction', and the term is better avoided, although it is still used in such terms as 'reaction propulsion'.

reaction propulsion *See* JET PROPULSION.

reactor 1. *See* NUCLEAR REACTOR. **2.** Any device, such as an inductor or capacitor, that introduces *reactance into an electrical circuit.

real gas A gas that does not have the properties of an *ideal gas. Its molecules have a nonzero size and there are forces between them (*see* EQUATION OF STATE).

real image *See* IMAGE.

real-is-positive convention (**real-positive convention**) A convention used in optical formulae relating to lenses and mirrors. In this convention, distances from optical components to real objects, images, and foci are taken as positive, whereas distances to virtual points are taken as negative.

Réaumur temperature scale A temperature scale in which the melting point of ice is taken as 0°R and the boiling point of water as 80°R. It was devised by René Antoine Réaumur (1683–1757).

recalescence A phenomenon that occurs during the cooling of iron and other ferromagnetic metals (*see* MAGNETISM) after they have been heated. When iron is heated to white heat and then allowed to cool, it abruptly evolves heat at a certain temperature. This evolution of heat, which slows down the cooling process and can lead to a brief reheating, is caused by an *exothermic reaction when the structure of the crystal changes. The temperature at which this occurs is called the **recalescence point**. For pure iron there are two recalescence points: at 780°C and 880°C. A reverse phenomenon causing cooling, called **decalescence**, occurs when ferromagnetic metals are heated.

reciprocal space A theoretical space used in the mathematical analysis of crystals in which there is a lattice called the **reciprocal lattice**. If the primitive translation vectors of the real (direct) lattice are $\boldsymbol{a}$, $\boldsymbol{b}$, $\boldsymbol{c}$ the primitive translation vectors $\boldsymbol{a'}$, $\boldsymbol{b'}$, $\boldsymbol{c'}$ of the reciprocal lattice are defined by $\boldsymbol{a'} = \boldsymbol{b} \times \boldsymbol{c}$, $\boldsymbol{b'} = \boldsymbol{c} \times \boldsymbol{a}$, $\boldsymbol{c'} = \boldsymbol{a} \times \boldsymbol{b}$. This definition means that every plane in a real lattice becomes a point in the reciprocal lattice. The concept of reciprocal space is very useful in X-ray crystallography and energy-band theory.

recombination era The period after the big bang when the universe was sufficiently cool for electrons and nuclei to form atoms for the first time. The recombination occurred about 300 000 years after the big bang. Up to this time, photons in the universe were scattered by free electrons and it is sometimes said that the universe was 'opaque'. With recombination, photons were able to travel without impedement, and the universe became 'transparent'. *See also* MICROWAVE BACKGROUND RADIATION.

recombination process The process in which a neutral atom or molecule is formed by the combination of a positive ion and a negative ion or electron; i.e. a process of the type:

$A^+ + B^- \rightarrow AB$ or $A^+ + e^- \rightarrow A$

In recombination, the neutral species formed is usually in an excited state, from which it can decay with emission of light or other electromagnetic radiation.

rectification 1. (in physics) The process of obtaining a direct current

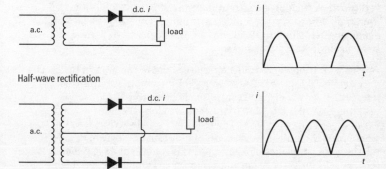

Half-wave rectification

Full-wave rectification

from an alternating electrical supply. *See* RECTIFIER. **2.** (in chemistry) The process of purifying a liquid by distillation.

rectifier An electrical device that allows more current to flow in one direction than the other, thus enabling alternating e.m.f.s to drive only direct current. The device most commonly used for rectification is the semiconductor *diode. In **half-wave rectification**, achieved with one diode, a pulsating current is produced. In **full-wave rectification** two diodes are used, one pair conducting during the first half cycle and the other conducting during the second half (see illustration). The full-wave rectified signal can be smoothed using a capacitor or an inductor. The **bridge rectifier** illustrated also gives full-wave rectification.

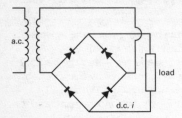

Bridge rectifier

red dwarf A dim and cool dwarf star (i.e. one that has a low *luminosity and a lower temperature than all other dwarf stars). The diameter of such stars is about half that of the sun and their surface temperature lies between 2000K and 3000K. It is thought that red dwarf stars are the most common type of star in the universe. The name red dwarf was suggested by the Danish astronomer Ejnar Hertzsprung (1873–1969).

A **red dwarf white star** is a star with a surface temperature less than 400K, which is about 10 000 times fainter than the sun. In these stars radiation from the surface has cooled the star very rapidly.

red giant A *giant star thought to be in the later stages of *stellar evolution. It has a surface temperature in the range 2000–3000 K and a diameter 10–100 times that of the sun. *See also* DEATH OF A STAR; HERTZSPRUNG–RUSSELL DIAGRAM.

redshift **1. (Doppler redshift)** A displacement of the lines in the spectra of certain galaxies towards the red end of the visible spectrum (i.e. towards longer wavelengths). It is usually interpreted as a *Doppler effect resulting from the recession of the galaxies along the line of sight (*see* EXPANSION OF THE UNIVERSE). The redshift is usually expressed as $\Delta\lambda/\lambda$, where $\Delta\lambda$ is the shift in wavelength of radiation of wavelength λ. For relatively low velocities of recession this is equivalent to v/c, where v is the relative velocity of recession and c is the speed of light. If very high velocities of recession are involved, a relativistic version of v/c is required

(*see* RELATIVISTIC SPEED). The redshift of spectral lines occurs in all regions of the electromagnetic spectrum; ultraviolet can be shifted into the visible region and visible radiation can be shifted into the infrared region.
2. (gravitational *or* **Einstein redshift)** A similar displacement of spectral lines towards the red caused not by a Doppler effect but by a high gravitational field. This type of redshift was predicted by Einstein and some astronomers believe that this is the cause of the large redshifts of *quasars, which can be as high as 3.35.

reflectance The ratio of the radiant flux reflected by a surface to that falling on it. This quantity is also known as the **radiant reflectance**. The radiant reflectance measured for a specified wavelength of the incident radiant flux is called the **spectral reflectance**.

reflecting telescope (reflector) *See* TELESCOPE.

reflection The return of all or part of a beam of particles or waves when it encounters the boundary between two media. The **laws of reflection** state: (1) that the incident ray, the reflected ray, and the normal to the reflecting interface at the point of incidence are all in the same plane; (2) that the angle of incidence equals the angle of reflection. *See also* MIRROR; REFLECTANCE; TOTAL INTERNAL REFLECTION.

refracting telescope (refractor) *See* TELESCOPE.

refraction The change of direction suffered by wavefront as it passes obliquely from one medium to another in which its speed of propagation is altered. The phenomenon occurs with all types of waves, but is most familiar with light waves. In optics the direction is changed in accordance with **Snell's law**, i.e. $n_1 \sin i = n_2 \sin r$, where i and r are the angles made by the incident beam of radiation and the refracted beam to the normal (an imaginary line perpendicular to the interface between the two media); n_1 and n_2 are the *refractive indices of the two media. This law is also known as one of the **laws of refraction**. The other law of refraction is that the incident ray, the refracted ray, and the normal at the point of incidence lie in the same plane. The change of direction results from a change in

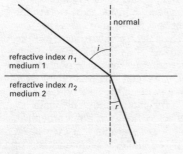

Refraction

the speed of propagation and the consequent change in wavelength (see illustration).

refractive index (refractive constant) Symbol n. The **absolute refractive index** of a medium is the ratio of the speed of electromagnetic radiation in free space to the speed of the radiation in that medium. As the refractive index varies with wavelength, the wavelength should be specified. It is usually given for yellow light (sodium D-lines; wavelength 589.3 nm). The **relative refractive index** is the ratio of the speed of light in one medium to that in an adjacent medium. *See also* REFRACTION.

refractivity A measure of the extent to which a medium will deviate a ray of light entering its surface. In some contexts it is equal to $(n - 1)$, where n is the *refractive index.

refractometer Any of various instruments for measuring the *refractive index of a substance or medium. An example is the **Pulfrich refractometer**, which is a glass block with a polished top, with a small cell on top of the block for liquid samples. A telescope rotating on a vertical circular scale is used to find the angle (α) between the top of the block and the direction in which the limiting ray (from incident light parallel to the top face) leaves the side of the block. If the refractive index of the block (n_g) is known, that of the liquid can be calculated using $n = \sqrt{(n_g^2 - \sin^2\alpha)}$.

refractory 1. Having a high melting point. Metal oxides, carbides, and silicides tend to be refractory materials, and are extensively used for lining furnaces, etc. **2.** A refractory material.

refrigerant *See* REFRIGERATION.

refrigeration The process of cooling a substance and of maintaining it at a temperature below that of its surroundings. In the domestic refrigerator a cycle of operations equivalent to those of a *heat pump is used.

The vapour-compression cycle

In this cycle (see illustration) a volatile liquid refrigerant, such as a fluorocarbon, is pumped through the cooling coils of the evaporator to the ice-making compartment of the refrigerator. Inside these coils the refrigerant evaporates, taking the latent heat required to effect the change of state from the inside of the ice-making compartment, causing the temperature in this compartment to fall. The vapour is then passed to an electrically driven compressor before entering the condenser, where the high-pressure gas is converted back to a liquid. The heat produced by this second change of state is given out, usually at the back of the refrigerator, to the room in which the refrigerator stands. The liquid refrigerant then enters a storage vessel before finally passing through an expansion valve to reduce its pressure prior to beginning the cycle again in the evaparator. This cycle is repeated over and over again until the temperature reaches the desired level (about 1–2°C in the food chamber

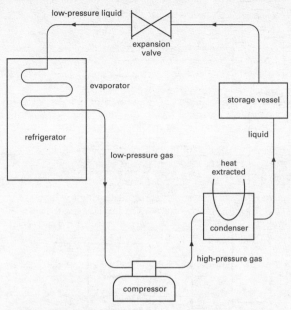

low-pressure liquid

expansion
valve

evaporator

storage vessel

refrigerator

liquid

low-pressure gas

heat
extracted

condenser

high-pressure gas

compressor

Vapour-compression cycle

of a domestic refrigerator and minus 15–18°C in the deep-freeze
compartment). The compressor is then switched off, and on again later,
by a thermostat to maintain a steady temperature. In order to transfer
heat from the cold interior to the warm surroundings without
contravening the second law of thermodynamics energy has to be
supplied to the cycle by the electric current that drives the compressor.

The vapour-absorption cycle

In this cycle there are no moving parts and energy is supplied as heat
either by an electric heater or a gas burner. The refrigerant is usually
ammonia, which is liberated from a water solution and moved through
the evaporator by a stream of hydrogen gas under pressure. Heat is
applied to the generator, raising the ammonia and water vapour to the
separator; the ammonia separates from the water and passes to the
condenser, where it cools and liquefies, giving off its latent heat to the
surroundings. The liquid ammonia is then mixed with hydrogen gas,
which carries it through the evaporator and helps in the process of
evaporation. Subsequently the hydrogen and ammonia vapour enter the
absorber, where the water returned from the separator dissolves the
ammonia before returning to the generator.

Small-scale refrigeration

These two systems account for all domestic and industrial refrigerators
including those in ships, trains, and refrigerated lorries. However,

Vapour-absorption cycle

small-scale refrigeration is sometimes achieved by means of the *Peltier effect at junctions for n-type and p-type semiconductors.

Regnault's method A technique for measuring gas density by evacuating and weighing a glass bulb of known volume, admitting gas at known pressure, and reweighing. The determination must be carried out at constant known temperature and the result corrected to standard temperature and pressure. The method is named after the French chemist Henri Victor Regnault (1810–78).

regression analysis (least-squares method) A method for calculating the best straight line for a linear equation with two variables x and y, for which x can be measured accurately and y has a random error associated with its measurement. The linear equation is written in the form $y = \alpha + \beta x$, where α and β are constants. The error is associated with a mean (see AVERAGE) and *variance. For a given x_i, the true value of y_i is given by $y_i = \alpha + \beta x_i$. The measured value Y_i of y_i is given by $Y_i = y_i + \varepsilon_i$, where ε_i is a random error. The relation between x and y is a straight line together with a random error. If the line $y = a + bx$ is considered then the quantity $d_i = a + bx_i - Y_i$ is the difference between the value of y_i on the line at x_i and Y_i. The method of least squares is used to find a and b, which minimize S, defined by

$$S = \sum_{i=1}^{n} d_i^2.$$

This is done using *partial derivatives. b is given by:

$$(\sum_{i=1}^{n} x_i Y_i - n\bar{x}\,\bar{Y})/(\sum_{i=1}^{n} x_i^2 - n\bar{x}^2),$$

where $\bar{x}$ and $\bar{Y}$ denote the mean values. a can be worked out from b by $a = \bar{Y} - b\bar{x}$. The line $y = a + bx$, with the values of a and b obtained by the method of least squares, is called the **regression line**.

relative aperture *See* APERTURE.

relative atomic mass (atomic weight; r.a.m.) Symbol A_r. The ratio of the average mass per atom of the naturally occurring form of an element to 1/12 of the mass of a carbon–12 atom.

relative density (r.d.) The ratio of the *density of a substance to the density of some reference substance. For liquids or solids it is the ratio of the density (usually at 20°C) to the density of water (at its maximum density). This quantity was formerly called **specific gravity**. Sometimes relative densities of gases are used; for example, relative to dry air, both gases being at s.t.p.

relative humidity *See* HUMIDITY.

relative molecular mass (molecular weight) Symbol M_r. The ratio of the average mass per molecule of the naturally occurring form of an element or compound to 1/12 of the mass of a carbon–12 atom. It is equal to the sum of the relative atomic masses of all the atoms that comprise a molecule.

relative permeability *See* PERMEABILITY.

relative permittivity *See* PERMITTIVITY.

relativistic mass The mass of a moving body as measured by an observer in the same frame of reference as the body. According to the special theory of *relativity the mass m of a body moving at speed v is given by $m = m_0/\sqrt{(1 - v^2/c^2)}$, where m_0 is its *rest mass and c is the speed of light. The relativistic mass therefore only differs significantly from the rest mass if its speed is a substantial fraction of the speed of light. If $v = c/2$, for example, the relativistic mass is 15% greater than the rest mass.

relativistic mechanics An extension of Newtonian mechanics that takes into account the theory of *relativity.

relativistic quantum theory *See* QUANTUM THEORY; QUANTUM FIELD THEORY.

relativistic speed A speed that is sufficiently large to make the mass of a body significantly greater than its *rest mass. It is usually expressed as a proportion of the speed of light. *See* RELATIVITY; RELATIVISTIC MASS.

relativity Two widely accepted theories proposed by Albert *Einstein to account for departures from Newtonian mechanics. The **special theory**, of 1905, refers to non-accelerated frames of reference, while the **general theory**, of 1915, extends to accelerated systems.

The special theory. For Galileo and Newton, all uniformly moving frames of reference (Galilean frames) are equivalent for describing the dynamics of moving bodies. There is no experiment in dynamics that can distinguish between a stationary laboratory and a laboratory that is moving at uniform velocity. Einstein's special theory of relativity takes this notion of equivalent frames one step further: he required all physical phenomena, not only those of dynamics, to be independent of the uniform motion of the laboratory.

When Einstein published his special theory, he realized that it could explain the apparent lack of experimental evidence for an ether, which was supposed to be the medium required for the propagation of electromagnetic waves (*see* MICHELSON–MORLEY EXPERIMENT). Einstein recognized that the existence of an ether would render invalid any equivalent relativity principle for electromagnetic phenomena; i.e. uniform movement of the laboratory through the ether would lead to measurable differences in the propagation of electromagnetic waves *in vacuo*. Since no experimental evidence of the ether was forthcoming, Einstein was encouraged to continue his search for a relativity principle that encompassed all physical phenomena. Since light is a physical phenomenon, its propagation *in vacuo* could not be used to distinguish between uniformly moving frames of reference. Therefore in all such frames the measured speed of light *in vacuo* must be the same.

This conclusion has some important consequences for the nature of space and time. In his popular exposition of 1916, Einstein illustrated these consequences with thought experiments. In one such experiment, he invites the reader to imagine a very long train travelling along an embankment with a constant velocity v in a given direction (see diagram).

Observers on the train use it as a rigid reference body, regarding all events with reference to the train. Einstein posed a simple question: Are two events, which are simultaneous relative to the railway embankment, also simultaneous relative to the observer on the train? For example, two lightning strokes strike the embankment at points *A* and *B* simultaneously with respect to the embankment, so that an observer at *M*

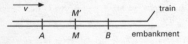

Einstein's thought experiment

(the mid-point of the line AB) will record no time lapse between them. However, the events A and B also correspond to positions A and B on the train. M' is the mid-point of the distance AB on the moving train. When the flashes occur, from the point of view of the embankment, M' coincides with M. However, M moves with speed v towards the right and therefore hastens towards the beam of light coming from B, while moving on ahead of the beam from A. The observer at M' would not agree with the observer at M on the simultaneity of the events A and B because the beam of light from B will be seen to be emitted before the beam of light at A.

At first sight there may seem to be a problem here. If the observer at M' is 'hastening towards the beam of light from B', is this not equivalent to saying that the beam of light is travelling towards M' at a combined speed of v + c, where c is the speed of light *in vacuo*?

The resolution of this problem is the basis of special relativity. According to Einstein, the moving observer at M' must measure the speed of light *in vacuo* to be c, since there can be no experiment that distinguishes the train's moving frame from any other Galilean frame. It is therefore the concept of time measurement that requires revision; that is, the time required for a particular event to occur with respect to the train cannot have the same duration as the same event when judged from the embankment.

This remarkable result also has implications for the measurement of spatial intervals. The measurement of a spatial interval requires the time coincidence of two points along a measuring rod. The relativity of simultaneity means that one cannot contend that an observer who traverses a distance x m per second in the train, traverses the same distance x m also with respect to the embankment in each second. In trying to include the law of propagation of light into a relativity principle, Einstein questioned the way in which measurements of space and time in different Galilean frames are compared. Place and time measurements in two different Galilean frames must be related by a transformation preserving the relativity principle that every ray of light has a velocity of transmission c relative to observers in both frames.

Transformations that preserve the relativity principle are called *Lorentz transformations. The form of these transformations looks complicated at first (see diagram). However, they arise from the simple requirement that there can be no experiment in dynamics or electromagnetism that will distinguish between two different Galilean frames of reference.

These transformations suggest that observers in the two frames will not agree on measurements of length made in the y-direction. Indeed, the duration of intervals of time cannot be agreed upon in the two frames. This is exactly what was suggested in Einstein's thought experiment on the train. Simple manipulations lead to the following formulae, which relate lengths and time intervals in the x', y', z' frame to their equivalent quantities in the x, y, z frame:

$$\Delta l = \Delta l' \, (1 - v^2/c^2)^{\frac{1}{2}}$$

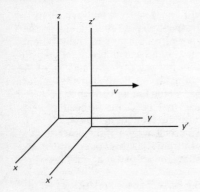

The frame (x', y', z') is moving uniformly with respect to (x, y, z) at a velocity v in the y direction.

Coordinates in the two frames are related as follows by the Lorentz transformations

$$y' = \frac{y - vt}{\sqrt{1 - \frac{v^2}{c^2}}}$$

$$t' = \frac{t - vy/c^2}{\sqrt{1 - \frac{v^2}{c^2}}}$$

$$x' = x$$
$$z' = z$$

Lorentz transformations

and

$$\Delta t = \Delta t' \, (1 - v^2/c^2)^{-\frac{1}{2}},$$

where Δl and Δt are respectively intervals in space and time. Motion therefore leads to a length contraction of the x', y', z' lengths with respect to the x, y, z lengths. Similarly, the equation relating the time intervals in the two frames leads to a time dilation in the x', y', z' system compared to the x, y, z system. From these expressions it is clear that the velocity c plays the part of a limiting velocity, which can neither be reached nor exceeded by any material body.

The velocity c is often said to be the limiting velocity for the transfer of information in the universe. Faster-than-light signals violate *causality when taken to their logical conclusions. The universe, therefore, according to the special theory of relativity, updates itself at the maximum speed of light c; that is, any local changes in the properties within a region of space are not communicated to the rest of the universe instantaneously. Rather, the universe is updated through a wave of reality, which emanates at speed c from the region in which the change took place.

The general theory. In his special theory, Einstein updated the Galilean principle of relativity by including electromagnetic phenomena. Galileo, and later Newton, were well aware that no experiment in the dynamics of moving bodies could distinguish between frames of reference moving relative to each other at constant velocity (Galilean frames). If two Galilean frames move with respect to each other at uniform velocity, no experiment could determine which frame was in absolute motion and which frame was at absolute rest.

This is the basic principle of Einstein's special theory of relativity. However, Einstein was not content with the apparent absolute status conferred to accelerating frames by the behaviour of bodies within them. Einstein sought a general principle of relativity that would require all frames of reference, whatever their relative state of motion, to be equivalent for the formulation of the general laws of nature. In his popular exposition of 1916, Einstein explains this by describing the experiences of an observer within a railway carriage that is decelerating. In his own words:

"At all events it is clear that the Galilean law does not hold with respect to the non-uniformly moving carriage. Because of this, we feel compelled at the present juncture to grant a kind of absolute physical reality to non-uniform motion, in opposition to the general principle of relativity."

Once again it was one of Galileo's observations that provided the starting point for the formulation of Einstein's ideas. Galileo observed that bodies moving under the sole influence of a *gravitational field acquire an acceleration that does not depend upon the material or the physical state of the body. Einstein realized that this property of gravitational fields implied equivalence between gravity and accelerating frames of reference. This equivalence, which became the basis of his general theory of relativity, is well illustrated by one of Einstein's thought experiments. Imagine an elevator, so far removed from stars and other large masses that there is no appreciable gravitational field. An observer inside the elevator is equipped with the appropriate apparatus and uses the elevator as a reference frame. Initially, if the elevator is a Galilean frame, the observer would feel weightless with only inferences from decorations inside the elevator to distinguish between 'up' and 'down'. However, if a rope attached to the top of the elevator were to be pulled with a constant acceleration of 9.81 m s^{-2}, the observer would detect this acceleration as a force reaction on the floor of the elevator. The experiences of the observer in the elevator are equivalent to the experiences of an observer in an elevator in the earth's gravitational field of strength 9.81 N kg^{-1}. Moreover, the force reaction at the feet of the observer in the accelerating frame is due to the observer's inertial *mass (the mass that represents the reluctance of the observer's body to accelerate under the influence of a force).

An observer in the earth-bound elevator would feel the same force reaction at the floor of the elevator, but for this observer the force is due to the influence of the earth's gravitational field on the observer's gravitational mass. Guided by this example, Einstein realized that his extension of the principle of relativity to include accelerations implies the equality of inertial and gravitational mass, which had been established experimentally by Roland Eötvös (1848–1919) in 1888.

These considerations have significant implications for the nature of space and time under the influence of a gravitational field. Another of Einstein's thought experiments illustrates these implications. Imagine a Galilean frame of reference K from which an observer A takes measurements of a non-Galilean frame K', which is a rotating disc inhabited by an observer B. A notes that B is in circular motion and experiences a centrifugal acceleration. This acceleration is produced by a force, which may be interpreted as an effect of B's inertial mass. However, on the basis of the general principle of relativity, B may contend that he is actually at rest but under the influence of a radially directed gravitational field.

A comparison of time-measuring devices placed at the centre and edge of the rotating disc would show a remarkable result. For although the devices would both be at rest with respect to K', the motion of the disc with respect to K would lead to a *time dilation at the edge with respect to measured time at the centre. It follows that the clock at the disc's periphery runs at a permanently slower rate than that at the centre, i.e. as observed from K. The same effect would be noted by an observer who is sitting next to the clock at the centre of the disc. Thus, on the disc, or indeed in any gravitational field, a timing device will run at different rates depending on where it is situated.

The measurement of spatial intervals on the rotating disc will also incur a similar lack of definition. Standard measuring rods placed tangentially around the circumference C of the disc will all be contracted in length due to relativistic length contraction with respect to K. However, measuring rods will not experience shortening in length, as judged from K, if they are applied across a diameter D. Dividing the circumference by the diameter would produce a surprising result from K's point of view. Normally such a quotient would have the value $\pi = 3.14159...$, but in this situation the quotient is larger. Euclidean geometry does not seem to hold in an accelerating frame, or indeed by the principle of relativity, within a gravitational field. Spaces in which the propositions of Euclid are not valid are sometimes called curved spaces. For example, the sum of the internal angles of a triangle drawn on a flat sheet of paper will be 180°; however, a triangle drawn on the curved surface of a sphere will not follow this Euclidean rule.

Einstein fully expected to see this effect in gravitational fields, such was his belief in the general principle of relativity. In fact, it was the effect gravitational fields have on the propagation of light that was heralded as the major verification of his general relativity. Einstein realized that rays of light would be perceived as curving in an accelerating frame. This led him to conclude that, in general, rays of light are propagated curvilinearly in gravitational fields. By means of photographs taken of stars during the solar eclipse of 29 May 1919, the existence of the deflection of starlight around the sun's mass was confirmed.

The mathematics required to describe the curvature of space in the presence of gravitational fields existed before Einstein had need for it, but it was essentially rediscovered by Einstein to solve his general relativistic problems. In general relativity, material bodies follow lines of shortest distance, called **geodesics**. The line formed by stretching an elastic band over a curved surface would be a geodesic on the curved surface. Light follows geodesics called **null-geodesics**. The motions of material bodies are therefore determined by the curvature of the space in the region through which they pass. However, it is the mass of the bodies that causes the curvature of the space in the first place, which demonstrates the elegant self-consistency of Einstein's general theory.

relaxation oscillator *See* OSCILLATOR.

relay An electrical or electronic device in which a variation in the current in one circuit controls the current in a second circuit. These devices are used in an enormous number of different applications in which electrical control is required. The simplest is the electromechanical relay in which the first circuit energizes an electromagnet, which operates a switch in a second circuit. The *thyratron gas-filled relay found many uses in the past but has now been largely replaced by the *thyristor solid-state relay.

reluctance Symbol R. The ratio of the magnetomotive force to the total magnetic flux in a magnetic circuit or component. It is measured in henries.

reluctivity The reciprocal of magnetic *permeability.

rem *See* RADIATION UNITS.

remanence (retentivity) The magnetic flux density remaining in a ferromagnetic substance when the saturating field is reduced to zero. *See* HYSTERESIS.

remote sensing The gathering and recording of information concerning the earth's surface by techniques that do not involve actual contact with the object or area under study. These techniques include photography (e.g. aerial photography), multispectral imagery, infrared imagery, and radar. Remote sensing is generally carried out from aircraft and, increasingly, satellites. The techniques are used, for example, in cartography (map making).

renewable energy sources Sources of energy that do not use up the earth's finite mineral resources. **Nonrenewable energy sources** are *fossil fuels and fission fuels (*see* NUCLEAR FISSION). Various renewable energy sources are being used or investigated. *See* GEOTHERMAL ENERGY; HYDROELECTRIC POWER; NUCLEAR FUSION; SOLAR ENERGY; TIDES; WIND POWER; WAVE POWER.

renormalization A procedure used in relativistic *quantum field theory to deal with the fact that in *perturbation theory calculations give rise to infinities beyond the first term. Renormalization was first used in *quantum electrodynamics, where the infinities were removed by taking the observed mass and charge of the electron as 'renormalized' parameters rather than the 'bare' mass and charge.

Theories for which finite results for all perturbation-theory calculations exist, by taking a finite number of parameters from experiment and using renormalization, are called **renormalizable**. Only certain types of quantum field theories are renormalizable. Theories that need an infinite number of parameters are said to be **nonrenormalizable** and are regarded as incomplete physical theories. The *gauge theories that describe the strong, weak, and electromagnetic interactions are renormalizable. The quantum theory of gravitational interactions is a nonrenormalizable theory, which perhaps indicates that gravity needs to be unified with other fundamental interactions before one can have a consistent quantum theory of gravity.

renormalization group A technique used to understand systems in which many length-scales are involved. Such systems include phase transitions, turbulence, polymers, many-electron systems, and the localization of electrons in disordered systems. The renormalization group has its origin in *quantum field theory, in which it is used to calculate how *coupling constants change with energy. The way in which this change with energy takes place involves the procedure of *renormalization and gives physical insight into this procedure.

reptation A motion describing the dynamics of a polymer in a highly entangled state, such as a network. Regarding the entangled state as a set of chains between crosslinks it is possible to regard the chain as being in a 'tube', with the tube being formed by topological constraints. The chain is longer than the tube so that the 'slack' of the chain moves through the tube, which causes the tube itself to change with time. This motion was called **reptation** (from the Latin *reptare*, to creep) by the French physicist P. G. de Gennes, who postulated it in 1971. Many experiments indicate that reptation dominates the dynamics of polymer chains when they are entangled.

resistance Symbol R. The ratio of the potential difference across an electrical component to the current passing through it. It is thus a measure of the component's opposition to the flow of electric charge. In general, the resistance of a metallic conductor increases with temperature, whereas the resistance of a *semiconductor decreases with temperature.

resistance thermometer (resistance pyrometer) A *thermometer that relies on the increase of electrical resistance of a metal wire with rising temperature, according to the approximate relationship $R = R_0(1 + aT + bT^2)$, where R is the resistance of the wire at temperature T and R_0 is the resistance of the wire at a reference temperature, usually 0°C; a and b are constants characteristic of the metal of the wire. The metal most frequently used is platinum and the platinum resistance coil is usually incorporated into one arm of a *Wheatstone bridge. The effect of the temperature change on the leads carrying current to the platinum coil is compensated by including a pair of dummy leads within the casing carrying the coil. *See also* THERMISTOR.

resistivity Symbol ρ. A measure of a material's ability to oppose the flow of an electric current. The resistivity of a material is given by RA/l, where R is the resistance of a uniform specimen of the material, having a length l and a cross-sectional area A. It is usually given at 0°C or 20°C and is measured in ohm metres. It was formerly known as **specific resistance**.

resistor A component in an electrical or electronic circuit that is present because of its electrical resistance. For electronic purposes many resistors are either wire-wound or consist of carbon particles in a ceramic binder. The ceramic coating carries a number or colour code indicating the value of the resistance. Some resistors can be varied manually by means of a sliding contact; others are markedly dependent on temperature or illumination.

resolution The separation of a vector quantity into two components, which are usually at right angles to each other. Thus, a force F acting on a body in a vertical plane at an angle θ to the horizontal can be resolved into a horizontal component $F\cos\theta$ and a vertical component $F\sin\theta$, both in the same plane as the original force.

resolving power A measure of the ability of an optical instrument to form separable images of close objects or to separate close wavelengths of radiation. The **chromatic resolving power** for any spectroscopic instrument is equal to $\lambda/\delta\lambda$, where $\delta\lambda$ is the difference in wavelength of two equally strong spectral lines that can barely be separated by the instrument and λ is the average wavelength of these two lines. For a telescope forming images of stars the **angular resolving power** is the smallest angular separation of the images; the **linear resolving power** is the linear separation of the images in the focal plane. In a telescope forming images of two stars, as a result of diffraction by the lens aperture each image consists of a bright central blob surrounded by light and dark rings. According to the **Rayleigh criterion** for resolution, the central ring of one image should fall on the first dark ring of the other. The angular resolving power in radians is then $1.22\lambda/d$, where d is the diameter of the objective lens in centimetres and λ is the wavelength of the light (usually taken as 560 nanometres). For microscopes, the resolving power is usually taken as the minimum distance between two points that can be separated. In both cases, the smaller the resolving power, the better the resolution; to avoid this apparent paradox the resolving power is now sometimes taken as the reciprocals of the quantities stated above.

resonance 1. An oscillation of a system at its natural frequency of vibration, as determined by the physical parameters of the system. It has the characteristic that large amplitude vibrations will ultimately result from low-power driving of the system. Resonance can occur in atoms and molecules, mechanical systems, and electrical circuits (*see* RESONANT CIRCUIT; RESONANT CAVITY). **2.** A very short-lived *elementary particle that can be regarded as an excited state of a more stable particle. Resonances decay by the strong interaction (*see* FUNDAMENTAL INTERACTIONS) in 10^{-24} second.

resonant cavity (cavity resonator) A closed space within a conductor in which an electromagnetic field can be made to oscillate at frequencies above those at which a *resonant circuit will operate. The resonant frequency of the oscillation will depend on the dimensions and the shape of the cavity. The device is used to produce microwaves (*see* KLYSTRON; MAGNETRON).

resonant circuit A reactive circuit (*see* REACTANCE) so arranged that it is capable of *resonance. In a **series resonant circuit** a resistor, inductor, and capacitor are arranged in series. Resonance occurs when the *impedance (Z) is a minimum and the current amplitude therefore a maximum. In a **parallel resonant circuit** the inductance and capacitance are in parallel and resonance (with minimal current amplitude) occurs at maximum impedance. The frequency at which resonance occurs is called the **resonant frequency**. In a series resonant circuit

$$Z = R + i[\omega L - 1/\omega C],$$

where $\omega = 2\pi f$ and f is the frequency, R is the resistance, L is the inductance, and C is the capacitance. At resonance, Z is a minimum and $\omega L = 1/\omega C$, i.e. the circuit behaves as if it is purely resistive. In the parallel circuit, resonance occurs when $R^2 + \omega^2 L^2 = L/C$, which in most cases also approximates to $\omega L = 1/\omega C$. Resonant circuits are widely used in *radio to select one signal frequency in preference to others.

rest energy The *rest mass of a body expressed in energy terms according to the relationship $E_0 = m_0 c^2$, where m_0 is the rest mass of the body and c is the speed of light.

restitution coefficient Symbol e. A measure of the elasticity of colliding bodies. For two spheres moving in the same straight line, $e = (v_2 - v_1)/(u_1 - u_2)$, where u_1 and u_2 are the velocities of bodies 1 and 2 before collision ($u_1 > u_2$) and v_1 and v_2 are the velocities of 1 and 2 after impact ($v_2 > v_1$). If the collision is perfectly elastic $e = 1$ and the kinetic energy is conserved; for an inelastic collision $e < 1$.

rest mass The mass of a body at rest when measured by an observer who is at rest in the same frame of reference. *Compare* RELATIVISTIC MASS.

resultant A *vector quantity that has the same effect as two or more other vector quantities of the same kind. *See* PARALLELOGRAM OF VECTORS.

retardation (deceleration) The rate of reduction of speed, velocity, or rate of change.

retardation plate A transparent plate of a birefringent material, such as quartz, cut parallel to the optic axis. Light falling on the plate at 90° to the optic axis is split into an ordinary ray and an extraordinary ray (*see* DOUBLE REFRACTION), which travel through the plate at different speeds. By cutting the plate to different thicknesses a specific phase difference can be introduced between the transmitted rays. In the **half-wave plate** a phase difference of π radians, equivalent to a path difference of half a wavelength, is introduced. In the **quarter-wave plate** the waves are out of step by one quarter of a wavelength.

retina The light-sensitive membrane that lines the interior of the eye. The retina consists of two layers. The inner layer contains nerve cells, blood vessels, and two types of light-sensitive cells (*rods and *cones). The outer layer is pigmented, which prevents the back reflection of light and consequent decrease in visual acuity. Light passing through the lens stimulates individual rods and cones, which generates nerve impulses that are transmitted through the optic nerve to the brain, where the visual image is formed.

retrograde motion 1. The apparent motion of a planet from east to west as seen from the earth against the background of the stars. **2.** The clockwise rotation of a planet, as seen from its north pole. *Compare* DIRECT MOTION.

retrorocket A small rocket motor that produces thrust in the opposite direction to a rocket's main motor or motors in order to decelerate it.

reverberation time The time taken for the energy density of a sound to fall to the threshold of audibility from a value 10^6 times as great; i.e. a fall of 60 decibels. It is an important characteristic of an auditorium. The optimum value is proportional to the linear dimensions of the auditorium.

reverberatory furnace A metallurgical furnace in which the charge to be heated is kept separate from the fuel. It consists of a shallow hearth on which the charge is heated by flames that pass over it and by radiation reflected onto it from a low roof.

reverse osmosis A method of obtaining pure water from water containing a salt, as in desalination. Pure water and the salt water are separated by a semipermeable membrane and the pressure of the salt water is raised above the osmotic pressure, causing water from the brine to pass through the membrane into the pure water. This process requires a pressure of some 25 atmospheres, which makes it difficult to apply on a large scale.

reversible process Any process in which the variables that define the state of the system can be made to change in such a way that they pass through the same values in the reverse order when the process is reversed. It is also a condition of a reversible process that any exchanges of energy, work, or matter with the surroundings should be reversed in direction and order when the process is reversed. Any process that does not comply with these conditions when it is reversed is said to be an **irreversible process**. All natural processes are irreversible, although some processes can be made to approach closely to a reversible process.

Reynolds number Symbol Re. A dimensionless number used in fluid dynamics to determine the type of flow of a fluid through a pipe, to design prototypes from small-scale models, etc. It is the ratio $v\rho l/\eta$, where v is the flow velocity, ρ is the fluid density, l is a characteristic linear dimension, such as the diameter of a pipe, and η is the fluid viscosity. In a smooth straight uniform pipe, laminar flow usually occurs if $Re < 2000$ and turbulent flow is established if $Re > 3000$. It is named after Osborne Reynolds (1842–1912).

rhe A unit of fluidity equal to the reciprocal of the *poise.

rheology The study of the deformation and flow of matter.

rheopexy The process by which certain thixotropic substances set more rapidly when they are stirred, shaken, or tapped. Gypsum in water is such a **rheopectic substance**.

rheostat A variable *resistor, the value of which can be changed without interrupting the current flow. In the common wire-wound rheostat, a sliding contact moves along the length of the coil of wire.

Richardson equation (Richardson–Dushman equation) *See* THERMIONIC EMISSION.

Richter scale A logarithmic scale devised in 1935 by C. F. Richter (1900–85) to compare the magnitude of earthquakes. The scale ranges from 0 to 10 and the Richter scale value is related to the logarithm of the amplitude of the ground motion divided by the period of the dominant wave, subject to certain corrections. On this scale a value of 2 can just be felt as a tremor and damage to buildings occurs for values in excess of 6. The largest shock recorded had a magnitude of 9.5.

Riemannian geometry A type of non-Euclidean geometry proposed by the German mathematician George Friedrich Bernhard Riemann (1826–66) in 1854. It is ideally suited to describing the curvature of space–time in general relativity theory.

rigidity modulus *See* ELASTIC MODULUS.

ring main 1. A domestic wiring system in which all the power outlets have their own individual fuses and in which the outlets are connected in parallel to a ring circuit, which starts and ends at a mains supply point. **2.** An electric supply main that is closed upon itself, thus forming a ring. If a power station supplies the ring at one point only, there are two independent paths for the current to take. Therefore if a fault in the ring occurs the section with the fault can be repaired without disrupting the supply of electricity to all consumers.

ripple 1. A surface wave on a liquid having so short a wavelength that the motion of the wave is dominated by *surface tension, i.e. the wavelength must be less than $2\pi(\gamma/\rho g)^{1/2}$, where γ is the surface tension and ρ the density of the liquid and g is the acceleration of free fall. In the case of water, the wavelength must be less than 1.7 cm. Patterns made in mud or sand by currents in the water that flows over them are called **ripple marks**. **2.** (in electricity) The alternating-current component of a direct-current power supply resulting from sources within the power supply.

RMS value *See* ROOT-MEAN-SQUARE VALUE.

robotics The study of the design, manufacture, and operation of **robots**, i.e. machines capable of being programmed to perform mechanical tasks and to move by automatic control. Robots are used in industry to perform tasks that are either repetitive or in dangerous environments. The concept of programmable robots dates back to the 1920s. As computers develop robots will be used for more and more intricate tasks, e.g. laser cutting.

Rochelle salt Potassium sodium tartrate tetrahydrate, $KNaC_4H_4O_6.4H_2O$. A colourless crystalline salt used for its piezoelectric properties.

Rochon prism An optical device consisting of two quartz prisms; the first, cut parallel to the optic axis, receives the light; the second, with the optic axis at right angles, transmits the ordinary ray without deviation but the extraordinary ray is deflected and can be absorbed by a screen. The device can be used to produce plane-polarized light and it can also be used with ultraviolet radiation.

rock crystal *See* QUARTZ.

rocket A space vehicle or projectile that is forced through space or the atmosphere by *jet propulsion and that carries its own propellants and oxidizers. It is therefore independent of the earth's atmosphere for lift, thrust, or oxygen and is the only known vehicle for travel outside the earth's atmosphere. Rocket motors (or rocket engines) are currently driven by solid or liquid chemical propellants, which burn in an oxidizer carried within the rocket. Typical liquid bipropellant combinations include liquid hydrogen with liquid oxygen for main engines and hydrazine with dinitrogen tetroxide oxidizer for smaller positioning rockets. Experimental rocket motors have also been tested using ionized gases and plasmas to provide thrust (*see also* ION ENGINE). The measure of a rocket motor's performance is its *specific impulse.

rocking-chair cell *See* INTERCALATION CELL.

rod A type of light-sensitive receptor cell present in the retinas of vertebrates. Rods contain the pigment rhodopsin and are essential for vision in dim light. *Compare* CONE.

roentgen The former unit of dose equivalent (*see* RADIATION UNITS). It is named after the discoverer of X-rays, W. K. Roentgen.

Roentgen, William Konrad (1845–1923) German physicist, who made many contributions to physics, the best known being his discovery of X-rays in 1895. For this work he was awarded the first Nobel Prize for physics in 1901.

rolling friction *Friction between a rolling wheel and the plane surface on which it is rotating. As a result of any small distortions of the two surfaces, there is a frictional force with a component, F_1, that opposes the motion. If N is the normal force, $F_r = N\mu_r$, where μ_r is called the **coefficient of rolling friction**.

ROM Read-only memory. A form of computer memory, fabricated from *integrated circuits, whose contents are permanently recorded at the time of manufacture. It is thus used to store data that never require modification. (The contents of **programmable ROM** (or **PROM**) are recorded in a separate process after manufacture.) Like *RAM it consists of an array of 'cells' to which there is direct and extremely rapid access.

root-mean-square value (RMS value) **1.** (in statistics) A typical value of a number (n) of values of a quantity ($x_1, x_2, x_3 \ldots$) equal to the square root of the sum of the squares of the values divided by n, i.e.

RMS value = $[\sqrt{(x_1{}^2 + x_2{}^2 + x_3{}^2 \ldots)/n}]$

2. (in physics) A typical value of a continuously varying quantity, such as an alternating electric current, obtained similarly from many samples taken at regular time intervals during a cycle. Theoretically this can be shown to be the **effective value**, i.e. the value of the equivalent direct current that would produce the same power dissipation in a given resistor. For a sinusoidal current this is equal to $I_m/\sqrt{2}$, where I_m is the maximum value of the current.

Rose's metal An alloy of low melting point (about 100°C) consisting of 50% bismuth, 25–28% lead, and 22–25% tin.

rot *See* CURL.

rotary converter A device for converting direct current to alternating current or one d.c. voltage to another. It consists of an electric motor coupled to a generator.

rotational motion The laws relating to the rotation of a body about an axis are analogous to those describing linear motion. The **angular displacement** (θ) of a body is the angle in radians through which a point or line has been rotated in a specified sense about a specified axis. The **angular velocity** (ω) is the rate of change of angular displacement with time, i.e. $\omega = d\theta/dt$, and the **angular acceleration** (α) is the rate of change of angular velocity, i.e. $\alpha = d\omega/dt = d^2\theta/dt^2$.

The equations of linear motion have analogous rotational equivalents, e.g.:

$\omega_2 = \omega_1 + \alpha t$

$\theta = \omega_1 t + \alpha t^2/2$

$$\omega_2^2 = \omega_1^2 + 2\theta\alpha$$

The counterpart of Newton's second law of motion is $T = I\alpha$, where T is the *torque causing the angular acceleration and I is the *moment of inertia of the rotating body.

rotor The rotating part of an electric motor, electric generator, turbine, etc. *Compare* STATOR.

r-process A *nucleosynthesis process in which a nucleus captures a large number of neutrons in rapid succession. The r-process (rapid process) can only occur when there are many neutrons available, as in a *supernova explosion. It is thought that many elements that are heavier than iron, such as gold, were formed in this way.

rubidium–strontium dating A method of dating geological specimens based on the decay of the radioisotope rubidium–87 into the stable isotope strontium–87. Natural rubidium contains 27.85% of rubidium–87, which has a half-life of 4.7×10^{10} years. The ratio $^{87}Rb/^{87}Sr$ in a specimen gives an estimate of its age (up to several thousand million years).

Rumford, Count (Benjamin Thompson; 1753–1814) American-born British physicist, who acted as a British spy during the American Revolution. As a result he was forced to flee in 1775, first to England and then to Munich. There he observed the boring of cannon barrels, which led him to his best-known proposition, that *friction produces heat. While in Munich he was made a count of the Holy Roman Empire. Returning to England in 1795, he helped to demolish the *caloric theory.

Russell–Saunders coupling (L–S coupling) A type of coupling in systems involving many *fermions. These systems include electrons in atoms and nucleons in nuclei, in which the energies associated with electrostatic repulsion are much greater than the energies associated with *spin–orbit coupling. *Multiplets of many-electron atoms with a low *atomic number are characterized by Russell–Saunders coupling. It is named after the US physicists Henry Norris Russell (1877–1957) and F. A. Saunders, who postulated this type of coupling to explain the spectra of many-electron atoms with low atomic number in 1925. The multiplets of heavy atoms and nuclei are better described by *j-j coupling or **intermediate coupling**, i.e. a coupling in which the energies of electrostatic repulsion and spin–orbit coupling are similar in size.

Rutherford, Ernest, Lord (1871–1937) New Zealand-born British physicist, who worked under Sir J. J. *Thomson at Cambridge University (1895–98). He then took up a professorship at McGill University, Canada, and collaborated with Frederick *Soddy in studying radioactivity. In 1899 he discovered *alpha particles and beta particles, followed by the discovery of *gamma radiation the following year. In 1905, with Soddy, he announced that radioactive *decay involves a series of transformations. He moved to Manchester University in 1907 and there, with Hans *Geiger and E. Marsden, devised the alpha-particle scattering experiment that led in 1911 to the discovery of the atomic nucleus (*see* RUTHERFORD SCATTERING). After moving to Cambridge in 1919 he achieved the artificial splitting of light atoms. In 1908 he was awarded the Nobel Prize for chemistry.

Rutherford and Royd's apparatus An apparatus designed by Ernest Rutherford and his assistant Royd in 1909 to confirm the nature of *alpha particles. The *specific charge of alpha particles was determined soon after their discovery. It was found to be the same as that of a doubly ionized helium atom (helium nucleus). Rutherford and Royd's apparatus (see diagram) was designed to allow alpha particles, which were products of a radon-gas decay, to pass a thin glass partition into a partially evacuated chamber X. Alpha particles trapped in X were then able to acquire electrons either by ionization of some traces of air or by collisions with glass. The apparatus was left for about a week before the gas trapped in X was analysed. Electrical discharge through the gas produced the characteristic spectrum of helium. Alpha particles were therefore confirmed to be helium nuclei.

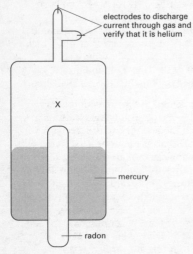

Rutherford and Royd's apparatus

rutherfordium Symbol Rf. A radioactive *transactinide element; a.n. 104. It was first reported in 1964 at Dubna, near Moscow, and in 1969 it was detected by A. Ghiorso and a team at Berkeley, California. It can be made by bombarding californium–249 nuclei with carbon–12 nuclei.

Rutherford scattering The scattering of *alpha particles by thin films of heavy metals, notably gold. The experiments, performed in 1909 by

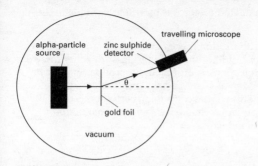

Rutherford scattering

Geiger and Marsden under Ernest Rutherford's direction, provided evidence for the hypothesis that atoms possessed a discrete nucleus. In Rutherford's apparatus, a narrow beam of alpha particles from a radon source (see diagram) fell upon a thin metal foil. A glass screen coated with zinc sulphide (which scintillates on absorbing alpha particles) was placed at the end of a travelling microscope and was used to detect scattered alpha particles. The travelling microscope could be rotated about the metal foil, and by counting the number of scintillations produced in various positions during equal intervals, the angular dependence of the scattering could be determined. Since the range of alpha particles in air is limited, the central chamber of the apparatus was evacuated. The majority of alpha particles suffered only small angles of deflection (θ). However, a very small number, about 1 in 8000, were deviated by more than 90°.

These findings led Rutherford to postulate that alpha particles deflected by angles greater than 90° had encountered a small intense positive charge of high inertia. Rutherford went on to propose in 1911 that an atom has a positively charged core or nucleus, which contains most of the mass of the atom and which is surrounded by orbiting electrons (*see* BOHR THEORY). Since very few alpha particles were scattered through large angles, it follows that the probability of a head-on collision with the nucleus is small. The nucleus therefore occupies a very small proportion of the atom's total volume. The nucleus of an atom is of the order of 10^{-15} m across, whereas the atomic radius is of the order of 10^{-10} m, a factor of 100 000. *See also* NUCLEUS.

Rydberg constant Symbol R. A constant that occurs in the formulae for atomic spectra and is related to the binding energy between an electron and a nucleon. It is connected to other constants by the relationship $R = \mu_0^2 m e^4 c^3 / 8 h^3$, where μ_0 is the magnetic constant (*see* PERMEABILITY), m and e are the mass and charge of an electron, c is the speed of light, and h is the *Planck constant. It has the value 1.097×10^7 m^{-1}. It is named after the Swedish physicist Johannes Robert Rydberg (1854–1919), who developed a formula for the spectrum of atoms, which is particularly simple in the case of hydrogen atoms.

Rydberg spectrum An absorption spectrum of a gas in the ultraviolet region, consisting of a series of lines that become closer together towards shorter wavelengths, merging into a continuous absorption region. The absorption lines correspond to electronic transitions to successively higher energy levels. The onset of the continuum corresponds to photoionization of the atom or molecule, and can thus be used to determine the ionization potential.

r

Sachs–Wolfe effect A phenomenon, predicted in 1967 by R. K. Sachs and A. M. Wolfe, in which density fluctuations associated with quantum-mechanical effects in the early universe caused 'ripples' in the cosmic microwave background radiation. It was first observed by the *COBE satellite and analysed in more detail by *WMAP.

sacrificial protection (cathodic protection) The protection of iron or steel against corrosion (*see* RUSTING) by using a more reactive metal. A common form is galvanizing (*see* GALVANIZED IRON), in which the iron surface is coated with a layer of zinc. Even if the zinc layer is scratched, the iron does not rust because zinc ions are formed in solution in preference to iron ions. Pieces of magnesium alloy are similarly used in protecting pipelines, etc.

sampling The selection of small groups of entities to represent a large number of entities in *statistics. In **random sampling** each individual of a population has an equal chance of being selected as part of the sample. In **stratified random sampling**, the population is divided into strata, each of which is randomly sampled and the samples from the different strata are pooled. In **systematic sampling**, individuals are chosen at fixed intervals; for example, every hundredth article on a production line. In **sampling with replacement**, each individual chosen is replaced before the next selection is made.

satellite 1. (natural satellite) A relatively small natural body that orbits a planet. For example, the earth's only natural satellite is the moon. **2.** (artificial satellite) A man-made spacecraft that orbits the earth, moon, sun, or a planet. Artificial satellites are used for a variety of purposes. **Communication satellites** are used for relaying telephone, radio, and television signals round the curved surface of the earth (*see* SYNCHRONOUS ORBIT). They are of two types: **passive satellites** reflect signals from one point on the earth's surface to another; **active satellites** are able to amplify and retransmit the signals that they pick up. **Astronomical satellites** are equipped to gather and transmit to earth astronomical information from space, including conditions in the earth's atmosphere, which is of great value in weather forecasting.

saturated 1. (of a ferromagnetic material) Unable to be magnetized more strongly as all the domains are orientated in the direction of the field. **2.** (of a solution) Containing the maximum equilibrium amount of solute at a given temperature. In a saturated solution the dissolved substance is in equilibrium with undissolved substance; i.e. the rate at which solute particles leave the solution is exactly balanced by the rate at which they dissolve. A solution containing less than the equilibrium amount is said to be **unsaturated**. One containing more than the equilibrium amount is **supersaturated**. Supersaturated solutions can be made by slowly cooling a saturated solution. Such solutions are metastable; if a small crystal seed is added the excess solute crystallizes out of solution. **3.** (of a vapour) *See* VAPOUR PRESSURE.

saturation 1. *See* COLOUR. **2.** *See* SUPERSATURATION.

sawtooth waveform A waveform in which the variable increases uniformly with time for a fixed period, drops sharply to its initial value, and then repeats the sequence periodically. The illustration shows the ideal waveform and the waveform generated by practical electrical circuits. Sawtooth generators are frequently used to provide a time base for electronic circuits, as in the *cathode-ray oscilloscope.

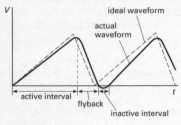

Sawtooth waveform

scalar potential A potential function in which the variable is a *scalar quantity. The potential functions of gravitation described by *Newton's law of gravitation and of *electrostatics are scalar potentials. In the case of a *magnetic field the potential function is a vector quantity and is therefore called a **vector potential**.

scalar product (dot product) The product of two vectors U and V, with components U_1, U_2, U_3 and V_1, V_2, V_3, respectively, given by:

$$U.V = U_1V_1 + U_2V_2 + U_3V_3.$$

It can also be written as $UV\cos\theta$, where U and V are the lengths of U and V, respectively, and θ is the angle between them. *Compare* VECTOR PRODUCT.

scalar quantity A quantity in which direction is either not applicable (as in temperature) or not specified (as in speed). *Compare* VECTOR.

scalar triple product *See* TRIPLE PRODUCT.

scalene Denoting a triangle having three unequal sides.

scaler (scaling circuit) An electronic counting circuit that provides an output when it has been activated by a prescribed number of input pulses.

A **decade scaler** produces an output pulse when it has received ten or a multiple of ten input pulses; a **binary scaler** produces its output after two input pulses.

scaling analysis The analysis of a physical problem in which the terms in the equation describing the problem are expressed in powers of dimensionless units. This enables the orders of magnitude of terms to be estimated, which allows simplifications to be made by dropping very small terms. Scaling analysis is used extensively in theories describing phase transitions, disordered solids, polymers, and turbulence.

scanning The process of repeatedly crossing a surface or volume with a beam, aerial, or moving detector in order to bring about some change to the surface or volume, to measure some activity, or to detect some object. The fluorescent screen of a television picture tube is scanned by an electron beam in order to produce the picture; an area of the sky may be scanned by the movable dish aerial of a radio telescope in order to detect celestial bodies, etc.

scanning electron microscope *See* ELECTRON MICROSCOPE.

scanning tunnelling microscope (STM) A type of microscope in which a fine conducting probe is held close to the surface of a sample. Electrons tunnel between the sample and the probe, producing an electrical signal. The probe is slowly moved across the surface and raised and lowered so as to keep the signal constant. A profile of the surface is produced, and a computer-generated contour map of the surface is produced. The technique is capable of resolving individual atoms, but works better with conducting materials. *See also* ATOMIC FORCE MICROSCOPE.

scattering of electromagnetic radiation The process in which electromagnetic radiation is deflected by particles in the matter through which it passes. In **elastic scattering** the photons of the radiation are reflected; i.e. they bounce off the atoms and molecules without any change of energy. In this type of scattering, known as **Rayleigh scattering** (after Lord *Rayleigh), there is a change of phase but no frequency change. In **inelastic scattering** and **superelastic scattering**, there is interchange of energy between the photons and the particles. Consequently, the scattered photons have a different wavelength as well as a different phase. Examples include the *Raman effect and the *Compton effect. *See also* RUTHERFORD SCATTERING; TYNDALL EFFECT.

schlieren photography A technique that enables density differences in a moving fluid to be photographed. In the turbulent flow of a fluid, for example, short-lived localized differences in density create differences of refractive index, which show up on photographs taken by short flashes of light as streaks (German: *Schliere*). Schlieren photography is used in wind-tunnel studies to show the density gradients created by turbulence and the shock waves around a stationary model.

Schmidt camera *See* TELESCOPE.

Schottky defect *See* DEFECT.

Schottky effect A reduction in the *work function of a substance when an external accelerating electric field is applied to its surface in a vacuum. The field reduces the potential energy of electrons outside the substance, distorting the potential barrier at the surface and causing *field emission. A similar effect occurs when a metal surface is in contact with a *semiconductor rather than a vacuum, when it is known as a **Schottky barrier**. The effect was discovered by the German physicist Walter Schottky (1886–1976).

Schrieffer, John *See* BARDEEN, JOHN.

Schrödinger, Erwin (1887–1961) Austrian physicist, who became professor of physics at Berlin University in 1927. He left for Oxford to escape the Nazis in 1933, returned to Graz in Austria in 1936, and then left again in 1938 for Dublin's Institute of Advanced Studies. He finally returned to Austria in 1956. He is best known for the development of *wave mechanics and the *Schrödinger equation, work that earned him a share of the 1933 Nobel Prize for physics with Paul *Dirac.

Schrödinger equation An equation used in wave mechanics (*see* QUANTUM MECHANICS) for the wave function of a particle. The time-independent Schrödinger equation is:

$$\nabla^2\psi + 8\pi^2 m(E - U)\psi/h^2 = 0$$

where ψ is the wave function, ∇^2 the Laplace operator (*see* LAPLACE EQUATION), h the Planck constant, m the particle's mass, E its total energy, and U its potential energy. It was devised by Erwin Schrödinger, who was mainly responsible for wave mechanics. It can be solved exactly for very simple systems, such as the harmonic oscillator and the hydrogen atom.

Schrödinger's cat A thought experiment introduced by Erwin Schrödinger in 1935 to illustrate the paradox in *quantum mechanics regarding the probability of finding, say, a subatomic particle at a specific point in space. According to Niels *Bohr, the position of such a particle remains indeterminate until it has been observed. Schrödinger postulated a sealed vessel containing a live cat and a device triggered by a quantum event, such as the radioactive decay of a nucleus. If the quantum event occurs, cyanide is released and the cat dies; if the event does not occur the cat lives. Schrödinger argued that Bohr's interpretation of events in quantum mechanics means that the cat could only be said to be alive or dead when the vessel has been opened and the situation inside it had been observed. This paradox has been extensively discussed since its introduction. It is currently thought that the concept of *decoherence might resolve this paradox in a satisfactory way.

Wigner's friend is a variation of the Schrödinger's cat paradox in which a friend of the physicist Eugene Wigner (1902–95) is the first to look inside the vessel. The friend will either find a live or dead cat. However, if Professor Wigner has both the vessel with the cat and the friend in a closed room, the state of mind of the friend (happy if there is a live cat but sad if there is a dead cat) cannot be determined in Bohr's interpretation of quantum mechanics until the professor has looked into the room although the friend has already looked at the cat. These paradoxes indicate the absurdity of the overstated roles of measurement and observation in Bohr's interpretation of quantum mechanics.

Schwarzschild radius A critical radius of a body of given mass that must be exceeded if light is to escape from that body. It is equal to $2GM/c^2$, where G is the gravitational constant, c is the speed of light, and M is the mass of the body. If the body collapses to such an extent that its radius is less than the Schwarzschild radius the escape velocity becomes equal to the speed of light and the object becomes a *black hole. The Schwarzschild radius is then the radius of the hole's event horizon. The concept was devised by the German astronomer Karl Schwarzschild (1873–1916). *See also* DEATH OF A STAR.

Schwinger effect An effect predicted to occur in *quantum electrodynamics in which electron–positron pairs are 'sucked out' of a vacuum by an *electric field. This effect, predicted in 1951 by the US physicist Julian Schwinger (1918–94), has never been observed since it is extremely small but it is thought that it could be observed with a sufficiently large electric field. A possibility is that it could occur in heavy ion reactions. The Schwinger effect can be thought of as analogous to *field emission in metals.

scintillation camera *See* GAMMA CAMERA.

scintillation counter A type of particle or radiation counter that makes use of the flash of light (scintillation) emitted by an excited atom falling back to its ground state after having been excited by a passing photon or particle. The scintillating medium is usually either solid or liquid and is used in connection with a *photomultiplier, which produces a pulse of current for each scintillation. The pulses are counted with a *scaler. In certain cases, a pulse-height analyser can be used to give an energy spectrum of the incident radiation.

sclerometer A device for measuring the hardness of a material by determining the pressure on a standard point that is required to scratch it or by determining the height to which a standard ball will rebound from it when dropped from a fixed height. The rebound type is sometimes called a **scleroscope**.

scotopic vision The type of vision that occurs when the *rods in the eye are the principal receptors, i.e. when the level of illumination is low. With scotopic vision colours cannot be identified. *Compare* PHOTOPIC VISION.

screen grid A wire grid in a tetrode or pentode *thermionic valve, placed between the anode and the *control grid to reduce the grid–anode capacitance. *See also* SUPPRESSOR GRID.

screw 1. A simple *machine effectively consisting of an inclined plane wrapped around a cylinder. The mechanical advantage of a screw is $2\pi r/p$, where r is the radius of the thread and p is the **pitch**, i.e. the distance between adjacent threads of the screw measured parallel to its axis. **2.** A symmetry element in a crystal lattice that consists of a combination of a rotation and a translation. *See also* GLIDE.

S-drop *See* STRANGE MATTER.

search coil A small coil in which a current can be induced to detect and measure a magnetic field. It is used in conjunction with a *fluxmeter.

Searle's bar An apparatus for determining the thermal conductivity of a bar of material. The bar is lagged and one end is heated while the other end is cooled, by steam and cold water respectively. At two points d apart along the length of the bar the temperature is measured using a thermometer or thermocouple. The conductivity can then be calculated from the measured temperature gradient.

second 1. Symbol s. The SI unit of time equal to the duration of 9 192 631 770 periods of the radiation corresponding to the transition between two hyperfine levels of the ground state of the caesium–133 atom. **2.** Symbol ″. A unit of angle equal to 1/3600 of a degree or 1/60 of a minute.

secondary cell A *voltaic cell in which the chemical reaction producing the e.m.f. is reversible and the cell can therefore be charged by passing a current through it. *See* ACCUMULATOR; INTERCALATION CELL. *Compare* PRIMARY CELL.

secondary colour Any colour that can be obtained by mixing two *primary colours. For example, if beams of red light and green light are made to overlap, the secondary colour, yellow, will be formed. Secondary colours of light are sometimes referred to as the pigmentary primary colours. For example, if transparent yellow and magenta pigments are overlapped in white light, red will be observed. In this case the red is a pigmentary secondary although it is a primary colour of light.

secondary emission The emission of electrons from a surface as a result of the impact of other charged particles, especially as a result of bombardment with (primary) electrons. As the number of secondary electrons can exceed the number of primary electrons, the process is important in *photomultipliers. *See also* AUGER EFFECT.

secondary winding The winding on the output side of a *transformer or *induction coil. *Compare* PRIMARY WINDING.

secular magnetic variation *See* GEOMAGNETISM.

sedimentation The settling of the solid particles through a liquid either to produce a concentrated slurry from a dilute suspension or to clarify a liquid containing solid particles. Usually this relies on the force of gravity, but if the particles are too small or the difference in density between the solid and liquid phases is too small, a *centrifuge may be used. In the simplest case the rate of sedimentation is determined by *Stokes' law, but in practice the predicted rate is rarely reached. Measurement of the rate of sedimentation in an *ultracentrifuge can be used to estimate the size of macromolecules.

Seebeck effect (thermoelectric effect) The generation of an e.m.f. in a circuit containing two different metals or semiconductors, when the junctions between the two are maintained at different temperatures. The magnitude of the e.m.f. depends on the nature of the metals and the difference in temperature. The Seebeck effect is the basis of the *thermocouple. It was named after T. J. Seebeck (1770–1831), who actually found that a magnetic field surrounded a circuit consisting of two metal conductors only if the junctions between the metals were maintained at different temperatures. He wrongly assumed that the conductors were magnetized directly by the temperature difference. *Compare* PELTIER EFFECT.

Seger cones (pyrometric cones) A series of cones used to indicate the temperature inside a furnace or kiln. The cones are made from different mixtures of clay, limestone, feldspars, etc., and each one softens at a different temperature. The drooping of the vertex is an indication that the known softening temperature has been reached and thus the furnace temperature can be estimated.

Segrè plot A plot of *neutron number (N) against *atomic number (Z) for all stable nuclides. The stability of nuclei can be understood qualitatively on the basis of the nature of the strong interaction (*see* FUNDAMENTAL INTERACTIONS) and the competition between this attractive force and the repulsive electrical force. The strong interaction is independent of electric charge, i.e. at any given separation the strong force between two neutrons is the same as that between two protons or between a proton or a neutron. Therefore, in the absence of the electrical repulsion between protons, the most stable nuclei would be those having equal numbers of neutrons and protons. The electrical repulsion shifts the balance to favour a greater number of neutrons, but a nucleus with too

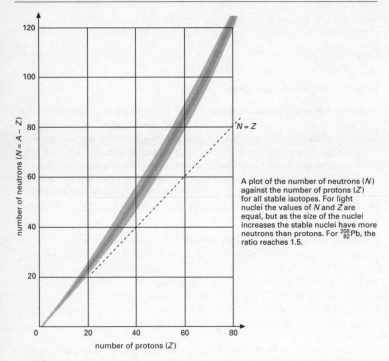

Segrè plot

A plot of the number of neutrons (N) against the number of protons (Z) for all stable isotopes. For light nuclei the values of N and Z are equal, but as the size of the nuclei increases the stable nuclei have more neutrons than protons. For $^{208}_{82}$Pb, the ratio reaches 1.5.

many neutrons is unstable, because not enough of them are paired with protons.

As the number of nucleons increases, the total energy of the electrical interaction increases faster than that of the nuclear interaction. The (positive, repulsive) electric potential energy of the nucleus increases approximately as Z^2, while the (negative, attractive) nuclear potential energy increases approximately as $N + Z$ with corrections for pairing effects. For large $N + Z$ values, the electrical potential energy per nucleon grows much faster than the nuclear potential energy per nucleon, until a point is reached where the formation of a stable nucleus is impossible. The competition between the electric and nuclear forces therefore accounts for the increase with Z of the neutron–proton ratio in stable nuclei as well as the existence of maximum values for $N + Z$ for stability. The plot is named after Emilio Segrè (1905–89). *See* BINDING ENERGY; DECAY; DRIPLINE; LIQUID-DROP MODEL.

seismic waves Vibrations propagated within the earth or along its surface as a result of an *earthquake or explosion. Earthquakes generate two types of body waves that travel within the earth and two types of surface wave. The body waves consist of primary (or longitudinal) waves that impart a back-and-forth motion to rock particles along their path. They travel at speeds between 6 km per second in surface rock and 10.4 km per second near the earth's core. Secondary (or transverse or shear) waves cause rock particles to move back and forth perpendicularly to their direction of propagation. They travel at between 3.4 km per second in surface rock and 7.2 km per second near the core.

The surface waves consist of Rayleigh waves (after Lord *Rayleigh, who predicted them) and Love waves (after A. E. Love). The Love waves displace particles perpendicularly to the direction of propagation and have no longitudinal or vertical components. They travel in the surface layer above a solid layer of rock with different elastic characteristics. Rayleigh waves travel over the surface of an elastic solid giving an elliptical motion to rock particles. It is these Rayleigh waves that have the strongest effect on distant seismographs.

seismograph An instrument that records ground oscillations, e.g. those caused by earthquakes, volcanic activity, and explosions. Most modern seismographs are based on the inertia of a delicately suspended mass and depend on the measurement of the displacement between the mass and a point fixed to the earth. Others measure the relative displacement between two points on earth. The record made by a seismograph is known as a **seismogram**.

seismology The branch of geology concerned with the study of earthquakes.

selection rules Rules that determine which transitions between different energy levels are possible in a system, such as an elementary particle, nucleus, atom, molecule, or crystal, described by quantum mechanics. Transitions cannot take place between any two energy levels. *Group theory, associated with the *symmetry of the system, determines which transitions, called **allowed transitions**, can take place and which transitions, called *forbidden transitions, cannot take place. Selection rules determined in this way are very useful in analysing the *spectra of quantum-mechanical systems.

selectron *See* SUPERSYMMETRY.

selenium cell Either of two types of *photoelectric cell; one type relies on the photoconductive effect, the other on the photovoltaic effect (*see* PHOTOELECTRIC EFFECT). In the photoconductive selenium cell an external e.m.f. must be applied; as the selenium changes its resistance on exposure to light, the current produced is a measure of the light energy falling on the selenium. In the photovoltaic selenium cell, the e.m.f. is generated within the cell. In this type of cell, a thin film of vitreous or metallic selenium is applied to a metal surface, a transparent film of another metal, usually gold or platinum, being placed over the selenium. Both types of cell are used as light meters in photography.

selenology The branch of astronomy concerned with the scientific study of the *moon.

self-energy In *classical physics the self-energy of a system is the contribution to the energy of the system resulting from the interaction between different parts of the system. In *quantum field theory the self-energy of a particle is the contribution to the energy of the particle due to virtual emission and absorption of particles. In the *many-body problem in *quantum mechanics the self-energy E_s of a particle is the difference $E_Q - E_B$, where E_Q is the energy of the *quasiparticle associated with the particle and E_B is the energy of the 'bare' particle. The particle interacts with its surrounding medium, which in turn acts back on the original particle.

self-exciting generator A type of electrical generator in which the field electromagnets are excited by current from the generator output.

self inductance *See* INDUCTANCE.

self-organization The spontaneous order arising in a system when certain parameters of the system reach critical values. Self-organization occurs in many systems in physics, chemistry, and biology; an example is the *Bénard cell. Self-organization can occur when a system is driven far from thermal *equilibrium. Since a self-organizing system is open to its environment, the second law of *thermodynamics is not violated by the formation of an ordered phase, as entropy can be transferred to the environment. Self-organization is related to the concepts of *broken symmetry, *complexity, nonlinearity, and nonequilibrium *statistical mechanics. Many systems that undergo transitions to self-organization can also undergo transitions to *chaos.

self-similarity *See* FRACTAL.

semiclassical approximation An approximation technique used to calculate quantities in quantum mechanics. This technique is called the semiclassical approximation because the *wave function is written as an asymptotic series with ascending powers of the Planck constant h, with the first term being purely classical. The semiclassical approximation is also known as the **Wentzel–Kramers–Brillouin (WKB) approximation**, named after its inventors Gregor Wentzel (1898–1978), Hendrik Anton Kramers (1894–1952), and Léon Brillouin (1889–1969), who invented it independently in 1926. The semiclassical approximation is particularly successful for calculations involving the *tunnel effect, such as *field emission and radioactive decay producing *alpha particles.

semiconductor A crystalline solid, such as silicon or germanium, with an electrical conductivity (typically 10^5–10^{-7} siemens per metre) intermediate between that of a conductor (up to 10^9 S m^{-1}) and an insulator (as low as 10^{-15} S m^{-1}). As the atoms in a crystalline solid are close together, the orbitals of their electrons overlap and their individual *energy levels are spread out into *energy bands. Conduction occurs in semiconductors as the result of a net movement, under the influence of an electric field, of electrons in the conduction band and empty states, called **holes**, in the valence band. A hole behaves as if it was an electron with a positive charge. Electrons and holes are known as the **charge carriers** in a semiconductor. The type of charge carrier that predominates in a particular region or material is called the **majority carrier** and that with the lower concentration is the **minority carrier**. An **intrinsic semiconductor** is one in which the concentration of charge carriers is a characteristic of the material itself; electrons jump to the conduction band from the valence band as a result of thermal excitation, each electron that makes the jump leaving behind a hole in the valence band. Therefore, in an intrinsic semiconductor the charge carriers are equally divided between electrons and holes. In **extrinsic semiconductors** the type of conduction that predominates depends on the number and valence of the impurity atoms present. Germanium and silicon atoms have a valence of four. If impurity atoms with a valence of five, such as arsenic, antimony, or phosphorus, are added to the lattice, there will be an extra electron per atom available for conduction, i.e. one that is not required to pair with the four valence electrons of the germanium or silicon. Thus extrinsic semiconductors doped with atoms of valence five give rise to crystals with electrons as majority carriers, the so-called **n-type conductors**. Similarly, if the impurity atoms have a valence of three, such as boron, aluminium, indium, or gallium, one hole per atom is created by an unsatisfied bond. The majority carriers are therefore holes, i.e. **p-type conductors**.

Semiconductor devices have virtually replaced thermionic devices, because they are several orders of magnitude smaller, cheaper in energy consumption, and more reliable. The basic structure for electronic semiconductor devices is the **semiconductor diode** (*see also* TRANSISTOR). This consists of a silicon crystal doped in such a way that half is *p*-type and half is *n*-type. At the junction between the two halves there is a depletion layer in which electrons from the *n*-type have filled holes from the *p*-type. This sets up a potential barrier, which tends to keep the remaining electrons in the *n*-region and the remaining holes in the *p*-region. However, if the *p*-region is biased with a positive potential, the height of the barrier is reduced; the diode is said to be forward biased, because the majority holes in the *p*-region can then flow to the *n*-region and majority electrons in the *n*-region flow to the *p*-region. When forward biased there is a good current flow across the barrier. On the other hand if the *p*-region is negatively biased, the height of the potential barrier is increased and there is only a small leakage current of minority electrons from the *p*-region able to flow to the *n*-region. Thus the *p*–*n* junction acts as an efficient rectifier, for which purpose it is widely used.

semiconductor diode *See* DIODE; SEMICONDUCTOR.

semiconductor laser A type of *laser in which semiconductors provide the excitation. The laser action results from electrons in the conduction band (*see* ENERGY BANDS) being stimulated to recombine with holes in the valence band. When this occurs the electrons give up the energy corresponding to the band gap. Materials, such as gallium arsenide, are suitable for this purpose. A junction between p-type and n-type semiconductors can be used with the light passing along the plane of the junction. Mirrors for the laser action are provided by the ends of the crystals. Semiconductor lasers can be as small as 1 mm in length.

semipermeable membrane A membrane that is permeable to molecules of the solvent but not the solute in *osmosis. Semipermeable membranes can be made by supporting a film of material (e.g. cellulose) on a wire gauze or porous pot.

series A sequence of terms each of which can be written in a form that is an algebraic function of its position in the series. For example, the *exponential series $1 + x + x^2/2! + x^3/3!$ has an nth term $x^n/n!$. The sum of all the terms from $n = 0$ to $n = \infty$ is written:

$$\sum_{n=0}^{\infty} = x^n/n!$$

This series has an infinite number of terms and is therefore called an **infinite series**. A **finite series** has a fixed number of terms. *See also* ASYMPTOTIC SERIES; CONVERGENT SERIES.

series circuits Circuits in which the circuit elements are arranged in sequence so that the same current flows through each of them in turn. For resistances in series, the total resistance is the sum of the individual resistances. For capacitors in series, the total capacitance, C, is given by $1/C = 1/C_1 + 1/C_2 + 1/C_3 \ldots$

series-wound machine *See* SHUNT.

sets Collections of objects or elements that have at least one characteristic in common. For example, the set X may consist of all the

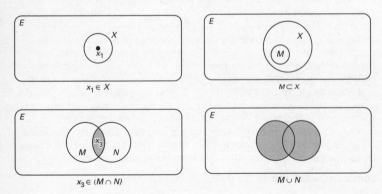

Sets

elements x_1, x_2, x_3, etc. This is written $\{x_1, x_2, x_3, \ldots\} = X$. A specific element in a set is characterized by $x_1 \in X$, meaning x_1 is a member of set X. A **subset** of set X, say M, would be written $M \subset X$, i.e. M is contained in X. If x_3 is a member of both subsets M and N, then $x_3 \in (M \cap N)$, i.e. x_3 belongs to the **intersection** of M and N. $M \cup N$ means the **union** of M and N. For example, if M consists of $\{1, 4, 5, 8\}$ and N consists of $\{2, 3, 4, 5\}$ then $M \cap N = \{4, 5\}$ and $M \cup N = \{1, 2, 3, 4, 5, 8\}$. In the diagram, the rectangle represents the universal set E, circles represent sets or subsets. These diagrams are called **Venn diagrams**, after John Venn (1834–1923), who invented them.

sextant An instrument used in navigation to measure the altitude of a celestial body. Originally it had an arc of 60° (one sixth of a circle, hence its name) but modern instruments have various angles. The sextant uses two mirrors: the horizon glass, in which only the lower half is silvered, and the index mirror, which can be rotated about an axis perpendicular to the plane of the instrument (see illustration). An arm attached to the index glass sweeps round the calibrated arc, from which angles are read. The instrument is aimed at the horizon and the index mirror rotated until the celestial object can also be seen through the telescope. After careful adjustment to make the image of the celestial body just touch the horizon, the angle is read off the graduated scale.

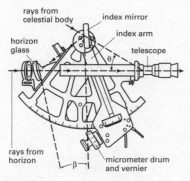

A sextant

shadow An area of darkness formed on a surface when an object intercepts the light falling on the surface from a source. In the case of a point source the shadow has a sharply defined outline. If the source has an appreciable size the shadow has two distinct regions; one of full-shadow, called the **umbra**, the other of half-shadow, called the **penumbra** (see illustration).

shearing force A force that acts parallel to a plane rather than perpendicularly, as with a tensile or compressive force. A **shear stress** requires a combination of four forces acting over (most simply) four sides of a plane and produces two equal and opposite couples. It is measured as

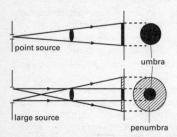

Shadows

the ratio of one shearing force to the area over which it acts, $F/(ab)$ in the diagram. The shear strain is the angular deformation, θ, in circular measure. The **shear modulus** is the ratio of the shear stress to the shear strain (*see also* ELASTIC MODULUS).

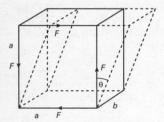

Shearing force

shear modulus *See* ELASTIC MODULUS; SHEARING FORCE.

shell model 1. *See* ATOM. **2.** A model of the atomic nucleus in which nucleons are assumed to move under the influence of a central field in shells that are analogous to atomic electron shells. The model provides a good explanation of the stability of nuclei that have *magic numbers and is successful in predicting other properties of many nuclei.

sherardizing The process of coating iron or steel with a zinc corrosion-resistant layer by heating the iron or steel in contact with zinc dust to a temperature slightly below the melting point of zinc. At a temperature of about 371°C the two metals amalgamate to form internal layers of zinc–iron alloys and an external layer of pure zinc. The process was invented by Sherard Cowper-Coles (d. 1935).

shielding 1. A barrier surrounding a region to exclude it from the influence of an energy field. For example, to protect a region from an electric field an earthed barrier is required; to protect it from a magnetic field a shield of high magnetic permeability is needed. **2.** A barrier used to surround a source of harmful or unwanted radiations. For example, the core of a *nuclear reactor is surrounded by a cement or lead shield to absorb neutrons and other dangerous radiation. **3.** (in atoms) The barrier provided by inner electron shells to the influence of nuclear charge on outer electrons. Shielding has an effect on ionic radius, as in the lanthanoids.

SHM *See* SIMPLE HARMONIC MOTION.

Shockley, William *See* BARDEEN, JOHN.

shock wave A very narrow region of high pressure and temperature formed in a fluid when the fluid flows supersonically over a stationary object or a projectile flying supersonically passes through a stationary fluid. A shock wave may also be generated by violent disturbances in a fluid, such as a lightning stroke or a bomb blast.

short-sightedness *See* MYOPIA.

shower *See* COSMIC RADIATION.

shunt An electrical resistor or other element connected in parallel with some other circuit or device, to take part of the current passing through it. For example, a shunt is used across the terminals of a galvanometer to increase the current that can pass through the system. A **shunt-wound** electric generator or motor is one in which the field winding is in parallel with the armature circuit. In a **series-wound** electrical machine the field coils and the armature circuit are in series.

sideband The band of frequencies above or below the frequency of the carrier wave in a telecommunications system within which the frequency components of the wave produced by *modulation fall. For example, if a carrier wave of frequency f is modulated by a signal of frequency x, the **upper sideband** will have a frequency $f + x$ and the **lower sideband** a frequency $f - x$.

sidereal day *See* DAY.

sidereal period The time taken for a planet or satellite to complete one revolution of its orbit measured with reference to the background of the stars. *See also* DAY; SYNODIC PERIOD; YEAR.

siemens Symbol S. The SI unit of electrical conductance equal to the conductance of a circuit or element that has a resistance of 1 ohm. 1 S = 10^{-1} Ω. The unit was formerly called the mho or reciprocal ohm. It is named after the German-born British engineer Sir William Siemens (1823–83).

sievert The SI unit of dose equivalent (*see* RADIATION UNITS). It is named after the Swedish physicist Rolf Sievert (1896–1966).

sigma particle A type of spin ½ *baryon. There are three types of sigma particles, denoted Σ^-, Σ^0, Σ^+, for the negatively charged, electrically neutral, and positively charged forms, respectively. The quark content of the sigma particles are Σ^- (dds), Σ^0 (dus), Σ^+ (uus), where d, u, and s denote down, up, and strange, respectively. The masses of the sigma particles are 1189.36 MeV (Σ^+), 1192.46 MeV (Σ^0), 1197.34 MeV (Σ^-); their average lifetimes are 0.8×10^{-10} s (Σ^+), 5.8×10^{-20} s (Σ^0), and 1.5×10^{-10} s (Σ^-).

signal The variable parameter that contains information and by which information is transmitted in an electronic system or circuit. The signal is often a voltage source in which the amplitude, frequency, and waveform can be varied.

signal-to-noise ratio A ratio of importance in electrical communication, especially in *radio and *television. Noise in this context refers to electrical disturbances that cause unwanted interference (crackling, hum, etc.) in radio and spoil the picture in television with 'snow-storms'. Some electrical noise is caused by external electrical equipment, such as electric motors. In addition, fluctuations in electric current occur because the current consists of electrons, which have random movements. The signal-to-noise ratio of a signal is defined as the ratio of the amplitude of the signal to the mean amplitude of the noise.

sign convention A set of rules determined by convention for giving plus or minus signs to distances in the formulae involving lenses and mirrors. The real-is-positive convention is now usually adopted. The **New Cartesian convention** is now not widely used. In this convention distances to the right of the pole are treated as positive, those to the left as negative. This system has the advantage of conforming to the sign convention used with Cartesian coordinates in mathematics and is therefore preferred by some for the more complicated calculations.

significant figures The number of digits used in a number to specify its accuracy. The number 6.532 is a value taken to be accurate to four significant figures. The number 7.3×10^3 is accurate only to three significant figures. Similarly 0.0732 is also only accurate to three significant figures. In these cases the zeros only indicate the order of magnitude of the number, whereas 7.065 is accurate to four significant figures as the zero in this case is significant in expressing the value of the number.

silicon chip A single crystal of a semiconducting silicon material, typically having millimetre dimensions, fabricated in such a way that it can perform a large number of independent electronic functions (*see* INTEGRATED CIRCUIT).

simple harmonic motion (SHM) A form of periodic motion in which a point or body oscillates along a line about a central point in such a way that it ranges an equal distance on either side of the central point and that its acceleration towards the central point is always proportional to its distance from it. One way of visualizing SHM is to imagine a point

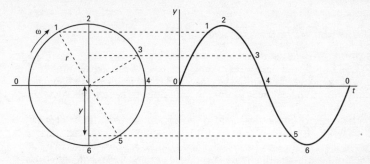

Simple harmonic motion

rotating around a circle of radius r at a constant angular velocity ω. If the distance from the centre of the circle to the projection of this point on a vertical diameter is y at time t, this projection of the point will move about the centre of the circle with simple harmonic motion. A graph of y against t will be a sine wave, whose equation is $y = r\sin\omega t$ (see diagram). *See also* PENDULUM.

simulation *See* MONTE CARLO SIMULATION.

simultaneity The condition in which two or more events occur at the same instant. In Newtonian physics any two events found to be simultaneous by one observer will also appear to be simultaneous to any other observer in uniform relative motion. In the theory of *relativity, however, this is not true. *See* TIME.

sine wave (sinusoidal wave) Any waveform that has an equation in which one variable is proportional to the sine of the other. Such a waveform can be generated by an oscillator that executes *simple harmonic motion.

single-sideband modulation *See* AMPLITUDE MODULATION.

singularity Somewhere in space–time where general relativity theory predicts that certain quantities, such as the curvature of space–time, become infinite. For example, singularities are predicted to occur at the big bang and at the centre of a black hole. The existence of singularities is usually regarded as a limitation of general relativity theory. It has been suggested that singularities will not occur in a full theory of *quantum gravity due to space–time being an emergent quantity in such a theory.

sintering The process of heating and compacting a powdered material at a temperature below its melting point in order to weld the particles together into a single rigid shape. Materials commonly sintered include metals and alloys, glass, and ceramic oxides. Sintered magnetic materials, cooled in a magnetic field, make especially retentive permanent magnets.

sinusoidal oscillator *See* OSCILLATOR.

sinusoidal wave *See* SINE WAVE.

siphon An inverted U-tube with one limb longer than the other. Liquid
will be transferred from a reservoir at the base of the shorter limb to the
end of the longer limb, provided that the U-tube is filled with liquid (see
illustration). The device is useful for emptying an inaccessible container,
such as a car's petrol tank.

The pressure (p_1) on the liquid at the base of the short limb (length h_1) is
$p - h_1k$, where p is the atmospheric pressure and k is a constant equal to
the product of the density of the liquid and the acceleration of free fall.
The pressure (p_2) on the liquid at the base of the long limb (length h_2) is
$p - h_2k$. Thus for fluid flow to occur through the tube, from short limb to
long limb, $p_1 > p_2$, and for this to occur $h_2 > h_1$. Thus if the limbs are of
equal length no flow will occur; it will only occur if the limb dipping into
the reservoir is shorter than the delivering limb.

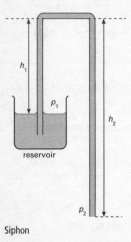

Siphon

Sisyphus effect *See* LASER COOLING.

SI units Système International d'Unités: the international system of
units now recommended for most scientific purposes. A coherent and
rationalized system of units derived from the *m.k.s. units, SI units have
now replaced *c.g.s. units and *Imperial units for many purposes. The
system has seven **base units** and two **dimensionless units** (formerly called
supplementary units) from which all other units are derived. There are 18
derived units with special names. Each unit has an agreed symbol (a
capital letter or an initial capital letter if it is named after a scientist,
otherwise the symbol consists of one or two lower-case letters). Decimal
multiples of the units are indicated by a set of prefixes; whenever possible

a prefix representing 10 raised to a power that is a multiple of three should be used. See Appendix.

skin depth Symbol δ. The depth of the surface region of a metal that an alternating current flows in. The variation of current flow, with the flow rapidly decreasing from the surface of the conductor to its centre, is called the **skin effect**. At room temperatures the skin depth is considerably greater than the mean free path of the electrons at very low temperatures. This makes the phenomenon, known as the **anomalous skin effect** in this situation, more complicated.

skip distance The minimum distance from the transmitter of a radio wave at which reception is possible by means of a sky wave (*see* RADIO TRANSMISSION). If a radio wave strikes the ionosphere at a small angle of incidence the wave passes through it and is not reflected. There is therefore a minimum angle of incidence at which reflection occurs for a given frequency. This leads to a region around a transmitter in which sky waves cannot be received. As the frequency of the transmission increases the minimum angle of incidence at which ionospheric reflection occurs becomes greater. Above about 4 megahertz there may be a region of several hundred kilometres around a transmitter, which is within the skip distance and in which ground waves are too attenuated to be effectively received. In this region no reception is possible.

sky wave *See* RADIO TRANSMISSION.

slepton *See* SUPERSYMMETRY.

slow neutron A neutron with a kinetic energy of less than 10^2 eV (10^{-17} joule). *See also* FAST NEUTRON; THERMALIZATION.

slug An f.p.s. unit of mass equal to the mass that will acquire an acceleration of 1 ft $\sec^{-2}$ when acted on by a force of one pound-force.

S-matrix theory A theory initiated by the German physicist Werner *Heisenberg in 1943 and developed extensively in the 1950s and 1960s to describe *strong interactions in terms of their scattering properties. S-matrix theory uses general properties, such as *causality in *quantum mechanics and the special theory of *relativity. The discovery of *quantum chromodynamics as the fundamental theory of strong interactions limited the use of S-matrix theory to a convenient way of deriving general results for scattering in *quantum field theories. *String theory, as a theory for *hadrons, originated in attempts to provide a more fundamental basis for S-matrix theory.

smectic *See* LIQUID CRYSTAL.

smelting The process of separating a metal from its ore by heating the ore to a high temperature in a suitable furnace in the presence of a reducing agent, such as carbon, and a fluxing agent, such as limestone. Iron ore is smelted in this way so that the metal melts and, being denser than the molten slag, sinks below the slag, enabling it to be removed from the furnace separately.

smoke A fine suspension of solid particles in a gas.

Snell's law *See* REFRACTION.

snowflake curve *See* FRACTAL.

Soddy, Frederick (1877–1956) British chemist, who worked with Ernest *Rutherford in Canada and William *Ramsay in London before finally settling in Oxford in 1919. His announcement in 1913 of the existence of *isotopes won him the 1921 Nobel Prize for physics.

sodium–sulphur cell A type of *secondary cell that has molten electrodes of sodium and sulphur separated by a solid electrolyte consisting of beta alumina (a crystalline form of aluminium oxide). When the cell is producing current, sodium ions flow through the alumina to the sulphur, where they form sodium polysulphide. Electrons from the sodium flow in the external circuit. The opposite process takes place during charging of the cell. Sodium–sulphur batteries have been considered for use in electric vehicles because of their high peak power levels and relatively low weight. However, some of the output has to be used to maintain the operating temperature (about 370°C) and the cost of sodium is high.

sodium-vapour lamp A form of *electric lighting that gives a yellow light as a result of the luminous discharge obtained by the passage of a stream of electrons between tungsten electrodes in a tube containing sodium vapour. To facilitate starting, the tube also contains some neon; for this reason, until the lamp is warm the neon emits a characteristic pink glow. As the sodium vaporizes, the yellow light predominates. Sodium-vapour lamps are widely used as street lights because of their high luminous efficiency and because the yellow light is less absorbed than white light by fog and mist. Low-pressure sodium lamps emit a characteristic yellow light; in high-pressure lamps the atoms are sufficiently close to each other to interact and broaden the spectral lines into the orange and green regions.

soft iron A form of iron that contains little carbon, has high relative permeability, is easily magnetized and demagnetized, and has a small hysteresis loss. Soft iron and other **soft ferromagnetic materials**, such as silicon steel, are used in making parts exposed to rapid changes of magnetic flux, such as the cores of electromagnets, motors, generators, and transformers.

By comparison, **hard ferromagnetic materials**, such as cobalt steel and various alloys of nickel, aluminium, and cobalt, have low relative permeability, are difficult to magnetize, and have a high hysteresis loss. They are used in making permanent magnets.

soft matter A general name given to noncrystalline condensed matter. This includes liquids, disordered solids (including glasses), liquid crystals, and random networks of polymers.

soft radiation Ionizing radiation of low penetrating power, usually used with reference to X-rays of long wavelength. *Compare* HARD RADIATION.

software *See* COMPUTER.

sol A *colloid in which small solid particles are dispersed in a liquid continuous phase.

solar cell An electric cell that uses the sun's radiation to produce usable electric current. Most solar cells consist of a single-crystal silicon p–n junction. When photons of light energy from the sun fall on or near the *semiconductor junction the electron–hole pairs created are forced by the electric field at the junction to separate so that the holes pass to the p-region and the electrons pass to the n-region. This displacement of free charge creates an electric current when a load is connected across the terminals of the device (see illustration). Individual silicon solar cells cannot be made with a surface area much in excess of 4000 mm^2 and the maximum power delivered by such a cell is approximately 0.6 W at about 0.5 V in full sun. The efficiency of such devices is about 15%. For practical use, therefore, solar cells have to be assembled in arrays. Panels of solar cells have been the exclusive source of power for satellites and space capsules. Their use on earth has been largely limited by their high cost, a reduction in the cost by a factor of 10 being required to make them competitive with other energy sources at present.

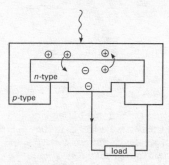

Silicon p–n junction

solar constant The rate at which solar energy is received per unit area at the outer limit of the earth's atmosphere at the mean distance between the earth and the sun. The value is 1.353 kW m^{-2}.

solar day *See* DAY.

solar energy The electromagnetic energy radiated from the sun. The tiny proportion (about 5×10^{-10} of the total) that falls on the earth is indicated by the *solar constant. The total quantity of solar energy falling on the earth in one year is about 4×10^{18} J, whereas the total annual energy consumption of the earth's inhabitants is only some 3×10^{14} J. The sun, therefore, could provide all the energy needed. The direct ways of making use of solar energy can be divided into thermal methods (*see* SOLAR HEATING) and nonthermal methods (*see* SOLAR CELL).

solar heating A form of domestic or industrial heating that relies on the direct use of solar energy. The basic form of **solar heater** is a thermal device in which a fluid is heated by the sun's rays in a collector (see illustration) and pumped or allowed to flow round a circuit that provides some form of heat storage and some form of auxiliary heat source for use when the sun is not shining. More complicated systems are combined heating-and-cooling devices, providing heat in the winter and air-conditioning in the summer. The simplest form of solar collector is the flat-plate collector, in which a blackened receiving surface is covered by one or more glass plates that acts like a greenhouse (*see* GREENHOUSE EFFECT) and traps the maximum amount of solar energy. Tubes attached to the receiving surface carry air, water, or some other fluid to which the absorbed heat is transferred. The whole panel is insulated at the back and can thus form part of the roof of a building. More sophisticated collectors focus the sun's rays using reflectors. *See also* SOLAR CELL.

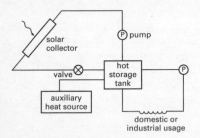

Typical solar heating system

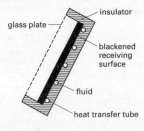

Flat plate solar collector

solar neutrino problem The problem that the number of neutrinos detected at the earth as being given off by the sun is far less than the number predicted by theories of nuclear processes in the sun. It is now thought that this is because some of the neutrinos given off by the sun are converted from electron neutrinos to the two other types of neutrinos en route to the earth.

solar parallax The angle subtended by the earth's equatorial radius at the centre of the sun at the mean distance between the earth and the sun (i.e. at 1 astronomical unit). It has the value 8.794 148 arc seconds.

solar prominence A cloud of gas that forms temporarily in the upper *chromosphere or inner *corona of the sun. It has a lower temperature but higher density than its surroundings and is observed as a bright projection.

solar system The sun, the nine major planets (Mercury, Venus, earth, Mars, Jupiter, Saturn, Uranus, Neptune, and Pluto) and their natural satellites, the asteroids, the comets, and meteoroids. Over 99% of the mass of the system is concentrated in the sun. The solar system as a whole moves in an approximately circular orbit about the centre of the Galaxy, taking about 2.2×10^8 years to complete its orbit. See Feature (p. 488).

solar wind A continuous outward flow of charged particles, mostly protons and electrons, from the sun's *corona into interplanetary space. The particles are controlled by the sun's magnetic field and are able to escape from the sun's gravitational field because of their high thermal energy. The average velocity of the particles in the vicinity of the earth is about 450 km s^{-1} and their density at this range is about 8×10^6 protons per cubic metre.

solar year *See* YEAR.

solder An alloy used to join metal surfaces. A **soft solder** melts at a temperature in the range 200–300°C and consists of a tin–lead alloy. The tin content varies between 80% for the lower end of the melting range and 31% for the higher end. **Hard solders** contain substantial quantities of silver in the alloy. **Brazing solders** are usually alloys of copper and zinc, which melt at over 800°C.

solenoid A coil of wire wound on a cylindrical former in which the length of the former is greater than its diameter. When a current is passed through the coil a magnetic field is produced inside the coil parallel to its axis. This field can be made to operate a plunger inside the former so that the solenoid can be used to operate a circuit breaker, valve, or other electromechanical device.

solid A state of matter in which there is a three-dimensional regularity of structure, resulting from the proximity of the component atoms, ions, or molecules and the strength of the forces between them. Solids can be crystalline or *amorphous. If a crystalline solid is heated, the kinetic energy of the components increases. At a specific temperature, called the **melting point**, the forces between the components become unable to contain them within the crystal structure. At this temperature, the lattice breaks down and the solid becomes a liquid.

solid angle Symbol Ω. The three-dimensional 'angle' formed by the vertex of a cone. When this vertex is the centre of a sphere of radius r and the base of the cone cuts out an area s on the surface of the sphere, the solid angle in *steradians is defined as s/r^2.

solid solution A crystalline material that is a mixture of two or more

SOLAR SYSTEM

The solar system is dominated by a star, the *sun. The sun's mass makes up more than 99% of the solar system's mass, and it is the gravitational attraction of the sun that holds the planets, asteroids, comets, and meteoroids in their orbits around it. Only the planetary satellites (such as the earth's moon) owe more allegiance to their parent planets, although the satellites too have to accompany the planets in their endless journey around the sun. The solar system itself also moves, rotating about the centre of the Milky Way galaxy about once every 2.2×10^8 years.

The solar system had its origin in a cloud of interstellar dust and gas that condensed around the proto-sun. The planets continued to grow by {accretion}, and by about 4.6 billion years ago the earth had formed. The inner, or terrestrial, planets – Mercury, Venus, earth, and Mars – are comparatively small. They are composed of rock and metal, with the metallic part forming a dense central core. Only the earth and Mars have natural satellites.

The outer, or giant, planets – Jupiter, Saturn, Uranus, and Neptune – have a rock-metal core surrounded by layers of solid, liquid, and gaseous hydrogen and helium. They are much farther from the sun and therefore much cooler than the inner planets, and have been able to retain gases of such low density. They have many satellites; Jupiter has at least 63, Saturn 34, Uranus 27, and Neptune 13. Pluto, the outermost planet, is small with one huge satellite.

The space between the orbits of Mars and Jupiter is occupied by thousands of *asteroids, or minor planets. Most of these are small, often irregularly-shaped chunks of rock, with perhaps only 150 of them more than 100 km across. Even smaller are hundreds of thousands of meteoroids, some no larger than grains of dust. Those that enter the earth's atmosphere and burn up as trails of light (shooting stars) are termed *meteors. The largest ones that reach the ground are called meteorites.

The last important members of the solar system are *comets. Often described as 'dirty snowballs', they consist of a nucleus of dust and ice a kilometre or two across, surrounded by a gaseous coma and with a tail that appears as the comet nears the sun. A dusty tail behind the comet is the source of many meteoroids.

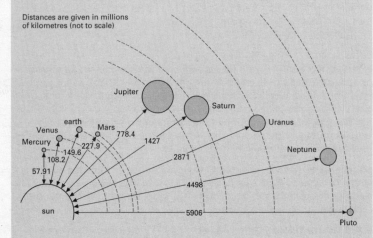

Distances are given in millions of kilometres (not to scale)

S

Jupiter

earth

Venus   Mars

Mercury

Saturn

Uranus

Neptune

778.4

227.9

1427

149.6

2871

108.2

4498

57.91

sun

5906

Pluto

components, with ions, atoms, or molecules of one component replacing some of the ions, atoms, or molecules of the other component in its normal crystal lattice. Solid solutions are found in certain alloys. For example, gold and copper form solid solutions in which some of the copper atoms in the lattice are replaced by gold atoms. In general, the gold atoms are distributed at random, and a range of gold–copper compositions is possible. At a certain composition, the gold and copper atoms can each form regular individual lattices (referred to as **superlattices**). Mixed crystals of double salts (such as alums) are also examples of solid solutions. Compounds can form solid solutions if they are isomorphous (*see* ISOMORPHISM).

solid-state detector *See* JUNCTION DETECTOR.

solid-state physics The study of the physical properties of solids, with special emphasis on the electrical properties of semiconducting materials in relation to their electronic structure. **Solid-state devices** are electronic components consisting entirely of solids (e.g. semiconductors, transistors, etc.) without heating elements, as in thermionic valves.

   Recently the term **condensed-matter physics** has been introduced to include the study of crystalline solids, amorphous solids, and liquids.

soliton A stable particle-like solitary wave state that is a solution of certain equations for propagation. Solitons are thought to occur in many areas of physics and applied mathematics, such as plasmas, fluid mechanics, lasers, optics, solid-state physics, and elementary-particle physics.

solstice **1.** Either of the two points on the *ecliptic midway between the *equinoxes, at which the sun is at its greatest angular distance north (**summer solstice**) or south (**winter solstice**) of the celestial equator. **2.** The time at which the sun reaches either of these points. The summer solstice occurs on June 21 and the winter solstice on December 21 in the northern hemisphere; the dates are reversed in the southern hemisphere.

solubility The quantity of solute that dissolves in a given quantity of solvent to form a saturated solution. Solubility is measured in kilograms per metre cubed, moles per kilogram of solvent, etc. The solubility of a substance in a given solvent depends on the temperature. Generally, for a solid in a liquid, solubility increases with temperature; for a gas, solubility decreases.

solubility product Symbol K_s. The product of the concentrations of ions in a saturated solution. For instance, if a compound A_xB_y is in equilibrium with its solution

$$A_xB_y(s) \leftrightarrow xA^+(aq) + yB^-(aq)$$

the equilibrium constant is

$$K_c = [A^+]^x[B^-]^y/[A_xB_y]$$

Since the concentration of the undissolved solid can be put equal to 1, the solubility product is given by

$$K_s = [A^+]^x[B^-]^y$$

The expression is only true for sparingly soluble salts. If the product of ionic concentrations in a solution exceeds the solubility product, then precipitation occurs.

solute The substance dissolved in a solvent in forming a *solution.

solution A homogeneous mixture of a liquid (the *solvent) with a gas or solid (the **solute**). In a solution, the molecules of the solute are discrete and mixed with the molecules of solvent. There is usually some interaction between the solvent and solute molecules. Two liquids that can mix on the molecular level are said to be **miscible**. In this case, the solvent is the major component and the solute the minor component. *See also* SOLID SOLUTION.

solvation The interaction of ions of a solute with the molecules of solvent. For instance, when sodium chloride is dissolved in water the sodium ions attract polar water molecules, with the negative oxygen atoms pointing towards the positive Na^+ ion. Solvation of transition-metal ions can also occur by formation of coordinate bonds, as in the hexaquocopper(II) ion $[Cu(H_2O)_6]^{2+}$. Solvation is the process that causes ionic solids to dissolve, because the energy released compensates for the energy necessary to break down the crystal lattice. It occurs only with polar solvents. Solvation in which the solvent is water is called **hydration**.

solvent A liquid that dissolves another substance or substances to form a *solution. **Polar solvents** are compounds such as water and liquid ammonia, which have dipole moments and consequently high dielectric constants. These solvents are capable of dissolving ionic compounds or covalent compounds that ionize (*see* SOLVATION). **Nonpolar solvents** are compounds such as ethoxyethane and benzene, which do not have permanent dipole moments. These do not dissolve ionic compounds but will dissolve nonpolar covalent compounds.

sonar *See* ECHO.

sonic boom A strong *shock wave generated by an aircraft when it is flying in the earth's atmosphere at supersonic speeds. This shock wave is radiated from the aircraft and where it intercepts the surface of the earth a loud booming sound is heard. The loudness depends on the speed and altitude of the aircraft and is lower in level flight than when the aircraft is undertaking a manoeuvre. The maximum increase of pressure in the shock wave during a transoceanic flight of a commercial supersonic transport (SST) is 120 Pa, equivalent to 136 decibels.

sonometer A device consisting essentially of a hollow sounding box with two bridges attached to its top. The string, fixed to the box at one end, is stretched between the two bridges so that the free end can run over a pulley and support a measured load. When the string is plucked the frequency of the note can be matched with that of another sound source, such as a tuning fork. It can be used to verify that the frequency (f) of a stretched string is given by $f = (1/2l)\sqrt{(T/m)}$, where l is the length of the string, m is its mass per unit length, and T is its tension.

Originally called the **monochord**, the sonometer was widely used as a tuning aid, but is now used only in teaching laboratories.

sorption *Absorption of a gas by a solid.

sorption pump A type of vacuum pump in which gas is removed from a system by absorption on a solid (e.g. activated charcoal or a zeolite) at low temperature.

sound A vibration in an elastic medium at a frequency and intensity that is capable of being heard by the human ear. The frequency of sounds lie in the range 20–20 000 Hz, but the ability to hear sounds in the upper part of the frequency range declines with age (*see also* PITCH). Vibrations that have a lower frequency than sound are called **infrasounds** and those with a higher frequency are called **ultrasounds**.

Sound is propagated through an elastic fluid as a longitudinal **sound wave**, in which a region of high pressure travels through the fluid at the *speed of sound in that medium. At a frequency of about 10 kilohertz the maximum excess pressure of a sound wave in air lies between 10^{-4} Pa and 10^3 Pa. Sound travels through solids as either longitudinal or transverse waves.

source The electrode in a field-effect *transistor from which electrons or holes enter the interelectrode space.

space 1. A property of the universe that enables physical phenomena to be extended into three mutually perpendicular directions. In Newtonian physics, space, time, and matter are treated as quite separate entities. In Einsteinian physics, space and time are combined into a four-dimensional continuum (*see* SPACE–TIME) and in the general theory of *relativity matter is regarded as having an effect on space, causing it to curve. **2. (outer space)** The part of the universe that lies outside the earth's atmosphere.

space group A *group formed by the set of all symmetry operations of a crystal lattice. This set consists of translations, rotations, and reflections and their combinations, such as *glide and *screw. It was discovered in the late 19th century that there are 230 possible space groups for a lattice in three dimensions. Space groups are used in the quantum theory of solids and in structure analysis in crystallography.

space probe An unmanned spacecraft that investigates features within the solar system. A **planetary probe** examines the conditions on or in the vicinity of one or more planets and a **lunar probe** is designed to obtain information about the moon. Probes are propelled by rocket motors and once out of the earth's gravitational field use their propulsion systems for course changes. Many are powered by panels of *solar cells, for both internal operation and radio communications.

space-reflection symmetry *See* PARITY.

space–time (space–time continuum) A geometry that includes the three dimensions and a **fourth dimension** of time. In Newtonian physics, space and time are considered as separate entities and whether or not events are simultaneous is a matter that is regarded as obvious to any competent observer. In Einstein's concept of the physical universe, based on a system of geometry devised by H. Minkowski (1864–1909), space and time are regarded as entwined, so that two observers in relative motion could disagree regarding the simultaneity of distant events. In Minkowski's geometry, an event is identified by a **world point** in a four-dimensional continuum.

spallation A type of nuclear reaction in which the interacting nuclei disintegrate into a large number of protons, neutrons, and other light particles, rather than exchanging nucleons between them. It is thought that most of the nuclei of light elements, such as boron, are made in this way.

spark *See* ELECTRIC SPARK.

spark chamber A device for detecting charged particles. It consists of a chamber, filled with helium and neon at atmospheric pressure, in which a stack of 20 to 100 plates are placed; the plates are connected alternately to the positive and negative terminals of a source of high potential (10 000 V or more). An incoming particle creates ion pairs in its track, causing the gas to become conducting and sparks to jump between the plates. The light from the sparks is focused to obtain stereoscopic photographs of the particles' tracks. It can also function as a counter (called a **spark counter**) when connected to suitable counting circuits. Some versions use crossed sets of parallel wires rather than plates; simple patterns may have a single wire near a plate, in the open atmosphere.

spark counter *See* SPARK CHAMBER.

special theory of relativity *See* RELATIVITY.

specific 1. Denoting that an extensive physical quantity (*see* EXTENSIVE VARIABLE) so described is expressed per unit mass. For example, the **specific latent heat** of a body is its latent heat per unit mass. When the extensive physical quantity is denoted by a capital letter (e.g. L for latent heat), the specific quantity is denoted by the corresponding lower-case letter (e.g. l for specific latent heat). **2.** In some older physical quantities the adjective 'specific' was added for other reasons (e.g. specific gravity, specific resistance). These names are now no longer used.

specific activity *See* ACTIVITY.

specific charge The ratio of the charge of an *elementary particle or other charged body to its mass.

specific gravity *See* RELATIVE DENSITY; SPECIFIC.

specific heat capacity *See* HEAT CAPACITY.

specific humidity *See* HUMIDITY.

specific impulse A measure of the thrust available from a rocket propellant. It is the ratio of the thrust produced to the fuel consumption.

specific intensity *See* PLANCK'S RADIATION LAW.

specific latent heat *See* LATENT HEAT.

specific resistance *See* RESISTIVITY; SPECIFIC.

specific surface The surface area of a particular substance per unit mass, expressed in $m^2 \ kg^{-1}$. It provides a measure of the surface area available for a process, such as adsorption, for a given mass of a powder or porous substance.

specific volume The volume of a substance per unit mass. The reciprocal of density, it has the units $m^3 \ kg^{-1}$.

spectral class (spectral type) A form of classification used for stars, based on their spectra. The **Harvard classification**, introduced in 1890 and modified in the 1920s, is based on the seven star types known as O, B, A, F, G, K, M:

  O hottest blue stars; ionized helium lines dominant
  B hot blue stars; neutral helium lines dominant, no ionized helium
  A blue blue-white stars; hydrogen lines dominant
  F white stars; metallic lines strengthen, hydrogen lines weaken
  G yellow stars; ionized calcium lines dominant
  K orange-red stars; neutral metallic lines dominant, some molecular
    bands
  M coolest red stars; molecular bands dominant

spectrochemical series A series of ligands arranged in the order in which they cause splitting of the energy levels of d orbitals in metal complexes (*see* LIGAND-FIELD THEORY). The series for some common ligands has the form:

  $CN^->NO_2^->NH_3>C_5H_5N>H_2O>OH^->F^->Cl^->Br^->I^-$

spectrograph *See* SPECTROSCOPE.

spectrometer Any of various instruments for producing a spectrum and measuring the wavelengths, energies, etc., involved. A simple type, for visible radiation, is a spectroscope equipped with a calibrated scale allowing wavelengths to be read off or calculated. In the X-ray to infrared region of the electromagnetic spectrum, the spectrum is produced by dispersing the radiation with a prism or diffraction grating (or crystal, in the case of hard X-rays). Some form of photoelectric detector is used, and the spectrum can be obtained as a graphical plot, which shows how the intensity of the radiation varies with wavelength. Such instruments are also called **spectrophotometers**. Spectrometers also exist for investigating the gamma-ray region and the microwave and radio-wave regions of the spectrum (*see* ELECTRON-SPIN RESONANCE; NUCLEAR MAGNETIC RESONANCE). Instruments for obtaining spectra of particle beams are also called spectrometers (*see* SPECTRUM; PHOTOELECTRON SPECTROSCOPY).

spectrophotometer *See* SPECTROMETER.

spectroscope An optical instrument that produces a *spectrum for visual observation. The first such instrument was made by R. W. Bunsen; in its simplest form it consists of a hollow tube with a slit at one end by which the light enters and a collimating lens at the other end to produce a parallel beam, a prism to disperse the light, and a telescope for viewing the spectrum (see illustration). In the **spectrograph**, the spectroscope is provided with a camera to record the spectrum. For a broad range of spectroscopic work, from the ultraviolet to the infrared, a diffraction grating is used instead of a prism. *See also* SPECTROMETER.

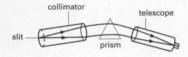

Spectroscope

spectroscopic binary *See* BINARY STARS.

spectroscopy The study of methods of producing and analysing *spectra using *spectroscopes, *spectrometers, spectrographs, and spectrophotometers. The interpretations of the spectra so produced can be used for chemical analysis, examining atomic and molecular energy levels and molecular structures, and for determining the composition and motions of celestial bodies (*see* REDSHIFT).

spectrum 1. A distribution of entities or properties arrayed in order of increasing or decreasing magnitude. For example, a beam of ions passed through a mass spectrograph, in which they are deflected according to their charge-to-mass ratios, will have a range of masses called a **mass spectrum**. A **sound spectrum** is the distribution of energy over a range of frequencies of a particular source. **2.** A range of electromagnetic energies arrayed in order of increasing or decreasing wavelength or frequency (*see* ELECTROMAGNETIC SPECTRUM). The **emission spectrum** of a body or substance is the characteristic range of radiations it emits when it is heated, bombarded by electrons or ions, or absorbs photons. The **absorption spectrum** of a substance is produced by examining, through

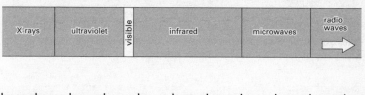

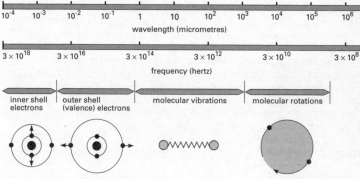

Spectrum: sources of electromagnetic spectra

the substance and through a spectroscope, a continuous spectrum of radiation. The energies removed from the continuous spectrum by the absorbing medium show up as black lines or bands. With a substance capable of emitting a spectrum, these are in exactly the same positions in the spectrum as some of the lines and bands in the emission spectrum.

Emission and absorption spectra may show a **continuous spectrum**, a **line spectrum**, or a **band spectrum**. A continuous spectrum contains an unbroken sequence of frequencies over a relatively wide range; it is produced by incandescent solids, liquids, and compressed gases. Line spectra are discontinuous lines produced by excited atoms and ions as they fall back to a lower energy level. Band spectra (closely grouped bands of lines) are characteristic of molecular gases or chemical compounds. *See also* SPECTROSCOPY.

speculum An alloy of copper and tin formerly used in reflecting *telescopes to make the main mirror as it could be cast, ground, and polished to make a highly reflective surface. It has now been largely replaced by silvered glass for this purpose.

speed The ratio of a distance covered by a body to the time taken. Speed is a *scalar quantity, i.e. no direction is given. Velocity is a *vector quantity, i.e. both the rate of travel and the direction are specified.

speed of light Symbol c. The speed at which electromagnetic radiation travels. The speed of light in a vacuum is $2.997\ 924\ 58 \times 10^8$ m s^{-1}. When light passes through any material medium its speed is reduced (*see*

REFRACTIVE INDEX). The speed of light in a vacuum is the highest speed attainable in the universe (*see* RELATIVITY; CERENKOV RADIATION). It is thought to be a universal constant and is independent of the speed of the observer. Since October 1983 it has formed the basis of the definition of the *metre.

speed of sound Symbol c or c_s. The speed at which sound waves are propagated through a material medium. In air at 20°C sound travels at 344 m s^{-1}, in water at 20°C it travels at 1461 m s^{-1}, and in steel at 20°C at 5000 m s^{-1}. The speed of sound in a medium depends on the medium's modulus of elasticity (E) and its density (ρ) according to the relationship $c = \sqrt{(E/\rho)}$. For longitudinal waves in a narrow solid specimen, E is the Young modulus; for a liquid E is the bulk modulus (*see* ELASTIC MODULUS); and for a gas $E = \gamma p$, where γ is the ratio of the principal specific *heat capacities and p is the pressure of the gas. For an ideal gas the relationship takes the form $c = \sqrt{(\gamma r T)}$, where r is the gas constant per unit mass and T is the thermodynamic temperature. This equation shows how the speed of sound in a gas is related to its temperature. This relationship can be written $c = c_0\sqrt{(1 + t/273)}$, where c_0 is the speed of sound in a particular gas at 0°C and t is the temperature in °C.

sphere The figure generated when a circle is rotated about a diameter. The volume of a sphere is $4\pi r^3/3$ and its surface area is $4\pi r^2$, where r is its radius. In Cartesian coordinates the equation of a sphere centred at the origin is $x^2 + y^2 + z^2 = r^2$.

spherical aberration *See* ABERRATION.

spherical harmonics The natural vibrations of an elastic sphere. Spherical harmonics may be used to describe the *normal modes of oscillation of spherically shaped objects. They have been used mathematically to model systems as diverse as the surface of the sun and the atomic nucleus. Each harmonic has two identifying indices that distinguish it from other spherical harmonic waveforms. On the surface of a vibrating sphere, certain **nodal circles** appear, where the surface is at rest. The number of these nodal circles for a given spherical harmonic is called its order n; n is one of the indices used to identify the normal mode. The second identification index corresponds to the number m of nodal circles, which pass through the poles of the vibrating sphere. A general property of spherical harmonics is that if a nodal circle does not pass through the poles of the sphere, then it must lie in a plane parallel to the sphere's equator. Consequently all nodal circles are lines of definite latitude or longitude.

spherical mirror *See* MIRROR.

spherical polar coordinates *See* POLAR COORDINATES.

spherometer An instrument for measuring the curvature of a surface. The usual instrument for this purpose consists of a tripod, the pointed

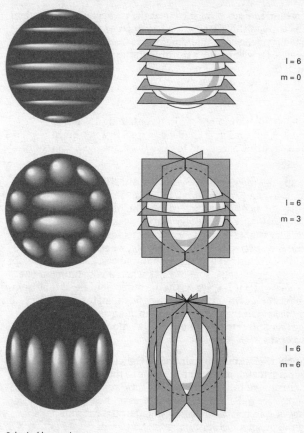

l = 6
m = 0

l = 6
m = 3

l = 6
m = 6

Spherical harmonics

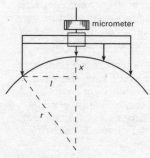

Spherometer

legs of which rest on the spherical surface at the corners of an equilateral triangle. In the centre of this triangle is a fourth point, the height of which is adjusted by means of a micrometer screw (see illustration). If the distance between each leg and the axis through the micrometer screw is l, and the height of the micrometer point above (or below) a flat surface is x, the radius (r) of the sphere is given by

$$r = (l^2 + x^2)/2x.$$

spiegel (spiegeleisen) A form of *pig iron containing 15–30% of manganese and 4–5% of carbon. It is added to steel in a Bessemer converter as a deoxidizing agent and to raise the manganese content of steel.

spin (intrinsic angular momentum) Symbol m_s or s. The part of the total angular momentum of a particle, atom, nucleus, etc., that is distinct from its orbital angular momentum. A molecule, atom, or nucleus in a specified energy level, or a particular elementary particle, has a particular spin, just as it has a particular charge or mass. According to *quantum theory, this is quantized and is restricted to multiples of $h/2\pi$, where h is the *Planck constant. Spin is characterized by a quantum number m_s. For example, for an electron $m_s = \pm\frac{1}{2}$, implying a spin of $+h/4\pi$ when it is spinning in one direction and $-h/4\pi$ when it is spinning in the other. Because of their spin, particles also have their own intrinsic *magnetic moments and in a magnetic field the spin of the particles lines up at an angle to the direction of the field, precessing around this direction. *See* NUCLEAR MAGNETIC RESONANCE.

spin glass An alloy of a small amount of a magnetic metal (0.1–10%) with a nonmagnetic metal, in which the atoms of the magnetic element are randomly distributed through the crystal lattice of the nonmagnetic element. Examples are AuFe and CuMn. Theories of the magnetic and other properties of spin glasses are complicated by the random distribution of the magnetic atoms.

spin network A quantum theory of space–time, first suggested by Roger Penrose in the 1960s, that classical space–time emerges from a network that is like a polymer network in rubber. The spin network picture has re-emerged from work on *quantum gravity in the 1990s. There is presently no observational or experimental evidence for spin networks.

spinodal curve A curve that separates a metastable region from an unstable region in the coexistence region of a binary fluid. Above the spinodal curve the process of moving towards equilibrium occurs by droplet nucleation, while below the spinodal curve there are periodic modulations of the *order parameter, which have a small amplitude at first (*see* SPINODAL DECOMPOSITION). The spinodal curve is not a sharp boundary in real systems as a result of fluctuations.

spinodal decomposition The process of moving towards equilibrium in a part of a phase diagram in which the *order parameter is conserved. Spinodal decomposition is observed in the quenching of binary mixtures. *See also* SPINODAL CURVE.

spinor A mathematical entity similar to a vector but having the property that it changes sign on each rotation through 360°. The wave function of a spin-½ particle, such as an electron, in relativistic quantum mechanics is an example of a spinor. Spinors have also been used extensively in general relativity theory.

spin–orbit coupling An interaction between the orbital angular momentum and the spin angular momentum of an individual particle, such as an electron. For light atoms, spin–orbit coupling is small so that *multiplets of many-electron atoms are described by *Russell–Saunders coupling. For heavy atoms, spin–orbit coupling is large so that multiplets of many-electron atoms are described by *j-j coupling. For medium-sized atoms the sizes of the energies associated with spin–orbit coupling are comparable to the sizes of energies associated with electrostatic repulsion between the electrons, the multiplets in this case being described as having **intermediate coupling**. Spin–orbit coupling is large in many nuclei, particularly heavy nuclei.

spin–statistics theorem A fundamental theorem of relativistic *quantum field theory that states that half-integer *spins can only be quantized consistently if they obey Fermi–Dirac statistics and even-integer spins can only be quantized consistently if they obey Bose–Einstein statistics (*see* QUANTUM STATISTICS). This theorem enables one to understand the result of quantum statistics that wave functions for bosons are symmetric and wave functions for fermions are antisymmetric. It also provides the foundation for the *Pauli exclusion principle. It was first proved by Pauli in 1940 and has been proved in several different ways since then.

spintronics A branch of technology that specifically makes use of the quantum-mechanical *spin of electrons in electronic devices. The subject is being developed very rapidly and is likely to lead to fast components that both store and process information and to devices used in *quantum computing.

spin wave (magnon) A *collective excitation associated with magnetic systems. Spin waves occur in both ferromagnetic and antiferromagnetic systems (*see* MAGNETISM).

spiral galaxy *See* GALAXY.

spontaneous emission The emission of a photon by an atom as it makes a transition from an excited state to the ground state. Spontaneous emission occurs independently of any external electromagnetic radiation; the transition is caused by interactions between atoms and vacuum fluctuations (*see* VACUUM STATE) of the quantized electromagnetic field.

The process of spontaneous emission, which cannot be described by nonrelativistic *quantum mechanics, as given by formulations such as the *Schrödinger equation, is responsible for the limited lifetime of an excited state of an atom before it emits a photon. *See also* QUANTUM THEORY OF RADIATION; EINSTEIN COEFFICIENTS; LASER.

spring balance A simple form of *balance in which a force is measured by the extension it produces in a helical spring. The extension, which is read off a scale, is directly proportional to the force, provided that the spring is not overstretched. The device is often used to measure the weight of a body approximately.

s-process A process in *nucleosynthesis in which a nucleus captures a neutron and will probably undergo beta decay before capturing another neutron. The s-process ('slow process') occurs if there is a steady but not large supply of neutrons from nuclear fusion reactions in stars. Many elements that are heavier than iron are produced by the s-process.

sputtering The process by which some of the atoms of an electrode (usually a cathode) are ejected as a result of bombardment by heavy positive ions. Although the process is generally unwanted, it can be used to produce a clean surface or to deposit a uniform film of a metal on an object in an evacuated enclosure.

square wave A train of rectangular voltage pulses that alternate between two fixed values for equal lengths of time. The time of transition between each fixed value is negligible compared to the duration of the fixed value. See diagram.

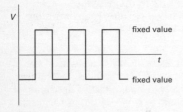

Square wave

squark *See* SUPERSYMMETRY.

stable equilibrium *See* EQUILIBRIUM.

stainless steel A form of *steel containing at least 11–12% of chromium, a low percentage of carbon, and often some other elements, notably nickel and molybdenum. Stainless steel does not rust or stain and therefore has a wide variety of uses in industrial, chemical, and domestic environments. A particularly successful alloy is the steel known as 18–8, which contains 18% Cr, 8% Ni, and 0.08% C.

standard cell A *voltaic cell, such as a *Clark cell or *Weston cell, used as a standard of e.m.f.

standard deviation A measure of the dispersion of data in statistics. For a set of values $a_1, a_2, a_3, \ldots, a_n$, the mean m is given by $(a_1 + a_2 + \ldots + a_n)/n$. The **deviation** of each value is the absolute value of the difference from the mean, i.e. $|m - a_1|$, etc. The standard deviation is the square root of the mean of the squares of these values, i.e.

$$\sqrt{[(|m - a_1|^2 + \ldots + |m - a_n|^2)/n]}$$

When the data is continuous the sum is replaced by an integral.

standard electrode An electrode (a half cell) used in measuring electrode potential. *See* HYDROGEN HALF CELL.

standard electrode potential *See* ELECTRODE POTENTIAL.

standard form A way of writing a number, especially a large or small number, in which only one integer appears before the decimal point, the value being adjusted by multiplying by the appropriate power of 10. For example, 236,214 would be written in the standard form as $2.362\ 14 \times 10^5$; likewise 0.006821047 would be written $6.821\ 047 \times 10^{-3}$. Note that in the standard form, commas are not used: the digits are grouped into threes and a space is left between groups.

standard model *See* ELEMENTARY PARTICLES.

standard state A state of a system used as a reference value in thermodynamic measurements. Standard states involve a reference value of pressure (usually one atmosphere, 101.325 kPa) or concentration (usually 1 M). Thermodynamic functions are designated as 'standard' when they refer to changes in which reactants and products are all in their standard and their normal physical state. For example, the standard molar enthalpy of formation of water at 298 K is the enthalpy change for the reaction

$$H_2(g) + \tfrac{1}{2}O_2(g) \rightarrow H_2O(l)$$

$$\Delta H^{\ominus}{}_{298} = -285.83 \text{ kJ mol}^{-1}.$$

Note the superscript $^{\ominus}$ is used to denote standard state and the temperature should be indicated.

standard temperature and pressure *See* S.T.P.

standing wave *See* STATIONARY WAVE.

star A self-luminous celestial body, such as the *sun, that generates nuclear energy within its core. Stars are not distributed uniformly throughout the universe, but are collected together in *galaxies. The age and lifetime of a star are related to its mass (*see* STELLAR EVOLUTION; HERTZSPRUNG–RUSSELL DIAGRAM).

star cluster A group of stars that are sufficiently close to each other for them to be physically associated. Stars belonging to the cluster are formed together from the same cloud of interstellar gas and have approximately the same age and initial chemical composition. Because of this, and since the stars in a given cluster are at roughly the same distance from earth, observations of star clusters are of great importance in studies of stellar evolution.

There are two types of star cluster. **Open** (or **galactic**) **clusters** are fairly loose systems of between a few hundred and a few thousand members. The stars in open clusters are quite young by astronomical standards (some as young as a few million years) and have relatively high abundances of heavy elements. **Globular clusters** are approximately spherical collections of between ten thousand and a million stars. These are very old (of order 10^{10} years) and have low heavy-element abundances.

Stark effect The splitting of lines in the *spectra of atoms due to the presence of a strong electric field. It is named after the German physicist Johannes Stark (1874–1957), who discovered it in 1913. Like the normal *Zeeman effect, the Stark effect can be understood in terms of the classical electron theory of Lorentz. The Stark effect for hydrogen atoms was also described by the *Bohr theory of the atom. In terms of *quantum mechanics, the Stark effect is described by regarding the electric field as a *perturbation on the quantum states and energy levels of an atom in the absence of an electric field. This application of perturbation theory was its first use in quantum mechanics.

stat- A prefix attached to the name of a practical electrical unit to provide a name for a unit in the electrostatic system of units, e.g. statcoulomb, statvolt. *Compare* AB-. In modern practice both absolute and electrostatic units have been replaced by *SI units.

state of matter One of the three physical states in which matter can exist, i.e. *solid, *liquid, or *gas. *Plasma is sometimes regarded as the fourth state of matter.

static electricity The effects produced by electric charges at rest, including the forces between charged bodies (*see* COULOMB'S LAW) and the field they produce (*see* ELECTRIC FIELD).

statics The branch of mechanics concerned with bodies that are acted upon by balanced forces and couples so that they remain at rest or in unaccelerated motion. *Compare* DYNAMICS.

stationary orbit *See* SYNCHRONOUS ORBIT.

stationary state A state of a system when it has an energy level permitted by *quantum mechanics. Transitions from one stationary state to another can only occur by the emission or absorption of an appropriate quanta of energy (e.g. in the form of photons).

stationary wave (**standing wave**) A form of *wave in which the profile of the wave does not move through the medium but remains stationary. This is in contrast to a **travelling** (or **progressive**) **wave**, in which the profile moves through the medium at the speed of the wave. A stationary wave results when a travelling wave is reflected back along its own path. In a stationary wave there are points at which the displacement is zero; these are called **nodes**. Points of maximum displacement are called **antinodes**. The distance between a node and its neighbouring antinode is one quarter of a wavelength. In a stationary wave all the points along the wave have different amplitudes and the points between successive nodes are in phase; in a travelling wave every point vibrates with the same amplitude and the phase of vibration changes for different points along its path.

statistical error *See* MEASUREMENT ERROR.

statistical mechanics The branch of physics in which statistical methods are applied to the microscopic constituents of a system in order to predict its macroscopic properties. The earliest application of this method was the 19th-century work by Maxwell and Boltzmann, which attempted to explain the thermodynamic properties of gases on the basis of the statistical properties of large assemblies of molecules.

In classical statistical mechanics, each particle is regarded as occupying a point in *phase space, i.e. to have an exact position and momentum at any particular instant. The probability that this point will occupy any small volume of the phase space is taken to be proportional to the volume. The Maxwell–Boltzmann law gives the most probable distribution of the particles in phase space.

With the advent of quantum theory, the exactness of these premises was disturbed (by the Heisenberg uncertainty principle). In the *quantum statistics that evolved as a result, the phase space is divided into cells, each having a volume h^f, where h is the Planck constant and f is the number of degrees of freedom of the particles. This new concept led to Bose–Einstein statistics, and for particles obeying the Pauli exclusion principle, to Fermi–Dirac statistics.

statistics The branch of mathematics concerned with the inferences that can be drawn from numerical data on the basis of probability. A **statistical inference** is a conclusion drawn about a population as a result of an analysis of a representative sample. *See* SAMPLING.

stator The stationary electromagnetic structure of an electric motor or electric generator. *Compare* ROTOR.

steady-state theory The cosmological theory that the universe has always existed in a steady state, that it had no beginning, will have no end, and has a constant mean density. To compensate for the observed *expansion of the universe this theory postulates that matter is created throughout the universe at a rate of about 10^{-10} nucleon per metre cubed per year as a property of space. Because it has failed to account for the *microwave background radiation or the evidence of evolution in the universe it has lost favour to the *big-bang theory. It was first proposed by Hermann Bondi (1919–), Thomas Gold (1920–2004), and Fred *Hoyle in 1948.

steam distillation A method of distilling liquids that are immiscible with water by bubbling steam through them. It depends on the fact that the vapour pressure (and hence the boiling point) of a mixture of two immiscible liquids is lower than the vapour pressure of either pure liquid.

steam engine A *heat engine in which the thermal energy of steam is converted into mechanical energy. It consists of a cylinder fitted with a piston and valve gear to enable the high-pressure steam to be admitted to the cylinder when the piston is near the top of its stroke. The steam forces the piston to the bottom of its stroke and is then exhausted from the cylinder usually into a condenser. The reciprocating motion of the piston is converted to rotary motion of the flywheel by means of a connecting rod, crosshead, and crank. The steam engine reached its zenith at the end of the 19th century, since when it has been replaced by the steam turbine and the internal-combustion engine. *See also* RANKINE CYCLE.

steam point The temperature at which the maximum vapour pressure of water is equal to the standard atmospheric pressure (101 325 Pa). On the Celsius scale it has the value 100°C.

steel Any of a number of alloys consisting predominantly of iron with varying proportions of carbon (up to 1.7%) and, in some cases, small quantities of other elements (**alloy steels**), such as manganese, silicon, chromium, molybdenum, and nickel. Steels containing over 11–12% of chromium are known as *stainless steels.

Carbon steels exist in three stable crystalline phases: **ferrite** has a body-

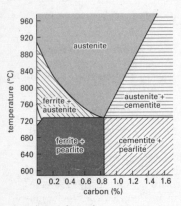

Phase diagram for steel

centred cubic crystal, **austenite** has a face-centred cubic crystal, and
cementite has an orthorhombic crystal. **Pearlite** is a mixture of ferrite and
cementite arranged in parallel plates. The phase diagram shows how the
phases form at different temperatures and compositions.

Steels are made by the *basic-oxygen process (L–D process), which has
largely replaced the *Bessemer process and the *open-hearth process, or
in electrical furnaces.

Stefan's law (Stefan–Boltzmann law) The total energy radiated per unit
surface area of a *black body in unit time is proportional to the fourth
power of its thermodynamic temperature. The constant of
proportionality, the **Stefan constant** (or **Stefan–Boltzmann constant**) has
the value 5.6697×10^{-8} J s^{-1} m^{-2} K^{-4}. The law was discovered by the
Austrian physicist Joseph Stefan (1853–93) and theoretically derived by
Ludwig *Boltzmann.

stellar evolution The changes that occur to a *star during its lifetime,
from birth to final extinction. A star is believed to form from a
condensation of interstellar matter, which collects either by chance or for
unexplained reasons, and grows by attracting other matter towards itself
as a result of its gravitational field. This initial cloud of cold contracting
matter, called a **protostar**, builds up an internal pressure as a result of its
gravitational contraction. The pressure raises the temperature until it
reaches $5–10 \times 10^6$ K, at which temperature the thermonuclear
conversion of hydrogen to helium begins. In our *sun, a typical star,
hydrogen is converted at a rate of some 10^{11} kg s^{-1} with the evolution of
some 6×10^{25} J s^{-1} of energy. It is estimated that the sun contains
sufficient hydrogen to burn at this rate for 10^{10} years and that it still has
half its life to live as a main-sequence star (see HERTZSPRUNG–RUSSELL
DIAGRAM). Eventually, however, this period of stability comes to an end,
because the thermonuclear energy generated in the interior is no longer
sufficient to counterbalance the gravitational contraction. The core, which
is now mostly helium, collapses until a sufficiently high temperature is
reached in a shell of unburnt hydrogen round the core to start a new
phase of thermonuclear reaction. This burning of the shell causes the
star's outer envelope to expand and cool, the temperature drop changes
the colour from white to red and the star becomes a **red giant** or a
supergiant if the original star was very large. The core now contracts,
reaching a temperature of 10^8 K, and the helium in the core acts as the
thermonuclear energy source. This reaction produces carbon, but a star of
low mass relatively soon runs out of helium and the core collapses into a
*white dwarf, while the outer regions drift away into space, possibly
forming a **planetary nebula**. Larger stars (several times larger than the
sun) have sufficient helium for the process to continue so that heavier
elements, up to iron, are formed. But iron is the heaviest element that can
be formed with the production of energy and when the helium has all
been consumed there is a catastrophic collapse of the core, resulting in a
*supernova explosion, blowing the outer layers away. The current theory

suggests that thereafter the collapsed core becomes a *neutron star or a *black hole, depending on its mass.

stellar wind The radial outflow of matter from the atmosphere of a very hot star. In the case of the sun, the stellar wind is called *solar wind. The rate at which the stellar matter is emitted varies with the lifetime of the star.

steradian Symbol sr. The supplementary *SI unit of solid angle equal to the *solid angle that encloses a surface on a sphere equal to the square of the radius of the sphere.

stere A unit of volume equal to 1 m^3. It is not now used for scientific purposes.

Stern–Gerlach experiment A classic experiment performed in 1921, which demonstrated the quantum mechanical nature of atomic magnetic moments. A beam of silver atoms entered a strong non-uniform magnetic field. Classical physics leads to the expectation that the beam should be broadened into a continuous band. However, in agreement with quantum mechanics, the beam was split into two separate beams. The experiment was performed by the German physicists Otto Stern (1888–1969) and Walter Gerlach (1889–1979).

stimulated emission *See* INDUCED EMISSION; LASER.

Stirling engine A heat engine consisting of a hot cylinder and a cold cylinder separated by a regenerator acting as a heat exchanger. The cylinders enclose oscillating pistons. Heat is applied externally to the hot cylinder, causing the working fluid within it to expand and drive the piston. The fluid is cooled in the regenerator before entering the cold cylinder, where it is compressed by the piston and driven back to be heated in the regenerator before entering the hot cylinder again. Stirling engines, which were invented by Robert Stirling (1790–1878) in 1816, are silent and efficient but costly to produce. They have found limited use; interest in them revived in the 1960s.

STM *See* SCANNING TUNNELLING MICROSCOPE.

stochastic process Any process in which there is a random element. Stochastic processes are important in *nonequilibrium statistical mechanics and *disordered solids. In a **time-dependent stochastic process**, a variable that changes with time does so in such a way that there is no correlation between different time intervals. An example of a stochastic process is *Brownian movement. Equations, such as the *Langevin equation and the *Fokker–Planck equation, that describe stochastic processes are called **stochastic equations**. It is necessary to use statistical methods and the theory of probability to analyse stochastic processes and their equations.

stokes Symbol St. A c.g.s. unit of kinematic viscosity equal to the ratio of the viscosity of a fluid in poises to its density in grams per cubic centimetre. 1 stokes = 10^{-4} m² s⁻¹. It is named after Sir George Gabriel Stokes.

Stokes, Sir George Gabriel (1819–1903) British physicist and mathematician, born in Ireland, who worked at Cambridge University all his life. He is best known for *Stokes' law, concerning the movement of objects in a fluid. A unit of *kinematic viscosity is named after him.

Stokes' law A law that predicts the frictional force F on a spherical ball moving through a viscous medium. According to this law $F = 6\pi r \eta v$, where r is the radius of the ball, v is its velocity, and η is the viscosity of the medium. The sphere accelerates until it reaches a steady terminal speed. For a falling ball, F is equal to the gravitational force on the sphere, less any upthrust. The law was discovered by Sir George Gabriel Stokes.

Stokes' parameters A set of four parameters, denoted s_0, s_1, s_2, s_3, that determine the state of *polarization of a wave of *monochromatic radiation from observations of the beam. The Stokes' parameters are quadratic in the field strength and can be determined by intensity measurements or by a *polarizer. The Stokes' parameters are not independent since they are related by $s_0^2 = s_1^2 + s_2^2 + s_3^2$. The Stokes' parameters are named after Sir George Gabriel Stokes, who introduced them in 1852. The determination of Stokes' parameters has astrophysical applications, which include gaining information concerning the mechanism of electromagnetic radiation from such objects as pulsars.

Stokes' theorem A theorem that is the analogue of the *divergence theorem for the *curl of a vector. Stokes' theorem states that if a surface S, which is smooth and simply connected (i.e. any closed curve on the surface can be contracted continuously into a point without leaving the surface), is bounded by a line L the vector $\mathbf{F}$ defined in S satisfies

$$\int_S \mathrm{curl}\mathbf{F}\cdot \mathrm{d}S = \int_L \mathbf{F}\cdot \mathrm{d}\mathbf{l}.$$

Stokes' theorem was stated by Sir George Gabriel Stokes as a Cambridge examination question, having been raised by Lord *Kelvin in a letter to Stokes in 1850.

stop A circular aperture that limits the effective size of a lens in an optical system. It may be adjustable, as the iris diaphragm in a camera, or have a fixed diameter, as the disc used in some telescopes.

stopping power A measure of the ability of matter to reduce the kinetic energy of a particle passing through it. The **linear stopping power**, $-\mathrm{d}E/\mathrm{d}x$, is energy loss of a particle per unit distance. The **mass stopping power**, $(1/\rho)\mathrm{d}E/\mathrm{d}x$, is the linear stopping power divided by the density (ρ) of the substance. The **atomic stopping power**, $(1/n)\mathrm{d}E/\mathrm{d}x = (A/\rho N)\mathrm{d}E/\mathrm{d}x$, is the energy loss per atom per unit area perpendicular to the particle's motion, i.e. n is the number of atoms in unit volume of the substance, N is the Avogadro number, and A is the relative atomic mass of the substance.

The relative stopping power is the ratio of the stopping power of a substance to that of a standard substance, usually aluminium, oxygen, or air.

storage ring A large evacuated toroidal ring forming a part of some particle accelerators. The rings are designed like *synchrotrons, except that they do not accelerate the particles circling within them but supply just sufficient energy to make up for losses (mainly *synchrotron radiation). The storage rings are usually built tangentially to the associated accelerator so that particles can be transferred accurately between them. At *CERN in Geneva, two interlaced storage rings are used, containing protons rotating in opposite directions. At the intersections very high collision energies (up to 1700 GeV) can be achieved.

s.t.p. Standard temperature and pressure, formerly known as **N.T.P.** (normal temperature and pressure). The standard conditions used as a basis for calculations involving quantities that vary with temperature and pressure. These conditions are used when comparing the properties of gases. They are 273.15 K (or 0°C) and 101 325 Pa (or 760.0 mmHg).

strain A measure of the extent to which a body is deformed when it is subjected to a *stress. The **linear strain** or **tensile strain** is the ratio of the change in length to the original length. The **bulk strain** or **volume strain** is the ratio of the change in volume to the original volume. The **shear strain** is the angular distortion in radians of a body subjected to a *shearing force. *See also* ELASTICITY; ELASTIC MODULUS.

strain gauge A device used to measure a small mechanical deformation in a body (*see* STRAIN). The most widely used devices are metal wires or foils or semiconductor materials, such as a single silicon crystal, which are attached to structural members; when the members are stretched under tensile *stress the resistance of the metal or semiconductor element increases. By making the resistance a component in a *Wheatstone-bridge circuit an estimate of the strain can be made by noting the change in resistance. Other types of strain gauge rely on changes of capacitance or the magnetic induction between two coils, one of which is attached to the stressed member.

strain hardening (**work hardening**) An increase in the resistance to the further plastic deformation of a body as a result of a rearrangement of its internal structure when it is strained, particularly by repeated stress. *See also* ELASTICITY.

strange attractor *See* ATTRACTOR.

strange matter Matter composed of up, down, and strange quarks (rather than the up and down quarks found in normal nucleons). It has been suggested that strange matter may have been formed in the *early universe, and that pieces of this matter (called **S-drops**) may still exist.

strangeness Symbol *s*. A property of certain elementary particles called hadrons (K-mesons and hyperons) that decay more slowly than would have been expected from the large amount of energy released in the process. These particles were assigned the quantum number *s* to account for this behaviour. For nucleons and other nonstrange particles *s* = 0; for strange particles *s* does not equal zero but has an integral value. In quark theory (*see* ELEMENTARY PARTICLES) hadrons with the property of strangeness contain a strange quark or its antiquark.

stratosphere *See* EARTH'S ATMOSPHERE.

stray capacitance A capacitance that arises in a circuit in an unintended way because of interactions between components of the circuit. Stray capacitance is minimized by placing these components as far away from each other as possible in the circuit.

streamline flow A type of fluid flow in which no *turbulence occurs and the particles of the fluid follow continuous paths, either at constant velocity or at a velocity that alters in a predictable and regular way (*see also* LAMINAR FLOW).

stress The force per unit area on a body that tends to cause it to deform (*see* STRAIN). It is a measure of the internal forces in a body between particles of the material of which it consists as they resist separation, compression, or sliding in response to externally applied forces. **Tensile stress** and **compressive stress** are axial forces per unit area applied to a body that tend either to extend it or compress it linearly. **Shear stress** is a tangential force per unit area that tends to shear a body. *See also* ELASTICITY; ELASTIC MODULUS.

string A one-dimensional object used in theories of elementary particles and in cosmology (**cosmic string**). **String theory** replaces the idea of a pointlike elementary particle (used in quantum field theory) by a line or loop (a closed string). States of a particle may be produced by standing waves along this string. The combination of string theory with supersymmetry leads to *superstring theory.

stroboscope A device for making a moving body intermittently visible in order to make it appear stationary. It may consist of a lamp flashing at regular intervals or a shutter that enables it to be seen intermittently. **Stroboscopic photography** is the taking of very short-exposure pictures of moving objects using an electronically controlled flash lamp.

strong interaction *See* FUNDAMENTAL INTERACTIONS.

structure formation in the universe The process that has produced a nonuniform distribution of matter throughout the universe, i.e. *large-scale structure, such as galaxies. In general, 'clumpiness' occurs when the matter is not homogeneous, i.e. the gravitational attraction between particles of matter ceases to be uniform. This can occur as a result of quantum fluctuations in the *early universe associated with inflation and the *Jeans instability in an expanding universe. There is some evidence for this hypothesis, based on measurements from *COBE and *WMAP. An alternative mechanism for structure formation is based on cosmic strings but observations are not in accord with the predictions of this theory. A complete quantitative theory of structure formation does not exist at present.

subatomic particle *See* ELEMENTARY PARTICLES.

subcritical *See* CRITICAL MASS; CRITICAL REACTION; MULTIPLICATION FACTOR.

sublimate A solid formed by sublimation.

sublimation A direct change of state from solid to gas.

submillimetre waves Electromagnetic radiation with wavelengths below one millimetre (and therefore frequencies greater than 300 gigahertz), extending to radiation of the far infrared. A source of submillimetre radiation is a medium pressure mercury lamp in quartz. Submillimetre waves can be detected by a *Golay cell.

subshell *See* ATOM.

subsonic speed A speed that is less than Mach 1 (*see* MACH NUMBER).

subtractive process *See* COLOUR.

sun The *star at the centre of the *solar system. A typical main-sequence dwarf star (*see* HERTZSPRUNG–RUSSELL DIAGRAM; STELLAR EVOLUTION), the sun is situated at a distance of some 149 600 000 km from earth. It has a diameter of about 1 392 000 km and a mass of 1.9×10^{30} kg. Hydrogen and helium are the primary constituents (about 75% hydrogen, 25% helium), with less than 1% of heavier elements. In the central core, some 400 000 km in diameter, hydrogen is converted into helium by thermonuclear reactions, which generate vast quantities of energy. This energy is radiated into space and provides the earth with all the light and heat necessary to have created and maintained life (*see* SOLAR CONSTANT). The surface of the sun, the *photosphere, forms the boundary between its opaque interior and its transparent atmosphere. It is here that *sunspots occur. Above the photosphere is the *chromosphere and above this the *corona, which extends tenuously into interplanetary space. *See also* SOLAR WIND.

sunspot A dark patch in the sun's *photosphere resulting from a localized fall in temperature to about 4000 K. Most spots have a central very dark umbra surrounded by a lighter penumbra. Sunspots tend to occur in clusters and to last about two weeks. The number of sunspots visible fluctuates over an eleven-year cycle – often called the **sunspot cycle**. The cause is thought to be the presence of intense localized magnetic fields, which suppress the convection currents that bring hot gases to the photosphere.

supercluster *See* GALAXY CLUSTER.

superconductivity The absence of measurable electrical resistance in certain substances at temperatures close to 0 K. First discovered in 1911 in mercury, superconductivity is now known to occur in some 26 metallic elements and many compounds and alloys. The temperature below which a substance becomes superconducting is called the **transition temperature** (or **critical temperature**). Compounds are now known that show superconductivity at liquid-nitrogen temperatures.

The theoretical explanation of the phenomenon was given by John *Bardeen, L. N. Cooper (1930–), and J. R. Schrieffer (1931–) in 1957 and is known as the **BCS theory**. According to this theory an electron moving through an elastic crystal lattice creates a slight distortion of the lattice as a result of Coulomb forces between the positively charged lattice and the negatively charged electron. If this distortion persists for a finite time it can affect a second passing electron. In 1956 Cooper showed that the effect of this phenomenon is for the current to be carried in superconductors not by individual electrons but by bound pairs of electrons, the **Cooper pairs**. The BCS theory is based on a *wave function in which all the electrons are paired. Because the total momentum of a Cooper pair is unchanged by the interaction between one of its electrons and the lattice, the flow of electrons continues indefinitely.

Superconducting coils in which large currents can circulate indefinitely can be used to create powerful magnetic fields and are used for this purpose in some particle accelerators and in other devices.

Superconductivity can also occur by a slightly more complicated mechanism than BCS theory in *heavy-fermion systems. In 1986, Georg Bednorz (1950–) and Karl Müller (1927–) found an apparently completely different type of superconductivity. This is called **high-temperature superconductivity**, since the critical temperature is very much higher than for BCS superconductors; some high-temperature superconductors have critical temperatures greater than 100 K. A typical high-temperature superconductor is $YBa_2Cu_3O_{1-7}$.

At the present time a theory of high-temperature superconductivity has not been established in spite of a great deal of effort, which is still going on. The BCS mechanism, and minor modifications of it, almost certainly do not apply. Some models of high-temperature superconductors have had some successes in explaining some of the properties of these materials, but a unified theory has not yet been produced.

supercooling 1. The cooling of a liquid to below its freezing point without a change from the liquid to solid state taking place. In this metastable state the particles of the liquid lose energy but do not fall into place in the lattice of the solid crystal. If the liquid is seeded with a small crystal, crystallization usually takes place and the temperature returns to the freezing point. This is a common occurrence in the atmosphere where water droplets frequently remain unfrozen at temperatures well below 0°C until disturbed, following which they rapidly freeze. The supercooled droplets, for example, rapidly freeze on passing aircraft forming 'icing', which can be a hazard. **2.** The analogous cooling of a vapour to make it supersaturated until a disturbance causes condensation to occur, as in the Wilson *cloud chamber.

supercritical See CRITICAL MASS; CRITICAL REACTION; MULTIPLICATION FACTOR.

superdeformed nucleus A nucleus that has an ellipsoidal shape, with the long axis about twice the size of the short axis. Such nuclei can be formed by fusing high-energy heavy ions with other nuclei. They lose energy by emitting a characteristic pattern of gamma rays as they revert to a more normal shape.

superficial expansivity See EXPANSIVITY.

superfluidity The property of liquid helium at very low temperatures that enables it to flow without friction. Both helium isotopes possess this property, but ^{4}He becomes superfluid at 2.172 K, whereas ^{3}He does not become superfluid until a temperature of 0.00093 K is reached. There is a basic connection between superfluidity and *superconductivity, so that sometimes a superconductor is called a charged superfluid.

supergiant The largest and most luminous type of star. They are formed from the most massive stars and are therefore very rare. They lie above the giants on the *Hertzsprung–Russell diagram. See also STELLAR EVOLUTION.

supergravity A *unified-field theory for all the known fundamental interactions that involves *supersymmetry. Supergravity is most naturally formulated as a *Kaluza–Klein theory in eleven dimensions. The theory contains particles of spin 2, spin 3/2, spin 1, spin 1/2, and spin 0. Although supersymmetry means that the infinities in the calculations are less severe than in other attempts to construct a quantum theory of gravity, it is probable that supergravity still contains infinities that cannot be removed by the process of *renormalization. It is thought by many physicists that to obtain a consistent quantum theory of gravity one has to abandon *quantum field theories, since they deal with point objects, and move to theories based on extended objects, such as *superstring theory and *supermembrane theory, and therefore that supergravity is not a complete theory of the fundamental interactions.

superheavy elements See SUPERTRANSURANICS.

superheterodyne receiver A widely used type of radio receiver in which the incoming radio-frequency signal is mixed with an internally generated signal from a local oscillator. The output of the mixer has a carrier frequency equal to the difference between the transmitted frequency and the locally generated frequency, still retains the transmitted modulation, and is called the **intermediate frequency** (IF). The IF signal is amplified and demodulated before being passed to the audio-frequency amplifier. This system enables the IF signal to be amplified with less distortion, greater gain, better selectivity, and easier elimination of noise than can be achieved by amplifying the radio-frequency signal.

super high frequency (SHF) A radio frequency in the range 3–30 gigahertz.

superlattice *See* SOLID SOLUTION.

supermembrane theory A unified theory of the *fundamental interactions involving *supersymmetry, in which the basic entities are two-dimensional extended objects (**supermembranes**). They are thought to have about the same length scale as superstrings (*see* SUPERSTRING THEORY), i.e. 10^{-35} m. At the present time there is no experimental evidence for supermembranes. Supermembranes and other higher-dimensional objects have been found as solutions in superstring theory.

supernova An explosive brightening of a star in which the energy radiated by it increases by a factor of 10^{10}. It takes several years to fade and while it lasts dominates the whole galaxy in which it lies. It is estimated that there could be a supernova explosion in the Milky Way every 30 years, although only six have actually been observed in the last 1000 years. A supernova explosion occurs when a star has burnt up all its available nuclear fuel and the core collapses catastrophically to form a neutron star (*see* STELLAR EVOLUTION). *Compare* NOVA.

superplasticity The ability of some metals and alloys to stretch uniformly by several thousand percent at high temperatures, unlike normal alloys, which fail after being stretched 100% or less. Since 1962, when this property was discovered in an alloy of zinc and aluminium (22%), many alloys and ceramics have been shown to possess this property. For superplasticity to occur, the metal grain must be small and rounded and the alloy must have a slow rate of deformation.

superposition principle A general principle of linear systems that when applied to wave phenomena asserts that the combined effect of any number of interacting waves at a point may be obtained by the algebraic summation of the amplitudes of all the waves at the point. For example, the superposition of two oscillations x_1 and x_2, both of frequency v, produces a disturbance of the same frequency. The amplitude and *phase angle of the resulting disturbance are functions of the component amplitudes and phases. Thus, if

$x_1 = a_1 \sin(2\pi v + \delta_1)$

and

$x_2 = a_2 \sin(2\pi v + \delta_2),$

the resultant disturbance, x, will be given by:

$x = A \sin(2\pi v + \Delta)$,

where amplitude A and phase angle Δ are both functions of a_1, a_2, δ_1, and δ_2.

supersaturated solution *See* SATURATED.

supersaturation 1. The state of the atmosphere in which the relative humidity is over 100%. This occurs in pure air where no condensation nuclei are available. Supersaturation is usually prevented in the atmosphere by the abundance of condensation nuclei (e.g. dust, sea salt, and smoke particles). **2.** The state of any vapour whose pressure exceeds that at which condensation usually occurs (at the prevailing temperature).

supersonic *See* MACH NUMBER.

superstring theory A unified theory of the *fundamental interactions involving supersymmetry, in which the basic objects are one-dimensional objects (**superstrings**). Superstrings are thought to have a length scale of about 10^{-35} m and, since very short distances are associated with very high energies, they should have energy scales of about 10^{19} GeV, which is far beyond the energy of any accelerator that can be envisaged.

Strings associated with bosons are only consistent as quantum theories in a 26-dimensional *space–time; those associated with fermions are only consistent as quantum theories in 10-dimensional space–time. It is thought that four macroscopic dimensions arise according to *Kaluza–Klein theory, with the remaining dimensions being 'curled up' to become very small.

One of the most attractive features of the theory of superstrings is that it leads to spin 2 particles, which are identified as *gravitons. Thus, a superstring theory automatically contains a quantum theory of the gravitational interaction. It is thought that superstrings are free of the infinities that cannot be removed by *renormalization, which plague attempts to construct a quantum field theory incorporating gravity. There is some evidence that superstring theory is free of infinities but not a complete proof yet.

Although there is no direct evidence for superstrings, some features of superstrings are compatible with the experimental facts of *elementary particles, such as the possibility of particles that do not respect *parity, as found in the weak interactions. Another attractive aspect of the theory is that it reveals the existence of particles that do not conserve parity, as in the weak interaction.

All the viable superstring theories and also supergravity theory are related by duality, with the unified theory underpinning this being an 11-dimensional theory called **M-theory**. Both the entropy of black holes and Hawking radiation have been explained in terms of superstring theory.

supersymmetry A *symmetry that can be applied to elementary particles so as to include both bosons and fermions. In the simplest supersymmetry theories, every boson has a corresponding fermion partner and every fermion has a corresponding boson partner. The boson partners of existing fermions have names formed by adding 's' to the beginning of the name of the fermion, e.g. **selectron**, **squark**, and **slepton**. The fermion partners of existing bosons have names formed by replacing '-on' at the end of the boson's name by '-ino' or by adding '-ino', e.g. **gluino**, **photino**, **wino**, and **zino**.

The infinities that cause problems in relativistic quantum field theories (*see* RENORMALIZATION) are less severe in supersymmetry theories because infinities of bosons and fermions can cancel one another out.

If supersymmetry is relevant to observed elementary particles then it must be a *broken symmetry, although there is no convincing evidence at present to show at what energy it would be broken. There is, in fact, no experimental evidence for the theory, although it is thought that it may form part of a unified theory of interactions. This would not necessarily be a *unified-field theory; the idea of strings with supersymmetry may be the best approach to unifying the four fundamental interactions (*see* SUPERSTRING THEORY).

supertransuranics (superheavy elements) A set of relatively stable elements with atomic numbers around 114 and mass numbers of about 298 predicted to exist beyond the *transuranic elements. These elements have not been observed, it is thought, because of the difficulties in synthesizing them rather than their instability. The existence of supertransuranic elements is postulated because of the stability associated with the closing of a proton or neutron shell (*see* SHELL MODEL) that occurs for *magic numbers. The number 114 is predicted to be a proton magic number and the number 184 is predicted to be a neutron magic number.

supplementary units *See* SI UNITS.

suppressor grid A wire grid in a pentode *thermionic valve placed between the *screen grid and the anode to prevent electrons produced by *secondary emission from the anode from reaching the screen grid.

surd A quantity that cannot be expressed as a *rational number. It consists of the root of an arithmetic member (e.g. $\sqrt{3}$), which cannot be exactly determined, or the sum or difference of such roots.

surface integral An integral taken over a surface that can involve vectors or scalars. If $\mathbf{V}(x,y,z)$ is a vector function defined in a region that contains the surface S and the *unit vector $\hat{\boldsymbol{n}}$ is defined as the (outward) normal to that surface, the surface integral is defined by $\int_s \mathbf{V} \cdot \hat{\boldsymbol{n}} \, dS = \int \mathbf{V} \cdot d\mathbf{S}$. This integral is the (outward) *flux of $\mathbf{V}$ through S. Surface integrals for flux occur in electricity and magnetism, an example being *Gauss' law. It is also possible to define the surface integral $\int_s \mathbf{V} \times d\mathbf{S}$ and for the scalar function $\phi(x,y,z)$ the surface integral $\int_s \phi \, d\mathbf{S}$.

surface tension Symbol γ. The property of a liquid that makes it behave as if its surface is enclosed in an elastic skin. The property results from intermolecular forces: a molecule in the interior of a liquid experiences interactions with other molecules equally from all sides, whereas a molecule at the surface is only affected by molecules below it in the liquid. The surface tension is defined as the force acting over the surface per unit length of surface perpendicular to the force. It is measured in newtons per metre. It can equally be defined as the energy required to increase the surface area isothermally by one square metre, i.e. it can be measured in joules per metre squared (which is equivalent to $N\,m^{-1}$).

The property of surface tension is responsible for the formation of liquid drops, soap bubbles, and meniscuses, as well as the rise of liquids in a capillary tube (**capillarity**), the absorption of liquids by porous substances, and the ability of liquids to wet a surface.

surfactant (surface active agent) A substance, such as a detergent, added to a liquid to increase its spreading or wetting properties by reducing its *surface tension.

susceptance Symbol B. The reciprocal of the *reactance of a circuit and thus the imaginary part of its *admittance. It is measured in siemens.

susceptibility 1. (magnetic susceptibility) Symbol χ_m. The dimensionless quantity describing the contribution made by a substance when subjected to a magnetic field to the total magnetic flux density present. It is equal to $\mu_r - 1$, where μ_r is the relative *permeability of the material. Diamagnetic materials have a low negative susceptibility, paramagnetic materials have a low positive susceptibility, and ferromagnetic materials have a high positive value. **2. (electric susceptibility)** Symbol χ_e. The dimensionless quantity referring to a *dielectric equal to $P/\varepsilon_0 E$, where P is the electric polarization, E is the electric intensity producing it, and ε_0 is the electric constant. The electric susceptibility is also equal to $\varepsilon_r - 1$, where ε_r is the relative *permittivity of the dielectric.

suspension A mixture in which small solid or liquid particles are suspended in a liquid or gas.

symmetry The set of invariances of a system. Upon application of a symmetry operation on a system, the system is unchanged. Symmetry is studied mathematically using *group theory. Some of the symmetries are directly physical. Examples include reflections and rotation for molecules and translation in crystal lattices. Symmetries can be **discrete** (i.e. have a finite number), such as the set of rotations for an octahedral molecule, or **continuous** (i.e. do not have a finite number), such as the set of rotations for atoms or nuclei. More general and abstract symmetries can occur, as in CPT invariance and in the symmetries associated with *gauge theories. *See also* BROKEN SYMMETRY; SUPERSYMMETRY.

synchrocyclotron A form of *cyclotron in which the frequency of the accelerating potential is synchronized with the increasing period of revolution of a group of the accelerated particles, resulting from their relativistic increase in mass as they reach *relativistic speeds. The accelerator is used with protons, deuterons, and alpha-particles.

synchronous Taking place at the same time, at the same rate, or with the same period. Many examples of synchronous events occur in physical science and engineering. A **synchronous computer** is a computer in which all its operations are controlled by one master clock.

synchronous motor *See* ELECTRIC MOTOR.

synchronous orbit (geosynchronous orbit) An orbit of the earth made by an artificial *satellite with a period exactly equal to the earth's period of rotation on its axis, i.e. 23 hours 56 minutes 4.1 seconds. If the orbit is inclined to the equatorial plane the satellite will appear from the earth to trace out a figure-of-eight track once every 24 hours. If the orbit lies in the equatorial plane and is circular, the satellite will appear to be stationary. This is called a **stationary orbit** (or **geostationary orbit** or **parking orbit**) and it occurs at an altitude of 35 900 km. Most communication satellites are in stationary orbits, with three or more spaced round the orbit to give worldwide coverage.

synchronous rotation The rotation of a natural satellite in which the period of rotation is equal to its orbital period. The moon, for example, is in synchronous rotation about the earth and therefore always presents the same face to the earth.

synchrotron A particle accelerator used to impart energy to electrons and protons in order to carry out experiments in particle physics and in some cases to make use of the *synchrotron radiation produced. The particles are accelerated in closed orbits (often circular) by radio-frequency fields. Magnets are spaced round the orbit to bend the trajectory of the particles and separate focusing magnets are used to keep the particles in a narrow beam. The radio-frequency accelerating cavities are interspersed between the magnets. The motion of the particles is automatically synchronized with the rising magnetic field, as the field strength has to increase as the particle energy increases; the frequency of the accelerating field also has to increase synchronously.

synchrotron radiation (magnetobremsstrahlung) Electromagnetic radiation that is emitted by charged particles moving at relativistic speeds in circular orbits in a magnetic field. The rate of emission is inversely proportional to the product of the radius of curvature of the orbit and the fourth power of the mass of the particles. For this reason, synchrotron radiation is not a problem in the design of proton *synchrotrons but it is significant in electron synchrotrons. The greater the circumference of a synchrotron, the less important is the loss of energy by synchrotron radiation. In *storage rings, synchrotron radiation is the principal cause of energy loss.

However, since the 1950s it has been realized that synchrotron radiation is itself a very useful tool for studying the structure of matter and many accelerator laboratories have research projects making use of the radiation on a secondary basis to the main high-energy research. The radiation used for these purposes is primarily in the ultraviolet and X-ray frequencies.

Much of the microwave radiation from celestial radio sources outside the Galaxy is believed to originate from electrons moving in curved paths in celestial magnetic fields; it is also called synchrotron radiation as it is analogous to the radiation occurring in a synchrotron.

synodic month (lunation) The interval between new *moons. It is equal to 29 days, 12 hours, and 44 minutes.

synodic period The mean time taken by any object in the solar system to move between successive returns to the same position, relative to the sun as seen from the earth. Since a planet is best observed at opposition the synodic period of a planet, S, is easier to measure than its *sidereal period, P. For inferior planets $1/S = 1/P - 1/E$; for superior planets $1/S = 1/E - 1/P$, where E is the sidereal period of the earth.

Système International d'Unités *See* SI UNITS.

systems software *See* APPLICATIONS SOFTWARE.

syzygy The situation that occurs when the sun, the moon (or a planet), and the earth are in a straight line. This occurs when the moon (or planet) is at *conjunction or *opposition.

S

tachometer An instrument for measuring angular speed, especially the number of revolutions made by a rotating shaft in unit time. Various types of instrument are used, including mechanical, electrical, and electronic devices. The widely used electrical-generator tachometer consists of a small generator in which the output voltage is a measure of the rate of rotation of the shaft that drives it.

tachyon A hypothetical particle that has a speed in excess of the *speed of light. According to electromagnetic theory, a charged particle travelling through a medium at a speed in excess of the speed of light in that medium emits *Cerenkov radiation. A charged tachyon would emit Cerenkov radiation even in a vacuum. No such particle has yet been detected. According to the special theory of *relativity, it is impossible to accelerate a particle up to the speed of light because its energy E, given by $E = mc^2/\sqrt{(1 - v^2/c^2)}$, would have to become infinite. The theory, however, does not forbid the existence of particles with $v > c$ (where c is the speed of light). In such cases the expression in the brackets becomes negative and the energy would be imaginary.

tandem generator A type of particle generator, essentially consisting of a *Van de Graaff generator that maintains one electrode at a high positive potential; this electrode is placed between two earthed electrodes. Negative ions are accelerated from earth potential to the positively charged electrode, where surplus electrons are stripped from the ions to produce positive ions, which are accelerated again from the positive electrode back to earth. Thus the ions are accelerated twice over by a single high potential. This tandem arrangement enables energies up to 30 MeV to be achieved.

tangent 1. A line that touches a curve or a plane that touches a surface. **2.** *See* TRIGONOMETRIC FUNCTIONS.

tangent galvanometer A type of galvanometer, now rarely used, in which a small magnetic needle is pivoted horizontally at the centre of a vertical coil that is adjusted to be parallel to the horizontal component of the earth's magnetic field. When a current I is passed through the coil, the needle is deflected so that it makes an angle θ with its equilibrium position parallel to the earth's field. The value of I is given by $I = (2Hr\tan\theta)/n$, where H is the strength of the earth's horizontal component of magnetizing force, r is the radius of the coil, and n is the number of turns in the coil. Although not now used for measuring current, the instrument provides a means of measuring the earth's magnetizing force.

tau particle *See* ELEMENTARY PARTICLES; LEPTON.

Taylor series The infinite power series of derivatives into which a function f(x) can be expanded, for a fixed value of the variable $x = a$:

$$f(x) = f(a) + f'(a)(x - a) + f''(a)(x - a)^2/2! + \ldots$$

When $a = 0$, the series formed is known as **Maclaurin's series**:

$$f(x) = f(0) + f'(0)x + f''(0)x^2/2! + \ldots$$

The series was discovered by Brook Taylor (1685–1731) and the special case was named after Colin Maclaurin (1698–1746).

tebi- *See* BINARY PREFIXES.

technicolour theory *See* HIGGS FIELD.

tektite A small black, greenish, or yellowish glassy object found in groups on the earth's surface and consisting of a silicaceous material unrelated to the geological formations in which it is found. Tektites are believed to have formed on earth as a result of the impact of meteorites.

telecommunications The study and application of means of transmitting information, either by wires or by electromagnetic radiation.

telephoto lens A camera lens used to produce an image of a distant object without the need for a lens of very long focal length (which would make the camera similar to a telescope). A telephoto lens consists of a combination of a converging lens and a weak diverging lens (nearest to the film). This combination of lenses has the same effect as a single lens with a focal length f_c, a long way from the film. The ratio f_c/f_b, where f_b is the distance between the film and the nearer lens, is called the **telephoto magnification**.

In shooting films and videos, the camera is often required to move from a distance shot to a close-up without the image losing focus. This can be achieved with a **zoom lens**, consisting of a combination of lenses that allow for a change of focal length, with the telephoto magnification varying in value from nearly one to a much higher value.

telescope An instrument that collects radiation from a distant object in order to produce an image of it or enable the radiation to be analysed. See Feature.

television The transmission and reception of moving images by means of radio waves or cable. The scene to be transmitted is focused onto a photoelectric screen in the television *camera. This screen is scanned by an electron beam. The camera produces an electric current, the instantaneous magnitude of which is proportional to the brightness of the portion of the screen being scanned. In Europe the screen is scanned by 625 lines and 25 such frames are produced every second. In the USA 525 lines and 30 frames per second are used. The picture signal so produced is used to modulate a VHF or UHF carrier wave and is transmitted with an

OPTICAL ASTRONOMICAL TELESCOPES

Optical astronomical telescopes fall into two main classes: **refracting telescopes** (or **refractors**), which use lenses to form the primary image, and **reflecting telescopes** (or **reflectors**), which use mirrors.

Refracting telescopes
The refracting telescopes use a converging lens to collect the light and the resulting image is magnified by the eyepiece. This type of instrument was first constructed in 1608 by Hans Lippershey (1587–1619) in Holland and developed in the following year as an astronomical instrument by Galileo, who used a diverging lens as eyepiece. The **Galilean telescope** was later improved by Johannes Kepler (1571–1630), who substituted a converging eyepiece lens. This form is still in use for small astronomical telescopes (the **Keplerian telescope**).

Reflecting telescopes
The first reflecting telescope was produced by Newton in 1668. This used a concave mirror to collect and focus the light and a small secondary mirror at an angle of 45° to the main beam to reflect the light into the magnifying eyepiece. This design is known as the **Newtonian telescope**. The **Gregorian telescope**, designed by James Gregory (1638–75), and the **Cassegrainian telescope**, designed by N. Cassegrain (*fl.* 1670s), use different secondary optical systems. The **coudé telescope** (French: angled) is sometimes used with larger instruments as it increases their focal lengths.

Catadioptic telescopes
Catadioptic telescopes use both lenses and mirrors. The most widely used astronomical instruments in this class are the **Maksutov telescope** and the **Schmidt camera**.

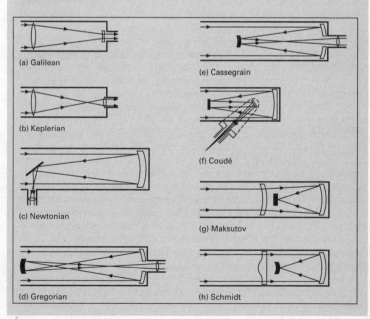

(a) Galilean
(b) Keplerian
(c) Newtonian
(d) Gregorian
(e) Cassegrain
(f) Coudé
(g) Maksutov
(h) Schmidt

independent sound signal, but with colour information (if any) incorporated into the brief gaps between the picture lines. The signals received by the receiving aerial are demodulated in the receiver; the demodulated picture signal controls the electron beam in a cathode-ray tube, on the screen of which the picture is reconstructed. *See also* COLOUR TELEVISION.

television tube *See* CATHODE RAYS.

temperament The way in which the intervals between notes on keyboard instruments are distributed throughout the scale to ensure that music in all keys sounds in tune. The problem can be illustrated by a piano keyboard. Taking a low C and a high C seven octaves above, the interval should be $2^7 = 128$. However, in passing through the cycle of 12 keys, each using as its fundamental the fifth of its predecessor, the interval between Cs becomes $(3/2)^{12} = 129.75$. The difference between 129.75 and 128 is known as the **comma of Pythagoras**. The **equal-temperament scale**, which has been in use since the time of J. S. Bach, distributes the comma of Pythagoras equally between the 12 intervals of the scale over seven octaves. Thus each fifth becomes $(128)^{1/12} = 1.4983$. All forms of temperament involve a measure of compromise; this system is now regarded as the best.

temperature The property of a body or region of space that determines whether or not there will be a net flow of heat into it or out of it from a neighbouring body or region and in which direction (if any) the heat will flow. If there is no heat flow the bodies or regions are said to be in **thermodynamic equilibrium** and at the same temperature. If there is a flow of heat, the direction of the flow is from the body or region of higher temperature. Broadly, there are two methods of quantifying this property. The empirical method is to take two or more reproducible temperature-dependent events and assign **fixed points** on a scale of values to these events. For example, the Celsius temperature scale uses the freezing point and boiling point of water as the two fixed points, assigns the values 0 and 100 to them, respectively, and divides the scale between them into 100 degrees. This method is serviceable for many practical purposes (*see* TEMPERATURE SCALES), but lacking a theoretical basis it is awkward to use in many scientific contexts. In the 19th century, Lord *Kelvin proposed a thermodynamic method to specify temperature, based on the measurement of the quantity of heat flowing between bodies at different temperatures. This concept relies on an absolute scale of temperature with an *absolute zero of temperature, at which no body can give up heat. He also used Sadi Carnot's concept of an ideal frictionless perfectly efficient heat engine (*see* CARNOT CYCLE). This Carnot engine takes in a quantity of heat q_1 at a temperature T_1, and exhausts heat q_2 at T_2, so that $T_1/T_2 = q_1/q_2$. If T_2 has a value fixed by definition, a Carnot engine can be run between this fixed temperature and any unknown temperature T_1, enabling T_1 to be calculated by measuring the values of q_1 and q_2. This concept remains the basis for defining **thermodynamic temperature**, quite independently of the nature of the working substance. The unit in which thermodynamic temperature is expressed is the *kelvin. In practice thermodynamic temperatures cannot be measured directly; they are usually inferred from measurements with a gas thermometer containing a nearly ideal gas. This is possible because another aspect of thermodynamic temperature is its relationship to the *internal energy of a given amount of substance. This can be shown most simply in the case of an ideal monatomic gas, in which the internal energy per mole (U) is equal to the total kinetic energy of translation of the atoms in one mole of the gas (a monatomic gas has no rotational or vibrational energy). According to *kinetic theory, the thermodynamic temperature of such a gas is given by $T = 2U/3R$, where R is the universal *gas constant.

temperature coefficient   A coefficient that determines the rate of change of some physical property with change in temperature. For example, the dependence of the resistance (R) of a material on the Celsius temperature t, is given by $R = R_0 + \alpha t + \beta t^2$, where R_0 is the resistance at 0°C and α and β are constants. If β is negligible, then α is the **temperature coefficient of resistance**.

temperature scales   A number of empirical scales of *temperature have been in use: the *Celsius scale is widely used for many purposes and in certain countries the *Fahrenheit scale is still used. These scales both rely on the use of **fixed points**, such as the freezing point and the boiling point of water, and the division of the **fundamental interval** between these two points into units of temperature (100 degrees in the case of the Celsius scale and 180 degrees in the Fahrenheit scale).

   However, for scientific purposes the scale in use is the **International Practical Temperature Scale (IPTS)**, which is designed to conform as closely as possible to thermodynamic temperature and is expressed in the unit of thermodynamic temperature, the *kelvin. The 1968 version of the table (known as IPTS-68) had 11 fixed points defined by both Celsius and thermodynamic temperatures. The most recent version (IPTS-90), introduced in 1990, has 16 fixed points with temperatures expressed in kelvins:

Triple point of hydrogen: 13.8033
Boiling point of hydrogen (33 321.3 Pa): 17.035
Boiling point of hydrogen (101 292 Pa): 20.27
Triple point of neon: 24.5561
Triple point of oxygen: 54.3584
Triple point of argon: 83.8058
Triple point of mercury: 234.3156
Triple point of water: 273.16 (0.01°C)
Melting point of gallium: 302.9146
Freezing point of indium: 429.7485
Freezing point of tin: 505.078
Freezing point of zinc: 692.677

Freezing point of aluminium: 933.473
Freezing point of silver: 1234.93
Freezing point of gold: 1337.33
Freezing point of copper: 1357.77

Methods for measuring intermediate temperatures between these fixed points are specified; for example, at low temperatures (0–5 K) they are measured by means of vapour-pressure determinations of ^{3}He and ^{4}He; at high temperatures (above 1234.93 K) a radiation pyrometer is used.

tempering The process of increasing the toughness of an alloy, such as steel, by heating it to a predetermined temperature, maintaining it at this temperature for a predetermined time, and cooling it to room temperature at a predetermined rate. In steel, the purpose of the process is to heat the alloy to a temperature that will enable the excess carbide to precipitate out of the supersaturated solid solution of *martensite and then to cool the saturated solution fast enough to prevent further precipitation or grain growth. For this reason steel is quenched rapidly by dipping into cold water.

temporary magnetism Magnetism in a body that is present when the body is in a magnetic field but that largely disappears when it is removed from the field.

tensile strength A measure of the resistance that a material offers to tensile *stress. It is defined as the stress, expressed as the force per unit cross-sectional area, required to break it.

tensimeter A form of differential manometer with two sealed bulbs attached to the limbs. It is used to measure the difference in vapour pressure between liquids sealed into the bulbs. If one liquid has a known vapour pressure (often water is used) that of the other can be determined.

tensiometer Any apparatus for measuring *surface tension.

tensor A quantity that has different values in different directions. A *vector is a special simple case of a tensor. Tensors are particularly useful in general relativity theory.

tera- Symbol T. A prefix used in the metric system to denote one million million times. For example, 10^{12} volts = 1 teravolt (TV).

terminal 1. The point at which electrical connection is made to a device or system. **2.** A device at which data is put into a *computer or taken from it.

terminal speed The constant speed finally attained by a body moving through a fluid under gravity when there is a zero resultant force acting on it. *See* STOKES' LAW.

terminator The boundary, on the surface of the moon or a planet, between the sunlit area and the dark area.

terrestrial magnetism *See* GEOMAGNETISM.

tertiary colour A colour obtained by mixing two *secondary colours.

tesla Symbol T. The SI unit of magnetic flux density equal to one weber of magnetic flux per square metre, i.e. 1 T = 1 Wb m^{-2}. It is named after Nikola Tesla (1870–1943), Croatian-born US electrical engineer.

Tesla coil A device for producing a high-frequency high-voltage current. It consists of a *transformer with a high turns ratio, the primary circuit of which includes a spark gap and a fixed capacitor; the secondary circuit is tuned by means of a variable capacitor to resonate with the primary. It was devised by Nikola Tesla. Tesla coils are commonly used to excite luminous discharges in glass vacuum apparatus, in order to detect leaks.

tetragonal *See* CRYSTAL SYSTEM.

tetrahedron A polyhedron with four triangular faces. In a **regular tetrahedron** all four triangles are congruent equilateral triangles. It constitutes a regular triangular *pyramid.

tetrode A *thermionic valve with a **screen grid** placed between the anode and the control grid of a *triode to reduce the capacitance between these two electrodes and so improve the valve's performance as an amplifier or oscillator at high frequencies. The screen grid is maintained at a fixed potential.

theodolite An optical surveying instrument for measuring horizontal and vertical angles. It consists of a sighting telescope, with crosshairs in the eyepiece for focusing on the target, which can be rotated in both the horizontal and vertical planes. It is mounted on a tripod and a spirit level is used to indicate when the instrument is horizontal. The angles are read off graduated circles seen through a second eyepiece in the instrument.

theoretical physics The study of physics by formulating and analysing theories that describe natural processes. Theoretical physics is complementary to the study of physics by experiment, which is called **experimental physics**. A large part of theoretical physics consists of analysing the results of experiments to see whether or not they obey particular theories. The branch of theoretical physics concerned with the mathematical aspects of theories in physics is called **mathematical physics**.

theory *See* LAWS, THEORIES, AND HYPOTHESES.

theory of everything A theory that provides a unified description of all known types of elementary particles, all known forces in the universe, and the origin and evolution of the universe. Some believe that *superstring theory is potentially a theory of everything. Others believe that it is impossible to formulate a theory of everything and that any such theory could only claim to be a theory for all forces, particles, and observations concerning the evolution of the universe known at the time. It is thought that a *quantum field theory or a *unified field theory cannot be a theory of everything. A theory of everything should explain the number of dimensions in the universe and why the number of observable dimensions is four.

therm A practical unit of energy defined as 10^5 British thermal units. 1 therm is equal to 1.055×10^8 joules.

thermal capacity *See* HEAT CAPACITY.

thermal conductivity *See* CONDUCTIVITY.

thermal diffusion The diffusion that takes place in a fluid as a result of a temperature gradient. If a column of gas is maintained so that the lower end is cooler than the upper end, the heavier molecules in the gas will tend to remain at the lower-temperature end and the lighter molecules will diffuse to the higher-temperature end. This property has been used in the separation of gaseous isotopes (*see* CLUSIUS COLUMN).

thermal equilibrium *See* EQUILIBRIUM.

thermal expansion *See* EXPANSIVITY.

thermal imaging The production of images by detecting, measuring, and recording the thermal (infrared) radiation emitted by objects. Applications of thermal imaging include mapping of the surface of the earth from the air, mapping the weather, and *thermography in medicine.

thermalization The reduction of the kinetic energy of neutrons in a thermal *nuclear reactor by means of a *moderator; the process of producing thermal neutrons.

thermal neutrons *See* MODERATOR; NUCLEAR REACTORS; THERMALIZATION.

thermal reactor *See* NUCLEAR REACTOR.

thermionic emission The emission of electrons, usually into a vacuum, from a heated conductor. The emitted current density, J, is given by the **Richardson** (or **Richardson–Dushman**) **equation**, i.e. $J = AT^2\exp(-W/kT)$, where T is the thermodynamic temperature of the emitter, W is its *work function, k is the Boltzmann constant, and A is a constant. Thermionic emission is the basis of the *thermionic valve and the *electron gun in cathode-ray tubes. The equation was derived by the British physicist Sir Owen Richardson (1879–1959) from classical statistical mechanics and later modified by the Russian-born US physicist Saul Dushman (1883–1954) using *quantum mechanics.

thermionics The branch of electronics concerned with the study and design of devices based on the emission of electrons from metal or metal-oxide surfaces as a result of high temperatures. The primary concern of thermionics is the design of *thermionic valves and the electron guns of cathode-ray tubes and other devices.

thermionic valve An electronic valve based on *thermionic emission. In such valves the cathode is either directly heated by passing a current through it or indirectly heated by placing it close to a heated filament. Directly heated cathodes are usually made of tungsten wire, whereas indirectly heated cathodes are usually coated with barium and strontium oxides. Most electronic valves are thermionic vacuum devices, although a few have cold cathodes and some are gas-filled (*see* THYRATRON). *See* DIODE; TRIODE; TETRODE; PENTODE.

thermistor An electronic device the resistance of which decreases as its temperature increases. It consists of a bead, rod, or disc of various oxides of manganese, nickel, cobalt, copper, iron, or other metals. Thermistors are used as thermometers, often forming one element in a resistance bridge. They are used for this purpose in such applications as bearings, cylinder heads, and transformer cores. They are also used to compensate for the increased resistance of ordinary resistors when hot, and in vacuum gauges, time-delay switches, and voltage regulators.

thermite A stoichiometric powdered mixture of iron(III) oxide and aluminium for the reaction:

$$2Al + Fe_2O_3 \rightarrow Al_2O_3 + 2Fe$$

The reaction is highly exothermic and the increase in temperature is sufficient to melt the iron produced. It has been used for localized welding of steel objects (e.g. railway lines) in the **Thermit process**. Thermite is also used in incendiary bombs.

thermochemistry The branch of physical chemistry concerned with heats of chemical reaction, heats of formation of chemical compounds, etc.

thermocouple A device consisting of two dissimilar metal wires or semiconducting rods welded together at their ends. A thermoelectric e.m.f. is generated in the device when the ends are maintained at different temperatures, the magnitude of the e.m.f. being related to the temperature difference. This enables a thermocouple to be used as a thermometer over a limited temperature range. One of the two junctions, called the **hot** or **measuring junction**, is exposed to the temperature to be measured. The other, the **cold** or **reference junction**, is maintained at a known reference temperature. The e.m.f. generated is measured by a suitable millivoltmeter or potentiometer incorporated into the circuit. *See* SEEBECK EFFECT; THERMOPILE.

thermodynamics The study of the laws that govern the conversion of energy from one form to another, the direction in which heat will flow, and the availability of energy to do work. It is based on the concept that in an isolated system anywhere in the universe there is a measurable quantity of energy called the *internal energy (U) of the system. This is the total kinetic and potential energy of the atoms and molecules of the system of all kinds that can be transferred directly as heat; it therefore excludes chemical and nuclear energy. The value of U can only be changed if the system ceases to be isolated. In these circumstances U can change by the transfer of mass to or from the system, the transfer of heat (Q) to or from the system, or by the work (W) being done on or by the system. For an adiabatic ($Q = 0$) system of constant mass, $\Delta U = W$. By convention, W is taken to be positive if work is done on the system and negative if work is done by the system. For nonadiabatic systems of constant mass, $\Delta U = Q + W$. This statement, which is equivalent to the law of conservation of energy, is known as the **first law of thermodynamics**.

All natural processes conform to this law, but not all processes conforming to it can occur in nature. Most natural processes are irreversible, i.e. they will only proceed in one direction (*see* REVERSIBLE PROCESS). The direction that a natural process can take is the subject of the **second law of thermodynamics**, which can be stated in a variety of ways. Rudolf *Clausius stated the law in two ways: "heat cannot be transferred from one body to a second body at a higher temperature without producing some other effect" and "the entropy of a closed system increases with time". These statements introduce the thermodynamic concepts of *temperature (T) and *entropy (S), both of which are parameters determining the direction in which an irreversible process can go. The temperature of a body or system determines whether heat will flow into it or out of it; its entropy is a measure of the unavailability of its energy to do work. Thus T and S determine the relationship between Q and W in the statement of the first law. This is usually presented by stating the second law in the form $\Delta U = T\Delta S - W$.

The second law is concerned with changes in entropy (ΔS). The **third law of thermodynamics** provides an absolute scale of values for entropy by stating that for changes involving only perfect crystalline solids at *absolute zero, the change of the total entropy is zero. This law enables absolute values to be stated for entropies.

One other law is used in thermodynamics. Because it is fundamental to, and assumed by, the other laws of thermodynamics it is usually known as the **zeroth law of thermodynamics**. This states that if two bodies are each in thermal equilibrium with a third body, then all three bodies are in thermal equilibrium with each other. *See also* ENTHALPY; FREE ENERGY.

thermodynamic temperature *See* TEMPERATURE.

thermoelectricity An electric current generated by temperature difference. *See* SEEBECK EFFECT. The converse effects, the *Peltier effect and the *Thomson effect, are also sometimes known as thermoelectric effects.

thermograph 1. A recording thermometer used in meteorology to obtain a continuous record of temperature changes over a period on a graph. **2.** A record so obtained. **3.** A record obtained by the technique of *thermography.

thermography A medical technique that makes use of the infrared radiation from the human skin to detect an area of elevated skin temperature that could be associated with an underlying cancer. The heat radiated from the body varies according to the local blood flow, thus an area of poor circulation produces less radiation. A tumour, on the other hand, has an abnormally increased blood supply and is revealed on the **thermogram** (or **thermograph**) as a 'hot spot'. The technique is used particularly in mammography, the examination of the infrared radiation emitted by human breasts in order to detect breast cancer.

thermoluminescence *Luminescence produced in a solid when its temperature is raised. It arises when free electrons and *holes, trapped in a solid as a result of exposure to ionizing radiation, unite and emit photons of light. The process is made use of in **thermoluminescent dating**, which assumes that the number of electrons and holes trapped in a sample of pottery is related to the length of time that has elapsed since the pottery was fired. By comparing the luminescence produced by heating a piece of pottery of unknown age with the luminescence produced by heating similar materials of known age, a fairly accurate estimate of the age of an object can be made.

thermoluminescent dating *See* THERMOLUMINESCENCE.

thermometer An instrument used for measuring the *temperature of a substance. A number of techniques and forms are used in thermometers depending on such factors as the degree of accuracy required and the range of temperatures to be measured, but they all measure temperature by making use of some property of a substance that varies with temperature. For example, **liquid-in-glass thermometers** depend on the expansion of a liquid, usually mercury or alcohol coloured with dye. These consist of a liquid-filled glass bulb attached to a partially filled capillary tube. In the **bimetallic thermometer** the unequal expansion of two dissimilar metals that have been bonded together into a narrow strip and coiled is used to move a pointer round a dial. The **gas thermometer**, which is more accurate than the liquid-in-glass thermometer, measures the variation in the pressure of a gas kept at constant volume. The **resistance thermometer** is based on the change in resistance of conductors or semiconductors with temperature change. Platinum, nickel, and copper are the metals most commonly used in resistance thermometers. *See also* PYROMETRY; THERMISTOR; THERMOCOUPLE.

thermonuclear reaction *See* NUCLEAR FUSION; THERMONUCLEAR REACTOR.

thermonuclear reactor (fusion reactor) A reactor in which *nuclear

Thermonuclear reactor: Tokamak principle

fusion takes place with the controlled release of energy. Although thermonuclear reactors do not yet exist, intense research in many parts of the world is being carried out with a view to achieving such a machine. There are two central problems in the creation of a self-sustaining thermonuclear reactor: heating the reacting nuclides to the enormous *ignition temperature (about 40×10^6 K for a deuterium–tritium reaction) and containing the reacting nuclides for long enough for the fusion energy released to exceed the energy required to achieve the ignition temperature (see LAWSON CRITERION). The two methods being explored are **magnetic containment** and **pellet fusion**.

In the closed magnetic-containment device the fusion *plasma is contained in a toroidal-shaped reactor, called a **Tokamak**, in which strong magnetic fields guide the charged plasma particles round the toroid without allowing them to contact the container walls (see illustration). In open-ended magnetic systems the plasma is trapped between magnetic mirrors (strong magnetic fields) at the two ends of a straight containment vessel.

In pellet fusion the fuel is a mixture of deuterium and tritium in pellet form (~ 1 mm). The fuel pellets are struck simultaneously by nanosecond pulses of high intensity radiation from several infrared lasers. The surface of the pellet vaporizes and a shock wave is generated, which increases the pellet's core density by up to 1000 times. The temperature of the core is raised to over 10^8 K. The plasma formed in this way is essentially confined to the small region of heating by its own inertia; for this reason the method is sometimes called **inertial confinement**. The heating occurs over such a short time (typically 1–2 nanoseconds) that the plasma has no time to disperse. Results with this type of equipment have been comparable to those achieved by magnetic confinement.

thermonuclear weapon *See* NUCLEAR WEAPONS.

thermopile A device used to detect and measure the intensity of radiant energy. It consists of a number of *thermocouples connected together in series to achieve greater sensitivity. The hot junctions of the thermocouples are blackened and exposed to the radiation to be detected or measured, while the cold junctions are shielded from the radiation. The thermoelectric e.m.f. generated enables the hot junction excess temperature to be calculated and the radiant intensity to be deduced. They are used in various applications, from a safety device that ceases to produce an electric current if a pilot light blows out to an instrument to measure the heat radiation received from the sun.

thermostat A device that controls the heating or cooling of a substance in order to maintain it at a constant temperature. It consists of a temperature-sensing instrument connected to a switching device. When the temperature reaches a predetermined level the sensor switches the heating or cooling source on or off according to a predetermined program. The sensing thermometer is often a *bimetallic strip that triggers a simple electrical switch. Thermostats are used for space-heating controls, in water heaters and refrigerators, and to maintain the environment of a scientific experiment at a constant temperature.

theta vacuum (θ vacuum) The vacuum state for non-Albelian gauge fields (in the absence of *fermion fields and *Higgs fields). In the theta vacuum there is an infinite number of *degenerate states with tunnelling (*see* TUNNEL EFFECT) between these states. This means that the theta vacuum is analogous to a Bloch function (*see* BLOCH'S THEOREM) in a crystal. This can be derived either as a general result or using instanton techniques. When a massless fermion field is present the tunnelling between states is completely suppressed. When there are fermion fields with a small mass tunnelling is much smaller than for pure gauge fields, but is not completely suppressed. The theta vacuum is the starting point of understanding the vacuum state in strongly interacting gauge theories, such as *quantum chromodynamics.

thixotropy *See* NEWTONIAN FLUID.

Thompson, Benjamin *See* RUMFORD, COUNT.

Thomson, Sir Joseph John (1856–1940) British physicist, who became a professor at Cambridge University in 1884. He is best known for his work on *cathode rays, which led to his discovery of the *electron in 1897. He went on to study the conduction of electricity through gases, and it is for this work that he was awarded the Nobel Prize for physics in 1906. His son, **Sir George Paget Thomson** (1892–1975), discovered *electron diffraction, for which he shared the 1937 Nobel Prize for physics with Clinton J. Davisson (1881–1958), who independently made the same discovery.

Thomson, Sir William *See* KELVIN, LORD.

Thomson effect (**Kelvin effect**) When an electric current flows through a conductor, the ends of which are maintained at different temperatures, heat is evolved at a rate approximately proportional to the product of the current and the temperature gradient. If either the current or the temperature gradient is reversed heat is absorbed rather than being evolved. It is named after Sir William Thomson (*see* KELVIN, LORD).

Thomson scattering The scattering of electromagnetic radiation by free charged particles, especially electrons, when the photon energy is small compared with the energy equivalent to the *rest mass of the charged particles. The energy lost by the radiation is accounted for by classical theory as a result of the radiation emitted by the charged particles when they are accelerated in the transverse electric field of the radiation. It is named after Sir J. J. *Thomson.

thorium Symbol Th. A grey radioactive metallic element belonging to the actinoids; a.n. 90; r.a.m. 232.038; r.d. 11.5–11.9 (17°C); m.p. 1740–1760°C; b.p. 4780–4800°C. It occurs in monazite sand in Brazil, India, and USA. The isotopes of thorium have mass numbers from 223 to 234 inclusive; the most stable isotope, thorium–232, has a half-life of 1.39 × 10^{10} years. It can be used as a nuclear fuel for breeder reactors as thorium–232 captures slow neutrons to breed uranium–233. Thorium dioxide (**thoria**, ThO_2) is used on gas mantles and in special refractories. The element was discovered by J. J. Berzelius (1779–1848) in 1829.

thorium series *See* RADIOACTIVE SERIES.

three-body problem *See* MANY-BODY PROBLEM.

threshold The minimum value of a parameter or variable that will produce a specified effect.

threshold frequency *See* PHOTOELECTRIC EFFECT.

thrust The propelling force generated by an aircraft engine or rocket. It is usually calculated as the product of the rate of mass discharge and the velocity of the exhaust gases relative to the vehicle.

thumb drive *See* USB DRIVE.

thunderstorm A convective storm accompanied by *lightning and

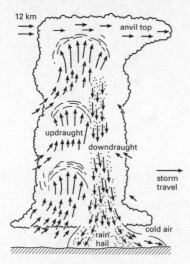

12 km

anvil top

updraught

downdraught

storm travel

cold air

rain hail

Cross section through a thunderstorm cell

thunder and a variety of weather conditions, especially heavy rain or hail, high winds, and sudden temperature changes. Thunderstorms originate when intense heating causes a parcel of moist air to rise, leading to instability and the development of cumulonimbus cloud – a towering cloud with a characteristic anvil-shaped top (see illustration). The exact mechanisms of thunderstorms are not fully understood. They occur most frequently in the tropics but are also common in the mid-latitudes.

thyratron A thermionic valve (usually a triode) that functions as a gas-filled relay. A positive pulse fed to a correctly biased thyratron causes a discharge to start and to continue until the anode voltage has been reduced. It has now been replaced by its solid-state counterpart, the silicon-controlled rectifier.

thyristor A silicon-controlled rectifier whose anode–cathode current is controlled by a signal applied to a third electrode (the gate) in much the same way as in a thyratron valve. It consists usually of a four-layer chip comprising three p–n junctions.

tidal energy *See* TIDES.

tides The regular rise and fall of the water level in the earth's oceans as a result of the gravitational forces between the earth, moon, and sun. The detailed theory of tides is complex. The moon is approximately twice as effective as the sun in causing tides. In illustration (a) the resultant gravitational forces between the moon and various points on the earth (solid lines) are shown as the vector sums of the tide-generating forces (broken lines) and a constant force (dotted lines) that is the same at all

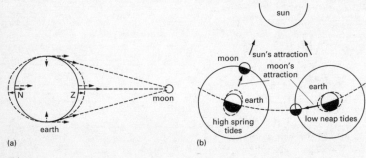

Tides

points on the earth and is equal to the moon's attraction on the earth's centre. The resultant force when the moon is in zenith (Z in the illustration) is greater than that at nadir (N) because Z is closer to the moon than N and the force is inversely proportional to the square of the distance according to *Newton's law of gravitation. Illustration (b) shows how at full and new moon the sun and moon act together to produce the high-range **spring tides**, while at quarter moon the forces are at right angles to each other causing the low-range **neap tides**.

The use of **tidal energy**, estimated at some 4×10^{18} J per annum at known tidal sites, dates back to medieval tidal mills. Modern tidal power stations use specially designed turbines, operated by tidal waters, to drive generators.

Tidal forces due to Jupiter have large effects on many of its moons.

timbre *See* QUALITY OF SOUND.

time A dimension that enables two otherwise identical events that occur at the same point in space to be distinguished (*see* SPACE–TIME). The interval between two such events forms the basis of time measurement. For general purposes, the earth's rotation on its axis provides the units of the clock (*see* DAY) and the earth's orbit round the sun (*see* YEAR) provides the units of the calendar. For scientific purposes, intervals of time are now defined in terms of the frequency of a specified electromagnetic radiation (*see* SECOND). *See also* TIME DILATION; TIME REVERSAL.

In physics, since the publication of the special theory of *relativity in 1905, Einstein has frequently been said to have abandoned the concept of absolute time. In this context absolute time is taken to mean "time that flows equably and independently of the state of motion of the observer". Time dilation effects and the collapse of absolute *simultaneity mean that absolute time in this sense cannot be applied to the measurement of an interval of time.

Although philosophers tend to describe Einstein's work on relativity as the beginning of a 20th century revolution in science, many of these 'revolutionary' concepts were not entirely original. In 1898, for example,

Jules Poincaré (1854–1912), the French mathematician, questioned the concept of absolute simultaneity commenting that "we have no direct intuition about the equality of two time intervals". Poincaré was also aware of the need to consider local time for a given observer. In 1904, he observed that clocks synchronized by light signals sent between observers in uniform relative motion "will not mark the true time", but, rather, "what one might call the local time".

A frequent misconception is that the theory of relativity removes absolute time from mechanics. This is true for the measurement of time as discussed above, but not for time itself. Newton's definition of absolute time is essentially a philosophical concept. Indeed, challenges to this concept in Newton's lifetime were usually made on philosophical, rather than experimental, grounds. Newton never claimed that one could measure absolute time; this absolute quantity had to be distinguished from the "sensible measures" used in "ordinary affairs".

In Einstein's view of the universe, descriptions of a physical phenomenon need to be fully relativistic, requiring *Lorentz transformations between the coordinates of systems in uniform relative motion. Contrary to popular belief, Newtonian mechanics was not based on absolute space and time and was fully relativistic, but in the Galilean sense; that is, *Galilean transformations were required between the coordinates of systems in uniform relative motion.

In considering simultaneity Einstein made use of a thought experiment (*see* RELATIVITY). As a result of this experiment in Einstein's view, the concept of absolute simultaneity has to be abandoned. His universe is causal, and in a causal universe, there is no such thing as simultaneity as there are no simultaneous events. Events have a definite order based on their causal sequence, which cannot be changed. This is what Newton meant by absolute time. Without making a direct statement, Einstein effectively introduced a third postulate in his theory of relativity; that no information can be transmitted faster than the speed of light. For both Newton and Einstein absolute time is really the absolute order of events, determined by causality, and not the measurement of time, which is the subject of ordinary observation.

timebase A voltage applied to the electron beam in a *cathode-ray tube at regular intervals so that the luminous spot on the screen is deflected in a predetermined manner. The timebase is usually designed to make the beam sweep the screen horizontally, the period during which the spot returns to its starting point (the **flyback**) being suppressed in some contexts (e.g. in television) although in a *cathode-ray oscilloscope the timebase can be put to more complicated uses.

time constant Symbol τ or T. A quantity that gives a measure of the rate of discharge of a capacitor through a circuit consisting of the capacitor and a resistor. It is given by CR seconds, where C is the capacitance in farads and R is the resistance in ohms. It is the time taken for the quantity to fall to $1/e$ of its original value.

time dilation (time dilatation) The principle, predicted by Einstein's special theory of *relativity, that intervals of time are not absolute but are relative to the motion of the observers. If two identical clocks are synchronized and placed side by side in an inertial frame of reference they will read the same time for as long as they both remain side by side. However, if one of the clocks has a velocity relative to the other, which remains beside a stationary observer, the travelling clock will show, to that observer, that less time has elapsed than the stationary clock. In general, the travelling clock goes more slowly by a factor $\sqrt{(1 - v^2/c^2)}$, when measured in a frame of reference travelling at a velocity v relative to another frame of reference; c is the speed of light. The principle has been verified in a number of ways; for example, by comparing the lifetimes of fast muons, which increase with the speed of the particles to an extent predicted by this factor.

time-lapse photography A form of photography used to record a slow process, such as plant growth. A series of single exposures of the object is made on film at predetermined regular intervals. The film produced can then be projected at normal ciné speeds and the process appears to be taking place at an extremely high rate.

time reversal Symbol T. The operation of replacing time t by time $-t$. The *symmetry of time reversal is known as **T invariance**. As with CP violation, **T violation** occurs in certain weak interactions, notably kaon decay. *See also* CP INVARIANCE; CPT THEOREM.

time sharing A means by which several jobs – data plus the programs to manipulate the data – share the processing time and other resources of a computer. A brief period is allocated to each job by the computer's operating system, and the computer switches rapidly between jobs. A *multiaccess system relies on time sharing.

time travel The act of moving to a different time by moving through space–time. Time travel is theoretically allowed in general relativity theory. In *superstring theory time travel is forbidden in some of the situations in which it is allowed in general relativity theory. However, there is no known general principle forbidding time travel in superstring theory. *See also* CHRONOLOGY PROTECTION CONJECTURE; WORMHOLE.

T invariance *See* TIME REVERSAL.

Tokamak *See* THERMONUCLEAR REACTOR.

tomography The use of X-rays to photograph a selected plane of a human body with other planes eliminated. The **CT** (**computerized tomography**) **scanner** is a ring-shaped X-ray machine that rotates through 180° around the horizontal patient, making numerous X-ray measurements every few degrees. The vast amount of information acquired is built into a three-dimensional image of the tissues under examination by the scanner's own computer. The patient is exposed to a dose of X-rays only some 20% of that used in a normal diagnostic X-ray.

topological invariant A geometrical quantity that remains unchanged by continuous deformations. Examples are the 'charges' of *magnetic monopoles and analogous quantities for instantons in *gauge theories.

topology The branch of geometry concerned with the properties of geometrical objects that are unchanged by continuous deformations, such as twisting or stretching. Mathematical approaches employing topology are of great importance in modern theories of the *fundamental interactions and in many other branches of physics.

top quark *See* ELEMENTARY PARTICLES.

torque (moment of a force or couple) The product of a force and its perpendicular distance from a point about which it causes rotation or *torsion. The unit of torque is the newton metre, a vector product, unlike the joule, also equal to a newton metre, which is a scalar product. A turbine produces a torque on its central rotating shaft. *See also* COUPLE.

torr A unit of pressure, used in high-vacuum technology, defined as 1 mmHg. 1 torr is equal to 133.322 pascals. The unit is named after Evangelista Torricelli (1609–47).

Torricellian vacuum The vacuum formed when a long tube, closed at one end and filled with mercury, is inverted into a mercury reservoir so that the open end of the tube is below the surface of the mercury. The pressure inside the Torricellian vacuum is the vapour pressure of mercury, about 10^{-3} torr. It is named after Evangelista Torricelli, who was responsible for the design of the mercury barometer.

torsion A twisting deformation produced by a *torque or *couple. A **torsion bar** is a form of spring in which one end of a bar is fixed and a torque is applied to the other end. Torsion bars are used in the suspension systems of motor vehicles.

torsion balance An instrument for measuring very weak forces. It consists of a horizontal rod fixed to the end of a vertical wire or fibre or to the centre of a taut horizontal wire. The forces to be measured are applied to the end or ends of the rod. The turning of the rod may be measured by the displacement of a beam of light reflected from a plane mirror attached to it. The best-known form is that used by Henry *Cavendish and later by Sir Charles Boys (1855–1944) to determine the *gravitational constant; in this form the balance is calibrated by determining the torsional coefficient of the suspension by treating the device as a torsional pendulum.

torus A solid generated by rotating a circle about an external line in its plane, also called an **anchor ring**. It has the shape of the inner tube of a tyre. If r is the radius of the rotating circle and R the distance between the centre of the circle and the axis of rotation, the volume of the torus is $2\pi^2 R r^2$ and the surface area is $4\pi^2 R r$. In Cartesian coordinates, if the z-axis is the axis of rotation, the equation of the torus is

$$[\sqrt{(x^2 + y^2)} - R]^2 + z^2 = r^2.$$

total internal reflection The total reflection of a beam of light at the interface of one medium and another medium of lower refractive index, when the angle of incidence to the second medium exceeds a specific **critical angle**.

If a beam of light passing through a medium A (say glass) strikes the boundary to a medium B of lower refractive index (say air) with a small angle of incidence i, part will be refracted, with an angle of refraction r, and part will be reflected (see illustration a). If i is increased it will reach a critical angle c, at which $r = 90°$ (see illustration b). If i is now increased further, no refraction can occur and all the light energy is reflected by the interface (see illustration c). This total internal reflection occurs when c (given by $n \sin c = 1$) is exceeded (n is the refractive index of A relative to B). The critical angle of optical glass is usually about 40° and total internal reflection is made use of by incorporating prisms in some optical instruments instead of mirrors.

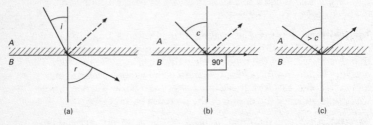

Total internal reflection

totality The period during a total *eclipse of the sun in which the view of the sun's surface from a point on the earth is totally obscured by the moon. The maximum duration of totality is 7.67 minutes, but it is usually less.

total-radiation pyrometer *See* PYROMETRY.

tracing (radioactive tracing) *See* LABELLING.

trajectory *See* PHASE SPACE.

transactinide elements Elements with an atomic number greater than 103, i.e. elements above lawrencium in the *periodic table. So far, elements up to 116 have been detected. Because of the highly radioactive and transient nature of these elements, there has been much dispute about priority of discovery and, consequently, naming of the elements. The International Union of Pure and Applied Chemistry (IUPAC) introduced a set of systematic temporary names based on affixes, as shown in the table below:

Affix	Number	Symbol
nil	0	n
un	1	u
bi	2	b
tri	3	t
quad	4	q
pent	5	p
hex	6	h
sept	7	s
oct	8	o
enn	9	e

Table of affixes

All these element names end in -ium. So, for example, element 109 in this system is called un + nil + enn + ium, i.e. unnilennium, and given the symbol u+n+e, i.e. Une.

One long-standing dispute was about the element 104 (rutherfordium), which has also been called kurchatovium (Ku). There have also been disputes between IUPAC and the American Chemical Union about element names.

In 1994 IUPAC suggested the following list:
mendelevium (Md, 101)
nobelium (No, 102)
lawrencium (Lr, 103)
dubnium (Db, 104)
joliotium (Jl, 105)
rutherfordium (Rf, 106)
bohrium (Bh, 107)
hahnium (Hn, 108)
meitnerium (Mt, 109)

The ACU favoured a different set of names:
mendelevium (Md, 101)
nobelium (No, 102)
lawrencium (Lr, 103)
rutherfordium (Rf, 104)
hahnium (Ha, 105)
seaborgium (Sg, 106)
nielsbohrium (Ns, 107)
hassium (Hs, 108)
meitnerium (Mt, 109)

A compromise list was adopted by IUPAC in 1997 and is generally accepted:
mendelevium (Md, 101)
nobelium (No, 102)

lawrencium (Lr, 103)
rutherfordium (Rf, 104)
dubnium (Db, 105)
seaborgium (Sg, 106)
bohrium (Bh, 107)
hassium (Hs, 108)
meitnerium (Mt, 109)

Element 110 was named as darmstadtium in 2003 and element 111 was named roentgenium in 2004. So far elements 112 (ununbium, Uub), 113 (ununtrium, Uut), 114 (ununquadium, Uuq), 115 (ununpentium, Uup), and 116 (ununhexium, Uuh) are not officially named. All these elements are unstable and have very short half-lives.

transcendental number A number that is not algebraic, such as π or e. A **transcendental function** is also nonalgebraic, such as a^x, $\sin x$, or $\log x$.

transducer A device for converting a nonelectrical signal, such as sound, light, heat, etc., into an electrical signal, or vice versa. Thus microphones and loudspeakers are **electroacoustic transducers**. An **active transducer** is one that can itself introduce a power gain and has its own power source. A **passive transducer** has no power source other than the actuating signal and cannot introduce gain.

transformer A device for transferring electrical energy from one alternating-current circuit to another with a change of voltage, current, phase, or impedance. It consists of a primary winding of N_p turns magnetically linked by a ferromagnetic core or by proximity to the secondary winding of N_s turns. The **turns ratio** (N_s/N_p) is approximately equal to V_s/V_p and to I_p/I_s, where V_p and I_p are the voltage and current fed to the primary winding and V_s and I_s are the voltage and current induced in the secondary winding, assuming that there are no power losses in the core. In practice, however, there are *eddy-current and *hysteresis losses in the core, incomplete magnetic linkage between the coils, and heating losses in the coils themselves. By the use of a *laminated core and careful design, transformers with 98% efficiency can be achieved.

transient A brief disturbance or oscillation in a circuit caused by a sudden rise in the current or e.m.f.

transistor A *semiconductor device capable of amplification in addition to rectification. It is the basic unit in radio, television, and computer circuits, having almost completely replaced the *thermionic valve. The **point-contact transistor**, which is now obsolete, was invented in 1948. It consists of a small germanium crystal with two rectifying point contacts attached to it; a third contact, called the **base**, makes a low-resistance nonrectifying (ohmic) connection with the crystal. Current flowing through the device between the point contacts is modulated by the signal fed to the base. This type of transistor was replaced by the **junction transistor**, which was developed in 1949–50. The **field-effect transistor**

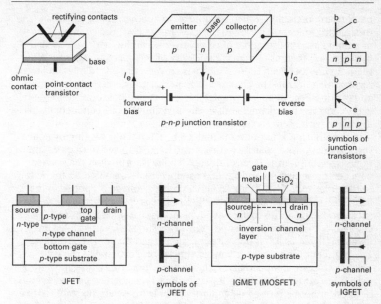

Transistors

(**FET**) was a later invention. **Bipolar transistors**, such as the junction transistor, depend on the flow of both majority and minority carriers, whereas in **unipolar transistors**, such as the FET, the current is carried by majority carriers only.

In the bipolar junction transistor, two p-type semiconductor regions are separated by a thin n-type region, making a p-n-p structure. Alternatively, an n-p-n structure can also be used. In both cases the thin central region is called the base and one outer region of the sandwich is called the **emitter**, the other the **collector**. The emitter–base junction is forward-biased and the collector–base junction is reverse-biased. In the p-n-p transistor, the forward bias causes holes in the emitter region to flow across the junction into the base; as the base is thin, the majority of holes are swept right across it (helped by the reverse bias), into the collector. The minority of holes that do not flow from the base to the collector combine with electrons in the n-type base. This recombination is balanced by a small electron flow in the base circuit. The diagram illustrates the (conventional) current flow using the **common-base** type of connection. If the emitter, base, and collector currents are I_e, I_b, and I_c, respectively, then $I_e = I_b + I_c$ and the current gain is I_c/I_b.

Field-effect transistors are of two kinds, the **junction FET** (**JFET** or **JUGFET**) and the **insulated-gate FET** (**IGFET**; also known as a **MOSFET**, i.e. metal-oxide-semiconductor FET). Both are unipolar devices and in both the current flows through a narrow **channel** between two electrodes (the

gate) from one region, called the **source**, to another, called the **drain**. The modulating signal is applied to the gate. In the JFET, the channel consists of a semiconductor material of relatively low conductivity sandwiched between two regions of high conductivity of the opposite polarity. When the junctions between these regions are reverse-biased, *depletion layers form, which narrow the channel. At high bias the depletion layers meet and pinch-off the channel completely. Thus the voltage applied to the two gates controls the thickness of the channel and thus its conductivity. JFETs are made with both n-type and p-type channels.

In the IGFET, a wafer of semiconductor material has two highly doped regions of opposite polarity diffused into it, to form the source and drain regions. An insulating layer of silicon dioxide is formed on the surface between these regions and a metal conductor is evaporated on to the top of this layer to form the gate. When a positive voltage is applied to the gate, electrons move along the surface of the p-type substrate below the gate, producing a thin surface of n-type material, which forms the channel between the source and drain. This surface layer is called an **inversion layer**, as it has opposite conductivity to that of the substrate. The number of induced electrons is directly proportional to the gate voltage, thus the conductivity of the channel increases with gate voltage. IGFETs are also made with both p-type and n-type channels. Because MOS devices cannot be formed on gallium arsenide (there are no stable native oxides of GaAs), metal semiconductor FETs (MESFET) devices are used. This makes use of Schottky barrier (*see* SCHOTTKY EFFECT) as the gate electrode rather than a semiconductor junction.

transition point (transition temperature) **1.** The temperature at which one crystalline form of a substance changes to another form. **2.** The temperature at which a substance changes phase. **3.** The temperature at which a substance becomes superconducting (*see* SUPERCONDUCTIVITY). **4.** The temperature at which some other change, such as a change of magnetic properties (*see also* CURIE POINT), takes place.

translation Motion of a body in which all the points in the body follow parallel paths.

translucent Permitting the passage of radiation but not without some scattering or diffusion. For example, frosted glass allows light to pass through it but an object cannot be seen clearly through it because the light rays are scattered by it. *Compare* TRANSPARENT.

transmission coefficient *See* TRANSMITTANCE.

transmission electron microscope *See* ELECTRON MICROSCOPE.

transmittance (transmission coefficient) The ratio of the energy of some form of radiation transmitted through a surface to the energy falling on it. The reciprocal of the transmittance is the **opacity**.

transmitter 1. The equipment used to generate and broadcast radio-frequency electromagnetic waves for communication purposes. In **transmitted-carrier transmission** it consists of a carrier-wave generator, a device for modulating the carrier wave in accordance with the information to be broadcast, amplifiers, and an aerial system. In **suppressed-carrier transmission**, the carrier component of the carrier wave is not transmitted; one *sideband (**single-sideband transmission**) or both sidebands (**double-sideband transmission**) are transmitted and a local oscillator in the receiver regenerates the carrier frequency and mixes it with the received signal to detect the modulating wave. **2.** The part of a telephone system that converts sound into electrical signals.

transparent Permitting the passage of radiation without significant deviation or absorption. *Compare* TRANSLUCENT. A substance may be transparent to radiation of one wavelength but not to radiation of another wavelength. For example, some forms of glass are transparent to light but not to ultraviolet radiation, while other forms of glass may be transparent to all visible radiation except red light. *See also* RADIOTRANSPARENT.

transponder A radio transmitter–receiver that automatically transmits signals when the correct interrogation is received. An example of a transponder is a radar beacon mounted on an aircraft (or missile) in which a receiver is tuned to the incoming radar frequency and a transmitter that radiates the received signal at a higher intensity.

transport coefficients Quantities that characterize transport in a system. Examples of transport coefficients include electrical and thermal *conductivity. One of the main purposes of nonequilibrium *statistical mechanics is to calculate such coefficients from first principles. It is difficult to calculate transport coefficients exactly for noninteracting systems and it is therefore necessary to use *approximation techniques and/or *model systems. A transport coefficient gives a measure for flow in a system. An **inverse transport coefficient** gives a measure of resistance to flow in a system. An example of an inverse transport coefficient is *resistivity.

transport number Symbol t. The fraction of the total charge carried by a particular type of ion in the conduction of electricity through electrolytes.

transport theory The theory of phenomena involving the transfer of matter or heat. The calculation of *transport coefficients and inverse transport coefficients, such as *conductivity and *viscosity, is an aim of transport theory. Calculations from first principles in transport theory start from *nonequilibrium statistical mechanics. Because of the difficulties involved in calculations in nonequilibrium statistical mechanics, transport theory uses approximate methods, including the *kinetic theory of gases and *kinetic equations, such as the *Boltzmann equation.

transuranic elements Elements with an atomic number greater than 92, i.e. elements above uranium in the periodic table.

transverse wave *See* WAVE.

travelling wave *See* WAVE.

travelling-wave accelerator *See* LINEAR ACCELERATOR.

Travers, Morris *See* RAMSAY, SIR WILLIAM.

triangle of vectors A triangle constructed so that each of its sides represents one of three coplanar *vectors acting at a point with no resultant. If the triangle is completed, with the sides representing the vectors in both magnitude and direction, so that there are no gaps between the sides, then the vectors are in equilibrium. If the three vectors are forces, the figure is called a **triangle of forces**; if they are velocities, it is a **triangle of velocities**.

triboelectricity *Static electricity produced as a result of friction.

tribology The study of friction, lubrication, and lubricants.

triboluminescence *Luminescence caused by friction; for example, some crystalline substances emit light when they are crushed as a result of static electric charges generated by the friction.

triclinic *See* CRYSTAL SYSTEM.

trigonometric functions Functions defined in terms of a right-angled triangle (see diagram) and widely used in the solution of many mathematical problems. They are defined as:

tangent of angle A, written $\tan A = a/b$

sine of angle A, written $\sin A = a/c$

cosine of angle A, written $\cos A = b/c$,

where a is the length of the side opposite the angle A, b is the length of the side opposite the angle B, and c is the hypotenuse of the triangle. The reciprocal functions are:

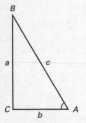

Trigonometric functions

cotangent of angle A, written $\cot A = 1/\tan A = b/a$

secant of angle A, written $\sec A = 1/\cos A = c/b$

cosecant of angle A, written $\operatorname{cosec} A = 1/\sin A = c/a$

triode A *thermionic valve with three electrodes. Electrons produced by the heated cathode flow to the anode after passing through the negatively biased *control grid. Small voltage fluctuations superimposed on the grid bias cause large fluctuations in the anode current. The triode was thus the first electronic device capable of amplification. Its role has now been taken over by the transistor, except where high power (radio-frequency transmitters producing more than 1kW in power) is required.

triple point The temperature and pressure at which the vapour, liquid, and solid phases of a substance are in equilibrium. For water the triple point occurs at 273.16 K and 611.2 Pa. This value forms the basis of the definition of the *kelvin and the thermodynamic *temperature scale.

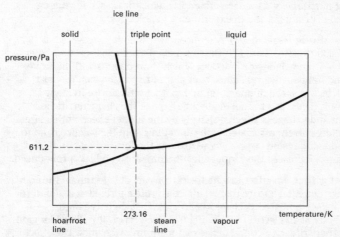

Triple point of water

triple product Either a *scalar product or a *vector product each having three components. A **scalar triple product** is obtained by multiplying three *vectors a, b, and c in the manner $a.(b \times c)$; the result is a scalar. If the three vectors represent the positions of three points with respect to the origin, the magnitude of the scalar triple product is the volume of the parallelepiped with corners at the three points and the origin. A **vector triple product** is obtained by multiplying three vectors a, b, and c in the manner $a \times (b \times c)$; the result is a vector. It also equals $(a.c)b - (a.b)c$ (but note it does not equal $(a \times b) \times c$).

tritiated compound *See* LABELLING.

tritium Symbol T. An isotope of hydrogen with mass number 3; i.e. the nucleus contains 2 neutrons and 1 proton. It is radioactive (half-life 12.3 years), undergoing beta decay to helium–3. Tritium is used in *labelling.

triton The nucleus of a tritium atom.

tropical year *See* YEAR.

troposphere *See* EARTH'S ATMOSPHERE.

tuned circuit A *resonant circuit in which there is a *resonance at a certain frequency. The process of altering the characteristics of the circuit so that it resonates, i.e. its *impedance is at its minimum, is called tuning the circuit. This is achieved either by altering the capacitance of the circuit, called **capacitive tuning**, or altering the inductance of the circuit, called **inductive tuning**.

tuning fork A metal two-pronged fork that when struck produces an almost pure tone of a predetermined frequency. It is used for tuning musical instruments and in experiments in acoustics.

tunnel diode (Esaki diode) A semiconductor diode, discovered in 1957 by L. Esaki (1925–), based on the *tunnel effect. It consists of a highly doped p–n semiconductor junction, which short circuits with negative bias and has negative resistance over part of its range when forward biased. Its fast speed of operation makes it a useful device in many electronic fields. In a tunnel diode the shape of the characteristic for current and voltage is affected sharply by the tunnel effect, with a large tunnelling current associated with tunnelling from the valence band to the conduction band (*see* ENERGY BANDS). As the forward bias on the junction is increased the tunnel effect becomes less and less pronounced.

tunnel effect An effect in which electrons are able to tunnel through a narrow *potential barrier that would constitute a forbidden region if the electrons were treated as classical particles. That there is a finite probability of an electron tunnelling from one classically allowed region to another arises as a consequence of *quantum mechanics. The effect is made use of in the *tunnel diode. Alpha decay (*see* ALPHA PARTICLE) is an example of a tunnelling process.

turbine A machine in which a fluid is used to produce rotational motion. The most widely used turbines are the **steam turbines** and **water turbines** that provide some 95% of the world's electric power (in the form of *turbogenerators) and the **gas turbines** that power all the world's jet-propelled aircraft. In the **impulse turbine** a high-pressure low-velocity fluid is expanded through stationary nozzles, producing low-pressure high-velocity jets, which are directed onto the blades of a rotor. The rotor blades reduce the speed of the jets and thus convert some of the fluid's kinetic energy into rotational kinetic energy of the rotor shaft. In the **reaction turbine** the discharge nozzles are themselves attached to the

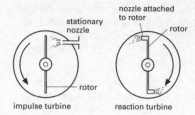

stationary nozzle

nozzle attached to rotor

rotor

rotor

impulse turbine

reaction turbine

Principle of the turbine

rotor. The acceleration of the fluid leaving the nozzles produces a force of reaction on the pipes, causing the rotor to move in the opposite direction to that of the fluid. (See illustrations.) Many turbines work on a combination of the impulse and reaction principles.

turbogenerator A steam turbine driving an electric generator. This is the normal method of generating electricity in power stations. In a conventional power station the steam is raised by burning a fossil fuel (coal, oil, or natural gas); in a nuclear power station the steam is raised by heat transfer from a nuclear reactor.

turbojet *See* JET PROPULSION.

turbulence A form of fluid flow in which the particles of the fluid move in a disordered manner in irregular paths, resulting in an exchange of momentum from one portion of a fluid to another. Turbulent flow takes over from *laminar flow when high values of the *Reynolds number are reached.

turns ratio *See* TRANSFORMER.

T violation *See* TIME REVERSAL.

tweeter A small loudspeaker capable of reproducing sounds of relatively high frequency, i.e. 5 kilohertz upwards. In high-fidelity equipment a tweeter is used in conjunction with a *woofer.

twinning A process in which two crystals of the same material form with orientations such that the crystals are related to each other by a symmetry operation. This may be either reflection in a plane (the **twinning plane**) or rotation about an axis (the **twinning axis**). The plane or axis is common to the two crystals.

twin paradox A paradox resulting from the special theory of *relativity; if one of a pair of twins remains on earth while the other twin makes a journey to a distant star at close to the speed of light and subsequently returns to earth, the twins will have aged differently. The twin remaining on earth will have aged considerably more than the twin who travelled to a star. This paradox can be explained by the geometry of Minkowski (*see* SPACE–TIME). The *world line of the twin who remained at home is denoted AB, while the world line of the other twin is the sum AC + CB, corresponding to the journeys to and from the star. Such distances in Minkowski space satisfy the inequality AB > AC + CB, thus demonstrating that the time experienced by the twin who remains on earth is greater than that experienced by the other twin.

Tyndall effect The scattering of light as it passes through a medium containing small particles. If a polychromatic beam of light is passed through a medium containing particles with diameters less than about one-twentieth of the wavelength of the light, the scattered light appears blue. This accounts for the blue appearance of tobacco smoke. At higher particle diameters, the scattered light remains polychromatic. It is named after John Tyndall (1820–93). *See also* SCATTERING OF ELECTROMAGNETIC RADIATION.

t

ultracentrifuge A high-speed centrifuge used to measure the rate of sedimentation of colloidal particles or to separate macromolecules, such as proteins or nucleic acids, from solutions. Ultracentrifuges are electrically driven and are capable of speeds up to 60 000 rpm.

ultrahigh frequency (UHF) A radio frequency in the range 3×10^9– 0.3×10^9 Hz; i.e. having a wavelength in the range 10 cm to 1 m.

ultramicroscope A form of microscope that uses the *Tyndall effect to reveal the presence of particles that cannot be seen with a normal optical microscope. Colloidal particles, smoke particles, etc., are suspended in a liquid or gas in a cell with a black background and illuminated by an intense cone of light that enters the cell from the side and has its apex in the field of view. The particles then produce diffraction-ring systems, appearing as bright specks on the dark background.

ultrasonic imaging The use of *ultrasonics to form an image or representation of an object (such as an internal part of the human body). Ultrasonic *tomography is used for these purposes.

In ultrasonic imaging, a probe containing a piezoelectric crystal acts both as the source and the detector of the ultrasonic waves, which are typically in a frequency range of 1–15 MHz. Pulses of ultrasound are transmitted into the body and are reflected at boundaries between different tissue types. By analysing the time lapse and absorption of the returning signals, positions and orientations of the underlying organs and tissues may be deduced.

The diagrams overleaf describe the main features of three main types of ultrasound scan employed in modern medicine. In the M-scan, ultrasound is transmitted into the patient's body. The delay is used to calculate the depth of the structure under examination. Electronics converts the received signal into a potential difference, which is fed to the X plates of a *cathode-ray oscilloscope. The *timebase of the oscilloscope becomes the time axis of the M-scan. Regular oscillations of the structure under the skin produces a wavy trace. The M-scan techniques are often used for monitoring heart movement. However, the viewing of the heart is limited by surrounding bone and air. Care has to be taken to avoid the ribs and lungs. These restrictions mean that ultrasound cardiography (UCG) requires considerable skill.

In the A-scan, the surfaces in the patient's body produce echoes, which are picked up by the probe and analysed. A block diagram of the electronics that finally produces the A-scan is shown on the right. The A-scan is a range-finding scan. The surfaces a, b, c, and d produce echoes of

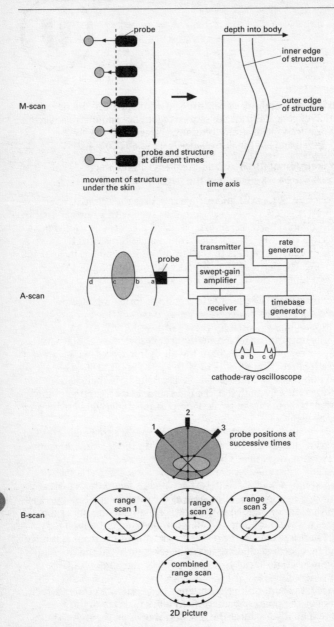

M-scan

probe

depth into body

inner edge
of structure

outer edge
of structure

probe and structure
at different times

movement of structure
under the skin

time axis

A-scan

probe

transmitter

rate
generator

swept-gain
amplifier

receiver

timebase
generator

d   c   b   a

a b   c d

cathode-ray oscilloscope

2

1

3 probe positions at
successive times

B-scan

range
scan 1

range
scan 2

range
scan 3

combined
range scan

2D picture

Ultrasonic imaging

the detected ultrasound. The depth of the surfaces are effectively shown by the time delay on the circular screen. The height of a spike which corresponds to a surface represents the size of the echo. The rate generator triggers the transmitter of ultrasound, the swept-gain amplifier, and the timebase simultaneously. The function of the swept-gain amplifier is to increase the return pulse amplification, since echoes returning from deeper structures will be more attenuated. The swept-gain amplifier therefore amplifies the received signals just enough to compensate for the attenuation. The output of the receiver is fed into the cathode-ray oscilloscope which displays the A-scan. A-scans are used when precise depth measurements are required, e.g. in echoencephalography, the determination of the midline of the brain.

In the B-scan, the ultrasound probe is moved slowly over the patient's body and a cathode-ray oscilloscope is able to store the echoes of a succession of range-finding scans displays them as bright spots. The depth is calculated by the delay of the return signal and the brightness of the spots indicates the strength of the return signal. The two-dimensional images produced by B-scans are ideal for monitoring the progress of a foetus in a womb or locating tumours or other anomalies in the liver, kidneys, and ovaries. B-scans are also invaluable in the examination of aneurysms (swollen parts of arteries).

Ultrasound can also be used to monitor the function of organs or blood flow. This application of ultrasound relies on the *Doppler effect. The transmitter is separate from the receiver. The ultrasound used in this technique is a continuous beam rather than the series of pulses used for imaging. The transmitted and received signals are electronically mixed and the output is filtered so that only the Doppler-shift frequency is amplified. Doppler-shift frequencies are often in the human hearing range so that the operators can 'hear' the moving structures on earphones.

Ultrasound monitoring using the Doppler effect is also used in monitoring blood flow. The Doppler probe is placed on the skin directly above a blood vessel at an angle θ to the horizontal. The blood, travelling in a direction parallel to the skin, reflects the transmitted waves and Doppler-shifts them by an amount Δf given by the formula:

$$\Delta f = (^2fv/c)\cos\theta,$$

where f is the frequency of the transmitted ultrasound, v is the speed of the blood flow, and c is the speed of sound in blood. A blockage due to a clot (thrombosis), or a constriction due to plaque build-up on the vessel walls (atheroma), is immediately apparent if there is a change in the Doppler shift frequency.

Ultrasound may also be delivered at large intensities to localized regions of the body for therapeutic purposes. **Diathermy** is the application of ultrasound to cause localized heating of tissue to relieve pain in the joints of arthritis sufferers. Typical treatments use intensities of several W/cm^2 for periods of a few minutes several times a week. Actual destruction of tissue is also possible using sufficiently high intensities of ultrasound (intensities exceeding 1000 W/cm^2). Ultrasound has been shown to destroy cancer cells, although some studies have also shown that it can in some cases stimulate growth of the tumour.

ultrasonics The study and use of pressure waves that have a frequency in excess of 20 000 Hz and are therefore inaudible to the human ear. **Ultrasonic generators** make use of the *piezoelectric effect, *ferroelectric materials, or *magnetostriction to act as transducers in converting electrical energy into mechanical energy. Ultrasonics are used in medical diagnosis, particularly in conditions such as pregnancy, in which X-rays could have a harmful effect. Ultrasonic techniques are also used industrially to test for flaws in metals, to clean surfaces, to test the thickness of parts, and to form colloids.

ultraviolet microscope A *microscope that has quartz lenses and slides and uses *ultraviolet radiation as the illumination. The use of shorter wavelengths than the visible range enables the instrument to resolve smaller objects and to provide greater magnification than the normal optical microscope. The final image is either photographed or made visible by means of an *image converter.

ultraviolet radiation (UV) Electromagnetic radiation having wavelengths between that of violet light and long X-rays, i.e. between 400 nanometres and 4 nm. In the range 400–300 nm the radiation is known as the **near ultraviolet**. In the range 300–200 nm it is known as the **far ultraviolet**. Below 200 nm it is known as the **extreme ultraviolet** or the **vacuum ultraviolet**, as absorption by the oxygen in the air makes the use of evacuated apparatus essential. The sun is a strong emitter of UV radiation but only the near UV reaches the surface of the earth as the ozone layer of the atmosphere absorbs all wavelengths below 290 nm. Ultraviolet radiation is classified in three ranges according to its effect on the skin. The ranges are:

UV-A (320–400 nm);

UV-B (290–320 nm);

UV-C (230–290 nm).

The longest-wavelength range, UV-A, is not harmful in normal doses and is used clinically in the treatment of certain skin complaints, such as psoriasis. It is also used to induce vitamin D formation in patients that are allergic to vitamin D preparations. UV-B causes reddening of the skin followed by pigmentation (tanning). Excessive exposure can cause severe blistering. UV-C, with the shortest wavelengths, is particularly damaging. It has been claimed that short-wavelength ultraviolet radiation causes skin cancer and that the risk of contracting this has been increased by the depletion of the ozone layer.

Most UV radiation for practical use is produced by various types of *mercury-vapour lamps. Ordinary glass absorbs UV radiation and therefore lenses and prisms for use in the UV are made from quartz.

umbra *See* SHADOW.

uncertainty principle (Heisenberg uncertainty principle; principle of indeterminism) The principle that it is not possible to know with unlimited accuracy both the position and momentum of a particle. This principle, discovered in 1927 by Werner *Heisenberg, is usually stated in the form: $\Delta x \Delta p_x \geq h/4\pi$, where Δx is the uncertainty in the x-coordinate of the particle, Δp_x is the uncertainty in the x-component of the particle's momentum, and h is the *Planck constant. An explanation of the uncertainty is that in order to locate a particle exactly, an observer must be able to bounce off it a photon of radiation; this act of location itself alters the position of the particle in an unpredictable way. To locate the position accurately, photons of short wavelength would have to be used. The high momenta of such photons would cause a large effect on the position. On the other hand, using photons of lower momenta would have less effect on the particle's position, but would be less accurate because of the longer wavelength.

underdamped *See* DAMPING.

undetermined multiplier *See* LAGRANGE MULTIPLIER.

uniaxial crystal A double-refracting crystal (*see* DOUBLE REFRACTION) having only one *optic axis.

unified-field theory A comprehensive theory that would relate the electromagnetic, gravitational, strong, and weak interactions (*see* FUNDAMENTAL INTERACTIONS) in one set of equations. In its original context the expression referred only to the unification of general *relativity and classical electromagnetic theory. No such theory has yet been found but some progress has been made in the unification of the electromagnetic and weak interactions (*see* ELECTROWEAK THEORY).

Einstein attempted to derive *quantum mechanics from unified-field theory, but it is now thought that any unified-field theory has to start with quantum mechanics. Attempts to construct unified-field theories, such as *supergravity and *Kaluza–Klein theory, have run into great difficulties. At the present time it appears that the framework of relativistic *quantum field theory is not adequate to give a unified theory for all the known fundamental interactions and elementary particles, and that one has to go to extended objects, such as superstrings. Unified-field theories and other fundamental theories, such as *superstring theory, are of great importance in understanding cosmology, particularly the *early universe. In turn cosmology puts constraints on unified-field theories. *See also* GRAND UNIFIED THEORY.

unified model (collective model) A single model of the atomic nucleus that combines the *shell model with the *liquid-drop model; i.e. the single-particle and collective aspect of nuclei. It is necessary to use the unified model to calculate the quadruple moment of a nonspherical nucleus and to get a fully quantitative description of *nuclear fission.

uniform field A field in which the value of the field strength is the same at all points. For example, a uniform electric field exists between two parallel charged plates. At the ends of the plates the field is non-uniform. Uniform magnetic fields may be produced by *Helmholtz coils.

union *See* SETS.

unit A specified measure of a physical quantity, such as length, mass, time, etc., specified multiples of which are used to express magnitudes of that physical quantity. For many scientific purposes previous systems of units have now been replaced by *SI units.

unitary transformation A transformation that has the form $O' = UOU^{-1}$, where O is an operator, U is a unitary *matrix and U^{-1} is its reciprocal, i.e. if the matrix obtained by interchanging rows and columns of U and then taking the complex conjugate of each entry, denoted U^+, is the inverse of U; $U^+ = U^{-1}$. The inverse of a unitary transformation is itself a unitary transformation. Unitary transformations are important in *quantum mechanics. In the *Hilbert space formulation of states in quantum mechanics a unitary transformation corresponds to a rotation of axes in the Hilbert space. Such a transformation does not alter the state vector, but a given state vector has different components when the axes are rotated.

unit cell The group of particles (atoms, ions, or molecules) in a *crystal system that is repeated in three dimensions in the *crystal lattice.

unit magnetic pole *See* MAGNETIC POLES.

unit vector A vector that has the magnitude 1. If $\boldsymbol{a}$ is any nonzero vector the unit vector in the direction of $\boldsymbol{a}$ is given by $\boldsymbol{a}/|\boldsymbol{a}|$ and is denoted $\hat{\boldsymbol{a}}$.

universal constant *See* FUNDAMENTAL CONSTANT.

universality *See* PHASE TRANSITIONS.

universal motor *See* ELECTRIC MOTOR.

universe All the matter, energy, and space that exists. *See* COSMOLOGY; EARLY UNIVERSE; HEAT DEATH OF THE UNIVERSE.

Unruh effect The phenomenon, predicted in 1976 by William Unruh, that an accelerating body would seem to be surrounded by particles at a nonzero temperature, with this temperature being proportional to the acceleration. The vacuum state of a nonaccelerating observer is different from that of an accelerating observer because of distortion of the zero-point fluctuations. **Unruh radiation** is associated with this effect. The effect itself is very small and has not been verified experimentally.

unstable equilibrium *See* EQUILIBRIUM.

upper atmosphere The upper part of the *earth's atmosphere above about 30 km, which cannot be reached by balloons.

upthrust *See* ARCHIMEDES' PRINCIPLE.

uranium Symbol U. A white radioactive metallic element belonging to the actinoids; a.n. 92; r.a.m. 238.03; r.d. 19.05 (20°C); m.p. 1132±1°C; b.p. 3818°C. It occurs as uraninite, from which the metal is extracted by an ion-exchange process. Three isotopes are found in nature: uranium–238 (99.28%), uranium–235 (0.71%), and uranium–234 (0.006%). As uranium–235 undergoes *nuclear fission with slow neutrons it is the fuel used in *nuclear reactors and *nuclear weapons; uranium was discovered by M. H. Klaproth (1743–1817) in 1789.

uranium–lead dating A group of methods of dating certain rocks that depends on the decay of the radioisotope uranium–238 to lead–206 (half-life 4.5×10^9 years) or the decay of uranium–235 to lead–207 (half-life 7.1×10^8 years). One form of uranium–lead dating depends on measuring the ratio of the amount of helium trapped in the rock to the amount of uranium present (since the decay $^{238}U \rightarrow {}^{206}Pb$ releases eight alpha-particles). Another method of calculating the age of the rocks is to measure the ratio of radiogenic lead (^{206}Pb, ^{207}Pb, and ^{208}Pb) present to nonradiogenic lead (^{204}Pb). These methods give reliable results for ages of the order 10^7–10^9 years. *See* DATING TECHNIQUES.

uranium series *See* RADIOACTIVE SERIES.

URL *See* WORLD WIDE WEB.

USB drive In general, any storage device that can be attached to a computer through a special type of connection (*u*niversal *s*erial *b*us connection). The term is particularly used for small portable storage devices typically sealed in plastic. Their physical size is reflected in the various names for this type of device – **thumb drive**, **pen drive**, **keyring drive**. They have capacities as high as 1 gigabyte and on most modern personal computers they can be recognized as an additional drive without the need for a special driver. Storage devices of this type are increasingly used for backup, data transfer, storage of photographs or MP3 files, etc.

UV *See* ULTRAVIOLET RADIATION.

u

vacancy *See* DEFECT.

vacuum A space in which there is a low pressure of gas, i.e. relatively few atoms or molecules. A **perfect vacuum** would contain no atoms or molecules, but this is unobtainable as all the materials that surround such a space have a finite *vapour pressure. In a **soft** (or **low**) **vacuum** the pressure is reduced to about 10^{-2} pascal, whereas a **hard** (or **high**) vacuum has a pressure of 10^{-2}–10^{-7} pascal. Below 10^{-7} pascal is known as an **ultrahigh vacuum**. *See also* VACUUM PUMP.

vacuum pump A pump used to reduce the gas pressure in a container. The normal laboratory rotary oil-seal pump can maintain a pressure of 10^{-1} Pa. For pressures down to 10^{-7} Pa a *diffusion pump is required. *Ion pumps can achieve a pressure of 10^{-9} Pa and a *cryogenic pump combined with a diffusion pump can reach 10^{-13} Pa.

vacuum state The ground state in a relativistic *quantum field theory. A vacuum state does not mean a state of nothing. Because one is dealing with *quantum mechanics, the vacuum state has a *zero-point energy, which gives rise to **vacuum fluctuations**. The existence of vacuum fluctuations has observable consequences in *quantum electrodynamics.

vacuum tube *See* THERMIONIC VALVE.

valence band *See* ENERGY BANDS.

valence electron An electron in one of the outer shells of an atom that takes part in forming chemical bonds.

valency (valence) The combining power of an atom or radical, equal to the number of hydrogen atoms that the atom could combine with or displace in a chemical compound (hydrogen has a valency of 1). It is equal to the ionic charge in ionic compounds; for example, in Na_2S, sodium has a valency of 1 (Na^+) and sulphur a valency of 2 (S^{2-}). In covalent compounds it is equal to the number of bonds formed; in CO_2 both carbon and oxygen have a valency of 2.

valve *See* THERMIONIC VALVE.

Van Allen belts (radiation belts) Belts that are sources of intense radiation surrounding the earth, consisting of high-energy charged particles trapped in the earth's magnetic field within which they follow roughly helical paths. They were discovered in 1958 by James Van Allen (1914–) as a result of radiation detectors carried by Explorer satellites.

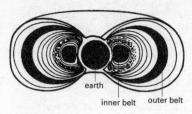

Van Allen belts

The lower belt, extending from 1000 to 5000 km above the equator, contains electrons and protons, while the upper belt, 15 000–25 000 km above the equator, contains mainly electrons (see illustration).

Van de Graaff accelerator *See* LINEAR ACCELERATOR.

Van de Graaff generator An electrostatic generator used to produce a high voltage, usually in the megavolt range. It consists of a large metal dome-shaped terminal mounted on a hollow insulating support. An endless insulating belt runs through the support from the base to a pulley within the spherical terminal. In the original type, charge is sprayed by point discharge from metal needles, held at a potential of about 10 kV, on to the bottom of the belt. A row of needles near the upper belt pulley removes the charge from the belt and passes it to the outer surface of the spherical terminal. The voltage achieved by the device is proportional to the radius of the spherical terminal. A typical device with a terminal having a radius of 1 m will produce about 1 MV. However, terminals can

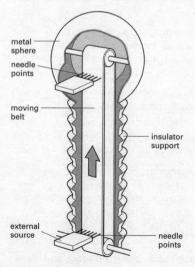

metal sphere

needle points

moving belt

insulator support

external source

needle points

Van de Graaff generator

be made smaller, for a given voltage, by enclosing the apparatus in nitrogen at a pressure of 10–20 atmospheres (1–2 MPa) to reduce sparking. Generators having a positive-ion source are fitted with an evacuated tube through which the particles can be accelerated for research purposes (*see* LINEAR ACCELERATOR). Machines having an electron source are used for various medical and industrial purposes. The generator was invented by R. J. Van de Graaff (1901–67).

Modern patterns of the generator have a chainlike belt of alternate links of metal and insulator. The metal links are charged by contact with a metal pulley, and discharge to the dome in the same way. This permits much higher current drain that the point discharge.

van der Waals' equation *See* EQUATION OF STATE.

van der Waals' force An attractive force between atoms or molecules, named after J. D. van der Waals (1837–1923). The force accounts for the term a/V^2 in the van der Waals equation (*see* EQUATION OF STATE). These forces are much weaker than those arising from valence bonds and are inversely proportional to the seventh power of the distance between the atoms or molecules. They are the forces responsible for nonideal behaviour of gases and for the lattice energy of molecular crystals. There are three factors causing such forces: (1) dipole–dipole interaction, i.e. electrostatic attractions between two molecules with permanent dipole moments; (2) dipole-induced dipole interactions, in which the dipole of one molecule polarizes a neighbouring molecule; (3) dispersion forces arising because of small instantaneous dipoles in atoms. A quantitative theory of van der Waals' force was given by F. London in 1930 in terms of quantum mechanics.

van't Hoff factor Symbol i. A factor appearing in equations for *colligative properties, equal to the ratio of the number of actual particles present to the number of undissociated particles. It was first suggested by Jacobus van't Hoff (1852–1911).

van't Hoff's isochore A thermodynamic equation devised by Jacobus van't Hoff for the variation of equilibrium constant with temperature:

$$(d \log_e K)/dT = \Delta H/RT^2,$$

where T is the thermodynamic temperature and ΔH the enthalpy of the reaction.

vapour density The density of a gas or vapour relative to hydrogen, oxygen, or air. Taking hydrogen as the reference substance, the vapour density is the ratio of the mass of a particular volume of a gas to the mass of an equal volume of hydrogen under identical conditions of pressure and temperature. Taking the density of hydrogen as 1, this ratio is equal to half the relative molecular mass of the gas.

vapour pressure The pressure exerted by a vapour. All solids and liquids give off vapours, consisting of atoms or molecules of the substances that have evaporated from the condensed forms. These atoms or molecules exert a vapour pressure. If the substance is in an enclosed space, the vapour pressure will reach an equilibrium value that depends only on the nature of the substance and the temperature. This maximum value occurs when there is a dynamic equilibrium between the atoms or molecules escaping from the liquid or solid and those that strike the surface of the liquid or solid and return to it. The vapour is then said to be a **saturated vapour** and the pressure it exerts is the **saturated vapour pressure**.

variable star A star that varies in brightness (*see* LUMINOSITY). This variation, which can be regular or irregular, can be due either to changes in internal conditions (an example of this being a **pulsating star** in which the variation in brightness is due to the star expanding and contracting) or due to external reasons, such as a star being eclipsed by another star (or stars). Changes in brightness of a star occur both in its early stages and close to its death. Extreme examples of variable stars are *nova and *supernova. Observations of variable stars provide important information concerning the mass, size, structure, temperature, and evolution of stars. There are more than 25 000 variable stars known, some of which can be seen with the naked eye. The time-scale from maximum to minimum brightness varies from less than a second to a few years.

variance A continuous distribution that has a probability density function f(x) has a variance σ^2 defined by

$$\sigma^2 = \int_{-\infty}^{\infty} (x - \mu)^2 \, f(x) dx,$$

where μ is defined by

$$\mu = \int_{-\infty}^{\infty} x f(x) dx.$$

If the distribution is discrete, with a probability function p(x), the variance σ^2 is defined by $\sigma^2 = \sum (x - \mu)^2 px$, where μ is defined by $\mu = \sum_x x px$.

variation *See* GEOMAGNETISM.

variational principle A principle involving the idea that a quantity of interest either be a minimum or a maximum. Many laws of physics are variational principles, examples being *Fermat's principle and the *principle of least action.

variometer 1. A variable inductor consisting of two coils connected in series and able to move relative to each other. It is used to measure inductance as part of an a.c. bridge. **2.** Any of several devices for detecting and measuring changes in the geomagnetic elements (*see* GEOMAGNETISM).

vector A quantity in which both the magnitude and the direction must be stated (*compare* SCALAR QUANTITY). Force, velocity, and field strength are examples of vector quantities. Note that distance and speed are scalar quantities, whereas displacement and velocity are vector quantities. Vector quantities must be treated by **vector algebra**; for example, the resultant of two vectors may be found by a *parallelogram of vectors. A (three-dimensional) vector **V** may be written in terms of components V_1, V_2, and V_3 along the x, y, and z axes (say) as $V_1\boldsymbol{i} + V_2\boldsymbol{j} + V_3\boldsymbol{k}$, where $\boldsymbol{i}$, $\boldsymbol{j}$, and $\boldsymbol{k}$ are **unit vectors** (i.e. vectors of unit length) along the x, y, and z axes. *See also* TRIANGLE OF VECTORS.

vector product (cross product) The product of two *vectors **U** and **V**, with components U_1, U_2, U_3 and V_1, V_2, V_3, respectively, given by:

$$\boldsymbol{U} \times \boldsymbol{V} = (U_2V_3 - U_3V_2)\boldsymbol{i} + (U_3V_1 - U_1V_3)\boldsymbol{j} + (U_1V_2 - U_2V_1)\boldsymbol{k}.$$

It is itself a vector, perpendicular to both **U** and **V**, and of length $UV\sin\theta$, where U and V are the lengths of **U** and **V**, respectively, and θ is the angle between them. *Compare* SCALAR PRODUCT.

vector space A set of *vectors for which an operation of addition is defined so that if $\boldsymbol{v}_1$ and $\boldsymbol{v}_2$ are vectors, the sum $\boldsymbol{v}_1 + \boldsymbol{v}_2$ is also a vector; an operation of scalar multiplication is defined so that if $\boldsymbol{v}$ is a vector and c is a *scalar quantity, the product $c\boldsymbol{v}$ is also a vector. *See also* HILBERT SPACE.

vector triple product *See* TRIPLE PRODUCT.

velocity Symbol $\boldsymbol{v}$. The rate of displacement of a body. It is the *speed of a body in a specified direction. Velocity is thus a *vector quantity, whereas speed is a scalar quantity.

velocity modulation *See* KLYSTRON.

velocity ratio (distance ratio) The ratio of the distance moved by the point of application effort in a simple *machine to the distance moved by the point of application load in the same time.

Venn diagram *See* SETS.

Venturi tube A device for mixing a fine spray of liquid with a gas or measuring a flow rate of a gas. It consists of two tapered sections of pipe joined by a narrow throat. The fluid velocity in the throat is increased and the pressure is therefore reduced. By attaching manometers to the three sections of the tube, the pressure drop can be measured and the flow rate through the throat can be calculated. In a carburettor, the petrol from the float chamber is made into a fine spray by being drawn through a jet into the low pressure in the throat of a Venturi tube, where it mixes with the air being drawn into the engine. The device was invented by G. B. Venturi (1746–1822).

vernier A short auxiliary scale placed beside the main scale on a measuring instrument to enable subdivisions of the main scale to be read accurately. The vernier scale is usually calibrated so that each of its divisions is 0.9 of the main scale divisions. The zero on the vernier scale is set to the observed measurement on the main scale and by noting which division on the vernier scale is exactly in line with a main scale division, the second decimal place of the measurement is obtained (see

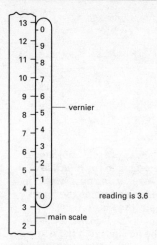

reading is 3.6

Vernier scale

illustration). The device was invented by Pierre Vernier (1580–1637) in about 1630.

very high frequency (VHF) A radio frequency in the range 3×10^8– 0.3 $\times 10^8$ Hz, i.e. having a wavelength in the range 1–10 m.

very low frequency (VLF) A radio frequency in the range 3×10^4– 0.3 $\times 10^4$ Hz, i.e. having a wavelength in the range 10–100 km.

Victor Meyer's method A method of measuring vapour density, devised by Victor Meyer (1848–97). A weighed sample in a small tube is dropped into a heated bulb with a long neck. The sample vaporizes and displaces air, which is collected over water and the volume measured. The vapour density can then be calculated.

virial equation A gas law that attempts to account for the behaviour of real gases, as opposed to an ideal gas. It takes the form

$$pV = RT + Bp + Cp^2 + Dp^3 + \ldots,$$

where B, C, and D are known as **virial coefficients**.

virtual image *See* IMAGE.

virtual particle *See* VIRTUAL STATE.

virtual state The state of the **virtual particles** that are exchanged between two interacting charged particles. These particles, called *photons, are not in the real state, i.e. directly observable; they are constructs to enable the phenomenon to be explained in terms of *quantum mechanics.

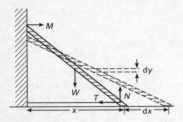

Virtual work

virtual work The imaginary work done when a system is subjected to infinitesimal hypothetical displacements. According to the **principle of virtual work**, the total work done by all the forces acting on a system in equilibrium is zero. This principle can be used to determine the forces acting on a system in equilibrium. For example, the illustration shows a ladder leaning against a wall, with the bottom of the ladder attached to the wall by a horizontal weightless string. The tension, T, in the string can be calculated by assuming that infinitesimal movement dx and dy take place as shown. Then by applying the principle of virtual work, Tdx + Wdy = 0. As dx and dy can be calculated from the geometry, T can be found.

viscometer An instrument for measuring the viscosity of a fluid. In the **Ostwald viscometer**, used for liquids, a bulb in a capillary tube is filled with the liquid and the time taken for the meniscus to reach a mark on the capillary, below the bulb, is a measure of the viscosity. The **falling-sphere viscometer**, based on *Stokes' law, enables the speed of fall of a ball falling through a sample of the fluid to be measured. Various other devices are used to measure viscosity.

viscosity A measure of the resistance to flow that a fluid offers when it is subjected to shear stress. For a *Newtonian fluid, the force, F, needed to maintain a velocity gradient, dv/dx, between adjacent planes of a fluid of area A is given by: $F = \eta A(dv/dx)$, where η is a constant, the coefficient of viscosity. In *SI units it has the unit pascal second (in the c.g.s. system it is measured in *poise). Non-Newtonian fluids, such as clays, do not conform to this simple model. *See also* KINEMATIC VISCOSITY.

visible spectrum The *spectrum of electromagnetic radiations to which the human eye is sensitive. *See* COLOUR.

visual binary *See* BINARY STARS.

visual-display unit (VDU) The part of a *computer system or word processor on which text or diagrams are displayed. It consists of a *cathode-ray tube and usually has its own input keyboard attached.

volt Symbol V. The SI unit of electric potential, potential difference, or e.m.f. defined as the difference of potential between two points on a conductor carrying a constant current of one ampere when the power dissipated between the points is one watt. It is named after Alessandro Volta.

Volta, Alessandro Giuseppe Antonio Anastasio (1745–1827) Italian physicist. In 1774 he began teaching in Como and in that year invented the *electrophorus. He moved to Pavia University in 1778. In 1800 he made the *voltaic cell, thus providing the first practical source of electric current (*see also* GALVANI, LUIGI). The SI unit of potential difference is named after him.

voltage Symbol V. An e.m.f. or potential difference expressed in volts.

voltage divider (potential divider; potentiometer) A resistor or a chain of resistors connected in series that can be tapped at one or more points to obtain a known fraction of the total voltage across the whole resistor or chain. In the illustration, V is the total voltage across the divider and v is required voltage, then

$v/V = R_2/(R_1 + R_2).$

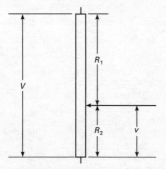

Voltage divider

voltaic cell (galvanic cell) A device that produces an e.m.f. as a result of chemical reactions that take place within it. These reactions occur at the surfaces of two electrodes, each of which dips into an electrolyte. The first voltaic cell, devised by Alessandro Volta, had electrodes of two different metals dipping into brine. *See* PRIMARY CELL; SECONDARY CELL.

voltaic pile An early form of battery, devised by Alessandro Volta, consisting of a number of flat *voltaic cells joined in series. The liquid electrolyte was absorbed into paper or leather discs.

voltameter (coulometer) 1. An electrolytic cell formerly used to measure quantity of electric charge. The increase in mass (m) of the cathode of the cell as a result of the deposition on it of a metal from a solution of its salt enables the charge (Q) to be determined from the relationship $Q = m/z$, where z is the electrochemical equivalent of the metal. **2.** Any other type of electrolytic cell used for measurement.

voltmeter An instrument used to measure voltage. *Moving-coil instruments are widely used for this purpose; generally a galvanometer is used in series with a resistor of high values (sometimes called a **multiplier**). To measure an alternating potential difference a rectifier must be included in the circuit. A moving-iron instrument can be used for either d.c. or a.c. without a rectifier. *Cathode-ray oscilloscopes are also used as voltmeters. The electronic **digital voltmeter** displays the value of the voltage in digits. The input is repeatedly sampled by the voltmeter and the instantaneous values are displayed.

volume Symbol V. The space occupied by a body or mass of fluid.

volume integral An integral over a volume, which can involve vectors and scalars. The volume element $dV = dxdydz$ is a scalar. For a vector function $\mathbf{F}(x,y,z)$ the volume integral is given by $\int_v \mathbf{F}dV$. For a scalar function $\phi(x,y,z)$ the volume integral is given by $\int_v \phi dV$.

von Laue, Max *See* LAUE, MAX VON.

V

wall effect Any effect resulting from the nature or presence of the inside wall of a container on the system it encloses.

Walton, Ernest *See* COCKCROFT, SIR JOHN DOUGLAS.

watt Symbol W. The SI unit of power, defined as a power of one joule per second. In electrical contexts it is equal to the rate of energy transformation by an electric current of one ampere flowing through a conductor the ends of which are maintained at a potential difference of one volt. The unit is named after James Watt (1736–1819).

wattmeter An instrument for measuring the power in watts in an alternating-current electric circuit. In a direct-current circuit, power is usually determined by separate measurements of the voltage and the current.

The **electrodynamic wattmeter** consists of two coils, one fixed (current) coil and one movable (potential) coil. The fixed coil carries the load current, and the movable coil carries a current proportional to the voltage applied to the measured circuit. The deflection of the needle attached to the movable coil indicates the power.

wave A periodic disturbance in a medium or in space. In a **travelling wave** (or **progressive wave**) energy is transferred from one place to another by the vibrations (*see also* STATIONARY WAVE). In a wave passing over the surface of water, for example, the water rises and falls as the wave passes but the particles of water on average do not move forward with the wave. This is called a **transverse wave** because the disturbances are at right angles to the direction of propagation. The water surface moves up and down while the waves travel across the surface of the water. Electromagnetic waves (see diagram) are also of this kind, with electric and magnetic fields varying in a periodic way at right angles to

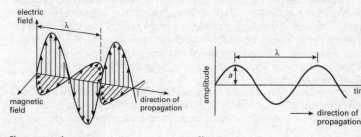

Electromagnetic waves   Sine wave

each other and to the direction of propagation. In sound waves, the air is alternately compressed and rarefied by displacements in the direction of propagation. Such waves are called **longitudinal waves**.

The chief characteristics of a wave are its **speed of propagation**, its **frequency**, its **wavelength**, and its **amplitude**. The speed of propagation is the distance covered by the wave in unit time. The frequency is the number of complete disturbances (cycles) in unit time, usually expressed in *hertz. The wavelength is the distance in metres between successive points of equal phase in a wave. The amplitude is the maximum difference of the disturbed quantity from its mean value.

Generally, the amplitude (a) is half the peak-to-peak value. There is a simple relationship between the wavelength (λ) and the frequency (f), i.e. $\lambda = c/f$, where c is the speed of propagation. The energy transferred by a progressive *sine wave (see diagram) is proportional to $a^2 f^2$. *See also* SIMPLE HARMONIC MOTION.

wave equation A partial differential equation of the form:

$$\nabla^2 u = (1/c^2)\partial^2 u/\partial t^2$$

where

$$\nabla^2 = \partial^2/\partial x^2 + \partial^2/\partial y^2 + \partial^2/\partial z^2$$

is the Laplace operator (*see* LAPLACE EQUATION). It represents the propagation of a wave, where u is the displacement and c the speed of propagation. *See also* SCHRÖDINGER'S EQUATION.

wave form The shape of a wave or the pattern representing a vibration. It can be illustrated by drawing a graph of the periodically varying quantity against distance for one complete wavelength. *See also* SINE WAVE.

wavefront A line or surface within a two- or three-dimensional medium through which waves are passing, being the locus of all adjacent points at which the disturbances are in phase. At large distances from a small source in a uniform medium, the fronts are small parts of a sphere of very large radius and they can be considered as plane. For example, sunlight reaches the earth with plane wavefronts.

wave function A function $\psi(x,y,z)$ appearing in *Schrödinger's equation in *quantum mechanics. The wave function is a mathematical expression involving the coordinates of a particle in space. If the Schrödinger equation can be solved for a particle in a given system (e.g. an electron in an atom) then, depending on the boundary conditions, the solution is a set of allowed wave functions (**eigenfunctions**) of the particle, each corresponding to an allowed energy level (**eigenvalue**). The physical significance of the wave function is that the square of its absolute value, $|\psi|^2$, at a point is proportional to the probability of finding the particle in a small element of volume, $dxdydz$, at that point. For an electron in an atom, this gives rise to the idea of atomic and molecular *orbitals.

wave guide A hollow tube through which microwave electromagnetic radiation can be transmitted with relatively little attenuation. They often have a rectangular cross section, but some have a circular cross section. In transverse electric (TE) modes the electric vector of the field has no component in the direction of propagation. In transverse magnetic (TM) modes, the magnetic vector has no such component.

wavelength *See* WAVE.

wave mechanics A formulation of *quantum mechanics in which the dual wave–particle nature (*see* COMPLEMENTARITY) of such entities as electrons is described by the *Schrödinger equation. Schrödinger put forward this formulation of quantum mechanics in 1926 and in the same year showed that it was equivalent to *matrix mechanics. Taking into account the *de Broglie wavelength, Schrödinger postulated a wave mechanics that bears the same relation to *Newtonian mechanics as physical optics does to geometrical optics (*see* OPTICS).

wavemeter A device for measuring the wavelength of electromagnetic radiation. For frequencies up to about 100 MHz a wavemeter consists of a tuned circuit with a suitable indicator to establish when resonance occurs. Usually the tuned circuit includes a variable capacitor calibrated to read wavelengths and resonance is indicated by a current-detecting instrument. At higher frequencies a cavity-resonator in a waveguide is often used. The cavity resonator is fitted with a piston, the position of which determines the resonant frequency of the cavity.

wave number Symbol k. The number of cycles of a wave in unit length. It is the reciprocal of the wavelength (*see* WAVE).

wave packet A superposition of waves with one predominant *wave number k, but with several other wave numbers near k. Wave packets are useful for the analysis of scattering in *quantum mechanics. Concentrated packets of waves can be used to describe localized particles of both matter and *photons. The Heisenberg *uncertainty principle can be derived from a wave-packet description of entities in quantum mechanics. The motion of a wave packet is in accord with the motion of the corresponding classical particle, if the potential energy change across the dimensions of the packet is very small. This proposition is known as **Ehrenfest's theorem**, named after the Dutch physicist Paul Ehrenfest (1880–1933), who proved it in 1927.

wave–particle duality The concept that waves carrying energy may have a corpuscular aspect and that particles may have a wave aspect; which of the two models is the more appropriate will depend on the properties the model is seeking to explain. For example, waves of electromagnetic radiation need to be visualized as particles, called *photons, to explain the *photoelectric effect while electrons need to be thought of as de Broglie waves in *electron diffraction. *See also* COMPLEMENTARITY; DE BROGLIE WAVELENGTH; LIGHT.

W

wave power The use of wave motion in the sea to generate energy. The technique used is to anchor a series of bobbing floats offshore; the energy of the motion of the floats is used to turn a generator. It has been estimated that there are enough suitable sites to generate over 100 GW of electricity in the UK.

wave theory *See* LIGHT.

wave vector A vector k associated with a *wave number k. In the case of free electrons the wave vector k is related to the momentum p in *quantum mechanics by $p = \hbar k$, where $\hbar$ is the rationalized *Planck constant. In the case of *Bloch's theorem, the wave vector k can only have certain values and can be thought of as a quantum number associated with the translational symmetry of the crystal.

W boson (W particle) Either of a pair of elementary particles (W^+ or W^-), classified as **intermediate vector bosons**, that are believed to transmit the weak interaction (*see* FUNDAMENTAL INTERACTIONS) in much the same way as photons transmit the electromagnetic interaction. They are not, however, massless like photons, and are believed to have a rest mass of the order of 10^{-25} kg (806 GeV). W bosons were discovered at CERN in 1983 with the expected mass. *See also* Z BOSON.

weak interaction *See* FUNDAMENTAL INTERACTIONS.

weakly interacting massive particle (WIMP) *See* MISSING MASS.

Web *See* WORLD WIDE WEB.

weber Symbol Wb. The SI unit of magnetic flux equal to the flux that, linking a circuit of one turn, produces in it an e.m.f. of one volt as it is reduced to zero at a uniform rate in one second. It is named after Wilhelm Weber.

Weber, Wilhelm Eduard (1804–91) German physicist, who became a professor at Göttingen. In 1833 he and Karl *Gauss built an electric telegraph between their laboratories. In 1843 Weber moved to Leipzig, where his main work was to develop a system of self-consistent elecrical units (as Gauss had already done for magnetism). Both systems were adopted in 1881. The SI unit of magnetic flux is named after him.

weight The gravitational force by which a body is attracted to the earth. *See* MASS.

weightlessness A condition of a body when it is an infinite distance from any other body. In practice the appearance of weightlessness occurs in space when the gravitational attraction of the earth on a body in space is equal to the centripetal force required by its orbital motion so that the body is effectively in free fall. Weightlessness can also be simulated for short periods in an aircraft flying a parabolic flight path, so that its occupants are again in free fall.

Weinberg–Salam model (WS model) *See* ELECTROWEAK THEORY.

Weston cell (cadmium cell) A type of primary *voltaic cell, which is used as a standard; it produces a constant e.m.f. of 1.0186 volts at 20°C. The cell is usually made in an H-shaped glass vessel with a mercury anode covered with a paste of cadmium sulphate and mercury(I) sulphate in one leg and a cadmium amalgam cathode covered with cadmium sulphate in the other leg. The electrolyte, which connects the two electrodes by means of the bar of the H, is a saturated solution of cadmium sulphate. In some cells sulphuric acid is added to prevent the hydrolysis of mercury sulphate. It is named after Edward Weston (1850–1936).

wet-and-dry bulb hygrometer *See* HYGROMETER.

Wheatstone, Sir Charles (1802–75) British physicist, who set up as a musical instrument-maker in London. He studied acoustics and optics, inventing a stereoscope in 1838. His most important work, done with William Cooke (1806–79), was the development of an electric telegraph, which they achieved in 1837. He gave his name to the *Wheatstone bridge, although he did not invent it.

Wheatstone bridge An electrical circuit for measuring the value of a resistance. In the illustration, R_1 is a resistance of unknown value, R_2 is a fixed resistance of known value, R_3 and R_4 are variable resistances with known values. When no current flows between A and B the bridge is said to be balanced, the galvanometer registers no deflection, and $R_1/R_2 = R_3/R_4$. R_1 can therefore be calculated. The Wheatstone bridge is used in various forms. In the **metre bridge**, a wire 1 metre long of uniform resistance is attached to the top of a board alongside a metre rule. A sliding contact is run along the wire, which corresponds to R_3 and R_4, until the galvanometer registers zero. Most practical forms use one or more rotary rheostats to provide the variation. The device was popularized though not invented by Sir Charles *Wheatstone.

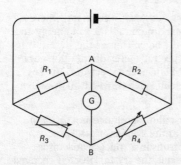

Wheatstone bridge

whistler A whistling noise that can be heard in a radio receiver following a lightning strike. The lightning produces long-wavelength electromagnetic waves, which follow the magnetic lines of force on the earth from the point where the lightning strikes to the antipodal point on the earth and back again several times.

white dwarf A compact stellar object that is supported against collapse under self-gravity by the *degeneracy pressure of electrons. White dwarfs are formed as the end products of the evolution of stars of relatively low mass (about that of the sun); high-mass stars may end up as *neutron stars or *black holes (*see* STELLAR EVOLUTION). White dwarfs consist of helium nuclei (and carbon and oxygen nuclei in the more massive cases) and a *degenerate gas of electrons. A typical white-dwarf density is 10^9 kg m^{-3}; white dwarf masses and radii are in the region of 0.7 solar masses and 10^3 km respectively. There is a maximum mass for white dwarfs, above which they are unstable to gravitational collapse – this is known as the *Chandrasekhar limit and is about 1.4 solar masses.

white hole A region of space–time that is the time reversal of a *black hole, meaning that matter would explode out of a white hole. Although white holes have sometimes been invoked to explain some violent events in the universe, there are strong theoretical arguments suggesting that they cannot exist.

white noise Random noise in which the energy per unit bandwidth is constant for the whole frequency range of interest.

Wiedemann–Franz law The ratio of the thermal conductivity of any pure metal to its electrical conductivity is approximately constant at a given temperature. The law is fairly well obeyed, except at low temperatures. It can be derived from the quantum theory of electrons in metals.

Wien formula *See* PLANCK'S RADIATION LAW.

Wien's displacement law For a *black body, $\lambda_m T$ = constant, where λ_m is the wavelength corresponding to the maximum radiation of energy and T is the thermodynamic temperature of the body. Thus as the temperature rises the maximum of the spectral energy distribution curve is displaced towards the short-wavelength end of the spectrum. The law was stated by the German physicist Wilhelm Wien (1864–1928). It can be derived as a special case of *Planck's radiation law.

Wigner energy Energy stored in a crystalline substance as a result of irradiation. This phenomenon is known as the **Wigner effect**. For example, some of the energy lost by neutrons in a *nuclear reactor is stored by the graphite moderator. As a result, the crystal lattice is changed and there is a consequent change in the physical dimensions of the moderator. It is named after the US physicist of Hungarian origin, Eugene Wigner (1902–95).

Wigner nuclides Pairs of isobars with odd nucleon numbers in which the atomic number and the neutron number differ by one. ^{3}H and ^{3}He are examples. They are named after Eugene Wigner.

Wigner's friend *See* SCHRÖDINGER'S CAT.

Wigner's rule A result concerning the *quantum statistics of systems with more than one fermion. If there are N fermions in the system and N is odd, then the total system is a fermion; but if N is even, the total system is a boson. Wigner's rule is named after Eugene Wigner, who discovered it in 1929. It was rediscovered independently by Paul Ehrenfest and Robert Oppenheimer in 1931.

Wilson, Charles Thomson Rees (1869–1959) British physicist, born in Scotland, who studied physics with J. J. *Thomson in Cambridge. His best-known achievement was the development of the *cloud chamber in 1911, for which he was awarded the 1927 Nobel Prize for physics.

Wilson cloud chamber *See* CLOUD CHAMBER.

WIMP Weakly interacting massive particle. *See* MISSING MASS.

Wimshurst machine A laboratory electrostatic generator. It consists of two insulating discs to which radial strips of metal foil are attached. After a few strips have been charged individually, the discs are rotated in opposite directions and the charge produced on the strips by induction is collected by metal combs or brushes. It was invented by J. Wimshurst (1836–1903).

window 1. A band of electromagnetic wavelengths that is able to pass through a particular medium with little reflection or absorption. For example, there is a **radio window** in the atmosphere allowing radio waves of wavelengths 5 mm to 30 m to pass through. This radio window enables *radio telescopes to be used on the surface of the earth. **2.** A period of time during which an event may occur in order to achieve a desired result. For example a **launch window** is the period during which a space vehicle must be launched to achieve a planned encounter.

wind power The use of winds in the earth's atmosphere to drive machinery, especially to drive an electrical generator. Practical land-based **wind generators** (**aerogenerators**) are probably capable of providing some 10^{20} J (10^{14} kW h) of energy per year throughout the world and interest in this form of renewable energy is increasing. The power, P, available to drive a wind generator is given by $P = kd^2v^3$, where k is the air density, d is the diameter of the blades, and v is the average wind speed. Wind farms now exist in many parts of the world; California, for example, has the capacity to produce over 1200 MW from wind energy.

wino *See* SUPERSYMMETRY.

Witten, Edward (1951–   ) US mathematical physicist who has been the

leading figure in the development of *superstring theory. He has also
made many important contributions to quantum field theory and
mathematics, notably *knot theory.

WMAP (Wilkinson Microwave Anisotopy Probe) A satellite launched in
2001 to study *microwave background radiation. In 2003 a full-sky picture
was obtained of the early universe (380 000 years after the big bang)
showing temperature fluctuations at high resolution. The results
supported the big-bang and inflation theories, and gave an age for the
universe of 13.7×10^9 years. They also indicated that about 70% of the
energy in the universe is *dark energy. WMAP is named in honour of the
US cosmologist and WMAP team member David Wilkinson, who died in
2002.

Wollaston prism A type of quartz prism for producing plane-polarized
light. It deviates the ordinary and extraordinary rays in opposite
directions by approximately the same amount. Like the *Rochon prism,
it can be used with ultraviolet radiation. It is named after the British
inventor and chemist William Hyde Wollaston (1766–1828).

Wood's metal A low-melting (71°C) alloy of bismuth (50%), lead (25%),
tin (12.5%), and cadmium (12.5%). It is used for fusible links in automatic
sprinkler systems. The melting point can be changed by varying the
composition. It is named after William Wood (1671–1730).

woofer A large loudspeaker designed to reproduce sounds of relatively
low frequency, in conjunction with a *tweeter and often a mid-range
speaker, in a high-fidelity sound reproducing system.

word A number of *bits, often 32, 48, or 64, processed by a computer as
a single unit.

work The work done by a force acting on a body is the product of the
force and the distance moved by its point of application in the direction
of the force. If a force $\mathbf{F}$ acts in such a way that the displacement $\mathbf{s}$ is in a
direction that makes an angle θ with the direction of the force, the work
done is given by: $W = \mathbf{F} \cdot \mathbf{s}\cos\theta$. Work is the scalar product of the force and
displacement vectors. It is measured in joules.

work function A quantity that determines the extent to which
thermionic or photoelectric emission will occur according to the
Richardson equation or Einstein's photoelectric equation. It is sometimes
expressed as a potential difference (symbol ϕ) in volts and sometimes as
the energy required to remove an electron (symbol W) in electronvolts or
joules. The former has been called the **work function potential** and the
latter the **work function energy**.

work hardening An increase in the hardness of metals as a result of
working them cold. It causes a permanent distortion of the crystal
structure and is particularly apparent with iron, copper, aluminium, etc.,

whereas with lead and zinc it does not occur as these metals are capable of recrystallizing at room temperature.

world line The history of a particle as represented in *space–time. The position of a particle at time t can be found by slicing space–time at time t and finding where the slice cuts the world line of the particle. The latter is straight if the particle moves uniformly and curved if it moves nonuniformly (i.e. there is acceleration). Light rays can be treated as the world lines of *photons. The world lines of particles under the influence of a *gravitational field are *geodesics in space–time. The world line of a photon near a star, such as the sun, is slightly bent as the light is deflected by the gravitational field of the sun.

World Wide Web (Web) A computer-based information service developed at CERN in the early 1990s. It is a hypermedia system (*see* HYPERTEXT) distributed over a large number of computer sites that allows users to view and retrieve information from 'documents' containing 'links'. It is accessed by a computer connected to the *Internet that is running a suitable program. Web documents may consist of text or other features, such as graphics, still or moving images, or audio clips. Within a document there will be material to be displayed and usually one or more links, which in a text document appear as highlighted words or phrases, or as icons. The links 'point' to other documents located elsewhere on the Web by means of a **URL** (universal resource locator), which contains information specifying, for example, the network address of the device holding the document and the local index entry for that document. Activating a link will result in the display of the requested document.

wormhole A solution of Einstein's equation for the general theory of *relativity that connects two distant regions in one asymptotically flat universe. These solutions have also been considered in *quantum gravity. There is no experimental evidence for workholes at present but their existence has been postulated in quantum gravity to explain why the *cosmological constant is very close to zero and in speculations concerning *time travel. Some of these applications of wormholes make use of the *tunnel effect, with calculations being performed using instanton techniques. Wormholes are predicted to have a very short lifetime.

W particle *See* W BOSON.

wrought iron A highly refined form of iron containing 1–3% of slag (mostly iron silicate), which is evenly distributed throughout the material in threads and fibres so that the product has a fibrous structure quite dissimilar to that of crystalline cast iron. Wrought iron rusts less readily than other forms of metallic iron and it welds and works more easily. It is used for chains, hooks, tubes, etc.

WS model Weinberg–Salam model. *See* ELECTROWEAK THEORY.

X-ray astronomy The study of *X-ray sources by rockets and balloons in the earth's atmosphere and by satellites beyond it. The first nonsolar X-ray source was detected during a rocket flight in 1962, and this observation heralded an entirely new branch of astronomy which developed rapidly with the availability of satellites in the 1970s.

X-ray crystallography The use of *X-ray diffraction to determine the structure of crystals or molecules. The technique involves directing a beam of X-rays at a crystalline sample and recording the diffracted X-rays on a photographic plate. The diffraction pattern consists of a pattern of spots on the plate, and the crystal structure can be worked out from the positions and intensities of the diffraction spots. X-rays are diffracted by the electrons in the atoms and if crystals of a compound are used, the electron density distribution in the ions or molecules can be determined. *See also* NEUTRON DIFFRACTION.

X-ray diffraction The diffraction of X-rays by a crystal. The wavelengths of X-rays are comparable in size to the distances between atoms in most crystals, and the repeated pattern of the crystal lattice acts like a diffraction grating for X-rays. Thus, a crystal of suitable type can be used to disperse X-rays in a spectrometer. X-ray diffraction is also the basis of X-ray crystallography. *See also* BRAGG'S LAW.

X-ray fluorescence The emission of *X-rays from excited atoms produced by the impact of high-energy electrons, other particles, or a primary beam of other X-rays. The wavelengths of the fluorescent X-rays can be measured by an X-ray spectrometer as a means of chemical analysis. X-ray fluorescence is used in such techniques as *electron-probe microanalysis.

X-rays Electromagnetic radiation of shorter wavelength than ultraviolet radiation produced by bombardment of atoms by high-quantum-energy particles. They were discovered by William *Roentgen in 1895. The range of wavelengths is 10^{-11} m to 10^{-9} m. Atoms of all the elements emit a characteristic **X-ray spectrum** when they are bombarded by electrons. The X-ray photons are emitted when the incident electrons knock an inner orbital electron out of an atom. When this happens an outer electron falls into the inner shell to replace it, losing potential energy (ΔE) in doing so. The wavelength λ of the emitted photon will then be given by $\lambda = ch/\Delta E$, where c is the speed of light and h is the Planck constant. *See also* BREMSSTRAHLUNG.

X-rays can pass through many forms of matter and they are therefore used medically and industrially to examine internal structures. X-rays are produced for these purposes by an *X-ray tube.

X-ray sources Sources of X-radiation from outside the solar system. Some 100 sources within the Galaxy have been observed as objects that emit most of their energy in the X-ray region of the electromagnetic spectrum and only a relatively small proportion of their energy in the visible spectrum. Many of these X-ray sources appear to be members of a binary system, consisting of one optically visible star and one very compact object; it is thought that the latter is either a *neutron star or (if very massive) a *black hole. Owing to the absorption of X-rays by the earth's atmosphere these X-ray sources are only visible by **X-ray telescopes** carried by space probes and satellites, although some high-energy X-rays can penetrate the upper atmosphere and are detectable by X-ray telescopes mounted on balloons.

X-ray spectrum *See* X-RAYS.

X-ray tube A device for generating *X-rays by accelerating electrons to a high energy by an electrostatic field and making them strike a metal target either in a tube containing a low-pressure gas or, as in modern tubes, in a high vacuum. The target is made from a heavy metal, usually tungsten, and is backed by a massive metal anode to conduct the heat away (see illustration showing a liquid-cooled copper anode). The electron beam is produced by heating the cathode by means of a white-hot tungsten filament. A transformer supplies the high voltage, often 100 kV, the tube acting as its own rectifier. On the half-cycles when the target is negative nothing happens. When the target becomes positive, the electrons bombarding it generate X-rays.

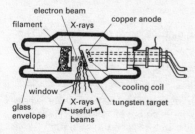

X-ray tube

Yagi aerial A directional aerial array widely used for television and *radio telescopes. It consists of one or two dipoles, a parallel reflector, and a series of closely spaced directors (0.15–0.25 wavelength apart) in front of the dipole. When used for reception this arrangement focuses the incoming signal on the dipole. For transmission, the output of the dipole is reinforced by the directors. It is named after Hidetsuga Yagi (1886–1976).

Yang–Mills theory *See* GAUGE THEORY.

yard The former Imperial standard unit of length. In 1963 the yard was redefined as 0.9144 metre exactly.

year The measure of time on which the calendar is based. It is the time taken for the earth to complete one orbit of the sun. The **calendar year** consists of an average of 365.25 mean solar days – three successive years of 365 days followed by one (leap) year of 366 days. The **solar year** (or **astronomical year**) is the average interval between two successive returns of the sun to the first point of Aries; it is 365.242 mean solar days. The **sidereal year** is the average period of revolution of the earth with respect to the fixed stars; it is 365.256 mean solar days. The **anomalistic year** is the average interval between successive perihelions; it is 365.259 mean solar days. *See also* EPHEMERIS TIME.

yield point *See* ELASTICITY.

yocto- Symbol y. A prefix used in the metric system to denote 10^{-24}. For example, 10^{-24} second = 1 yoctosecond (ys).

yotta- Symbol Y. A prefix used in the metric system to denote 10^{24}. For example, 10^{24} joules = a yottajoule (YJ).

Young, Thomas (1773–1829) British physician and physicist, who was a child prodigy and could speak 14 languages before he was 19. His early researches concerned the eye and vision, but he is best known for establishing the wave theory of *light (1800–04) and explaining the phenomenon of *interference (1807). The Young modulus of elasticity is named after him (*see* ELASTIC MODULUS).

Young modulus of elasticity *See* ELASTIC MODULUS.

Young's slits *See* INTERFERENCE.

Z boson An electrically neutral elementary particle, Z^0, which – like *W bosons – is thought to mediate the weak interactions in the *electroweak theory. The Z^0 boson was discovered at CERN in 1983 and has a mass of about 90 GeV as had been predicted from theory.

Zeeman effect The splitting of the lines in a spectrum when the source of the spectrum is exposed to a magnetic field. It was discovered in 1896 by the Dutch physicist Pieter Zeeman (1865–1943). In the **normal Zeeman effect** a single line is split into three if the field is perpendicular to the light path or two lines if the field is parallel to the light path. This effect can be explained by classical electromagnetic principles in terms of the speeding up and slowing down of electrons in atoms due to the applied field and in terms of the old quantum theory of Bohr and Sommerfeld. The **anomalous Zeeman effect** is a complicated splitting of the lines into several closely spaced lines, so called because it does not agree with classical predictions or the old quantum theory. This effect is explained by quantum mechanics in terms of electron spin.

Zeilinger's principle The principle that any elementary system carries just one bit of information. This principle was put forward by the Austrian physicist Anton Zeilinger in 1999 and subsequently developed by him to derive several aspects of quantum mechanics.

Zener diode A type of semiconductor diode, consisting of a p–n junction with high doping concentrations on either side of the junction. It acts as a rectifier until the applied reverse voltage reaches a certain value, the **Zener breakdown voltage**, when the device becomes conducting. This effect occurs as a result of electrons being excited directly from the valence band into the conduction band (*see* ENERGY BANDS). Zener diodes are used in voltage-limiting circuits; they are named after the US physicist Clarence Melvin Zener (1905–93).

zenith The point on the *celestial sphere that lies directly above an observer. *Compare* NADIR.

zepto- Symbol z. A prefix used in the metric system to denote 10^{-21}. For example, 10^{-21} second = 1 zeptosecond (zs).

zero-point energy The energy remaining in a substance at the *absolute zero of temperature (0 K). This is in accordance with quantum theory, in which a particle oscillating with simple harmonic motion does not have a stationary state of zero kinetic energy. Moreover, the *uncertainty principle does not allow such a particle to be at rest at exactly the centrepoint of its oscillations.

zeroth law of thermodynamics *See* THERMODYNAMICS.

zetta- Symbol Z. A prefix used in the metric system to denote 10^{21}. For example, 10^{21} joules = 1 zettajoule (ZJ).

zinc chloride cell *See* DRY CELL.

zino *See* SUPERSYMMETRY.

zodiac A band that passes round the *celestial sphere, extending 9° on either side of the *ecliptic. It includes the apparent paths of the sun, moon, and planets (except Pluto). The band is divided into the twelve **signs of the zodiac**, each 30° wide. These signs indicate the sun's position each month in the year and were named by the ancient Greeks after the **zodiacal constellations** that occupied the signs some 2000 years ago. However, as a result of the *precession of the equinoxes the constellations have since moved eastwards by over 30° and no longer coincide with the signs.

zodiacal light A faint luminous glow in the sky that can be observed on a moonless night on the western horizon after sunset or on the eastern horizon before sunrise. It is caused by the scattering of sunlight by dust particles in interplanetary space.

zone refining A technique used to reduce the level of impurities in certain metals, alloys, semiconductors, and other materials. It is based on the observation that the solubility of an impurity may be different in the liquid and solid phases of a material. To take advantage of this observation, a narrow molten zone is moved along the length of a specimen of the material, with the result that the impurities are segregated at one end of the bar and the pure material at the other. In general, if the impurities lower the melting point of the material they are moved in the same direction as the molten zone moves, and vice versa.

zoom lens *See* TELEPHOTO LENS.

zwitterion (ampholyte ion) An ion that has a positive and negative charge on the same group of atoms. Zwitterions can be formed from compounds that contain both acid groups and basic groups in their molecules. For example, aminoethanoic acid (the amino acid glycine) has the formula $H_2N.CH_2.COOH$. However, under neutral conditions, it exists in the different form of the zwitterion $^+H_3N.CH_2.COO^-$, which can be regarded as having been produced by an internal neutralization reaction (transfer of a proton from the carboxyl group to the amino group). Aminoethanoic acid, as a consequence, has some properties characteristic of ionic compounds; e.g. a high melting point and solubility in water. In acid solutions, the positive ion $^+H_3NCH_2COOH$ is formed. In basic solutions, the negative ion $H_2NCH_2COO^-$ predominates. The name comes from the German *Zwitter*, meaning hermaphrodite.

Z

Appendix 1. The Greek alphabet

Letters		Name
A	α	alpha
B	β	beta
Γ	γ	gamma
Δ	δ	delta
E	ε	epsilon
Z	ζ	zeta
H	η	eta
Θ	θ	theta
I	ι	iota
K	κ	kappa
Λ	λ	lambda
M	μ	mu
N	ν	nu
Ξ	ξ	xi
O	o	omicron
Π	π	pi
P	ρ	rho
Σ	σ	sigma
T	τ	tau
Υ	υ	upsilon
Φ	φ	phi
X	χ	chi
Ψ	ψ	psi
Ω	ω	omega

Appendix 2. SI units

TABLE 2.1 Base and dimensionless SI units

Physical quantity	Name	Symbol
length	metre	m
mass	kilogram	kg
time	second	s
electric current	ampere	A
thermodynamic temperature	kelvin	K
luminous intensity	candela	cd
amount of substance	mole	mol
*plane angle	radian	rad
*solid angle	steradian	sr

*dimensionless units

TABLE 2.2 Derived SI units with special names

Physical quantity	Name of SI unit	Symbol of SI unit
frequency	hertz	Hz
energy	joule	J
force	newton	N
power	watt	W
pressure	pascal	Pa
electric charge	coulomb	C
electric potential difference	volt	V
electric resistance	ohm	Ω
electric conductance	siemens	S
electric capacitance	farad	F
magnetic flux	weber	Wb
inductance	henry	H
magnetic flux density (magnetic induction)	tesla	T
luminous flux	lumen	lm
illuminance	lux	lx
absorbed dose	gray	Gy
activity	becquerel	Bq
dose equivalent	sievert	Sv

TABLE 2.3 Decimal multiples and submultiples to be used with SI units

Submultiple	Prefix	Symbol	Multiple	Prefix	Symbol
10^{-1}	deci	d	10	deca	da
10^{-2}	centi	c	10^2	hecto	h
10^{-3}	milli	m	10^3	kilo	k
10^{-6}	micro	μ	10^6	mega	M
10^{-9}	nano	n	10^9	giga	G
10^{-12}	pico	p	10^{12}	tera	T
10^{-15}	femto	f	10^{15}	peta	P
10^{-18}	atto	a	10^{18}	exa	E
10^{-21}	zepto	z	10^{21}	zetta	Z
10^{-24}	yocto	y	10^{24}	yotta	Y

TABLE 2.4 Conversion of units to SI units

From	To	Multiply by
in	m	2.54×10^{-2}
ft	m	0.3048
sq. in	m^2	6.4516×10^{-4}
sq. ft	m^2	9.2903×10^{-2}
cu. in	m^3	1.63871×10^{-5}
cu. ft	m^3	2.83168×10^{-2}
l(itre)	m^3	10^{-3}
gal(lon)	l(itre)	4.546 09
miles/hr	$m\ s^{-1}$	0.477 04
km/hr	$m\ s^{-1}$	0.277 78
lb	kg	0.453 592
$g\ cm^{-3}$	$kg\ m^{-3}$	10^3
lb/in^3	$kg\ m^{-3}$	$2.767\ 99 \times 10^4$
dyne	N	10^{-5}
poundal	N	0.138 255
lbf	N	4.448 22
mmHg	Pa	133.322
atmosphere	Pa	$1.013\ 25 \times 10^5$
hp	W	745.7
erg	J	10^{-7}
eV	J	$1.602\ 10 \times 10^{-19}$
kW h	J	3.6×10^6
cal	J	4.1868

Appendix 3. Fundamental constants

Constant	Symbol	Value in SI units
acceleration of free fall	g	$9.806\ 65$ m s^{-2}
Avogadro constant	L, N_A	$6.022\ 1367(36) \times 10^{23}$ mol^{-1}
Boltzmann constant	$k = R/N_A$	$1.380\ 658(12) \times 10^{-23}$ J K^{-1}
electric constant	ε_0	$8.854\ 187\ 817 \times 10^{-12}$ F m^{-1}
electronic charge	e	$1.602\ 177\ 33(49) \times 10^{-19}$ C
electronic rest mass	m_e	$9.109\ 3897(54) \times 10^{-31}$ kg
Faraday constant	F	$9.648\ 5309(29) \times 10^4$ C mol^{-1}
gas constant	R	$8.314\ 510(70)$ J K^{-1} mol^{-1}
gravitational constant	G	$6.672\ 59(85) \times 10^{-11}$ m^3 kg^{-1} s^{-2}
Loschmidt's constant	N_L	$2.686\ 763(23) \times 10^{25}$ m^{-3}
magnetic constant	μ_0	$4\pi \times 10^{-7}$ H m^{-1}
neutron rest mass	m_n	$1.674\ 9286(10) \times 10^{-27}$ kg
Planck constant	h	$6.626\ 0755(40) \times 10^{-34}$ J s
proton rest mass	m_p	$1.672\ 6231(10) \times 10^{-27}$ kg
speed of light	c	$2.997\ 924\ 58 \times 10^8$ m s^{-1}
Stefan–Boltzmann constant	σ	$5.670\ 51(19) \times 10^{-8}$ W m^{-2} K^{-4}

Appendix 4. The solar system

Planet	Equatorial diameter (km)	Mean distance from sun (10^6 km)	Sidereal period	
Mercury	4879.4	57.91	86.70	days
Venus	12 103.6	108.21	221.46	days
Earth	12 756.3	149.6	0.999	years
Mars	6794	227.94	677.0	days
Jupiter	142 985	778.41	11.86	years
Saturn	120 536	1426.72	29.42	years
Uranus	51 118	2870.97	83.75	years
Neptune	49 528	4498.25	163.72	years
Pluto	2390	5906.38	248.02	years

Appendix 5. The electromagnetic spectrum

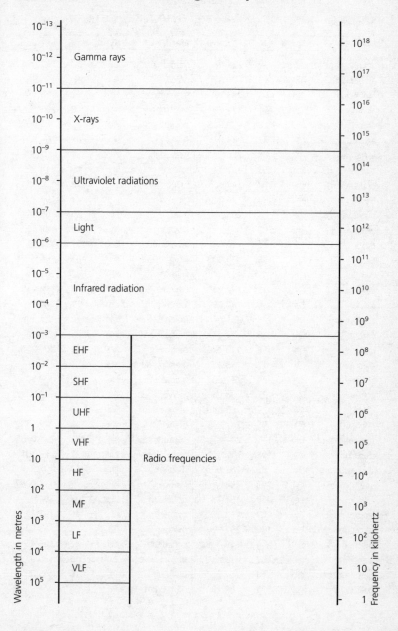

Appendix 6. The chemical elements

R.a.m. values with asterisk denote mass number of the most stable known isotope

Element	Symb	a.n.	r.a.m.	Element	Symb	a.n.	r.a.m.
actinium	Ac	89	227*	germanium	Ge	32	72.59
aluminium	Al	13	26.98	gold	Au	79	196.967
americium	Am	95	243*	hafnium	Hf	72	178.49
antimony	Sb	51	121.75	hassium	Hs	108	265*
argon	Ar	18	39.948	helium	He	2	4.0026
arsenic	As	33	74.92	holmium	Ho	67	164.93
astatine	At	85	210*	hydrogen	H	1	1.008
barium	Ba	56	137.34	indium	In	49	114.82
berkelium	Bk	97	247*	iodine	I	53	126.9045
beryllium	Be	4	9.012	iridium	Ir	77	192.20
bismuth	Bi	83	208.98	iron	Fe	26	55.847
bohrium	Bh	107	262*	krypton	Kr	36	83.80
boron	B	5	10.81	lanthanum	La	57	138.91
bromine	Br	35	79.909	lawrencium	Lr	103	256*
cadmium	Cd	48	112.41	lead	Pb	82	207.19
caesium	Cs	55	132.905	lithium	Li	3	6.939
calcium	Ca	20	40.08	lutetium	Lu	71	174.97
californium	Cf	98	251*	magnesium	Mg	12	24.305
carbon	C	6	12.011	manganese	Mn	25	54.94
cerium	Ce	58	140.12	meitnerium	Mt	109	266*
chlorine	Cl	17	35.453	mendelevium	Md	101	258*
chromium	Cr	24	52.00	mercury	Hg	80	200.59
cobalt	Co	27	58.933	molybdenum	Mo	42	95.94
copper	Cu	29	63.546	neodymium	Nd	60	144.24
curium	Cm	96	247*	neon	Ne	10	20.179
darmstadtium	Ds	110	271*	neptunium	Np	93	237.0482
dubnium	Db	105	262*	nickel	Ni	28	58.70
dysprosium	Dy	66	162.50	niobium	Nb	41	92.91
einsteinium	Es	99	254*	nitrogen	N	7	14.0067
erbium	Er	68	167.26	nobelium	No	102	254*
europium	Eu	63	151.96	osmium	Os	76	190.2
fermium	Fm	100	257*	oxygen	O	8	15.9994
fluorine	F	9	18.9984	palladium	Pd	46	106.4
francium	Fr	87	223*	phosphorus	P	15	30.9738
gadolinium	Gd	64	157.25	platinum	Pt	78	195.09
gallium	Ga	31	69.72	plutonium	Pu	94	244*

Element	Symb	a.n.	r.a.m.	Element	Symb	a.n.	r.a.m.
polonium	Po	84	210*	tantalum	Ta	73	180.948
potassium	K	19	39.098	technetium	Tc	43	98*
praseodymium	Pr	59	140.91	tellurium	Te	52	127.60
promethium	Pm	61	145	terbium	Tb	65	158.92
protactinium	Pa	91	231.036	thallium	Tl	81	204.39
radium	Ra	88	226.0254	thorium	Th	90	232.038
radon	Rn	86	222*	thulium	Tm	69	168.934
rhenium	Re	75	186.2	tin	Sn	50	118.69
rhodium	Rh	45	102.9	titanium	Ti	22	47.9
roentgenium	Rg	111	272*	tungsten	W	74	183.85
rubidium	Rb	37	85.47	ununbium	Uub	112	285*
ruthenium	Ru	44	101.07	ununtrium	Uut	113	284*
rutherfordium	Rf	104	261*	ununquadium	Uuq	114	289*
samarium	Sm	62	150.35	ununpentium	Uup	115	288*
scandium	Sc	21	44.956	ununhexium	Uuh	116	292*
seaborgium	Sg	106	263*	uranium	U	92	238.03
selenium	Se	34	78.96	vanadium	V	23	50.94
silicon	Si	14	28.086	xenon	Xe	54	131.30
silver	Ag	47	107.87	ytterbium	Yb	70	173.04
sodium	Na	11	22.9898	yttrium	Y	39	88.905
strontium	Sr	38	87.62	zinc	Zn	30	65.38
sulphur	S	16	32.06	zirconium	Zr	40	91.22

Appendix 7. Useful websites

AAAS www.sciencemag.org
Access to the *Science* magazine of the American Association for the Advancement of Science.

American Institute of Physics www.aip.org
A large site including *Physics News Update*, which is a digest of physics news items arising from physics meetings, physics journals, newspapers and magazines, and other news sources.

CERN www.cern.ch
The official website of the European Laboratory for Particle Physics.

Fermilab www.fnal.gov
The official website of the Fermi National Accelerator Laboratory.

Institute of Physics www.iop.org
The website of the Institute of Physics. It provides a number of free online services, including search facilities and physics portals.

IUPAC http://iupac.chemsoc.org/dhtml_home.html
The official home page of the International Union of Pure and Applied Chemistry, which is responsible for the naming of elements.

NASA www.nasa.gov/home
The website of the American National Aeronautics and Space Administration. The site contains a large amount of information on the solar system and space science.

National Institute of Standards and Technology www.nist.gov
This site provides a large amount of information about units and constants.

Nature Magazine Online www.nature.com
An online weekly journal that offers news articles and features, complete reference works online, and information on the latest science research.

New Scientist www.newscientist.com
A popular news and archive site for all branches of science.

Nobel Foundation www.nobel.se
The site lists all Nobel prize winners and includes extensive articles on the nature and significance of their work.

Physical Sciences Resource Center http://psrc.aapt.org
A site run by the American Physical Society. It contains a large collection of information and resources for the physical sciences.

Physics Central www.physicscentral.com
A site run by the American Association of Physics Teachers. It contains a searchable collection of articles on all branches of physics.

Scientific American www.sciam.com
A popular science news site containing selected recent articles.

Webelements www.webelements.com
A periodic table at the University of Sheffield linked to a very comprehensive database of the elements and their compounds.

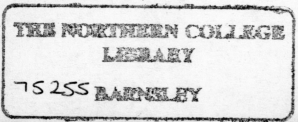